# Basic Topology 3

**Professor Mahima Ranjan Adhikari (1944–2021)**

Mahima Ranjan Adhikari

# Basic Topology 3

## Algebraic Topology and Topology of Fiber Bundles

 Springer

Mahima Ranjan Adhikari
Institute for Mathematics, Bioinformatics,
Information Technology and Computer
Science (IMBIC)
Kolkata, West Bengal, India

ISBN 978-981-16-6552-3        ISBN 978-981-16-6550-9   (eBook)
https://doi.org/10.1007/978-981-16-6550-9

Mathematics Subject Classification: 55-XX, 22-XX, 14F35, 14Fxx, 18F15

This Springer imprint is published by the registered company Springer Nature Singapore Pte Ltd.
The registered company address is: 152 Beach Road, #21-01/04 Gateway East, Singapore 189721,
Singapore

*Dedicated to*

*Prof. B. Eckmann and Prof. P. J. Hilton*

*for their fruitful interaction with them*

*during my academic visit to*

*ETH, Zurich, Switzerland*

*in 2003*

# Preface

Algebraic topology is an important branch of topology that utilizes algebraic tools to study topological problems. The main aim of algebraic topology is to construct algebraic invariants on topological spaces to convert topological problems into algebraic problems to have a better chance of a solution. For example, the algebraic invariants stem from homotopy theory are the homotopy groups $\pi_n(X, x_0)$ of a pointed topological space $(X, x_0)$ for $n \geq 1$. The other basic algebraic invariants are homology and cohomology groups of a space. Their constructions are much more complicated than those of homotopy groups. Fortunately, computations of homology and cohomology groups are easier than those of homotopy groups. In algebraic topology, algebraic invariants classify topological spaces up to homeomorphism. It is found that this classification is usually in most cases up to homotopy equivalence. It tries to measure degrees of connectivity using homology, cohomology and homotopy groups.

Several approaches have been used in topology to associate a topological space with a number of algebraic objects, for instance, groups, rings, etc. The basic idea of algebraic topology is the correspondence (or functor) associating a collection of certain algebraic objects to a collection of topological spaces and continuous mappings of spaces to corresponding homomorphisms. This functorial approach allows us to transform topological problem into the corresponding algebraic one. The solvability of this 'derived' algebraic problem in many cases implies the solvability of the initial topological problems.

The contents of Volume 3 are expanded into seven chapters and discuss geometric topology and manifolds by using algebraic topology.

Chapter 1 provides a background on algebra, topology and Lie groups to facilitate a smooth study of this volume.

Chapter 2 officially inaugurates a study of algebraic topology by conveying the basic concepts of homotopy and fundamental groups born through the work of H. Poincaré in his land-marking *Analysis situs*, Paris, 1895, and also discusses higher homotopy groups constructed in 1935 by H. Hurewicz (1904–1956) in his paper (Hurewicz, 1935), which are natural generalizations of fundamental groups. Homotopy theory studies those properties of topological spaces and continuous maps which are invariants under homotopic maps, called *homotopy invariants*. Finally, this chapter presents some interesting applications of homotopy, fundamental and higher homotopy groups in analysis, geometry, algebra, matrix theory, atmospheric science, vector field and extension problems and some others.

Chapter 3 conveys the basic concepts of homology theory starting from its invention by Heny H. Poincaré (1854–1912) in 1895 to the approach formulating axiomatization of homology, announced in 1952 by S. Eilenberg (1915–1998) and N. Steenrod (1910–1971), now known as *Eilenberg and Steenrod axioms*. This approach simplifies the proofs of many results by escaping avoidable difficulties to promote active learning in homology and cohomology theories, which is the most important contribution to algebraic topology after the invention homotopy and homology by Poincaré in 1895. This functorial approach facilitates in variety of cases to solve topological problems through the solvability of corresponding algebraic problems. The motivation of the study of algebraic topology comes from the study of geometric properties of topological spaces from the algebraic viewpoint.

Chapter 4 discusses the topology of fiber bundles starting with general theory of bundles and continues its study to Chap. 5. The topology of fiber bundles has created general interest and promises for more work because it is involved of interesting applications of topology to other areas such as algebraic topology, geometry, physics and gauge groups. The theory of fiber bundles was first recognized during the period 1935–1940 through the work of H. Whitney (1907–1989), H. Hopf (1894–1971) and E. Stiefel (1909–1978), J. Feldbau (1914–1945) and some others. A *fiber bundle* is a bundle with an additional structure derived from the action of a topological group on the fibers. A fiber bundle is a locally trivial fibration having covering homotopy property.

Chapter 5 continues the study topology of fiber bundles from the viewpoint of homotopy theory. Covering spaces provide tools to study the fundamental groups. Fiber bundles provide likewise tools to study higher homotopy groups (which are generalizations of fundamental groups). The notion of fiber spaces is the most fruitful generalization of covering spaces. The importance of fiber spaces was realized during 1935–1950 to solve several problems relating to homotopy and homology.

Chapter 6 studies geometric topology primarily, which studies manifolds and their embeddings in other manifolds. A particularly active area is low-dimensional topology, which studies manifolds of four or fewer dimensions. This includes knot theory, the study of mathematical knots. It proves more theorems and conveys further applications of topological concepts and results discussed in earlier chapters with a view to understand the beauty, power and scope of the subject topology. Moreover, it provides alternative proofs of some results proved in the previous chapters such as Brouwer–Poincaré theorem, Van Kampen theorem and Borsuk–Ulam theorem for any finite dimension. It proves Ham Sandwich theorem and Lusternik–Schnirelmann theorem.

Chapter 7 conveys the history of emergence of the concepts leading to the development of algebraic topology as a subject with their motivations.

The list of chapters shows that the book covers a wide range of topics. Some are more technical than others, but the reader without a great deal of technical knowledge should still find most of the text accessible. Avoiding readymade proofs of some theorems, their statements have appeared in this book in the form of problems providing an opportunity for exploration of the topics of the book along with somewhat diverging from the basic thrust of the book, since solving problems plays a key role in the study of mathematics.

The book is a clear exposition of the basic ideas of topology and conveys a straightforward discussion of the basic topics of topology and avoids unnecessary definitions and terminologies. Each chapter starts with highlighting the main results of the chapter with motivation and is split into several sections which discuss related topics with some degree of thoroughness and ends with exercises of varying degrees of difficulties, which not only impart an additional information about the text covered previously but also introduce a variety of ideas not treated in the earlier texts with certain references to the interested readers for more study. All these constitute the basic organizational units of the book.

The present book, together with the authors' two other Springer books, *Basic Modern Algebra with Applications* (M. R. Adhikari and Avishek Adhikari) and *Basic Algebraic Topology and its Applications* (M. R. Adhikari), will form a unitary module for the study of modern algebra, general and algebraic topology with applications in several areas.

The author acknowledges Higher Education Department of the Government of West Bengal for sanctioning the financial support to the Institute for Mathematics, Bioinformatics and Computer Science (IMBIC), toward writing this book vide order no. 432(Sanc)/EH/P/SE/SE/1G-17/07 dated August 29, 2017, and also to IMBIC, the University of Calcutta, Presidency University, Kolkata, India, and Moulana Abul Kalam Azad University of Technology, West Bengal, for providing the infrastructure toward implementing the scheme.

The author is indebted to the authors of the books and research papers listed in the Bibliography at the end of each chapter and are very thankful to Professors P. Stavrions (Greece), Constantine Udriste (Romania), Akira Asada (Japan) and Avishek Adhikari (India) and also to the reviewers of the manuscript for their scholarly suggestions for improvement of the book. We are thankful to Md. Kutubuddin Sardar for his cooperation towards the typesetting of the manuscript and to many UG and PG students of Presidency University and Calcutta University, and many other individuals who have helped in proofreading the book. Authors apologize to those whose names have been inadvertently not entered. Finally, the author acknowledges, with heartfelt thanks, the patience and sacrifice of the long-suffering family of the author, especially Minati Adhikari, Dr. Shibopriya Mitra Adhikari, and the beloved grand son Master Avipriyo Adhikari.

Kolkata, India                                    Mahima Ranjan Adhikari
June 2021

# A Note on Basic Topology—Volumes 1–3

The topic 'topology' has become one of the most exciting and influential fields of study in modern mathematics, because of its beauty and scope. The aim of this subject is to make a qualitative study of geometry in the sense that if one geometric object is continuously deformed into another geometrical object, then these two geometric objects are considered topologically equivalent, called *homeomorphic*. Topology starts where sets have some cohesive properties, leading to define continuity of functions.

The series of three books on *Basic Topology* is a project book funded by the Government of West Bengal, which is designed to introduce many variants of a basic course in topology through the study of point-set topology, topological groups, topological vector spaces, manifolds, Lie groups, homotopy and homology theories with an emphasis of their applications in modern analysis, geometry, algebra and theory of numbers:

Topics in topology is vast. The range of its basic topics is distributed among different topological subfields such as general topology, topological algebra, differential topology, combinatorial topology, algebraic topology and geometric topology. Each volume of the present book is considered as a separate textbook that promotes active learning of the subject highlighting elegance, beauty, scope and power of topology.

## Basic Topology—Volume 1: Metric Spaces and General Topology

This volume majorly studies metric spaces and general topology. It considers the general properties of topological spaces and their mappings. The special structure of a metric space induces a topology having many applications of topology in modern analysis, geometry and algebra. The texts of Volume I are expanded into eight chapters.

## Basic Topology—Volume 2: Topological Groups, Topology of Manifolds and Lie Groups

This volume considers additional structures other than topological structures studied in Volume 1 and links topological structure with other structures in a compatible way to study topological groups, topological vector spaces, topological and smooth manifolds, Lie groups and Lie algebra and also gives a complete classification of closed surfaces without using the formal techniques of homology theory. Volume 2 contains five chapters.

## Basic Topology—Volume 3: Algebraic Topology and Topology of Fiber Bundles

This volume mainly discusses algebraic topology and topology of fiber bundles. The main aim of topology is to classify topological spaces up to homeomorphism. To achieve this goal, algebraic topology constructs algebraic invariants and studies topological problems by using these algebraic invariants. Because of its beauty and scope, algebraic topology has become an essential branch of topology. Algebraic topology is an important branch of topology that utilizes algebraic tools to study topological problems. Its basic aim is to construct algebraic invariants that classify topological spaces up to homeomorphism. It is found that this classification, usually in most cases, is up to homotopy equivalence.

This volume conveys a coherent introduction to algebraic topology formally inaugurated by H. Poincaré (1854–1912) in his land-marking *Analysis situs*, Paris, 1895, through his invention of fundamental group and homology theory, which are topological invariants. It studies Euler characteristic, the Betti number and also certain classic problems such as the Jordan curve theorem. It considers higher homotopy groups and establishes links between homotopy and homology theories, axiomatic approach to homology and cohomology inaugurated by Eilenberg and Steenrod. It studies the problems of converting topological and geometrical problems to algebraic one in a functorial way for better chance for solution.

This volume also studies geometric topology and manifolds by using algebraic topology. The contents of Volume 3 are expanded into seven chapters.

Just after the concept of homeomorphisms is clearly defined, the subject of topology begins to study those properties of geometric figures which are preserved by homeomorphisms with an eye to classify topological spaces up to homeomorphism, which stands the ultimate problem in topology, where a geometric figure is considered to be a point set in the Euclidean space $\mathbf{R}^n$. But this undertaking becomes hopeless, when there exists no homeomorphism between two given topological spaces. The concepts of topological properties and topological invariants play key tools in such problems:

(a)    The concepts of topological properties, such as, compactness and connect-edness, introduced in general topology, solve this problem in very few cases (studied in *Basic Topology*, Vol. 1).

(b)    On the other hand, the subjects algebraic topology and differential topology (studied in Volume 2) were born to solve the problems of impossibility in many cases with a shift of the problem by associating *invariant objects* in the sense that homeomorphic spaces have the same object (up to equivalence), called *topological invariants.* Initially, these objects were integers and subsequent research reveals that more fruitful and interesting results can be obtained from the algebraic invariant structures such as groups and rings. For example, homology and homotopy groups are very important algebraic invariants which provide strong tools to study the structure of topological spaces.

# Contents

# About the Author

**Mahima Ranjan Adhikari,** Ph.D., M.Sc. (Gold Medalist), is the founder president of the Institute for Mathematics, Bioinformatics, Information Technology and Computer Science (IMBIC), Kolkata, India. He is a former professor at the Department of Pure Mathematics, University of Calcutta, India. His research papers are published in national and international journals of repute, including the *Proceedings of American Mathematical Society*. He has authored nine textbooks and is the editor of two, including: *Basic Modern Algebra with Applications* (Springer, 2014), *Basic Algebraic Topology and Applications* (Springer, 2016), and *Mathematical and Statistical Applications in Life Sciences and Engineering* (Springer, 2017).

Twelve students have been awarded Ph.D. degree under his guidance on various topics such as algebra, algebraic topology, category theory, geometry, analysis, graph theory, knot theory and history of mathematics. He has visited several universities and research institutions in India, USA, UK, Japan, China, Greece, Sweden, Switzerland, Italy, and many other counties on invitation. A member of the American Mathematical Society, Prof. Adhikari is on the editorial board of several journals of repute. He was elected as the president of the Mathematical Sciences Section (including Statistics) of the 95th Indian Science Congress, 2008. He has successfully completed research projects funded by the Government of India.

# Chapter 1
# Prerequisite Concepts of Algebra, Topology, Manifold and Category Theory

This chapter conveys a few basic concepts of algebra, topology, manifold and category theory for smooth study of Volume 3 of the present book series. For detailed study of the concepts and results given in this chapter, [Adhikari and Adhikari, 2003, 2006, 2014], [Adhikari, 2016], [Dugundji, 1966], [Simmons, 1963], [Adhikari and Adhikari, Volumes 1 and 2, 2022a, 2022b], [Alexandrov, 1979], [Borisovich et al. 1985], [Bredon, 1983], [Chatterjee et al. 2002], [MacLane, 1971], [Williard and Stephen, 1970] and some other references are given in Bibliography.

## 1.1 Some Basic Concepts on Algebraic Structures and their Homomorphisms

A group, ring, vector space, module or any algebraic system is defined as a nonempty set endowed with special algebraic structures. A homomorphism (transformation) is a function preserving the specific structures of the algebraic systems.

### 1.1.1 Groups and Fundamental Homomorphism Theorem

This subsection presents some basic concepts of group (abstract) theory which are subsequently used.

**Definition 1.1.1** Let $G$ be a group. A subset $C(G)$ of $G$ defined by

$$C(G) = \{g \in G : gx = xg \text{ for all } x \in G\}$$

forms a subgroup of $G$, called the **center** of $G$.

© The Author(s), under exclusive license to Springer Nature Singapore Pte Ltd. 2022
M. R. Adhikari, *Basic Topology 3*,
https://doi.org/10.1007/978-981-16-6550-9_1

**Definition 1.1.2** Let $H$ be a subgroup of group $G$. Then for any $x \in G$,

(i) the set
$$xH = \{xh : h \in H\}$$

is called a **left coset** of $H$ in $G$ and

(ii) the set
$$Hx = \{hx : h \in H\}$$

is called a **right coset** of $H$ in $G$.

**Definition 1.1.3** A subgroup $H$ of a group $G$ is said to be a **normal subgroup** of $G$, if
$$xH = Hx, \ \forall x \in G.$$

**Definition 1.1.4** (*Quotient group*) Let $G$ be group, $N$ be a normal subgroup of $G$ and $G/N$ be the set of all cosets of $N$ in $G$. Then $G/N$ is a group under the composition

$$xN \circ yN = xyN, \ \forall \ xN \text{ and } yN \text{ in } G/N,$$

called the **factor group (or quotient group)** of $G$ by $N$, denoted by $G/N$.

**Definition 1.1.5** Given an additive abelian group $G$, and a normal subgroup $N$ of $G$, the factor group $G/N$ is defined under the group operation

$$(x + N) + (y + N) = (x + y)N.$$

This quotient group is sometimes called a **difference group.**

**Definition 1.1.6** Let $G$ and $K$ be two groups. A map $f : G \to K$ is said to be a homomorphism if
$$f(xy) = f(x)f(y), \ \forall x, y \in G.$$

**Remark 1.1.7** A homomorphism of groups maps identity element into the identity element and inverse element into the inverse element. Special homomorphisms carry special names.

**Definition 1.1.8** A homomorphism $f : G \to K$ of groups is said to be

(i) a **monomorphism** if $f$ is injective;
(ii) an **embedding** if $f$ is a monomorphism;
(iii) an **epimorphism** if $f$ is surjective;
(iv) an **isomorphism** if $f$ is bijective;
(v) an isomorphism of $G$ onto itself is called an **automorphism** of $G$ and
(vi) a homomorphism of $G$ into itself is called an **endomorphism.**

**Remark 1.1.9** Two isomorphic groups are considered as replicas of each other, because they have the identical algebraic properties. So two isomorphic groups are identified and are considered the same group (up to isomorphism).

**Definition 1.1.10** The kernel of a homomorphism $f : G \to K$ of groups, denoted by $kerf$ is defined by

$$kerf = \{x \in G : f(x) = e_k\},$$

where $e_k$ is the identity element of $K$.

**Theorem 1.1.11** *Let $f : G \to K$ be a group homomorphism. Then $kerf$ is a normal subgroup of $G$. Conversely, if $N$ is a normal subgroup of $G$, then the map*

$$\pi : G \to K/N, \ x \mapsto xN$$

*is an epimorphism with $N$ as its kernel.*

**Corollary 1.1.12** *Let $N$ be a subgroup of a given group $G$. Then $N$ is a normal subgroup of $G$ if it is the kernel of some homomorphism.*

**Theorem 1.1.13** (First Isomorphism Theorem) *Let $f : G \to K$ be a homomorphism of groups. Then $f$ induces an isomorphism*

$$\tilde{f} : G/kerf \to Imf, \ x \, kerf \mapsto f(x).$$

**Corollary 1.1.14** *Let $f : G \to K$ be an epimorphism of groups. Then the groups $G/kerf$ and $K$ are isomorphic.*

**Example 1.1.15** Let $G = GL(n, \mathbf{R})$ be the general linear group over $\mathbf{R}$, $\mathbf{R}^*$ be the multiplicative group of nonzero real numbers and

$$det : G \to \mathbf{R}^*, \ M \mapsto det \, M$$

be the determinant function and $N = \{M \in GL(n, \mathbf{R}) : detM = 1\}$. Then

(i) $N$ is a normal subgroup of $G$ and
(ii) the groups $GL(n, \mathbf{R})/N$ and $\mathbf{R}^*$ are isomorphic by Corollary 1.1.14.

**Definition 1.1.16** (*Commutator subgroup*) Given a group $G$ and a pair of elements $g, h \in G$, the **commutator** of $g$ and $h$ denoted by $[g, h]$ is the element

$$[g, h] = ghg^{-1}h^{-1}.$$

The subgroup $K$ of $G$ generated by the set $S = \{ghg^{-1}h^{-1} : g, h \in G\}$ is called the **commutator subgroup** of $G$, and it consists of all finite products of commutators of $G$.

**Theorem 1.1.17** *Let $G$ be a group and $K$ be the commutator subgroup of $K$. Then*

*(i)   K is a normal subgroup of G;*
*(ii)  the quotient group G/K is always commutative;*
*(iii) the group G is commutative if its commutator subgroup K is the trivial group.*

### 1.1.2  Fundamental Theorem of Algebra

This subsection states the fundamental theorem of algebra and its proof is given in Chap. 2 by using the tools of homotopy theory. The completeness of the field $\mathbf{C}$ of complex numbers follows directly from the fundamental theorem of algebra.

**Definition 1.1.18** A field $F$ is said to be **algebraically closed or complete** if every polynomial ring $f(x)$ of degree $n$ $(n \geq 1)$ over $F$, has a root in $F$.

**Example 1.1.19** The field $\mathbf{C}$ is algebraically closed. It follows from fundamental theorem of Algebra 2.23.1.

**Theorem 1.1.20** (Fundamental Theorem of Algebra) *Every nonconstant polynomial with coefficients in the field* $\mathbf{C}$ *has a root in* $\mathbf{C}$.

### 1.1.3  Modules and Exact Sequences

**Definition 1.1.21** A module $M$ over a commutative ring $R$ with nonzero identity element $e$ is an additive abelian group $M$ together with an external law of composition

$$m : R \times M \to V,$$

the image of $(\alpha, x)$ under $m$ abbreviated $\alpha x$, such that the following conditions are satisfied:

(i)   $ex = x$;
(ii)  $\alpha(x + y) = \alpha x + \alpha y$;
(iii) $(\alpha + \beta)x = \alpha x + \beta x$;
(iv)  $(\alpha\beta)x = \alpha(\beta x)$
      $\forall x, y \in M, \alpha, \beta \in R$.

**Example 1.1.22** (i) Every additive abelian group is a module over the ring $(\mathbf{Z})$.
(ii) Every module over a field $F$ is a vector space.
(iii) Every ideal of a ring $R$ is a module over $R$.

**Remark 1.1.23** The definitions of submodules and quotient modules are analogous to the definitions corresponding to subgroups and quotient groups.

**Definition 1.1.24**  Let $M$ and $N$ be two modules over the same ring $R$. Then a group homomorphism $f$ is said to be an **$R$-module homomorphism** if

$$f(x+y) = f(x) + f(y), \ f(rx) = rf(x), \ \text{for all } x, y \text{ in } M \text{ and } r \text{ in } R.$$

**Definition 1.1.25**  A sequence of modules or groups and their homomorphisms

$$\cdots \to M_n \xrightarrow{f_n} M_{n+1} \xrightarrow{f_{n+1}} M_{n+2} \xrightarrow{f_{n+2}} M_{n+3} \to \cdots$$

(i)  is said to be exact at $M_{n+1}$ if $Im\, f_n = ker\, f_{n+1}$ and
(ii)  is said to be **exact** if it is exact at every $M_{n+1}$ in the sense that

$$Im\, f_n = ker\, f_{n+1}, \ \forall n.$$

**Remark 1.1.26**  There exist many results that compare groups or modules of an exact sequence, but only some of them are given which are used in the book. As every abelian group is a module over $\mathbf{Z}$, every result valid for modules is also true for abelian groups.

**Proposition 1.1.27**  *Consider an exact sequence of four modules or abelian groups*

$$\{0\} \xrightarrow{f} M \xrightarrow{g} N \xrightarrow{h} \{0\},$$

*where the end module or group $\{0\}$ is the trivial module or group. Then $g$ is an isomorphism.*

**Proposition 1.1.28**  *Consider an exact sequence of five groups*

$$\{0\} \to M \xrightarrow{f} N \xrightarrow{g} P \to \{0\},$$

*where group $\{0\}$ is the trivial group. If $h : P \to N$ is a homomorphism such that $g \circ h$ is the identity map on $P$ and $N$ is abelian, then the groups*

$$N \cong M \oplus P.$$

## 1.2   Some Basic Concepts of Topology

This subsection communicates some basic concepts of topology. For their detailed study, see Basic Topology, Volume 1.

## *1.2.1  Homeomorphic Spaces*

This subsection communicates the concept of a homeomorphism in topology which is a basic concept in topology. This concept is analogous to the concept of an isomorphism between algebraic objects such as groups or rings. Every homeomorphism is a bijective map that preserves topological structure involved. So, every classification problem in topology involves classification of topological spaces up to homeomorphism. Its precise definition is formulated in Definition 1.2.1.

**Definition 1.2.1** Let $X$ and $Y$ be two topological spaces. A map

$$f : X \to Y$$

is said to be **continuous** if

$$f^{-1}(U) \subset X$$

is an open set in $X$ for every open set $U$ in $Y$.

**Definition 1.2.2** A continuous map $f : X \to Y$ between topological spaces $X$ and $Y$ is said to be a **homeomorphism** if $f$ is bijective and

$$f^{-1} : Y \to X$$

is also continuous.

**Remark 1.2.3** If $f : X \to Y$ is a homeomorphism, then both the maps $f$ and $f^{-1}$ are continuous in the sense that

(i)  the map $f$ not only sends points of $X$ to points of $Y$ in a (1-1) manner,

(ii)  but $f$ also sends open sets of $X$ to open sets of $Y$ in a (1-1) manner.

This implies that $X$ and $Y$ are topologically the same in the sense that a topological property enjoyed by $X$ is also enjoyed by $Y$ and conversely.

**Example 1.2.4** If the map

$$f : X \to Y$$

is a homeomorphism, then $X$ is compact (or connected) if $Y$ is compact (or connected), because compactness and connectedness properties are topological in the sense that they are shared by homeomorphic spaces.

**Example 1.2.5** Let $X$ and $Y$ be topological spaces. A bijective map

$$f : X \to Y$$

is not necessarily continuous. For example, let $\mathbf{R}$ be the set of real numbers endowed its usual topology $\sigma$ and $\mathbf{R}_l$ be the same set endowed with the lower topology. Then the identity map

$$f : \mathbf{R} \to \mathbf{R}_l, x \mapsto x$$

is a bijection, but it is not continuous.

**Example 1.2.6** (i) The open interval $(0, 1) \subset \mathbf{R}$ and the real line space $\mathbf{R}$ with usual topology are homeomorphic spaces.

(ii) The open ball $B = \{x = (x_1, x_2) \in \mathbf{R}^2 : \|x\| < 1\} \subset \mathbf{R}^2$ with Euclidean topology is homeomorphic to the whole plane $\mathbf{R}^2$.

(iii) The open square $A = \{(x, y) \in \mathbf{R}^2 : 0 < \langle x, y \rangle < 1\} \subset \mathbf{R}^2$ with Euclidean topology is homeomorphic to the open ball $B$ defined in (ii).

(vi) The cone $A = \{(x, y, z) \in \mathbf{R}^3 : x^2 + y^2 = z^2, z > 0\}$ with Euclidean topology is homeomorphic to the plane $\mathbf{R}^2$.

(v) Consider the $n$-sphere $S^n = \{x \in \mathbf{R}^{n+1} : \|x\| = 1, n \geq 1\}$, endowed with Euclidean topology having its north pole

$$N = (0, 0, \ldots, 1) \in \mathbf{R}^{n+1}$$

and its south pole

$$S = (0, 0, \ldots, -1) \in \mathbf{R}^{n+1}.$$

(a) The stereographic projection

$$f : S^n - N \to \mathbf{R}^n, x \mapsto \frac{1}{1 - x_{n+1}}(x_1, x_2, \ldots, x_n),$$

for every $x = (x_1, x_2, \ldots, x_{n+1}) \in S^n - N$ is a homeomorphism.

(b) The space $S^n - S$ is homeomorphic to $S^n - N$.

(vi) A circle minus (deleted) any of its point is homeomorphic to a line segment, and a closed arc is homeomorphic to a closed line segment.

**Theorem 1.2.7** *Consider the $n$-cube $\mathbf{I}^n = \{(x_1, x_2, \ldots, x_n) \in \mathbf{R}^n : 0 \leq x_i \leq 1\}$. Its interior*

$$\text{Int } \mathbf{I}^n = \{(x_1, x_2, \ldots, x_n) \in \mathbf{R}^n : 0 < x_i < 1\}$$

*and its boundary $\partial \mathbf{I}^n = \dot{\mathbf{I}}^n = \mathbf{I}^n - \text{Int } \mathbf{I}^n$. Then*

*(i) $\mathbf{I}^n$ is homeomorphic to the n-ball $B^n$ in $\mathbf{R}^n$ and*

*(ii) under this homeomorphism, $\partial \mathbf{I}^n = \mathbf{I}^n - \text{Int } \mathbf{I}^n$ corresponds to the $(n - 1)$-sphere $S^{n-1}$.*

## 1.2.2 Connectedness and Locally Connectedness

This subsection conveys the concepts of connectedness and locally connectedness.

**Definition 1.2.8** A topological space $X$ is said to be connected if the only sets which are both open and closed are $\emptyset$ and $X$.

**Remark 1.2.9** Connectedness of topological spaces is an important topological property and is characterized by the following theorem.

**Theorem 1.2.10** *A topological space $X$ is connected if it is not the union of two disjoint nonempty open sets.*

**Definition 1.2.11** Let $X$ be a topological space and $x$ be a point of $X$. Then $X$ is said to be locally connected at $x$, if for every open set $V$ containing x there exists a connected open set $U$ with $x \in U \subset V$. The space $X$ is said to be locally connected if it is locally connected at $x$ for all $x \in X$.

**Example 1.2.12** The Euclidean space $\mathbf{R}^n$ is connected and locally connected for all $n \geq 1$.

**Remark 1.2.13** The continuous image of a locally connected space may not be locally connected.

**Definition 1.2.14** A path in a topological space $X$ is a continuous map $f : \mathbf{I} \to X$ from the closed unit interval $\mathbf{I}$ to $X$.

**Definition 1.2.15** A topological space $X$ is said to be path connected, if any two points of $X$ can be joined by a path.

**Remark 1.2.16** A path-connected space is connected. A connected open subset of a Euclidean space is path connected.

**Example 1.2.17** For $n > 0$, the $n$-sphere $S^n$ is path-connected.

**Definition 1.2.18** A topological space $X$ is said to be locally path connected if for each $x \in X$, and each nbd $U$ of $x$, there is a path-connected nbd $V$ of $x$ which is contained in $U$.

**Example 1.2.19** The following spaces in real analysis are connected.

(i) The space $\mathbf{R}$ of real numbers;
(ii) Any interval in $\mathbf{R}$;
(iii) $\mathbf{R}^n$;
(iv) Any ball or cube in $\mathbf{R}^n$;
(v) The continuous image of a connected space is connected.

### 1.2.3 Compactness, Locally Compactness and Paracompactness

This subsection conveys the concept of compactness which is used throughout the book and that of paracompactness which is specially used in the classification of vector bundles.

**Definition 1.2.20** An open covering of a topological space $X$ is a family $\{U_i\}$ of open sets of $X$, whose union is the whole set $X$.

**Definition 1.2.21** A topological space $X$ is said to be **compact** if every open covering of $X$ has a finite subcovering.

**Remark 1.2.22** This means that from any open covering $\{U_i\}$ of a compact space $X$, we can choose finitely many indices $i_j, j = 1, 2, \ldots, n$ such that $\bigcup_{j=1}^{n} U_{i_j} = X$. If $X$ is a compact space, every sequence of points $x_n$ of $X$ has a convergent subsequence, which means, every subsequence $x_{n_1}, x_{n_2}, \ldots, x_{n_t}, \ldots$, converges to a point of $X$. For metric spaces, this condition is equivalent to compactness.

**Proposition 1.2.23** *A compact subspace of a Hausdorff topological space $X$ is closed in $X$ and every closed subspace of a compact space is compact.*

**Definition 1.2.24** A topological space $X$ is said to be **locally compact** if each of its points has a compact neighborhood (nbd).

**Example 1.2.25** Any compact space, the space $\mathbf{R}^n$, any discrete space, any closed subset of a locally compact space are locally compact spaces. On the other hand, the space $\mathbf{Q}$ of rational numbers is not locally compact.

**Definition 1.2.26** A topological space $X$ is said to be a Baire space if intersection of each countable family of open dense sets in $X$ is dense in $X$.

**Example 1.2.27** Every locally compact Hausdorff space is a Baire space.

**Definition 1.2.28** A topological space $X$ is said to be **compactly generated** if $X$ is a Hausdorff space and each subset $A$ of $X$ satisfying the property that $A \cap C$ is closed for every compact subset $C$ of $X$ is itself closed.

**Remark 1.2.29** If $X$ and $Y$ are two topological spaces such that $X$ is locally compact and $Y$ is compactly generated, then their Cartesian product is compactly generated.

**Definition 1.2.30** A topological space $X$ is said to be **paracompact** if every open covering of $X$ has a locally finite subcovering of $X$.

**Example 1.2.31** (i) $\mathbf{R}^n$ is paracompact.
(ii) Every closed subspace of a paracompact space is paracompact; but a subspace of a paracompact space is not necessarily paracompact.

**Definition 1.2.32**  A topological space $X$ is said to be **countably compact** if every countable open covering of $X$ has a finite subcovering.

**Theorem 1.2.33**  (Cantor's intersection theorem) *A topological space $X$ is countably compact if every descending chain of nonempty closed sets of $X$ has a nonempty intersection.*

### 1.2.4  Weak Topology

This subsection communicates the concept of weak topology which is utilized to construct some important topological spaces in this book.

**Definition 1.2.34**  Let

$$X_1 \subset X_2 \subset X_3 \subset \cdots$$

be a chain of closed inclusions of topological spaces. Then its set union $\bigcup_{i \geq 1} X_i$, which is the union of the sets $X_i$ defines a topology by declaring a subset $K \subset \bigcup_{i \geq 1} X_i$ to be closed if its intersection $K \cap X_i$ is closed in $X_i$ for all $i \geq 1$. This topology is known as **union topology**. It is also called **weak topology** with respect to the subspaces.

**Example 1.2.35**  (i)  The infinite sphere $S^{\infty} = \bigcup_{n=0}^{\infty} S^n$ has the weak topology.

(ii)  The infinite projective space $\mathbf{R}P^{\infty} = \bigcup_{n=0}^{\infty} \mathbf{R}P^n$ has the weak topology.

(iii)  The infinite complex projective space $\mathbf{C}P^{\infty} = \bigcup_{n=0}^{\infty} \mathbf{C}P^n$ has the weak topology.

## 1.3  Partition of Unity and Lebesgue Lemma

This section conveys the concept of 'partition of unity' and states Lebesgue lemma with Lebesgue number. A partition of unity subordinate to a given open covering of a topological space is an important concept in mathematics.

### 1.3.1  Partition of Unity

**Definition 1.3.1**  Let $\mathcal{U} = \{U_j : j \in \mathbf{A}\}$ be an open covering of a topological space $X$. A **partition of unity** subordinate to $\mathcal{U}$ consists of a family of functions

$$\{f_j : X \to \mathbf{I} : j \in \mathbf{A}\}$$

such that

(i) $f_j|_{(X-U_j)} = 0$ for all $j$ and each $x \in X$ has a nbd $V$ with the property $f_j|_V = 0$, except for a finite number of indices $j$, and

(ii) $\sum_j f_j(x) = 1$ for all $x \in X$.

**Remark 1.3.2** The sum $\sum_j f_j(x)$ is always a finite sum.

**Remark 1.3.3** Paracompactness of a topological space can be characterized with the help of partition of unity.

**Theorem 1.3.4** *A topological space X is paracompact if every open covering $\mathcal{U}$ of X admits a partition of unity subordinate to $\mathcal{U}$.*

**Proof** See [Dugundji, 1966]. ❏

### 1.3.2 Lebesgue Lemma and Lebesgue Number

Lebesgue lemma is used to prove many important results. This lemma is also called Lebesgue covering lemma.

**Lemma 1.3.5** (Lebesgue) *Let X be a compact metric space. Given an open covering $\{U_\alpha : \alpha \in \mathbf{A}\}$ of X, there exists a real number $\delta > 0$, called **Lebesgue number** of $\{U_\alpha\}$ having the property that every open ball of radius less than $\delta$ lies in some element of $\{U_\alpha\}$.*

## 1.4 Separation Axioms, Urysohn Lemma and Tietze Extension Theorem

This section imposes certain conditions on the topology to obtain some particular classes of topological spaces initially used by P.S. Alexandroff (1896–1982) and H. Hopf (1894–1971). Such spaces are important objects in algebraic topology. Moreover, this section presents Urysohn Lemma and Tietze Extension theorem which are used in this book.

**Definition 1.4.1** A topological space $(X, \tau)$ is said to be a

(i) $T_1$-space (due to Frechet) if for every pair of distinct points $p, q$ in $X$, there exist two open sets $U, V$ such that

$$p \in U, q \in V, p \notin V, \text{ and } q \notin U.$$

In other words, every pair of distinct points is weakly separated in $(X, \tau)$:
equivalently, for every pair of distinct points $p, q$ in $X$, there exist a neighborhood
of $p$ which does not contain $q$ and a neighborhood of $q$ which does not contain
$p$.

(ii) Hausdorff space (due to Hausdorff) if any two distinct points are strongly separated in $(X, \tau)$:
equivalently, distinct points have disjoint neighborhoods.

(iii) Regular space (due to Vietoris) if any closed set $F$ and any point $p \notin F$ are
always strongly separated in $(X, \tau)$.

(iv) Normal space (due to Tietze) if any two disjoint closed sets are strongly separated in $(X, \tau)$, equivalently, each pair of disjoint closed sets have disjoint
neighborhoods.

**Remark 1.4.2**  It is not true that a nonconstant real-valued continuous function can
always be defined on a given space. But on normal spaces there always exist nonconstant real-valued continuous functions. Urysohn lemma characterizes normal spaces
by real-valued continuous functions.

**Lemma 1.4.3**  (Urysohn ) *A topological space $(X, \tau)$ is normal if and only if every
pair of disjoint closed sets $P, Q$ in $(X, \tau)$ are separated by a continuous real-valued
function $f$ on $(X, \tau)$, such that*

$$f(x) = \begin{cases} 0, \text{ for all } x \in P \\ 1, \text{ for all } x \in Q \end{cases}$$

*and*

$$0 \leq f(x) \leq 1 \text{ for all } x \in X.$$

## 1.5  Function Spaces

This section introduces the concept of function spaces topolozied by the compact
open topology. Function spaces play an important role in topology and geometry.

**Definition 1.5.1**  (*Compact open topology*)  Let $X$ and $Y$ be topological spaces and
$Y^X$ (or $F(X, Y)$) be the set of all continuous functions $f : X \to Y$. Then a topology,
called compact open topology, can be endowed on $F(X, Y)$ by taking a subspace for
the topology of all sets of the form

$$V_{K,U} = \{f \in F(X, Y) : f(K) \subset U\},$$

where $K \subset X$ is compact and $U \subset Y$ is open.
Let $E : Y^X \times X \to Y$, $(f, x) \mapsto f(x)$ be the evaluation map. Then given a function
$h : Z \to Y^X$, the composite

$$\psi : Z \times X \xrightarrow{h \times 1_d} Y^X \times X \xrightarrow{E} Y$$

i.e., $\psi = E \circ (h \times 1_d) : Z \times X \to Y$ is a function.

**Theorem 1.5.2** (Theorem of Exponential Correspondence) *Let X be a locally compact Hausdorff space and Y, Z be topological spaces. Then a function $f : Z \to Y^X$ is continuous if*

$$E \circ (f \times 1_d) : Z \times X \to Y$$

*is continuous.*

**Theorem 1.5.3** (Exponential Law) *Let X be a locally compact Hausdorff space, Z be a Hausdorff space, and Y be a topological space. Then the function*

$$\psi : (Y^X)^Z \to Y^{Z \times X}$$

*defined by $\psi(f) = E \circ (f \times 1_d)$ is a homeomorphism.*

**Proposition 1.5.4** *If X is a compact Hausdorff space and Y is metricized by a metric d, then the space $Y^X$ is metricized by the metric $d'$ defined by*

$$d'(f, g) = \sup\{d(f(x), g(x)) : x \in X\}.$$

## 1.6  Manifolds

This section defines manifolds which form an important class of geometrical objects in topology. An $n$-manifold is a Hausdorff topological space which looks locally like Euclidean $n$-space $\mathbf{R}^n$, but not necessarily globally. A local Euclidean structure to manifold by introducing the concept of a chart is utilized to use the conventional calculus of several variables. Due to linear structure of vector spaces, for many applications in mathematics and in other areas it needs generalization of metrizable vector spaces, maintaining only the local structure of the latter. On the other hand, every manifold can be considered as a (in general nonlinear) subspace of some vector space. Both aspects are used to approach the theory of manifolds. Since dimension of a vector space is a locally defined property, a manifold has a dimension.

**Definition 1.6.1** An $n$-dimensional (topological) manifold or an $n$-**manifold** $M$ is a Hausdorff space with a countable basis such that each point of $M$ has a neighborhood homeomorphic to an open subset of $\mathbf{R}^n$. An one-dimensional manifold is called a **curve**, and a two-dimensional manifold is called a **surface**.

**Example 1.6.2** $S^2$, torus, $\mathbf{R}P^2$ are examples of surfaces.

**Remark 1.6.3** All manifolds $M$ in this book are assumed to be paracompact to ensure that $M$ is a separable metric space.

**Definition 1.6.4** An $n$-dimensional differentiable manifold or a smooth manifold $M$ is a Hausdorff topological space having a countable open covering $\{U_1, U_2, \ldots\}$ such that

DM(1)   for each $U_i$, there is a homeomorphism $\psi_i : U_i \to V_i$, where $V_i$ is an open disk in $\mathbf{R}^n$;

DM(2)   if $U_i \cap U_j \neq \emptyset$, the homeomorphism $\psi_{ji} = \psi_j \circ \psi_i^{-1} : \psi_i(U_i \cap U_j) \to \psi_j (U_i \cap U_j)$ is a differentiable map (or smooth maps) between open subsets of $\mathbf{R}^n$.

$(U_i, \psi_i)$ is called a **local chart** of $M$, and $\{(U_i, \psi_i)\}$ is a set of local charts of $M$.

**Definition 1.6.5** Let $M$ and $N$ be two smooth manifolds and $f : M \to N$ be a map. A chart $(\psi, U)$ for $M$ is said to be **adapted to** $f$ **by a chart** $(\phi, V)$ for $N$ if $f(U) \subset V$. Then the map

$$\phi \circ f \circ \psi^{-1} : \psi(U) \to \phi(V)$$

is well-defined, and it is called the **local representation** of $f$ at the point $x \in U$ in the given charts. The map $f$ is said to be differentiable (or smooth ) at $x$ if it has a local representation at $x$ which is differentiable (or smooth). This is well-defined because a local representation is a map between open sets in Euclidean spaces.

**Example 1.6.6** $\mathbf{R}^n, S^n, RP^n$ are $n$-dimensional differentiable manifolds.

**Example 1.6.7** $CP^n$ is a $2n$-dimensional differentiable manifold.

**Definition 1.6.8** A Hausdorff space $M$ is called an $n$-dimensional **manifold with boundary** $(n \geq 1)$ if each point of $M$ has a neighborhood homeomorphic to the open set in the subpace of $\mathbf{R}^n$.

**Example 1.6.9** The $n$-dimensional disk $D^n$ is an $n$-manifold with boundary.

**Remark 1.6.10** Let $\mathcal{S} = \{(U_i, \psi_i)\}$ be a set of local charts of a differentiable manifold $M$. Then $\mathcal{S}$ is said to be a differentiable structure on $M$. Every subset of $\mathcal{S}$ which satisfies $M = \cup U_i$, **DM(1)** and **DM(2)** is called a basis for the differential structure $\mathcal{S}$.

## 1.7   Topological Groups and Lie Groups

This section recalls the introductory concepts of topological groups and also Lie groups (studied in **Basic Topology: Volume 2**) of the present book series. Lie groups provide special topological groups which are also smooth manifolds satisfying compatibility conditions laid down in Definition 1.7.14. The detailed study of topological groups and Lie groups is available in **Basic Topology: Volume 2**

## 1.7.1 Topological Groups: Definitions and Examples

This subsection illustrates the concept of topological groups with examples. The basic concept of a topological group is that it is an abstract group endowed with a topology such that the multiplication and inverse operations are both continuous. This concept was accepted by mathematicians in the early 1930s.

**Definition 1.7.1** A nonempty set $G$ is said to be a **topological group** if it satisfies the following axioms:

**TG(1)**: $G$ is algebraically a group;

**TG(2)**: $G$ is topologically a Hausdorff space; i.e., every pair of distinct points of $G$ are strongly separated by disjoint open sets;

**TG(3)**: group multiplication $m : G \times G \to G, (x, y) \mapsto xy$ is continuous, where the topology on $G \times G$ is endowed with product topology;

**TG(4)**: group inversion $v : G \to G, x \to x^{-1}$ is continuous.

**Remark 1.7.2** Some authors do not assume 'Hausdorff property' for a topological group. The conditions **TG(3)** & **TG(4)** in Definition 1.7.1 are equivalent to the single condition

**TG(5)**: the map $\psi : G \times G \to G, (x, y) \mapsto xy^{-1}$ is continuous.

**Example 1.7.3** $\mathbf{R}^n$ (under usual addition) and $S^1 = \{z \in \mathbf{C} : |z| = 1\}$ (under usual multiplication of complex numbers) are important examples of topological groups.

**Example 1.7.4** (i) The general real linear group GL $(n, \mathbf{R})$ of all invertible $n \times n$ matrices over $\mathbf{R}$ is an important topological group, which is neither compact nor connected.

(ii) GL $(n, \mathbf{C})$ is the set of all $n \times n$ nonsingular matrices with complex entries, called the general complex linear group. GL $(n, \mathbf{C})$ is a topological group with $\dim_{\mathbf{C}}$ GL $(n, \mathbf{C}) = n^2$. It is not compact.

(iii) The set U$(n, \mathbf{C}) = \{M \in$ GL $(n, \mathbf{C}) : MM^* = I\}$ forms a subgroup of GL $(n, \mathbf{C})$, where $M^*$ is the transpose of the complex conjugate of $M$. This subgroup is compact.

(iv) For the sympletic group $SU(n, \mathbf{H}) = \{A \in$ GL $(n, \mathbf{H}) : AA^* = I\}$ which is an quaternionic analogue, see Exercise 1.9 of Sect. 1.9.

**Definition 1.7.5** Let $G$ and $K$ be two topological groups. A homomorphism $f : G \to K$ is a continuous map such that $f$ is a group homomorphism in abstract sense. An isomorphism $f : G \to K$ between two topological groups is a homeomorphism and is also a group homomorphism between $G$ and $K$.

**Example 1.7.6** The special orthogonal group SO$(2, \mathbf{R})$ and the circle group $S^1$ are both topological groups. These two topological groups are isomorphic because there exists an isomorphism of topological groups

$$\psi : SO(2, \mathbf{R}) \to S^1, \quad \begin{pmatrix} \cos\theta & -\sin\theta \\ \sin\theta & \cos\theta \end{pmatrix} \mapsto e^{i\theta}.$$

## 1.7.2  Actions of Topological groups and Orbit Spaces

This subsection communicates the concept of topological group actions on topological spaces which provide many geometrical objects such as real and complex
projective spaces, torus, lens spaces, etc., as orbit spaces obtained by specifying
actions of topological groups.

**Definition 1.7.7** Let $G$ be a topological group with identity element $e$ and $X$ be a
topological space. If $G \times X$ has the product topology, then $G$ **is said to act on** $X$
**from the left** if there is continuous map $\mu : G \times X \to X$, the image $\mu(g, x)$, denoted
by $g(x)$ or $gx$ such that

(i)  $e(x) = x$  for all $x \in X$;
(ii)  $(gk)(x) = g(k(x))$,  for all $x \in X$, and $g, k \in G$.

The group $G$ is then called a  **topological transformation group**  of $X$ relative
to the group action $\mu$. It is sometimes denoted by the triple $(G, X, \mu)$. A topological
space $X$ endowed with a left $G$-action on $X$ is said to be a **left $G$-space. A right
action and a right $G$-space are defined in an analogous way.**

There is a one-to-one correspondence between the left and right $G$-structures on
$X$. So it is sufficient to study only one of them according to the situation.

**Example 1.7.8**  For the general linear group $GL(n, \mathbf{R})$, the Euclidean $n$-space $\mathbf{R}^n$ is
a left $GL(n, \mathbf{R})$-space under usual multiplication of matrices.

**Definition 1.7.9** Let $X$ be a left $G$-space. Two given elements $x, y$ in $X$ are called
**$G$-equivalent**, if $y = g(x)$ (*i.e.*, $y = gx$) for some $g \in G$. The relation of being $G$-
equivalent is an equivalence relation on $X$ and the corresponding quotient space
$X mod\ G$ endowed with quotient topology induced from $X$ (i.e., the largest topology
such that the projection map $p : X \to X mod\ G, x \mapsto G(x)$ is continuous, where
$G(x) = \{g(x) : g \in G\}$ is called the  **orbit space of**  $x \in G$. For an element $x \in X$,
$G(x)$ is called the orbit of $x$ and the subgroup $G_x$ of $G$ defined by $G_x = \{g \in G :
g(x) = x\}$ is called the **stabilizer or isotropy group at**  $x$ of the corresponding group
action.

**Example 1.7.10** (Geometrical)  The actions of a given topological group on the
same topological space may be different. For example, let $T$ be the torus in $\mathbf{R}^3$
obtained by rotating the circle $C : (x - 3)^2 + z^2 = 1$ about the $z$-axis and $\mathbf{Z}_2 = <
h >$, generated by $h$. Consider the three actions

(i)
$$\psi_1 : \mathbf{Z}_2 \times T \to T, \ (h, (x, y, z)) \mapsto (x, -y, -z),$$

which geometrically represents rotation of $T$ through an angle $180°$ about the
$x$-axis;

(ii)
$$\psi_2 : \mathbf{Z}_2 \times T \to T, \ (h, (x, y, z)) \mapsto (-x, -y, z),$$

which geometrically represents rotation of $T$ through an angle $180^o$ about the $z$-axis;

(iii)
$$\psi_3 : \mathbf{Z}_2 \times T \to T, \ (h, (x, y, z)) \mapsto (-x, -y, -z)$$

which geometrically represents reflection of $T$ about the origin.

Each of $\psi_1$, $\psi_2$ and $\psi_3$ is a homeomorphism of $T$ of order 2 with the sphere, torus and Klein bottle as the resulting orbit spaces, respectively.

**Proposition 1.7.11** *Let $X$ be a left $G$-space and $g \in G$ be an arbitrary point. Then the map*

$$\psi_g : X \to X, \ x \mapsto g(x) \ (= gx)$$

*is a homeomorphism.*

**Proof** $\psi_g$ is continuous for every $g \in G$, since the group action is continuous. So, $\psi_g$ and $\psi_{g^{-1}}$ are two continuous maps such that $\psi_g \circ \psi_{g^{-1}} = I_X = \psi_{g^{-1}} \circ \psi_g$, and hence, it follows that $\psi_g$ is a homeomorphism. ❑

**Remark 1.7.12** Let $X$ be a *left $G$-space* and **homeo(X)** be the set of homeomorphisms $\psi_g : X \to X$ for all $g \in G$. Then **homeo(X)** $= \{\psi_g : g \in G\}$ is a group under usual composition of mappings. Proposition 1.7.13 shows that this group is closely related to the group action of $G$ on $X$.

**Proposition 1.7.13** *Let $X$ be a left $G$-space. Then the map*

$$f : G \to \mathbf{homeo(X)}, \ g \mapsto \psi_g$$

*is a group homomorphism.*

**Proof** Since for each $g \in G$, $\psi_g$ is a homeomorphism, the map $f$ is well defined. Again, since for $g, k \in G$, $f(gk) = \psi_{gk} = f(g)f(k)$ asserts that $f$ is a group homomorphism. ❑

## 1.7.3 Lie Groups and Examples

This subsection communicates the concept of Lie groups which are continuous transformation groups. Lie groups occupy a vast territory of topological groups carrying a differentiable structure and play a key role in the study of topology, geometry and physics. S. Lie developed his theory of continuous transformations with an eye to investigate differential equations. The basic ideas of his theory appeared in his paper (Lie 1880) published in 'Math. Ann., Vol 16, 1880.' A Lie group is a topological group possessing the structure of a smooth manifold on which the group operations are smooth functions. Its mathematical formulation is given in Definition 1.7.14.

**Definition 1.7.14** A topological group $G$ with identity element $e$ is said to be a **real Lie group** if

(i) $G$ is a real differentiable manifold;
(ii) the group multiplication

$$m : G \times G \rightarrow G, \; (x, y) \mapsto xy$$

and the group inversion

$$v : G \rightarrow G, \; x \mapsto x^{-1}$$

are both differentiable.

**Definition 1.7.15** A topological group $G$ is said to be a **complex Lie group** if

(i) $G$ is a complex manifold;
(ii) the group multiplication

$$m : G \times G \rightarrow G, (x, y) \mapsto xy$$

and the group inversion
$$v : G \rightarrow G, \; x \mapsto x^{-1}$$

are both holomorphic.

**Definition 1.7.16** The **dimension of a Lie group** is defined to be its dimension as a manifold.

**Example 1.7.17** Examples of Lie groups are plenty.

(i) The real line $\mathbf{R}$ is a Lie group under usual addition of real numbers.
(ii) $\mathbf{R}^+ = \{x \in \mathbf{R} : x > 0\}$ is a Lie group under usual multiplication of real numbers.
(iii) $\mathbf{R}^2$ is a Lie group under pointwise addition given by

$$(x_1, y_1) + (x_2, y_2) = (x_1 + x_2, \; y_1 + y_2).$$

(iv) Every finite dimensional vector space $V$ over $\mathbf{R}$ is a Lie group, since the map

$$f : V \times V \rightarrow V, (x, y) \mapsto x + y$$

is linear and hence differentiable. Similarly, the map $g : V \rightarrow V, x \rightarrow -x$ is smooth. In particular, the additive group $\mathbf{R}^n$ with its standard structure as a differentiable manifold is a Lie group.

## 1.8   Category, Functor and Natural Transformation

This section conveys the basic concepts of **category theory,** which provides a convenient language to unify several mathematical results and unifies many basic concepts of mathematics in an accessible way. This language born through the work of S. Eilenberg (1913–1998) and S. MacLane (1900–2005) during 1942–1945 is used throughout the present book. This section conveys the introductory concepts of category, functor and natural transformation in the language of category theory.

### *1.8.1   Introductory Concept of Category*

**Definition 1.8.1**   **A category** $\mathcal{C}$ consists of

(i) a certain family of objects $X, Y, Z, \ldots$ usually denoted by $ob(\mathcal{C})$;
(ii) for every ordered pair of objects $X, Y \in ob(\mathcal{C})$, a set of morphisms from $X$ to $Y$, denoted by $[X, Y]$ is specified;
(iii) for every ordered triple of objects $X, Y, Z \in ob(\mathcal{C})$ and any pair of morphisms $f \in [X, Y], g \in [Y, Z]$, their composition, denoted by $g \circ f \in [X, Z]$ is defined with the following properties:
(iv) **(associativity)**: if $f \in [X, Y]$, $g \in [Y, Z]$, $h \in [Z, K]$, then $h \circ (g \circ f) = (h \circ g) \circ f \in [X, K]$;
(v) **(existence of identity morphism)**: for every object $Y \in ob(\mathcal{C})$, there is a morphism $1_Y$, called identity morphism with the property: for every $f \in [X, Y]$, $g \in [Y, Z]$, both the equalities $1_Y \circ f = f$ and $g \circ 1_Y = g$ hold.

**Remark 1.8.2**   The identity morphism in a category is unique for each of its object.

**Example 1.8.3**   (i) The family of all sets and their mappings form a category, denoted by $\mathcal{S}et$, where objects are all sets, morphisms are all possible mappings between them and the composition is the usual composition of functions.
(ii) The family of all groups and their homomorphisms form a category, denoted by $\mathcal{G}rp$, where objects are all groups, morphisms are all possible homomorphisms between them and the composition is the usual composition of mappings.
(iii) The family of all abelian groups and their homomorphisms form a category, denoted by $\mathcal{A}b$, where objects are all abelian groups, morphisms are all possible homomorphisms between them and the composition is the usual composition of mappings.
(iv) The family of all vector spaces and their linear transformations form a category, denoted by $\mathcal{V}ect$.
(v) The family of all topological spaces and their continuous maps form a category, denoted by $\mathcal{T}op$.
(vi) The family of all pointed topological spaces and their base point preserving continuous maps form a category, denoted by $\mathcal{T}op_*$.

**Remark 1.8.4**  The concepts of bijective mappings of sets, isomorphism of groups, rings or vector spaces, topological spaces and so on can be unified through the concept of an equivalence in category theory.

**Definition 1.8.5**  A morphism $f \in [X, Y]$ in a category $C$ is said to be an **equivalence** if there is a morphism $g \in [Y, X]$ in the category $C$ such that

$$f \circ g = 1_Y, \ g \circ f = 1_X.$$

If $f$ is an equivalence in $C$, then $g$ is also an equivalence in $C$ and the objects $X$ and $Y$ are said to be equivalent.

**Example 1.8.6**  The equivalences and equivalent objects in the following categories are specified:

(i)  In the category $Set$, equivalences are bijective mappings and equivalent objects are precisely the sets having the same cardinality.
(ii)  In the category $Grp$, equivalences are group isomorphisms and equivalent objects are precisely isomorphic groups.
(iii)  In the category $Top$, equivalences are homeomorphisms and equivalent objects are precisely homeomorphic topological spaces.
(iv)  In the category $Ring$, equivalences are ring isomorphisms and equivalent objects are precisely isomorphic rings.

### 1.8.2   Introductory Concept of Functors

A functor is a natural mapping from one category to the other in the sense that it preserves the identity morphism and composites of well-defined morphisms. It plays a key role in converting a problem of one category to the problem of other category to have a better chance for solution. Important examples of functors from the topological viewpoint are available in Chaps. 2, 3 and 5.

**Definition 1.8.7**  Given two categories $C_1$ and $C_2$, **a covariant functor**

$$F : C_1 \rightarrow C_2, X \rightarrow F(X), f \mapsto F(f)$$

from category $C_1$ to the category $C_2$, consists of

(i)  an **object function** which assigns to every object $X \in C_1$ an object $F(X) \in C_2$ and
(ii)  a **morphism function** which assigns to every morphism

$$f \in [X, Y]$$

in the category $C_1$, a morphism

$$F(f) \in [F(X), F(Y)]$$

in the category $C_2$ such that

(iii)  $F(1_X) = 1_{F(X)}$ for every identity morphism $1_X$;

(iv)  for morphisms $f \in [X, Y], g \in [Y, Z]$ in the category $C_1$, the equality

$$F(g \circ f) = F(g) \circ F(f)$$

holds in the category $C_2$.

A contravariant functor is defined dually and is formulated in Definition 1.8.8.

**Definition 1.8.8**  Given two categories $C_1$ and $C_2$, **a contravariant functor**

$$F : C_1 \to C_2, X \to F(X), f \mapsto F(f)$$

from category $C_1$ to the category $C_2$, consists of

(i)  an object function which assigns to every object $X \in C_1$ an object $F(X) \in C_2$ and

(ii)  a morphism function which assigns to every morphism $f \in [X, Y]$ in the category $C_1$, a morphism

$$F(f) \in [F(Y), F(X)]$$

in the category $C_2$ such that

(iii)  $F(1_X) = 1_{F(X)}$;

(iv)  for morphisms $f \in [X, Y], g \in [Y, Z]$ in the category $C_1$, the equality

$$F(g \circ f) = F(f) \circ F(g)$$

holds in the category $C_2$.

**Example 1.8.9**  (i)  There is a covariant functor $F : \mathcal{G}rp \to \mathcal{S}et$ whose object function assigns to every group $G$ its underlying set $|G|$ and to every group homomorphism $f : G \to K$ in the category $\mathcal{G}rp$ to its corresponding underlying set function $f : |G| \to |K|$ in the category $\mathcal{S}et$. This functor is known as forgetful functor as it forgets the group structure.

(ii)  Let $V$ be an $n$-dimensional vector space over $\mathbf{R}$ and $V^d = L(V, \mathbf{R})$ be the set of all linear transformations $T : V \to \mathbf{R}$. Then it is also an $n$-dimensional vector space over $\mathbf{R}$, called the dual space of $V$, denoted by $V^d$. To each linear transformation $T : V_1 \to V_2$, its dual map $T^d : V_2^d \to V_1^d$ is defined by $T^d(\alpha)(x) = \alpha(T(x)), \forall x \in V_1, \forall \alpha \in V_2^d$. If $\mathcal{V}^n$ be the category of all $n$-dimensional vector space over $\mathbf{R}$ and their linear transformations, then the above results assert that there is a covariant functor

$$F : \mathcal{V}^n \to \mathcal{V}^n, V \mapsto V^d \ \forall \ objects \ V \ \in \mathcal{V}^n, \ and \ T \mapsto T^d \ \forall \ morphisms \ T \in \mathcal{V}^n$$

i.e., the object function maps $V$ to $V^d$ and morphism function maps $T$ to $T^d$ in the category $\mathcal{V}^n$ to itself.

(iii) Given an object $A \in \mathcal{C}$, there is a covariant functor

$$h_C : \mathcal{C} \to \mathcal{C}$$

whose object function assigns to every object $X \in \mathcal{C}$ to the object $h_C(X) = [C, X]$ (the set of all morphisms from the object $C$ to the object $X$ in the category $\mathcal{C}$) and the morphism function assigns to every morphism $f \in [X, Y]$ in $\mathcal{C}$, the morphism

$$h_C(f) : h_C(X) \to h_C(Y), g \mapsto f \circ g$$

in the category $\mathcal{C}$.

(iv) Given an object $A \in \mathcal{C}$, there is a contravariant functor

$$h^C : \mathcal{C} \to \mathcal{C}$$

whose object function assigns to every object $X \in \mathcal{C}$ to the object $h^C(X) = [X, C]$ (the set of all morphisms from the object $X$ to the object $C$ in the category $\mathcal{C}$) and the morphism function assigns to every morphism $f \in [X, Y]$ in $\mathcal{C}$, the morphism

$$h^C(f) : h^C(Y) \to h^C(X), g \mapsto g \circ f$$

in the category $\mathcal{C}$.

**Proposition 1.8.10**  *Let $\mathcal{C}_1$ and $\mathcal{C}_2$ be two categories and $F : \mathcal{C}_1 \to \mathcal{C}_2$ be a functor. If $X$ and $Y$ are two objects equivalent in the category $\mathcal{C}_1$, then the corresponding objects $F(X)$ and $F(Y)$ are also equivalent objects in the category $\mathcal{C}_2$.*

**Proof**  Let $X$ and $Y$ be two equivalent objects in the category $\mathcal{C}_1$. Then there exist two morphisms $f \in [X, Y]$, $g \in [Y, X]$ in the category $\mathcal{C}_1$ such that $f \circ g = 1_Y$ and $g \circ f = I_X$. Then the result follows from the definition of a functor $F$ (may be covariant or contravariant). ❑

**Corollary 1.8.11**  *Let $X$ and $Y$ be two objects in a category $\mathcal{C}_1$ and $F$ be a functor from the category $\mathcal{C}_1$ to the category $\mathcal{C}_2$ such that the objects $F(X)$ and $F(Y)$ are not equivalent. Then the objects $X$ and $Y$ are also not equivalent.*

**Remark 1.8.12**  The Corollary 1.8.11 is applied to show that the groups $\mathbf{Z}_5$ and $\mathbf{Z}_9$ are not isomorphic because the forgetful functor assigns to their underlying sets of different cardinalities. But the cardinal number of the underlying sets of two groups may be the same but the groups may not be isomorphic. For example, the groups $S_3$ and $\mathbf{Z}_6$ are not isomorphic but the cardinal number of their underlying sets is the same.

### 1.8.3   Introductory Concept of Natural Transformation

The subsection defines natural transformation between two functors. Natural transformation is an important concept needed to compare functors with each other of the same variance (i.e., when they are either both covariant or both contravariant).

**Definition 1.8.13** Let $\mathcal{C}_1$ and $\mathcal{C}_2$ be two categories and $F_1, F_2 : \mathcal{C}_1 \to \mathcal{C}_2$ be two functors of the same variance. **A natural transformation**

$$\psi : F_1 \to F_2$$

is a function which assigns to every object $X$ in the category $\mathcal{C}_1$ a morphism $\psi(X)$ such that for every morphism $f \in [X, Y]$ in $\mathcal{C}_1$, the following equality holds:

$$\psi(Y) \circ F_1(f) = F_2(f) \circ \psi(X)$$

(if both $F_1$ and $F_2$ are covariant functors)
   or

$$\psi(X) \circ F_1(f) = F_2(f) \circ \psi(Y)$$

(if both $F_1$ and $F_2$ are contravariant functors). In addition, if for each object $X$ in the category $\mathcal{C}_1$, the natural transformation $\psi(X)$ is an equivalence in the category $\mathcal{C}_2$, then $\psi$ is said to be a **natural equivalence.**

**Example 1.8.14** Given a commutative ring $R$, let $Mod_R$ be the category of all $R$-modules and their $R$-homomorphisms. Given a fixed $R$-module $A$, there is a covariant functor $\pi_A$ from the category $Mod_R$ to itself which assigns to an $R$-module $M$ the $R$-module $[A, M]$ (the module of all $R$-module homomorphisms from $A$ to $M$) and for a morphism $f : M \to N$ in the category $Mod_R$ the morphism $\pi_A(f)$ is defined by

$$\pi_A(f) : [A, M] \to [A, N] : g \mapsto f \circ g.$$

Then $\pi_A$ is a covariant functor. Similarly, for each fixed $R$-module $A$ there is a contravariant functor $\pi^A$ defined from the category $Mod_R$ to itself. Given a morphism $f \in [X, Y]$ in the category $Mod_R$, a natural transformation $f^* : \pi_Y \to \pi_X$ is defined by

$$f^*(M) : \pi_Y(M) \to \pi_X(M), \; h \mapsto h \circ f, \; \forall M \text{ in } Mod_R.$$

**Remark 1.8.15** For topological applications of category theory, see Chaps. 2, 3, 5 and 6.

## 1.9  Exercises

As solving exercises plays an essential role of learning mathematics, various types of exercises are given in this section. They form an integral part of the book series.

1. Show that the circle group $S^1 = \{z \in \mathbf{C} : |z| = 1\}$ in the complex plane is a Lie group (This group is written as $U(1, \mathbf{C})$ or simply as $U(1)$).

2. Prove that the general linear group GL $(n, \mathbf{H})$ over the quaternions $\mathbf{H}$ is a topological group but it is not compact.
   [Hint : In absence of a determinant function in this case, use the result that GL $(n, \mathbf{H})$ is an open subset of an Euclidean space.]

3. Show that the special real linear group

$$SL(n, \mathbf{R}) = \{X \in GL\ (n, \mathbf{R}) : \det X = 1\}$$

   is a noncompact connected topological group and it is a real Lie group with dimension $n^2 - 1$.
   [ Hint : Use the result that $SL(n, \mathbf{R})$ is a subgroup of GL $(n, \mathbf{R})$. It is a hypersurface of GL $(n, \mathbf{R})$. ]

4. Prove that the special complex linear group

$$SL(n, \mathbf{C}) = \{X \in GL\ (n, \mathbf{C}) : \det X = 1\}$$

   is a noncompact connected topological group and is a complex Lie group of dimension $n^2 - 1$.
   [Hint : $SL(n, \mathbf{C})$ is a subgroup of GL $(n, \mathbf{C})$. ]

5. Prove that the orthogonal group

$$O(n, \mathbf{R}) = \{A \in GL\ (n, \mathbf{R}) : AA^t = I = A^tA\}$$

   is a compact nonconnected topological group and it is a real Lie group with dimension $\frac{n(n-1)}{2}$.

6. Show that the special orthogonal group

$$SO(n, \mathbf{R}) = O(n, \mathbf{R}) \cap SL(n, \mathbf{R})$$

   is a real compact connected topological group and it is a real Lie group with dimension $\frac{n(n-1)}{2}$.

7. Prove that the general (complex) linear group $GL(n, \mathbf{C})$ is a topological group and it is a connected, noncompact complex Lie group with dimension $n^2$.

8. Show that the unitary group

$$U(n, \mathbf{C}) = \{A \in GL\ (n, \mathbf{C}) : AA^* = A^*A = I\}$$

is a connected compact topological group, and it is a real Lie group with dimension $n^2$, where $A^*$ denotes the conjugate transpose of $A$ (conjugate means reversal of all the imaginary components).
[ Hint : It is a subgroup of GL $(n, \mathbf{C})$. It is not a complex submanifold of GL $(n, \mathbf{C})$. It can be embedded as a subgroup of GL $(2n, \mathbf{R})$ .]

9. Let $SU(n, \mathbf{C})$ denote the special unitary group defined by

$$SL(n, \mathbf{C}) = U(n, \mathbf{C}) \cap SL(n, \mathbf{C}).$$

Show that the group

$$SU(2, \mathbf{C}) = \{A = \begin{pmatrix} z & w \\ -\bar{w} & \bar{z} \end{pmatrix} : z, w \in \mathbf{C} \text{ and } |z|^2 + |w|^2 = 1\}$$

is isomorphic to $S^3$.
[Hint. Use the form of $A$.]

10. The quaternionic analog of orthogonal unitary groups is the sympletic group

$$SU(n, \mathbf{H}) = \{A \in GL(n, \mathbf{H}) : AA^* = I\},$$

where $A^*$ denotes the quaternionic conjugate transpose of $A$. Prove that it is a compact topological group.

11. Show that the 3-dimensional projective space $\mathbf{R}P^3$ and SO(3, $\mathbf{R}$) are homeomorphic.

12. Let $G$ be a Lie group. Prove the following statements:

(i) the right translation $\mathcal{R} : G \times G \to G, (a, g) \mapsto R_a(g) = ga$ is a free and transitive action and

(ii) left translation $\mathcal{L} : G \times G \to G, (a, g) \mapsto L_a(g) = ag$ is also a free and transitive action.

13. Let $G$ be a Lie group and $M$ be a smooth manifold. Prove the following statements:

(i) the isotropy group $G_x$ of any point $x \in M$ is a Lie subgroup of $G$;

(ii) if $G$ acts freely on $M$, then the isotropy group $G_x$ of any point $x \in M$ is trivial.

14. Show that the orthogonal group O $(n + 1, \mathbf{R})$ acts on $\mathbf{R}P^n$ transitively from left.

15. Show that orthogonal group O $(n, \mathbf{R})$ acts transitively on the Grassmann manifold (see Chap. 5) $G_{n,r}(r \leq n)$.

16. Show that the special orthogonal group SO $(n, \mathbf{R})$ acts transitively on the Stiefel manifold (see Chap. 5) $V_{n,r} = V_r(\mathbf{R}^n), (r \leq n)$.

# References

Adhikari A, Adhikari MR. Basic topology, vol. 1: metric spaces and general topology. India: Springer; 2022a.

Adhikari A, Adhikari MR, Basic topology, vol. 2: topological groups, topology of manifolds and lie groups. India: Springer; 2022b.

Adhikari MR, Adhikari. A. Basic modern algebra with applications. New Delhi, New York, Heidelberg: Springer; 2014.

Adhikari MR. Basic algebraic topolgy and its applications. India: Springer; 2016.

Adhikari MR, Adhikari A, Groups, rings and modules with applications. Hyderabad: Universities Press; 2003.

Adhikari MR, Adhikari A. Textbook of linear algebra: an introduction to modern algebra. New Delhi: Allied Publishers; 2006.

Alexandrov PS, Introduction to set theory and general topology. Moscow; 1979.

Borisovich YU, Bliznyakov N, Izrailevich YA, Fomenko T. Introduction to topology. Moscow: Mir Publishers; 1985.

Bredon GE. Topology and geometry. New York: Springer; 1983.

Chatterjee BC, Ganguly S, Adhikari MR. A textbook of topology. New Delhi: Asian Books; 2002.

Dugundji J. Topology. Newton, MA: Allyn & Bacon; 1966.

MacLane S. Categories for the working mathematician. Springer, New York (1971)

Simmons G.Introduction to topology and modern analysis. McGraw Hill York; 1963.

Williard S General topology. Addision- Wesley; 1970.

# Chapter 2
# Homotopy Theory: Fundamental Group and Higher Homotopy Groups

This chapter officially inaugurates **homotopy theory** to begin a study of algebraic topology by conveying the basic concepts of **homotopy and fundamental groups** born through the work of H. Poincaré (1854–1912) in his land-marking 'Analysis Situs,' Paris, 1895, and also discusses **higher homotopy groups** constructed in 1935 by H. Hurewicz (1904–1956) in his paper [Hurewicz, 1935], which are natural generalizations of fundamental groups. Homotopy theory studies those properties of topological spaces and continuous maps which are invariants under homotopic maps, called **homotopy invariants.** Finally, this chapter presents some interesting applications of homotopy, fundamental and higher homotopy groups in analysis, geometry, algebra, matrix theory, atmospheric science, vector field and extension problems and some others.

Just after the concept of homeomorphisms is clearly defined, the subject of topology begins to study those properties of geometric figures which are preserved by homeomorphisms with an eye to classify topological spaces up to homeomorphism, which stands as the ultimate problem in topology, where a geometric figure is considered to be a point set in the Euclidean space $\mathbf{R}^n$. But this undertaking becomes hopeless when there exists no homeomorphism between the two given topological spaces. For example, the problem is whether the Euclidean plane $\mathbf{R}^2$ and the punctured Euclidean plane $\mathbf{R}^2 - \{(0, 0)\}$ are homeomorphic or not. It is difficult to solve such a problem by the concepts of topological properties such as compactness and connectedness, studied in **Basic Topology, Volume 1** of the present book series. So, it has become necessary to search for an alternative technique, which is created in algebraic topology.

A basic problem in topology is to classify continuous maps between topological spaces up to homeomorphism. This problem is known as **classification problems of topological spaces,** and it aims to investigate whether given two topological spaces are homeomorphic or not. To solve such a problem, either we have to find an explicit expression of a homeomorphism between them or we have to show that no such

© The Author(s), under exclusive license to Springer Nature Singapore Pte Ltd. 2022
M. R. Adhikari, *Basic Topology 3*,
https://doi.org/10.1007/978-981-16-6550-9_2

homeomorphism exists. To solve the problem of impossibility (when there exists no such homeomorphism), the problem is shifted in many cases to algebra by associating **invariant objects** in the sense that homeomorphic spaces have the same algebraic object (up to equivalence). These algebraic objects are well known as **topological invariants.** They are also called **algebraic invariants.** Initially, these objects were integers such as **Euler characteristic** of a polyhedron. But subsequent research reveals that more fruitful and interesting results can be obtained from the algebraic invariant structures such as groups and rings. For example, homology and homotopy groups are very important algebraic invariants that provide strong tools to study the structure of topological spaces.

On the other hand, it is a natural problem **in homotopy theory** to classify continuous maps between topological spaces up to homotopy: Two continuous maps from one topological space to other are homotopic if one map can be continuously deformed into the other map. This classification by an equivalence relation leads to the concepts of **the fundamental groups** and the **higher homotopy groups,** which are powerful topological invariants to solve many basic problems of topology.

**The basic aim of algebraic topology** is to devise ways to assign to every topological space an algebraic object and to every continuous map from a topological space to other a homomorphism between the corresponding algebraic objects in a functorial way. This functorial approach facilitates in a variety of cases to solve topological problems through the possible sovability of corresponding algebraic problems. This technique defines topological invariants, which are also algebraic invariants. They convert topological problems into algebraic ones to have a better chance for solution, which develop another branch of topology, known as **algebraic topology.** This branch is one of the most important and powerful creations in mathematics which uses algebraic tools to study topological spaces. The twentieth century witnessed its greatest development. For example, in classical mechanics, a natural topology (Euclidean topology) can be endowed on the **configuration space** and the **phase space** of a system, which provides a qualitative study of the system. A nonvanishing vector field on a nonempty subset $X$ of the Euclidean space $\mathbf{R}^n$ and a flow on it establish a close connection between topology and analysis. This chapter also studies **topological dynamics,** which is the study of flows and gives an **abstract form of differential equations.**

**Historically**, B. Riemann (1826–1866) made an extensive work generalizing the concept of a surface to higher dimensions. He studied a special class of surfaces, now called **Riemann surfaces**. While investigating the 3-dimensional and higher-dimensional manifolds (topological) in 1895, **Henri Poincaré** in his **'Analysis Situs'** formally introduced the concepts of homotopy, fundamental group, homology groups and Betti numbers. His monumental work embodied in his 'Analysis Situs' provided tools to solve problems on system of differential equations, and his research establishes a surprising connection between analysis and topology (see Topological Dynamics of Sect. 2.18.5).

Poincaré himself remarked in 1912: **'Geometers usually distinguish two kinds of geometry, the first of which they qualify as metric and the second is projective. ··· But it is a third geometry from which quantity is completely excluded and** which is purely quantitative; this is analysis situs. In this discipline, two figures are equivalent whenever one can pass from one to the other by a continuous deformation, whatever else the law of this deformation may be, it must be continuous. Thus a circle is equivalent to an ellipse or even to an arbitrary closed curve, but it not equivalent to a straight line segment since the segment is not closed. A sphere is equivalent to a convex surface; it is not equivalent to a torus since there is a hole in a torus and in a sphere there is not.'

Poincaré also posed a conjecture in 1904 known as '**Poincaré conjecture.**' One form his conjecture says: Is a compact $n$-manifold homotopically equivalent to $S^n$ homeomorphic to $S^n$? For $n = 3$, G. Perelman (1966–) proved this conjecture in 2003 by using Ricci flow. For other values of $n$, it was solved by others before 1994 (see Chap. 7). Finally, this chapter communicates some interesting applications of homotopy theory to algebra, matrix theory, atmospheric science, vector field and extension problems. Algebraic topology also includes homology theory invented by Poincaré in 1895 and cohomology theory (its dual theory) born thereafter, which are mainly studied in Chapter 3.

For this chapter, the books [Adhikari, 2016], [Armstrong , 1993], [Hatcher, 2002], [Hu, 1959], [Massey, 1967], [Patterson, 1959], [Rotman, 1988], [Spanier, 1966], [Whitehead, 1978] and some others are referred in bibliography.

## 2.1  Motivation of the Study of Algebraic Topology

This section conveys the motivation of the study of combinatorial or algebraic topology. This motivation comes from the study of geometric properties of topological spaces from the algebraic viewpoint by assigning algebraic objects such as groups modules and rings to topological spaces and assigning to continuous maps between topological space homomorphisms between the corresponding algebraic objects in a functorial way. The concepts of fundamental groups, homology and homotopy groups were born through this study. More precisely, combinatorial or algebraic topology was born in the 1890s (at the turn of the nineteenth century to the twentieth century) through the remarkable work of Henri Poincaré in his 'Analysis Situs' dealing with the theory of integral calculus in higher dimensions and dividing a topological space into geometric elements corresponding to the vertices, edges and faces of polyhedra, and their higher-dimensional analogues. His monumental work inaugurated homotopy and homology theories with the invention of fundamental group and homology groups which are basic topological invariants including a generalization of the Euler characteristic (see Sect. 2.7.3). The development of the methods of this branch of topology concurred with the set-topology was facilitated by set-theoretic approach of G. Cantor during 1874–1895 and F. Hausdorff during 1900–1910. Motivation of construction of fundamental group is available in Sect. 2.9, that of higher homotopy groups is available in Sect. 2.20, and that of homology groups is available in Chap. 3.

**Historically**, a systematic study of algebraic topology as a subject began with precise formulations and correct proofs at the turn of the nineteenth to twentieth century (1885–1904) through the work of Henri Poincare in his 'Analysis Situs,' Paris, 1895. But his deep insight did not attract mathematicians sufficiently until the 1920s, when the situation began to change with applications in many mathematical theories. For example, the importance of homotopy invented by Poincaré was first established by H. Hopf (1895–1971) in 1835 with his discovery of a new continuous map, now known as **Hopf map.** The exponential growth of algebraic topology in both theory and applications has been found since 1940. Algebraic topologists consider H. Poincaré as founder and H. Hopf as co-founder of algebraic topology.

There is a natural question:

1. What is the main problem of topology?
2. Why we study algebraic topology?

1. **Answer**: The main problem of topology is to solve classification problems of topological spaces by using topological properties such as compactness and connectedness properties discussed in **Basic Topology, Volume 1** of the present book series. But the methods applied in general topology to solve the classification problems are workable at some very restrictive situations. For example, the Euclidean line $\mathbf{R}$ and Euclidean space $\mathbf{R}^3$ are not homeomorphic, because deleting any point from $\mathbf{R}$ keeps the remaining space disconnected; on the other hand, deleting any point from $\mathbf{R}^3$ keeps the remaining space connected. But this technique fails to examine whether higher-dimensional Euclidean spaces such as $\mathbf{R}^{100}$ and $\mathbf{R}^{1000}$ are homeomorphic or not. But algebraic topology provides powerful tools to solve this problem in a quick way.

2. **Answer**: A characteristic of a topological space which is shared by homeomorphic spaces is called a **topological invariant** in the sense that it is an invariant which is preserved by a homeomorphism. The main objective in algebraic topology is to create powerful tools to invade topological problems. They are called topological invariants and are algebraic in nature. Fundamental groups, higher homotopy groups, homology and cohomology groups (ring) are central topics of the study in algebraic topology, and they are utilized in classification of topological spaces up to homeomorphism. On the other hand, the motivation of combinatorial topology (former name of algebraic topology) is to study a topological space by representing it as a union of simple pieces with a specified arrangement, called combinatorial, such that the properties of the original space depend on how the split pieces are arranged. As algebraic topology solves many problems of mathematics and beyond it by utilizing the tools of algebraic topology. This shows the beauty and scope of the subject (see Chap. 7) of this volume.

Throughout this book, $\mathbf{I} = [0, 1]$ denotes the closed unit interval with the topology induced on $\mathbf{I}$ by the natural topology of the real line space $\mathbf{R}$.

## 2.2 Homotopy: Introductory Concepts

This section starts studying algebraic topology by addressing introductory concepts of homotopy theory, which form a key part of algebraic topology. The term homotopy coined by M. Dehn (1878–1952) and P. Heegaard (1871–1948) in 1907 is now commonly used. The importance of homotopy lies on the fact that the most of the topological (algebraic) invariants are also homotopy invariants. A homotopy is a relation between continuous maps of topological spaces, but not between their subspaces. In the mapping space $C(X, Y)$ (endowed with compact open topology) of continuous maps from a topological space $X$ to a topological space $Y$, a continuous deformation of one mapping into the other is considered as a path $H_t$ in $C(X, Y)$, which starts from $f$ and ends at $g$. A path in $Y$ is generally studied by the particular choice of $X$ in $C(X, Y)$.

Historically, L. E. J. Brouwer (1881–1967) gave the precise definition of continuous deformation by using the concept of homotopy of continuous maps. The aim of homotopy is to make a qualitative study of geometry in the sense that if one geometric object is continuously deformed into another geometrical object, then these two geometric objects are considered topologically equivalent, called homeomorphic. For example, the geometric objects such as a circle, an ellipse and a square are topologically equivalent (though their shapes are different). So, one of the aims of topology is to determine whether two given geometric objects are homeomorphic or not. To solve such a problem, it is necessary to find an expression of a homeomorphism between them or to show that no such homeomorphism exists, specially by using topological invariants in algebraic topology.

### 2.2.1 Homotopy of Continuous Maps

This subsection formalizes the intuitive idea of continuous deformation of a continuous map between two topological spaces by introducing the concept of homotopy with illustrative examples. A homotopy is a family of continuous mappings parameterized by the unit interval. Like homeomorphism classes, homotopy classes are generated by homotopy whose elements are continuous maps.

**Definition 2.2.1** Two continuous maps $\alpha, \beta : X \to Y$ from a topological space $X$ to a topological space $Y$ are said to be **homotopic**, symbolized $\alpha \simeq \beta$, if there exists a continuous map

$$H : X \times \mathbf{I} \to Y$$

such that $H(x, 0) = \alpha(x)$ and $H(x, 1) = \beta(x)$, $\forall\, x \in X$, where the topology on $X \times \mathbf{I}$ is the product topology. The map $H$ is said to be a **homotopy** between $\alpha$ and $\beta$, abbreviated $H : \alpha \simeq \beta$. In particular, if $\beta$ is a constant map, then the map $\alpha$ is said to be **nullhomotopic**.

**Fig. 2.1**  Straight-line
homotopy between $f_1$ and $f_2$

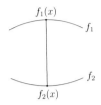

*Example 2.2.2*  Given two continuous maps $f_1, f_2 : X \to \mathbf{R}^2$, the map

$$H : X \times \mathbf{I} \to \mathbf{R}^2, \ (x, t) \mapsto (1 - t)f_1(x) + tf_2(x), \ \forall \ x \in X, \ \forall \ t \in \mathbf{I}$$

is a homotopy from $f_1$ to $f_2$, called a **straight-line homotopy or linear homotopy**,
depicted in Fig. 2.1.

**Geometrically**, the map $H$ shifts the point $f_1(x)$ to the point $f_2(x)$ along the
straight-line segment joining $f_1(x)$ and $f_2(x)$ as shown in Fig. 2.1. So, the map $H$ is
called a straight-line homotopy.

*Example 2.2.3*  Let $f, g : \mathbf{R} \to \mathbf{R}^2, x \mapsto (x, x^3), \ (x, e^x)$ be two maps. Then, the
map

$$H : \mathbf{R} \times \mathbf{I} \to \mathbf{R}^2, \ (x, t) \mapsto (x, te^x + (1 - t)x^3)$$

is homotopy from $f$ to $g$.

*Example 2.2.4*  Let $I_X, \ c : \mathbf{R}^n \to \mathbf{R}^n$ be two maps such that $I_X : X \to X, x \mapsto x$ is
the identity map on $X$ and $c : \mathbf{R}^n \to \mathbf{R}^n, x \mapsto \mathbf{0} = (0.0, \ldots, 0) \in \mathbf{R}^n$ is a constant
map. Then, the map

$$H : \mathbf{R}^n \times \mathbf{I} \to \mathbf{R}^n, \ (x, t) \mapsto tx$$

is a continuous map such that $H : c \simeq I_X$. This shows that the identity map $I_X$ is
nullhomotopic and hence $\mathbf{R}^n$ is a contractible space with $H$ as a contraction mapping
in the sense of Definition 2.4.9. Again, the maps $F$ and $G$ defined by

$$F : \mathbf{R}^n \times \mathbf{I} \to \mathbf{R}^n, \ (x, t) \mapsto (1 - t^2)x$$

and

$$G : \mathbf{R}^n \times \mathbf{I} \to \mathbf{R}^n, \ (x, t) \mapsto (1 - t)x$$

are also different homotopies between the maps $I_X$ and $c$ which imply that homotopy
between continuous maps is not unique.

*Remark 2.2.5*  (Geometrical Interpretation of Homotopy) Since Example 2.2.4
shows that homotopy between two continuous maps is not unique, it asserts that there
exist different ways of deforming a continuous map to another continuous map. Two
continuous maps $f, g : X \to Y$ are said to be homotopic if $f$ can be continuously

deformed into $g$ by a continuous 1-parameter family of maps $H_t : X \to Y$, given by $H_t(x) = H(x, t)$ with the property that $H_0 = f$ and $H_1 = g$ $\forall$ $x \in X$, $\forall$ $t \in I$. Here, $H_t$ forms a continuous family of maps in the sense that $H$ is continuous with respect to both $x$ and $t$ as a function from the product space $X \times I$ to the space $Y$. If $t$ is regarded as measuring time, then at $t = 0$, $H_0 = f$ and at $t = 1$, $H_1 = g$. Again as $t$ increases from 0 to 1, the map $f$ is continuously deformed into the map $g$. In particular, if $H$ is a contraction mapping of the topological space $X$ to the point $x_0 \in X$, then $H$ shrinks the whole space $X$ continuously into the point $x_0$.

Pasting Lemma 2.2.6 is a key result in proving continuity of a certain class of maps. For example, it is used to prove Theorem 2.2.7.

**Lemma 2.2.6** (Pasting or Gluing Lemma) Let $X$ be a topological space and $A$, $B$ be closed subsets in $X$ such that $X = A \cup B$. Given a topological space $Y$, if $f_1 : A \to Y$ and $f_2 : B \to Y$ are continuous maps such that $f_1(x) = f_2(x)$, $\forall$ $x \in A \cap B$, then the map

$$f : X \to Y, x \mapsto \begin{cases} f_1(x), & \text{if } x \in A \\ f_2(x), & \text{if } x \in B \end{cases}$$

is continuous.

**Proof** The map $f : X \to Y$ defined in this lemma is a well-defined unique map such that $f|_A = f_1$ and $f|_B = f_2$. We show that $f$ is continuous. Let $K$ be a closed set in $Y$. Then, $f^{-1}(K) = (A \cup B) \cap f^{-1}(K) = (A \cap f^{-1}(K)) \cup (B \cap f^{-1}(K)) = f_1^{-1}(K) \cup f_2^{-1}(K)$. Since each of the maps $f_1$ and $f_2$ is continuous, $f_1^{-1}(K)$ and $f_2^{-1}(K)$ are both closed in $X$. This implies that $f^{-1}(K)$ being the union of two closed sets is closed in $X$. This proves that $f$ is continuous. □

Theorem 2.2.7 proves that the relation of being homotopic '$\simeq$' of continuous maps $f, g : X \to Y$ is an equivalence relation on the set $\mathcal{C}(X, Y)$ of all continuous maps from $X$ to $Y$.

**Theorem 2.2.7** *Let $X$ and $Y$ be two topological spaces and $\mathcal{C}(X, Y)$ be the set of all continuous maps from $X$ to $Y$. Then, the relation of being homotopic '$\simeq$' is an equivalence relation on the set $\mathcal{C}(X, Y)$.*

**Proof** (i) '$\simeq$' is **reflexive.**: A homotopy

$$F : X \times I \to Y, \ (x, t) \mapsto f(x) \ \forall \ f \in \mathcal{C}(X, Y)$$

shows $f \simeq f$, $\forall f \in \mathcal{C}(X, Y)$, and hence, the relation '$\simeq$' is reflexive.
(ii) '$\simeq$' is **symmetric** : If $f, g \in \mathcal{C}(X, Y)$ are maps such that $f \simeq g$, then there exists a homotopy $F : f \simeq g$ such that

$$F(x, 0) = f(x), \ \text{and} \ F(x, 1) = g(x), \ \forall x \in X.$$

Consider the map $H : X \times I \to Y$, $(x, t) \mapsto F(x, 1 - t)$, which is continuous, since it is the composite of continuous maps

$$X \times I \to X \times I, \quad (x, t) \mapsto (x, 1 - t)$$

and
$$X \times I \to Y, \quad (x, t) \mapsto F(x, t).$$

Then, $H : g \simeq f$. It shows that the relation '$\simeq$' is symmetric.

(iii) '$\simeq$' is **transitive**: Finally, if $f, g, h \in C(X, Y)$ are maps such that $f \simeq g$, $g \simeq h$, then there exist homotopies $F : f \simeq g$ and $H : g \simeq h$. Consider the map

$$G : X \times I \to Y, \quad (x, t) \mapsto \begin{cases} F(x, 2t), & 0 \le t \le 1/2 \\ H(x, 2t - 1), & 1/2 \le t \le 1. \end{cases}$$

By using pasting Lemma 2.2.6, it follows that $G$ is continuous. Moreover, $G(x, 0) = F(x, 0) = f(x)$ and $G(x, 1) = H(x, 1) = h(x)$, $\forall\ x \in X$. It implies that $G : f \simeq h$ and hence the relation '$\simeq$' is transitive.

Consequently, the relation '$\simeq$' is an **equivalence relation** on the set $C(X, Y)$. $\square$

**Definition 2.2.8** The quotient set $C(X, Y)/\simeq$, abbreviated, $[X, Y]$, consists of all homotopy classes $[f]$ of maps $f \in C(X, Y)$.

**Remark 2.2.9** The set $[X, Y]$ plays a key role in algebraic topology, and it provides some algebraic structures on this set by specifying the spaces $X$ or $Y$. Fundamental group defined in Sect. 2.16 and higher homotopy groups defined in Sect. 2.20 are their outstanding examples.

The homotopy category $\mathcal{H}tp^*$ of pointed topological spaces given in Definition 2.2.10 plays a key role in algebraic topology for representation of topological invariants and many concepts in topology in the language of category theory in a unified way. For example, see functorial property of $\pi_1$ in Sect. 2.11 and functorial property of $\pi_n$ in Sect. 2.20.

**Definition 2.2.10** Topological spaces and their continuous maps form a category denoted by $\mathcal{T}op$, and the corresponding homotopy classes of continuous maps form a category denoted by $\mathcal{H}tp$. This category is called the **homotopy category**; i.e., $\mathcal{H}tp$ is the category whose objects are topological spaces and mor $(X, Y) = [X, Y]$, where the composition of maps is consistent with homotopies. In particular, pointed topological spaces and their base point preserving continuous maps form a category denoted by $\mathcal{T}op_*$ and the corresponding homotopy classes of their base point preserving continuous maps form a category, called the **homotopy category of pointed spaces,** denoted by $\mathcal{H}tp_*$.

### 2.2.2 Homotopy Equivalence

This subsection communicates the concept of homotopy equivalence, which generalizes the concept of homeomorphism ($\approx$) and gives rise to homotopically equivalent ($\simeq$) spaces which are natural generalization of homeomorphic spaces.

**Definition 2.2.11** Let $X$ and $Y$ be two topological spaces.

(i) Two continuous maps $f : X \to Y$ and $g : Y \to X$ are said to be homotopy equivalences if $f \circ g \simeq 1_Y$ and $g \circ f \simeq 1_X$, where $1_X$ and $1_Y$ are identity maps on $X$ and $Y$, respectively. Each of $f$ and $g$ is called a **homotopy equivalence** with homotopy inverse of each other.

(ii) The topological spaces $X$ and $Y$ are said to be **homotopy equivalent** denoted by $X \simeq Y$, if there exists a homotopy equivalence $f : X \to Y$. The relation $\simeq$ on the set of topological spaces is an equivalence relation. If $X \simeq Y$, then $X$ and $Y$ are also called topological spaces in the same **homotopy type.**

**Example 2.2.12** Let $\mathbf{R}^n$ be the Euclidean space with its origin $\mathbf{0} = (0, 0, \ldots, 0)$. Then, $\mathbf{R}^n \simeq \{\mathbf{0}\}$.

**Example 2.2.13** Let $n$ be a nonnegative integer and $\mathbf{0} = (0, 0, \ldots, 0)$ be the origin of the Euclidean $(n + 1)$-space $\mathbf{R}^{n+1}$. Then, the inclusion map

$$i : S^n \hookrightarrow \mathbf{R}^{n+1} - \{\mathbf{0}\}$$

is a homotopy equivalence by Proposition 2.27.5.

**Example 2.2.14** An isomorphism in the homotopy category $\mathcal{H}tp$ is a homotopy equivalence.

**Example 2.2.15** (Homeomorphic spaces are homotopy equivalent but its converse is not true). If two topological spaces $X$ and $Y$ are homotopy equivalent, it is a natural question: whether they are homeomorphic or not. The answer is negative. Two homeomorphic spaces are homotopy equivalent by Definition 2.2.11, but its converse is not necessarily true. In support, consider the examples,

(i) The spaces $\mathbf{R}^n \simeq \{\mathbf{0}\}$ but they are not homeomorphic.
(ii) The spaces $\mathbf{D}^n \simeq \{\mathbf{0}\}$ but they are not homeomorphic.
(iii) $X$ is the topological space $S^1$ together with the line segment $\mathbf{I}_0 = \{(x, 0) \in \mathbf{R}^2 : 0 \leq x \leq 1\}$ obtained by joining the point $(1, 0) \in S^1$ with the point $(2, 0)$, having the subspace topology inherited from $\mathbf{R}^2$. Then, $X \simeq S^1$ but they are not homeomorphic.

The homeomorphic spaces are of the same **topological type**. The homotopy equivalent spaces are said to have the same homotopy type. Proposition 2.2.16 asserts that the homeomorphic spaces have the same homotopy type, but the converse is not true by Example 2.2.15.

**Proposition 2.2.16** *Let X and Y be two homeomorphic spaces. Then, they have the same homotopy type.*

**Proof** By hypothesis, $X$ and $Y$ are two homeomorphic spaces. Then, there exists a homeomorphism $f : X \to Y$. Let $g = f^{-1} : Y \to X$ be the inverse map of $f$. Then, $g$ is a continuous map such that

$$g \circ f = f^{-1} \circ f = 1_X \simeq 1_X \text{ and } f \circ g = f \circ f^{-1} = 1_Y \simeq 1_Y.$$

This implies that $f$ is a homotopy equivalence. It proves that $X$ and $Y$ are of the same homotopy type.                                                                                □

Proposition 2.2.17 proves that every continuous map homotopic to a homotopy equivalence is also a homotopy equivalence.

**Proposition 2.2.17** *Let $f : X \to Y$ be a continuous map such that it is homotopic to a continuous map $g : X \to Y$. If $f$ is a homotopy equivalence, then $g$ is also a homotopy equivalence.*

**Proof** By hypothesis, $f : X \to Y$ is a continuous map such that $f \simeq g$. Suppose that $f$ is a homotopy equivalence. Then, there exists a continuous map $h : Y \to X$ such that

$$h \circ f \simeq 1_X \text{ and } f \circ h \simeq 1_Y.$$

By hypothesis, $f \simeq g$. Then, it follows that $f \circ h \simeq g \circ h$ and hence $1_Y \simeq g \circ h \Rightarrow g \circ h \simeq 1_Y$. Similarly, $h \circ g \simeq 1_X$. This proves that $g$ is a homotopy equivalence.  □

**Definition 2.2.18** A continuous map $f \in C(X, Y)$ is said to be an **inessential map** if $f \simeq c$ for some constant map $c : X \to Y$. Otherwise, $f$ is said to be an **essential map.** An inessential map is also nullhomotopic.

***Example 2.2.19***  (i)  The identity map $f : \mathbf{I} \to \mathbf{I}, t \mapsto t$ is homotopic to the constant map $f : \mathbf{I} \to \mathbf{I}, t \mapsto 0$. The homotopy

$$H : \mathbf{I} \times \mathbf{I} \to (t, s) \mapsto (1 - s)t$$

asserts that $f$ is inessential.
(ii)  The identity map $1_{S^1} : S^1 \to S^1$ is an essential map by Corollary 2.19.2.

## 2.3  Homotopy Classes of Continuous Maps

This section continues the study of the homotopy classes of maps formulated in Definition 2.2.8 and describes the quotient set $[X, Y] = C(X, Y)/\simeq$, which precisely consists of all homotopy classes $[f]$ of maps $f \in C(X, Y)$. If we keep the space $X$

fixed and vary $Y$ in the set $[X, Y]$, then this set is an invariant of the homotopy type of $Y$ in the sense that if $Z$ is any topological space such that $Y \simeq Z$, then the sets $[X, Y]$ and $[X, Z]$ are equivalent from the set-theoretic viewpoint; i.e., there exists a bijective map $\psi : [X, Y] \to [X, Z]$. Analogous result holds for pairs of topological spaces and also for pointed topological spaces. Many homotopy invariants are found by specializing the set $[X, Y]$ by particular choice of the spaces $X$ and $Y$. This section works in the category *Top* and in the category *Top*$_*$.

There are two problems that arise naturally in this section:

1. Corresponding to a given pointed topological space $Y$, does there exist a natural product in the set $[X, Y]$ making the set $[X, Y]$ a group for every pointed topological spaces $X$? A positive solution of this problem is available in Sect. 2.3.2.
2. Corresponding to a given pointed topological space $X$, does there exist a natural product in the set $[X, Y]$ making the set $[X, Y]$ a group for every pointed topological space $Y$?

To study these problems, this section works in the homotopy category *Htp*$_*$ of pointed topological spaces and their base point preserving continuous maps. This means that the set $[X, Y]$ is the set of morphisms from $X$ to $Y$ in the homotopy category *Htp*$_*$ of pointed topological spaces and their homotopy classes. This asserts the set $[X, Y]$ depends on the homotopy types of the spaces $X$ and $Y$.

## 2.3.1   Homotopy Sets

This subsection studies homotopy sets $[X, Y]$ and proves some special properties of such sets.

**Definition 2.3.1** (*Induced maps*) Given two continuous maps $f : X \to Y$ and $g : Y \to Z$, define maps

$$f^* : [Y, Z] \to [X, Z], \; [g] \mapsto [g \circ f]$$

and

$$g_* : [X, Y] \to [X, Z], \; [f] \mapsto [g \circ f].$$

The maps $f^*$ and $g_*$ are well defined, since the homotopy of $g \circ f$ depends precisely on the homotopy classes of $f$ and $g$. The maps $f^*$ and $g_*$ are called maps induced by $f$ and $g$, respectively.

Theorem 2.3.2 proves properties of induced maps from the viewpoint of homotopy.

**Theorem 2.3.2** *Given three pointed topological spaces $X$, $Y$ and $Z$, if $f : Y \to Z$ is a base point preserving continuous map, then $f$ induces a map*

$$f_* : [X, Y] \to [X, Z], \ [\alpha] \mapsto [f \circ \alpha]$$

*such that*

(i) *If $f \simeq h : Y \to Z$, then $f_* = h_*$; i.e., homotopic maps induce the same map.*

(ii) *If $1_Y : Y \to Y$ is the identity map, then $1_{Y*}$ is the identity map; i.e., the identity map induces the identity map.*

(iii) *If $g : Z \to W$ is another base point preserving continuous map, then $(g \circ f)_* = g_* \circ f_*$; i.e., the induced map preserves the composites.*

**Proof** The map

$$f_* : [X, Y] \to [X, Z], \ [\alpha] \mapsto [f \circ \alpha]$$

is well defined, because $f_*$ is independent of the choice of the representatives of the classes in the sense that if $\alpha \simeq \beta$, then $f \circ \alpha \simeq f \circ \beta$.

(i) If $f \simeq h : Y \to Z$, then $h_*([\alpha]) = [h \circ \alpha] = [f \circ \alpha] = f_*([\alpha]) \ \forall \ [\alpha] \in [X, Y]$ asserts that $h_* = f_*$.

(ii) $1_{Y*} : [X, Y] \to [X, Y], [\alpha]) \mapsto [1_Y \circ \alpha] = [\alpha], \ \forall \ [\alpha] \in [X, Y]$ asserts that $1_{Y*}$ is the identity map.

(iii) $(g \circ f)_* : [X, Y] \to [X, W], [\alpha] \mapsto [(g \circ f) \circ \alpha] = [g \circ (f \circ \alpha)] = (g_* \circ f_*)[\alpha], \ \forall \ [\alpha] \in [X, Y]$ asserts that $(g \circ f)_* = g_* \circ f_*$. $\qquad\square$

**Corollary 2.3.3** *Let $f : Y \to Z$ be a homotopy equivalence. Then, its induced map*

$$f_* : [X, Y] \to [X, Z]$$

*is bijective for any topological space $X$.*

**Proof** By hypothesis, $f : Y \to Z$ is a homotopy equivalence. Then, there exists a continuous map $g : Z \to Y$ such that $g \circ f \simeq 1_Y$ and $f \circ g \simeq 1_Z$. Hence, the corollary follows from Theorem 2.3.2. $\qquad\square$

**Corollary 2.3.4** *For every pair of homotopy equivalent spaces $Y$ and $Z$, there exists a bijection*

$$\psi : [X, Y] \to [X, Z]$$

*for any topological space $X$.*

**Proof** By hypothesis, $Y \simeq Z$. Then, $\exists$ a homotopy equivalence $f : Y \to Z$. Hence, it follows that

$$f_* = \psi : [X, Y] \to [X, Z]$$

is a bijection for any topological space $X$ by Corollary 2.3.3. $\qquad\square$

Theorem 2.3.5 expresses the above results in the language of the category theory.

**Theorem 2.3.5**  *Let* $\mathcal{T}op_*$ *be the category of pointed topological spaces,* $\mathcal{H}tp_*$ *be the homotopy category of pointed topological spaces and* $\mathcal{S}et$ *be the category of sets and their maps. Then for every object* $X \in \mathcal{T}op_*$, *there exists a covariant functor*

$$\pi_X : \mathcal{H}tp_* \to \mathcal{S}et,$$

*whose object function assigns to every object* $Y \in \mathcal{H}tp_*$ *the set* $[X, Y] \in \mathcal{S}et$ *and morphism function assigns to every morphism* $[f] \in \mathcal{H}tp_*$, *where* $f : Y \to Z \in \mathcal{T}op_*$, *the morphism* $f_* \in \mathcal{S}et$ *defined by*

$$\pi_x[f] = f_* : [X, Y] \to [X, Z], [\alpha] \mapsto [f \circ \alpha].$$

Theorem 2.3.6 formulates the dual result of Theorem 2.3.2.

**Theorem 2.3.6**  *Every base point preserving continuous map* $f : Y \to Z$ *induces a map*

$$f^* : [Z, X] \to [Y, X], \ [\alpha] \mapsto [\alpha \circ f]$$

*for any pointed space X such that*

(i)   *If* $f \simeq h$, *then* $f^* = h^*$.
(ii)  *If* $1_Y : Y \to Y$ *is the identity map, then* $1_{Y^*}$ *is the identity map.*
(iii) *If* $g : Z \to W$ *is another base point preserving continuous map, then* $(g \circ f)^* = f^* \circ g^*$.

**Proof**  Similar to the proof of Theorem 2.3.2.                          □

**Corollary 2.3.7**  *If* $f : Y \to Z$ *is a homotopy equivalence, then its induced map*

$$f^* : [Z, X] \to [Y, X]$$

*is bijective for any pointed topological space X.*

**Proof**  By hypothesis, $f : Y \to Z$ is a homotopy equivalence. Then, there exists a continuous map $g : Z \to Y$ such that $g \circ f \simeq 1_Y$ and $f \circ g \simeq 1_Z$. Hence, the corollary follows from Theorem 2.3.6.                          □

**Corollary 2.3.8**  *For every pair of homotopy equivalent spaces Y and Z, there exists a bijection*

$$f^* : [Z, X] \to [Y, X]$$

*for any pointed topological space X.*

**Proof**  By hypothesis, $Y$ and $Z$ are homotopy equivalent spaces. Then, there exist continuous maps $f : X \to Y$ and $g : Y \to X$ such that $g \circ f \simeq 1_X$, and $f \circ g \simeq 1_Y$. Hence, the corollary follows from Theorem 2.3.6.                          □

Theorem 2.3.9 formulates the dual result of Theorem 2.3.5.

**Theorem 2.3.9** *Let* $Top_*$ *be the category of pointed topological spaces,* $Htp_*$ *be the homotopy category of pointed topological spaces and* $Set$ *be the category of sets and their maps. Then for every object* $X \in Top_*$, *there exists a contravariant functor*

$$\pi^X : Htp_* \to Set.$$

**Proof** Proceed as in Theorem 2.3.5. □

Theorem 2.3.10 proves the converses of Corollaries 2.3.3 and 2.3.7.

**Theorem 2.3.10** *Let* $f : Y \to Z$ *is a base point preserving continuous map.*

(i) *If* $f_* : [X, Y] \to [X, Z]$ *is a bijective map for all pointed topological spaces* $X$, *then* $f$ *is a homotopy equivalence.*

(ii) *If* $f^* : [Z, X] \to [Y, X]$ *is a bijective map for all pointed topological spaces* $X$, *then* $f$ *is a homotopy equivalence.*

**Proof** (i) By hypothesis, for the particular choice of $X = Z$, it follows that $f_* : [Z, Y] \to [Z, Z]$ is a bijection. Hence, there exists a continuous map $g : Z \to Y$ such that $f_*([g]) = [1_Z]$. This shows that $f \circ g \simeq 1_Z$. Similarly, it follows that $g \circ f \simeq 1_Y$. This proves that $f$ is a homotopy equivalence.

(ii) Proceed as in (i). □

## 2.3.2 Homotopy Classes of Maps to a Topological Group

This subsection is dedicated to the study of continuous maps from an arbitrary pointed topological space to a topological group from the viewpoint of homotopy and proves that given a pointed topological space $X$ and a topological group $G$, the set $[X, G]$ admits a group structure. The essential feature which is retained in a topological group $G$ is a continuous multiplication with a unit. Every topological group $G$ with identity element $e$ as base point is an object in the category $Top_*$.

**Definition 2.3.11** A topological group $G$ is a Hausdorff topological space together with group operations such that

TG(1)  Group multiplication $m : G \times G \to G, (x, y) \mapsto xy$ is continuous.
TG(2)  Group inversion inv $: G \to G, x \mapsto x^{-1}$ is continuous.

**Definition 2.3.12** A topological group is said to be commutative if its continuous multiplication is commutative.

**Theorem 2.3.13** *Let* $G$ *be a topological group with identity element* $e$ *as base point. Then, the set* $[X, G]$ *admits a group structure for every topological space* $X \in Top_*$.

**Proof** Let $f, g : X \to G$ be any two base point preserving continuous maps. Define their pointwise multiplication

$$f \cdot g : X \to G, x \mapsto f(x)g(x),$$

where the right-side multiplication is the usual group multiplication $\mu$ of the topological group $G$. In the language of mapping, the product $f \cdot g$ is the composite map

$$f \cdot g = \mu \circ (f \times g) \circ \Delta : X \xrightarrow{\Delta} X \times X \xrightarrow{f \times g} G \times G \xrightarrow{\mu} G,$$

where $\Delta : X \to X \times X$, $x \mapsto (x, x)$ is the diagonal map. Carry this product $f \cdot g$ over the homotopy classes by the rule $[f] \circ [g] = [f \cdot g]$. This product on $[X, G]$ is well defined by Exercise 15 of Sect. 2.28. Hence, the group structure of $([X, G], \circ)$ follows from the corresponding properties of the topological group $G$. $\square$

**Corollary 2.3.14** *Let $G$ be a topological group. Then, every base point preserving continuous map $f : X \to Y$ induces a group homomorphism $f^* : [Y, G] \to [X, G]$.*

**Proof** By hypothesis, $G$ is a topological group and $f : X \to Y$ is a base point preserving continuous map. Then, $[X, G]$ and $[Y, G]$ are groups by Theorem 2.3.13. Define

$$f^* : [Y, G] \to [X, G], \ [\beta] \mapsto [\beta \circ f].$$

Then, $f^*$ is well defined and it is a group homomorphism. $\square$

**Corollary 2.3.15** *If $G$ is a commutative topological group, then the set $[X, G]$ admits a commutative group structure for every pointed topological space $X$.*

**Proof** Proceed as in Theorem 2.3.13. $\square$

The above behavior of a topological group $G$ with respect to the abstract group $[X, G]$ for every pointed topological space $X$ is formulated in Theorem 2.3.16 in the language of the category theory.

**Theorem 2.3.16** *Let $\mathcal{H}tp_*$ be the homotopy category of pointed topological spaces and $\mathcal{G}rp$ be the category of abstract groups and their homomorphisms. Then, every topological group $G$ determines a contravariant functor*

$$\pi^G : \mathcal{H}tp_* \to \mathcal{G}rp.$$

**Proof** The object function assigns to every object $X$ in the category $\mathcal{H}tp_*$, the group $[X, G]$, which is an object in the category $\mathcal{G}rp$. This assignment is well defined by Theorem 2.3.13. The morphism function assigns to every morphism $[f] \in \mathcal{H}tp_*$ of the homotopy class of the base point preserving continuous map $f : X \to Y$, the group homomorphism

$$f^* : [Y, G] \to [X, G], \quad [\beta] \mapsto [\beta \circ f],$$

which is a morphism in the category $\mathcal{G}rp$. This assignment is well defined by Corollary 2.3.14. Hence, by using Theorem 2.3.9, it follows that $\pi^G : \mathcal{H}tp_* \to \mathcal{G}rp$ is a contravariant functor. □

**Corollary 2.3.17** *Let $X$ be a pointed topological space such that $X \simeq G$ for some topological group $G$. Then, there is a natural equivalence between the contravariant functors $\pi^X$ and $\pi^G$.*

**Proof** By hypothesis, $X \simeq G$ and hence the corollary follows from Theorem 2.3.16. □

**Remark 2.3.18** Corollary 2.3.17 asserts that the functor $\pi^G$ can be considered as a functor on the category of groups.

**Example 2.3.19** Consider the commutative topological group $S^1 = \{z \in \mathbf{C} : |z| = 1\}$ under usual multiplication of complex numbers. Then, $[X, S^1]$ is a commutative group and if $f : X \to Y$ is a base point preserving continuous map, then $f^* : [Y, S^1] \to [X, S^1]$ is a homomorphism of groups.

**Example 2.3.20** Consider the noncommutative topological group $S^3$ consisting of the set of unit quaternions

$$S^3 = \{q = x + iy + jz + kw : x, y, z, w \in \mathbf{R} \text{ and } x^2 + y^2 + z^2 + w^2 = 1\}$$

under usual multiplication of quaternions of norm 1. Then, $[X, S^3]$ is a noncommutative group. Moreover, if $f : X \to Y$ is a base point preserving continuous map , then $f^* : [Y, S^3] \to [X, S^3]$ is group homomorphism.

### 2.3.3    Homotopy Between Smooth Maps on Manifolds

This section studies homotopy between smooth maps on manifolds defined in Chapter 3 of **Basic Topology, Volume 2** of the present series of books. The motivation of this study comes from the observation that in topology, there are certain properties of a continuous map which are not changed if the map is deformed in a smooth manner. Every smooth map is continuous, and the concept of smooth homotopy is borrowed from the usual concept of homotopy between continuous maps. Definition 2.2.1 formulates the concept of homotopy of continuous maps between topological spaces. In an analogous way, Definition 2.3.21 formulates the concept of homotopy of smooth maps between smooth manifolds.

**Definition 2.3.21** Let $M$ and $N$ be two smooth manifolds and $f, g : M \to N$ be two smooth maps. Then, the smooth maps $f$ and $g$ are said to be **smoothly homotopic** if there exists a smooth map

$$H : M \times \mathbf{R} \to N : H(x, 0) = f(x) \text{ and } H(x, 1) = g(x), \ \forall x \in M.$$

This gives rise to a **family of smooth maps**

$$H_t : M \to N, x \mapsto H(x, t), \ \forall t \in \mathbf{R}.$$

The smooth map $H$ is called a **smooth homotopy or homotopy between smooth maps** $f$ and $g$. In particular, if the map $H$ is just continuous, the maps $f$ and $g$ are said to be **continuously homotopic** or simply homotopic in the sense of in Definition 2.2.1. In general, $\{H_t\}$ is said to be a **smooth family of mappings** for some indexing set $\mathbf{A} \subset \mathbf{R}^n$ if there exists a smooth map $H$ such that

$$H : X \times \mathbf{A} \to Y, \ (x, t) \mapsto H_t(x).$$

Corresponding to above $H$, define a smooth map

$$\tilde{H} : M \times \mathbf{R} \to N : \tilde{H}(x, t) = f(x), \ \forall t \leq 0 \text{ and } \tilde{H}(x, t) = g(x), \ \forall t \geq 1.$$

The map $\tilde{H}$ is called the **normalized homotopy** corresponding to the smooth homotopy $H$.

Bump function formulated in Definition 2.3.22 is an important smooth function in differential topology. For example, it is used to study smooth homotopy and normalized homotopy (see Definition 2.3.21).

**Definition 2.3.22** A smooth function $\mathcal{B} : \mathbf{R} \to \mathbf{R}$ is said to be **bump function** if it satisfies the conditions

$$\mathcal{B}(t) = \begin{cases} 0, & \text{for all } t \leq 0 \\ 1, & \text{for all } t \geq 1 \end{cases}$$

and

$$0 < \mathcal{B}(t) < 1 \text{ for all } t \text{ such that } 0 < t < 1.$$

**Example 2.3.23** Given a smooth homotopy $H$ formulated in Definition 2.3.21, the smooth map $\tilde{H}$ defined by

$$\tilde{H}(x, t) = H(x, \mathcal{B}(t))$$

is its corresponding normalized homotopy. Moreover, the map

$$F : M \times \mathbf{R} \times \mathbf{R} \to N, \ (x, s, t) \mapsto H(x, (1 - t)s + t\mathcal{B}(s))$$

defines a smoothly homotopy map between $H$ and $\tilde{H}$.

**Remark 2.3.24 Geometrically,** two smooth maps are smoothly homotopic if one can be deformed to the other through smooth maps. For a continuous homotopy $F$ between two continuous maps $f, g : M \to N$, the map $F$ is taken as a continuous map from $M \times \mathbf{I} \to N$; on the other hand, for a smooth homotopy $F$ between smooth maps $f, g : M \to N$, the map $F$ is taken as a smooth map from $M \times \mathbf{R} \to N$. The technical reason is that if the manifold $M$ has a boundary, then the product space $M \times \mathbf{I}$ is not a smooth manifold.

**Proposition 2.3.25** *Let $\mathcal{F}(M, N)$ be the set of all smooth maps $f : M \to N$. Then, the smooth homotopy $\mathcal{H}$ is an equivalence relation on $\mathcal{F}(M, N)$.*

**Proof** Let $\mathcal{H}$ be a smooth homotopy relation on $\mathcal{F}(M, N)$.

(i) Then, the relation $\mathcal{H}$ on $\mathcal{F}(M, N)$ is reflexive and symmetric by Definition 2.3.21.
(ii) To prove that the relation $\mathcal{H}$ on $\mathcal{F}(M, N)$ is transitive, consider any three smooth maps $f, g, h \in \mathcal{F}(M, N)$ such that $f$ is smooth homotopic to $g$ and $g$ is smooth homotopic to $h$. Let $F$ be a normalized smooth homotopy between $f$ and $g$ and $G$ be a normalized smooth homotopy between $g$ and $h$. Define

$$\psi : M \times \mathbf{R} \to N, \ (x, t) \mapsto \begin{cases} F(x, 3t), & \text{if } t \leq 1/2 \\ G(x, 3t - 2), & \text{if } t \geq 1/2. \end{cases}$$

This implies that the map $\psi$ is well defined and it is a smooth map, because $F$ and $G$ are smooth maps and $\psi(x, t) = g(x), \ \forall t \in [1/3, 2/3]$. This asserts that $\psi$ is a normalized homotopy between $f, h \in \mathcal{F}(M, N)$ and hence it is proved that the relation $\mathcal{H}$ is also transitive.

Consequently, the relation $\mathcal{H}$ is an **equivalence relation** on $\mathcal{F}(M, N)$. □

**Definition 2.3.26** Let $M$ and $N$ be two smooth manifolds and $f, g : M \to N$ be two smooth maps. If $d$ is a metric on $N$ and $\delta : M \to \mathbf{R}$ is a positive-valued real function, then $g$ is said to be a $\delta$-approximation to $f$, if

$$d(f(x), g(x)) < \delta(x), \ \forall x \in M.$$

**Proposition 2.3.27** *Continuously homotopic smooth maps are also smoothly homotopic.*

**Proof** Let $f, g : M \to N$ be two smooth maps such that they are continuously homotopic and $F : M \times \mathbf{R} \to N$ be a normalized continuous homotopy between them. If $\mathbf{K} = (-\infty, 0] \cup [1, \infty)$, then $F$ is smooth on the closed set $M \times \mathbf{K}$, because, for $F|_{M \times (-\infty,0]} = f$ and $F|_{M \times [1,\infty)} = g$, there exists a positive continuous function $\delta : M \times \mathbf{R} \to \mathbf{R}$ by the smoothing theorem (see Basic Topology, Volume 2) such that

(i) $F$ can be $\delta$-approximated by a smooth map $H : M \times \mathbf{R} \to N$.
(ii) $F = H|_{M \times K}$.                                                                      □

## 2.4  Retraction, Contraction and Deformation Retraction

This section addresses the concepts of retraction, contraction, deformation retraction, weak and strong deformation retractions and studies them by using homotopy theory.

### 2.4.1  Retraction and Deformation Retraction

This subsection introduces the concepts of retraction and deformation retraction which are needed for our future study.

**Definition 2.4.1** Let $X$ be a topological space and $A$ be a subspace of $X$, with inclusion $i : A \hookrightarrow X$.

(i) The subspace $A$ is said to be a **retract of $X$,** if there exists a continuous map $r : X \to A$ with the property that $r \circ i = 1_A$. If such a map $r$ exists, then $r : X \to A$ is said to be a **retraction** of $X$ to $A$.
(ii) Additionally, if $i \circ r \simeq 1_X$, then $r$ is called a **deformation retraction** of $X$ to $A$ and $A$ is called a **deformation retract of $X$**.

**Remark 2.4.2** The retraction $r$ formulated in Definition 2.4.1 can be expressed in Fig. 2.2 making the triangle commutative in the sense that $r \circ i = 1_A$.

An interesting family of retractions is given in Proposition 2.4.3 and that of deformation retraction is available in Proposition 2.27.6 for $n \geq 1$.

**Proposition 2.4.3** *The $n$-disk $\mathbf{D}^n = \{x \in \mathbf{R}^n : ||x|| \leq 1\}$ is a retract of $\mathbf{R}^n$.*

**Proof** Let $i : \mathbf{D}^n \hookrightarrow \mathbf{R}^n$ be the inclusion map. Then, the continuous map

$$r : \mathbf{R}^n \to \mathbf{D}^n, x \mapsto \begin{cases} x, & \text{if } ||x|| \leq 1 \\ \frac{x}{||x||}, & \text{if } ||x|| \geq 1 \end{cases},$$

**Fig. 2.2** Diagram
representing retraction
$r : X \to A$

is a retraction. This shows that $\mathbf{D}^n$ is a retract of $\mathbf{R}^n$.                    □

**Remark 2.4.4** (Geometrical interpretation of retraction): The map $r$ defined in Proposition 2.4.3 shows that

(i) $r$ fixes every point $x \in \mathbf{D}^n$.
(ii) $r$ moves every point $x$ in $\mathbf{R}^n - \mathbf{D}^n$ along the straight line from the origin to $x$ onto the boundary $\partial \mathbf{D}^n \ (\approx S^{n-1})$ of $\mathbf{D}^n$.

**Proposition 2.4.5** *Let* $X = \mathbf{R}^n - \{0\}$ *be the punctured Euclidean space and* $S^{n-1}$ *be* $(n-1)$-*sphere. Then,* $S^{n-1}$ *is a deformation retract of* $X$.

**Proof** Let $i : S^{n-1} \hookrightarrow X$ be the inclusion map. Define a continuous map

$$r : X \to S^{n-1}, \ x \mapsto \frac{x}{||x||}$$

and another map

$$H : X \times I \to X, \ (x, t) \mapsto (1-t)x + \frac{tx}{||x||}.$$

Then, $H$ is a continuous map such that

$$H(x, 0) = x; \ H(x, 1) = \frac{x}{||x||} = (i \circ r)(x), \ \forall x \in X.$$

This shows that $H : 1_X \simeq i \circ r$ and hence $i \circ r \simeq 1_X$. Moreover, $r \circ i = 1_{S^{n-1}}$. Hence, $S^{n-1}$ is a deformation retract of $X$.                    □

**Proposition 2.4.6** *The retraction* $r : \mathbf{R}^n \to \mathbf{D}^n$ *formulated in Proposition 2.4.3 is also a deformation retraction.*

**Proof** To show this, define a map

$$F : \mathbf{R}^n \times I \to \mathbf{R}^n, \ (x, t) \mapsto \begin{cases} (1-t)x + tx/||x||, & \text{if } ||x|| \geq 1 \\ x, & \text{if } ||x|| < 1 \end{cases}$$

Then, $F$ is a continuous map such that $F : 1_{\mathbf{R}^n} \simeq i \circ r$. Since $r$ is a retraction, it proves that $r$ is a deformation retraction.

   **Geometrically,** $F$ fixes every point strictly inside $\mathbf{D}^n$ and shifts other points of $\mathbf{R}^n$ linearly from $x$ to $r(x)$ along the straight line joining the points $x$ and origin $(0, 0, \ldots, 0) \in \mathbf{R}^n$.                    □

### 2.4.2 Weak Retraction, and Weak and Strong Deformation Retractions

This subsection introduces the concepts of weak retraction, and weak and strong deformation retractions by imposing specific conditions on the inclusion map $i : A \hookrightarrow X$. These concepts are also used in our future study.

**Definition 2.4.7** Let $X$ be a topological space and $A$ be a subspace of $X$. If $i : A \hookrightarrow X$ is the inclusion map, then

(i) $A$ is said to be a **weak retract** of $X$, if there exists a continuous map $r : X \to A$ with the property that $r \circ i \simeq 1_A$. If such $r$ exists, then $r : X \to A$ is said to be a **weak retraction** of $X$ to $A$.

(ii) The subspace $A \subset X$ is said to be a **weak deformation retract** of $X$ if the inclusion map $i$ is a homotopy equivalence.

(iii) The subspace $A \subset X$ is said to be a strong deformation retract of $X$, if there exists a retraction $r : X \to A$ with the property

$$1_X \simeq i \circ r \operatorname{rel} A.$$

If $H : 1_X \simeq i \circ r \operatorname{rel} A$, then the homotopy $H$ is said to be a **strong deformation retraction** of $X$ to $A$.

*Example 2.4.8* (i) Every retraction is a weak retraction, but its converse is not true. For example, the comb space $X \subset \mathbf{I}^2$ is a weak retract of $\mathbf{I}^2$ but not a retract of $\mathbf{I}^2$. The comb space $X = (\mathbf{I} \times 0) \cup (0 \times \mathbf{I}) \cup \{1/n \times \mathbf{I} : n = 1, 2, \ldots, \}$ is a subspace of the Euclidean space $\mathbf{R}^2$ (see Fig. 2.4.22). It is contractible in the sense of Definition 2.4.9, but it is not a retract of $\mathbf{R}^2$ and the point $(0, 1)$ is not a strong deformation retract of $X$.

(ii) The $n$-sphere $S^n$ is strong deformation retract of the punctured Euclidean plane $\mathbf{R}^{n+1} - \{0\}$ with

$$H : (\mathbf{R}^{n+1} - \{0\}) \times \mathbf{I} \to \mathbf{R}^{n+1} - \{0\}, \quad (x, t) \mapsto (1 - t)x + \frac{tx}{||x||}$$

a strong deformation retraction.

### 2.4.3 Contractible Spaces

This subsection studies contractible spaces which are precisely those topological spaces that are homotopy equivalent to a one-point space, and hence, all contractible spaces have the homotopy type of a topological space deformable to a single point.

**Definition 2.4.9** A topological space $X$ is called **contractible** if the identity map $1_X : X \to X$ is nullhomotopic, i.e., if there exists a homotopy $H : 1_X \simeq c$ for some constant map $c : X \to X$, $x \mapsto x_0 \in X$. The homotopy $H$ is then called a **contraction mapping (or a contraction)** of the topological space $X$ to the point $x_0$.

*Example 2.4.10* Every convex subspace $X$ of the Euclidean space $\mathbf{R}^n$ is contractible. To show it, consider the continuous map

$$H : X \times \mathbf{I} \to X, (x, t) \mapsto (1 - t)x + tx_0, x, x_0 \in X, t \in \mathbf{I}.$$

If $c : X \to X$, $x \mapsto x_0$ is the constant map, then $H : 1_X \simeq c$ and hence the space $X$ is contractible with $H$ a contraction of $X$ to the point $x_0 \in X$. Since $\mathbf{R}^n$, $\mathbf{D}^n$, $\mathbf{I}$ are convex subspaces of $\mathbf{R}^n$, it follows by the first part that all of them are contractible spaces.

*Remark 2.4.11* (Geometrical interpretation of contraction) Let $X$ be a contractible space. Then, there exists a contraction $H : 1_X \simeq c$ of the topological space $X$ to the point $x_0 \in X$, which is geometrically interpreted as a continuous deformation of the space $X$ which shrinks ultimately the whole space $X$ into the point $x_0$. This asserts that the space $X$ is contracted to a point of $X$ by $H$.

**Proposition 2.4.12** *Let $X$ be any contractible space. Then, it is path connected.*

**Proof** By hypothesis, $X$ is contractible to a point $x_0 \in X$. If $c : X \to X$, $x \mapsto x_0$ is a constant map, then there exists a homotopy $H : 1_X \simeq c$. Given any point $p \in X$, define a path $g$ in $X \times \mathbf{I}$

$$g : \mathbf{I} \to X \times \mathbf{I}, \ t \mapsto (p, t).$$

Then, $\beta = H \circ g : \mathbf{I} \to X$ is a path in $X$ from the point $p$ to the point $x_0$. Since $p$ and $x_0$ are arbitrary points of $X$, it follows that $X$ is path connected. $\qquad\square$

**Corollary 2.4.13** *Every contractible space is connected.*

**Proof** Since every path-connected space is connected, the corollary follows from Proposition 2.4.12. $\qquad\square$

Proposition 2.4.14 characterizes contractibility of a topological space $X$ in terms of continuous maps from an arbitrary space to the space $X$.

**Proposition 2.4.14** *Let $X$ be a topological space. It is contractible iff given a continuous map $f : Y \to X$ from an arbitrary topological space $Y$, the map $f$ is homotopic to a constant map $c : Y \to X$, $y \mapsto x_0$.*

**Proof** First suppose that $X$ is contractible. Then, $1_X \simeq c$. Hence for an arbitrary continuous map $f : Y \to X$, the maps $1_X \circ f \simeq c \circ f$. Since $c \circ f : Y \to X$, $y \mapsto x_0$ is a constant map, it follows that $f$ is homotopic to a constant map. Conversely, suppose that given a continuous map $f : Y \to X$ from an arbitrary topological space $Y$, the map $f$ is homotopic to a constant map $c : Y \to X$, $y \mapsto x_0$. Then taking in particular, $Y = X$ and $f = 1_X : X \to X$, it follows that $X$ is contractible. $\qquad\square$

**Corollary 2.4.15**  *Any two continuous maps from an arbitrary space to a contractible space are homotopic.*

**Proof**  Let $X$ be a contractible space, $Y$ be an arbitrary space and $f, g : Y \to X$ be any two continuous maps. If $c : X \to X : x \mapsto x_0$, then by contractibility of $X$, it follows that $1_X \circ f \simeq c \circ f$ and $1_X \circ g \simeq c \circ g$. This implies that $f = 1_X \circ f \simeq c \circ f = c \circ g \simeq 1_X \circ g = g$. This proves that $f \simeq g$.  ☐

**Corollary 2.4.16**  *The identity map $1_X : X \to X$ of any contractible space $X$ is homotopic to any constant map of $X$ to itself.*

**Proof**  Let $X$ be a contractible space. Using Corollary 2.4.15, it is proved in particular that the identity map $1_X : X \to X$ is homotopic to any constant map of $X$ to itself.☐

Theorem 2.4.17 proves that contractible spaces are precisely the topological spaces which are homotopy equivalent to a one-point space. In other words, this proves that all contractible spaces have the homotopy type of any topological space deformable to a single point.

**Theorem 2.4.17**  *Let $X$ be a topological space. Then, $X$ is contractible iff it is of the same homotopy type of a one-point space $P = \{p\}$.*

**Proof**  Let $P = \{p\}$ be a one-point space. First suppose that the spaces $X \simeq P$; i.e., they are of the same homotopy type. Then, there exist continuous maps $f : X \to P$ and $g : P \to X$ such that

$$g \circ f \simeq 1_X \text{ and } f \circ g \simeq 1_P.$$

Suppose that $g(p) = x_0 \in X$, $c : X \to x_0$ and $H : 1_X \simeq g \circ f$. Then, $c$ is a constant map such that $g \circ f = c$ and hence $1_X \simeq c$. It proves that $X$ is contractible. Next, suppose that $X$ is contractible. Then, the identity map $1_X : X \to X$ is homotopic to some constant map $c : X \to X, x \mapsto x_0 \in X$ under some homotopy $H : 1_X \simeq c$. Hence, it follows that $c \circ i = 1_P$ and $H : 1_X \simeq i \circ c$. This proves that $X \simeq P$.  ☐

Corollary 2.4.18 proves that two contractible spaces are in the same homotopy type, and any continuous map between contractible spaces is a homotopy equivalence.

**Corollary 2.4.18**  *Let $X$ and $Y$ be any two contractible spaces. Then, $X$ and $Y$ are in the same homotopy type and any continuous map $f : X \to Y$ is a homotopy equivalence.*

**Proof**  Let $P$ is a one-point space. By hypothesis, $X$ and $Y$ are two contractible spaces. Then, $X \simeq P$ and $Y \simeq P$ by Theorem 2.4.17 and hence it follows by symmetry and transitivity of the relation $\simeq$ that $X \simeq Y$. This proves the first part of the corollary. For the second part, suppose that $X \simeq Y$. Then, there exists a homotopy equivalence $f : X \to Y$. If $g : X \to Y$ is an arbitrary continuous map, then $f \simeq g$ by Corollary 2.4.15.  ☐

**Definition 2.4.19** Let $X$ be a topological space and $c : X \to X, x \mapsto p \in X$ be a constant map. It is called **contractible to the point** $p$ relative to the subset $P = \{p\}$ if there exists a homotopy

$$H : X \times \mathbf{I} \to X \quad \text{such that } H : 1_X \simeq c \text{ rel } P.$$

**Theorem 2.4.20** *Let $X$ be a topological space contractible to a point $p \in X$ relative to the one-point space $P = \{p\}$. Then for every nbd $U$ of $p$ in $X$, there is a nbd $V \subset U$ of the point $p$ with the property that every point in $V$ can be joined to the point $p$ by a path lying entirely in $U$.*

**Proof** The theorem is proved by using the compactness property of $\mathbf{I}$. If $c : X \to X, x \mapsto p$, then under the given condition, there exists a homotopy $H : 1_X \simeq c$ rel $P$. This implies that the line $\{p\} \times \mathbf{I}$ is mapped by $H$ to the point $p \in X$. If $U$ is a nbd of $p$ in $X$, then from the continuity of the homotopy $H$, it follows that there exist nbd $V_t(p)$ of $p \in X$ for each $t \in \mathbf{I}$ and a nbd $W_t$ of $t$ in $\mathbf{I}$ such that

$$H(V_t(p) \times W_t) \subset U.$$

By using the compactness of $\mathbf{I}$, it follows that the open covering

$$\{W_t : t \in \mathbf{I}\}$$

of $\mathbf{I}$ has a finite subcovering

$$W_{t_1}, W_{t_2}, \ldots, W_{t_n}$$

such that

$$H(V_{t_i}(p) \times W_{t_i}) \subset U, \; \forall \; i = 1, 2, \ldots, n.$$

This asserts that $V(p) = \bigcap_{i=1}^{n} V_{t_i}(p)$ is a nbd of $p$ in $X$ such that $H(V(p) \times \mathbf{I}) \subset U$. For $x \in V(p)$, the image $H(V(p) \times \mathbf{I}) \subset U$, and hence it follows that the point $x$ can be joined to the point $p$ by a path which lies entirely in $U$. $\square$

**Example 2.4.21** The **comb space** is a subspace $X$ of the Euclidean plane $\mathbf{R}^2$ defined by

$$X = (\mathbf{I} \times 0) \cup (0 \times \mathbf{I}) \cup \{1/n \times \mathbf{I} : n = 1, 2, \ldots, \}.$$

**Geometrically,** the comb space $X$ consists of horizontal line segment $\mathcal{L}$ joining the point $(0, 0)$ to the point $(1, 0)$ and the vertical closed unit line segments standing on the points $(1/n, 0)$ for every $n \in \mathbf{N}$ together with the vertical line segment joining the points $(0, 0)$ with the point $(0, 1)$ as shown in Fig. 2.3. It is an important example of a contractible space by Proposition 2.4.22.

**Proposition 2.4.22** *Let $X$ be the comb space. Then, it is contractible but it is not contractible relative to the point $\{(0, 1)\}$.*

**Fig. 2.3**  Comb space $X$

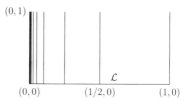

**Proof** **$X$ is contractible**:   Let $\mathcal{L}$ be the horizontal line segment joining the point $(0, 0)$ to the point $(1, 0)$ with $i : \mathcal{L} \hookrightarrow X$ be the inclusion map and $p : X \to \mathcal{L}$, $(x, y) \mapsto (x, 0)$ be the projection map. Then, $p \circ i = 1_{\mathcal{L}}$ (identity map on $\mathcal{L}$). Define the map

$$H : X \times \mathbf{I} \to X, ((x, y), t) \mapsto (x, (1 - t)y).$$

Then, $H : 1_X \simeq i \circ p$ and $p : X \to \mathcal{L}$ is a homotopy equivalence and hence $X \simeq \mathcal{L}$. Since $\mathcal{L} \approx \mathbf{I}$ (homeomorphic), it follows that $\mathcal{L} \simeq \mathbf{I}$. Since the space $\mathbf{I}$ is contractible space, it is of the same homotopy type of a one-point space and hence the spaces $\mathbf{I}$ and $\mathcal{L}$ are of the same homotopy type of one-point space. This implies that the space $X$ is of the same homotopy type of one-point space. This proves that the comb space $X$ is contractible by Theorem 2.4.17.

   **$X$ is not contractible relative to the point $\{(0, 1)\}$**:   To show it, take any small nbd $U$ of the point $(0, 1) \in X$ that contains an infinite number of path components of $X$. Let $\mathbf{D}_1^2$ be the open disk with center $(0, 1)$ and radius $\frac{1}{2}$. Consider the nbd $V = \mathbf{D}_1^2 \cap X$ of the point $(0, 1)$ in $X$. Then, $V$ has no nbd of the type $U$ such that each of its points can be joined to the point $(0, 1)$ by a path lying entirely in $V$. This proves that the comb space $X$ is not contractible relative to the point $\{(0, 1)\}$ by Theorem 2.4.20.                                                                            ☐

**Example 2.4.23**   The concept of relative homotopy is stronger than that of homotopy in the sense that if $A$ be a subspace of $X$ and $f, g : X \to Y$ are two continuous maps such that $f \simeq g$ rel $A$. Then $f \simeq g$. But its converse is not true. For example, consider the comb space $X$, its identity map $1_X : X \to X$, $x \mapsto x$ and the constant map $c : X \to X$, $(x, y) \mapsto (0, 1)$. Since $X$ is contractible, the maps $1_X \simeq c$ by Corollary 2.4.16. But by Proposition 2.4.22, the comb space $X$ is not contractible relative to $\{(0, 1)\}$.

## 2.5   Homotopy Extension Property: Retraction and Deformation Retraction

Homotopy extension property (HEP) is an important concept in topology to solve many problems in homotopy theory. For example, it is used in Chapter 5 to study cofibration of a continuous map. In this chapter, it is proved in Theorem 2.5.5 that

**Fig. 2.4** Homotopy
extension property of (X,A)
w.r.t Y

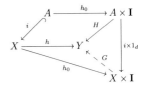

the concepts of weak retraction and retraction coincide under HEP. It is also proved
in Theorem 2.5.2 that HEP is a property in the homotopy category.

**Definition 2.5.1** A pair $(X, A)$ of topological spaces is said to have the **homotopy
extension property (HEP)** with respect to a topological space $Y$ if for any two
continuous maps $h : X \to Y$ and $H : A \times \mathbf{I} \to Y$ such that $H(x, 0) = h(x)$, $\forall x \in A$,
there exists a continuous map

$$G : X \times \mathbf{I} \to Y : G(x, 0) = h(x), \ x \in X \text{ and } G|_{A \times \mathbf{I}} = H.$$

In the mapping diagram, it is represented in Fig. 2.4, where $h_0 : X \to X \times \mathbf{I}$, $x \mapsto$
$(x, 0)$ and the dotted arrow represents the map $G$.

Theorem 2.5.2 asserts that the homotopy extension problem is a problem in the
homotopy category.

**Theorem 2.5.2** *Let the pair $(X, A)$ of spaces have the homotopy extension property
with respect to the space $Y$ and $f, g : A \to Y$ be two homotopic maps. Then, $f$ has
a continuous extension over $X$ iff $g$ has also a continuous extension over $X$.*

***Proof*** By hypothesis, $f \simeq g : A \to Y$. Then, there exists a homotopy

$$H : A \times \mathbf{I} \to Y : H(x, 0) = f(x) \text{ and } H(x, 1) = g(x), \ \forall \ x \in A.$$

First suppose that $f : A \to Y$ has a continuous extension $h : X \to Y$. Then,
$H(x, 0) = f(x) = h(x)$, $\forall \ x \in A$. As $(X, A)$ has the HEP with respect to $Y$, there
exists a continuous map

$$G : X \times \mathbf{I} \to Y : G|_{A \times \mathbf{I}} = H$$

and hence the diagram in Fig. 2.4 is commutative. This asserts the existence of the
map $G$. Define a map
$$k : X \to Y, x \mapsto G(x, 1).$$

Then, $k$ is a continuous extension of $g$ over $X$, because $k(a) = G(a, 1) = H(a, 1) =$
$g(a)$, $\forall \ a \in A$. Next suppose that $g : A \to Y$ has a continuous extension $k : X \to Y$.
Then proceeding as above, it is also proved that $f$ has also a continuous extension
$h : X \to Y$. $\qquad\qquad\square$

**Remark 2.5.3** Theorem 2.5.2 asserts that the homotopy extension property is a property in the homotopy category in the sense that a continuous map $f : A \to Y$ can or cannot be extended over $X$ is a property depending on the homotopy class of that map $f$.

**Remark 2.5.4** Example 2.4.8 shows that the concepts of weak retraction and retraction are different. Does there exist conditions under which these two concepts coincide? Theorem 2.5.5 gives its positive answer by providing suitable conditions under which the concepts of weak retraction and retraction coincide.

**Theorem 2.5.5** *Let the pair $(X, A)$ of spaces have the homotopy extension property with respect to the subspace $A$. Then, $A$ is a weak retract of $X$ iff $A$ is a retract of $X$.*

**Proof** By hypothesis, the pair $(X, A)$ of spaces have the homotopy extension property with respect to the subspace $A$ and $i : A \hookrightarrow X$ is the inclusion map. First suppose that $A \subset X$ is a retract of $X$ and $r : X \to A$ is a retraction. Then, $r \circ i = 1_A$ implies that $r \circ i \simeq 1_A$. This proves that $A$ is a weak retract of $X$. Conversely, suppose that $r : X \to A$ is a weak retraction. Then, $r \circ i \simeq 1_A$ asserts that there exists a homotopy

$$H : A \times \mathbf{I} \to A : H(x, 0) = r(x) \quad \text{and} \quad H(x, 1) = 1_A(x) = x.$$

Again as $(X, A)$ has the HEP with respect to $A$, there exists a continuous map $G : X \times \mathbf{I} \to A$ extending $H : A \times I \to A$. Hence, $G(x, 0) = r(x)$, $\forall\ x \in X$ and $G|_{A \times I} = H$. Define a map

$$\tilde{r} : X \to A, \ x \mapsto G(x, 1).$$

Then, $A$ is a retract of $X$ with retraction $\tilde{r}$.

$\square$

**Definition 2.5.6** Let $X$ be a topological space and $A$ be a subspace of $X$.

(i) A homotopy $D : A \times \mathbf{I} \to X$ is said to be a **deformation** of $A$ in $X$, if $D(x, 0) = x$, $\forall x \in A$.

(ii) Additionally, if $D(x, 1) \in Y \subset A$, $\forall x \in A$, then $D$ is said to be deformation of $A$ into $Y$ and $A$ is said to be **deformable** in $X$ into $Y$.

(iii) The inclusion map $i : A \subset X$ is said to have a **right homotopy inverse** $f : X \to A$, if there exists a continuous map

$$D : X \times \mathbf{I} \to X \text{ such that } D : 1_X \simeq i \circ f.$$

Theorem 2.5.7 characterizes deformability of a topological space in its subspace in terms of right homotopy inverse of the inclusion map.

**Theorem 2.5.7** *Let $X$ be a topological space and $A$ be a subspace of $X$. Then, $X$ is deformable into $A$ iff the inclusion map $i : A \hookrightarrow X$ has a right homotopy inverse.*

**Proof** By hypothesis, $X$ is a topological space and $A$ be a subspace of $X$ with the inclusion map $i : A \hookrightarrow X$. First suppose that $X$ is deformable into the subspace $A$ of $X$. Hence, there exists a continuous map

$$D : X \times \mathbf{I} \to X : D(x, 0) = x \text{ and } D(x, 1) \in A \subset X, \ \forall \ x \in X.$$

Define

$$f : X \to A : (i \circ f)(x) = D(x, 1), \ \forall \ x \in X.$$

This implies that $D : 1_X \simeq i \circ f$ and hence it is proved that $i$ has a right homotopy inverse.

Next suppose that $i : A \subset X$ has a right homotopy inverse $f : X \to A$. Then, $1_X \simeq i \circ f$. Hence, there exists a continuous map

$$D : X \times \mathbf{I} \to X \text{ such that } D : 1_X \simeq i \circ f.$$

This implies that

$$D(x, 0) = 1_X(x) = x \text{ and } D(x, 1) = (i \circ f)(x) \in A \subset X, \ \forall \ x \in X.$$

This proves that $X$ is deformable into $A$.                                                   □

**Remark 2.5.8** (Geometrical interpretation of retraction $D$) The map $D : X \times \mathbf{I} \to X$ formulated in Definition 2.5.6 starts with the identity map $1_X : X \to X$, then moves every point of $X$ continuously, including the points of $A$ and finally, pushes each point into a point of $A$. If, in particular, a topological space $X$ is deformable into a point $x_0 \in X$, then $X$ is contractible, and conversely, if $X$ is contractible then it is deformable into one of its points.

**Proposition 2.5.9** *The pair* $(\mathbf{D}^n, S^{n-1})$ *has the homotopy extension property.*

**Proof** Let $X$ be a given space and $H : S^{n-1} \times \mathbf{I} \to X$ and $f : \mathbf{D}^n \to X$ be any two continuous maps such that $H(x, 0) = f(x), \ \forall x \in S^{n-1}$. Consider the map

$$G : \mathbf{D}^n \times \mathbf{I} \to X : (x, t) \mapsto \begin{cases} f(x/(1 - t/2)), & \text{if } ||x|| \leq 1 - t/2 \\ H(x/||x||, 2(||x|| - 1 + t/2)), & \text{if } ||x|| \geq 1 - t/2. \end{cases}$$

Since $X$ is an arbitrary topological space, it follows that $(\mathbf{D}^n, S^{n-1})$ has the homotopy extension property.                                                   □

## 2.6  Paths and Homotopy of Paths

This section studies the concepts of paths and homotopy of paths in arbitrary topological spaces.

### 2.6.1  Paths in a Topological Space

This subsection defines a path in a topological space with illustrative examples.

**Definition 2.6.1**  Given a topological space $X$, a continuous map $\alpha : \mathbf{I} \to X$ such that $\alpha(0) = x_0 \in X$ and $\alpha(1) = x_1 \in X$ is said to be a **path** in $X$ from the point $x_0$ to the point $x_1$. The points $x_0$ and $x_1$ are called the **initial and terminal points** of the path $\alpha$ in $X$, respectively.

**Example 2.6.2**  Given two points $x_0, x_1 \in \mathbf{R}^2$, the map $\alpha : \mathbf{I} \to \mathbf{R}^2$, $t \mapsto (1 - t)x_0 + tx_1$ is a path in $\mathbf{R}^2$ from $x_0$ to $x_1$. On the other hand, the map $\beta : \mathbf{I} \to \mathbf{R}^2$, $t \mapsto tx_0 + (1 - t)x_1$ is a path in $\mathbf{R}^2$ from $x_1$ to $x_0$.

**Example 2.6.3**  Let $(X, \tau)$ be a topological space and $x_0 \in X$. Then, a constant map $c : \mathbf{I} \to X$, $x \mapsto x_0$ is said to be a **constant path or null path** in $X$ at $x_0$.

### 2.6.2  Homotopy of Paths

This subsection considers homotopy of a special class of paths $\alpha : \mathbf{I} \to X$ in a topological space $X$.

**Definition 2.6.4**  Let $X$ be a topological space and $\alpha, \beta : \mathbf{I} \to X$ be two paths in $X$. Then, they are said to be **path homotopic** if they have the same initial point and the same terminal point and there exists a continuous map $H : \mathbf{I} \times \mathbf{I} \to X$ with the conditions

$$H(t, 0) = \alpha(t), \;\; H(t, 1) = \beta(t), \;\; \forall t \in \mathbf{I}$$

and

$$H(0, s) = x_0, \;\; H(1, s) = x_1, \;\; \forall s \in \mathbf{I}.$$

The map $H$ is called a **path homotopy** between the paths $\alpha$ and $\beta$ as shown in Fig. 2.5, and it is represented by

$$H : \alpha \underset{p}{\simeq} \beta.$$

**Remark 2.6.5**  In Definition 2.6.4

(i) The first condition asserts that $H$ is a homotopy between $\alpha$ and $\beta$.
(ii) The second condition asserts that for each $t \in \mathbf{I}$, the path $t \mapsto H(t, s)$ is a path in $X$ from $x_0$ to $x_1$.

**Fig. 2.5**  Path homotopy
between $\alpha$ and $\beta$

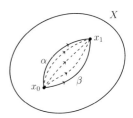

Geometrically, the map

$$H : \alpha \underset{p}{\simeq} \beta$$

provides a continuous way of deforming the path $\alpha$ to the path $\beta$ keeping the end
points of the paths remained fixed during the deformation as shown in Fig. 2.5.

**Theorem 2.6.6**  *Let $X$ be a topological space and $\mathcal{P}(X)$ be the set of all paths in $X$
with the same initial point $x_0$ and the same terminal point $x_1$. Then, the path homotopy
relation '$\underset{p}{\simeq}$' on $\mathcal{P}(X)$ is an equivalence relation.*

***Proof***  Let $\alpha_1, \alpha_2, \alpha_3 \in \mathcal{P}(X)$ be paths in $X$ such that $\alpha_1(0) = \alpha_2(0) = \alpha_3(0) = x_0$
and $\alpha_1(1) = \alpha_2(1) = \alpha_3(1) = x_1$.

(i)  **Reflexivity**: Let $\alpha \in \mathcal{P}(X)$ be an arbitrary path. Define a map

$$H : \mathbf{I} \times \mathbf{I} \to X, \ (t, s) \mapsto \alpha(t).$$

Since $H = \alpha \circ p_1$ is the composite of the projection map $p_1 : \mathbf{I} \times \mathbf{I} \to \mathbf{I}, (t, s) \mapsto$
$t$ onto the first factor and the map $\alpha : \mathbf{I} \to X$, which is continuous, it follows
that the map $H$ is also continuous. Moreover,

$$H(t, 0) = \alpha(t), \ H(t, 1) = \alpha(t), \ \forall \ t \in \mathbf{I}, \ H(0, s) = x_0, H(1, s) = x_1, \ \forall \ s \in \mathbf{I}.$$

This implies that $H : \alpha \underset{p}{\simeq} \alpha, \ \forall \ \alpha \in \mathcal{P}(X)$. This shows that the path homotopy
relation is reflexive.

(ii)  **Symmetry**: Let $\alpha_1 \underset{p}{\simeq} \alpha_2$ and $H : \alpha_1 \underset{p}{\simeq} \alpha_2$. Define a map

$$F : \mathbf{I} \times \mathbf{I} \to X, \ (t, s) \mapsto H(t, 1 - s).$$

Then, $F$ is a continuous map such that

$$F(t, 0) = H(t, 1) = \alpha_2(t), \ and \ F(t, 1) = H(t, 0) = \alpha_1(t), \ \forall \ t \in \mathbf{I}$$

and

$$F(0, s) = H(0, 1 - s) = x_0, \ F(1, s) = H(1, 1 - s) = x_1.$$

**Fig. 2.6**  Path (linear)
homotopy $H : \alpha \simeq_p \beta$

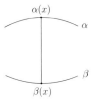

Hence, $F : \alpha_2 \simeq_p \alpha_1$. This asserts that this relation is symmetric.

(iii)  **Transitivity**: Let $\alpha_1 \simeq_p \alpha_2, f_2 \simeq_p \alpha_3$. Using the homotopies $G : \alpha_1 \simeq_p \alpha_2,$
and $H : \alpha_2 \simeq_p \alpha_3$ and pasting lemma, construct a homotopy $F : \alpha_1 \simeq_p \alpha_3$ (see
Theorem 2.2.7 ).

This proves that the path homotopy relation '$\simeq_p$' on $\mathcal{P}(X)$ is an equivalence
relation.                                                                                    □

**Definition 2.6.7**  Given a topological space $X$, the quotient set obtained by the path
homotopy relation on $\mathcal{P}(X)$ denoted by $\mathcal{P}(X)/\simeq_p$ is called the set of **path homotopy
classes** of paths in $X$.

***Example 2.6.8***  Let $\alpha, \beta : X \to \mathbf{R}^2$ be two continuous maps. Define a linear (continuous ) map

$$H : X \times \mathbf{I} \to, \mathbf{R}^2, \ (x, t) \mapsto (1 - t)\alpha(x) + t\beta(x).$$

Then, $H : \alpha \simeq \beta$.

**Geometrically,** $H$ shifts the point $\alpha(x)$ to the point $\beta(x)$ along the straight-line
segment joining $\alpha(x)$ and $\beta(x)$, as shown in Fig. 2.6. The map $H$ is called a **linear
homotopy.**

### 2.6.3   Homotopy of Loops Based at a Point

This subsection studies a particular class of paths, called loops. More precisely,
given a topological space $X$, an arbitrary chosen point $x_0 \in X$, called a base point,
the homotopy classes of loops in $X$ based at $x_0$ are studied in this subsection, which
leads to the concept of the fundamental group of $X$ based at the point $x_0$.

**Definition 2.6.9**  A path $\alpha : \mathbf{I} \to X$ is called a **loop** in $X$ based at $x_0 \in X$ if $\alpha(0) =$
$\alpha(1) = x_0$. If $\dot{\mathbf{I}} = \{0, 1\}$, then it is a subspace of $\mathbf{I}$. A loop $\alpha$ in $X$ based at $x_0$ is a
continuous map $\alpha : (\mathbf{I}, \dot{\mathbf{I}}) \to (X, x_0)$. In particular, the constant map $\delta : \mathbf{I} \to X, t \mapsto$
$x_0, \forall t \in \mathbf{I}$, is called a **constant loop or a null loop** in $X$ at $x_0$.

**Definition 2.6.10** Let $\alpha, \beta : (\mathbf{I}, \dot{\mathbf{I}}) \to (X, x_0)$ be two loops in $X$ based at the point $x_0$. Then, $\alpha$ and $\beta$ are said to be **homotopic relative to the subspace** $\dot{\mathbf{I}} = \{0, 1\}$ of $\mathbf{I}$ denoted by $\alpha \simeq \beta$ rel $\dot{\mathbf{I}}$, if there exists ( $\exists$ ) a continuous map $H : \mathbf{I} \times \mathbf{I} \to \mathbf{X}$ such that

$$H(t, 0) = \alpha(t),\ \forall\ t \in \mathbf{I},\ H(t, 1) = \beta(t),\ \forall\ t \in \mathbf{I},\ and\ H(0, s) = H(1, s) = x_0,\ \forall\ s \in \mathbf{I}.$$

**Remark 2.6.11 Geometrically**, the continuous map $H$ given in Definition 2.6.10 sends the square $\mathbf{I} \times \mathbf{I}$ into $X$ so that the bottom of the square is mapped by $\alpha$, the top of the square is mapped by $\beta$ and vertical sides of the square are mapped onto the point $x_0 \in X$. It shows intuitively that if $t$ represents time, then the path $\alpha$ is continuously deformed to $\beta$ throughout the unit interval.

**Theorem 2.6.12** *Given a topological space $(X, \tau)$, let $\mathcal{L}(X, x_0)$ be the set of all loops $\alpha : (\mathbf{I}, \dot{\mathbf{I}}) \to (X, x_0)$ in $X$ based at the point $x_0 \in X$. Then, the relation $\alpha \simeq \beta$ rel $\dot{I}$ on $\mathcal{L}(X, x_0)$ is an equivalence relation.*

**Proof** It follows from Theorem 2.6.6. □

**Remark 2.6.13** The quotient set $\mathcal{L}(X, x_0)/ \simeq$ of homotopy classes of loops relative to $\dot{\mathbf{I}} = \{0, 1\}$, denoted by $\pi_1(X, x_0)$, admits a group structure, called **the fundamental group** of the pointed topological space $(X, x_0)$ (see Theorem 2.9.10). It is a very important topological invariant and is studied in Sect. 2.9.

## 2.7 Euler Characteristic

This section considers geometric objects such as points, lines and surfaces in $\mathbf{R}^3$ through Euler characteristic, which is an important topological invariant (integral), but it is different from the powerful topological properties such as compactness and connectedness studied in Chap. 5, **Basic Topology, Volume 1** of the present series of books. Euler characteristic invented by L. Euler (1703–1783) in 1752, an integral invariant, is the first topological invariant which distinguishes nonhomeomorphic spaces. The search of other topological invariants led to invention of fundamental group (see Sect. 2.9) and homology group (see Chap. 3) by Henri Poincaré in 1895. If the $X$ and $Y$ are two homotopy equivalent spaces, then the characteristics $\kappa(X) = \kappa(Y)$. Moreover, $\kappa(X) \in \mathbf{Z}$, which is an algebraic object. Euler characteristic establishes a relation between geometry and algebra through topology. A generalization of Euler characteristic through homology theory is available in Chap. 3.

### 2.7.1 Simplexes and Polyhedra

This subsection develops some machinery to compute Euler characteristic and fundamental group of a vast family of topological spaces by using the familiar concepts of simplexes and polyhedra. It is widely used in homology theory developed in Chapter 3. **Historically,** algebraic topology was born through the work of H. Poincaré based on the idea of dividing a topological space into geometric elements corresponding to the vertices, edges and faces of polyhedra, and their higher-dimensional analogues. Such investigation presents many topological invariants including the Euler characteristic.

*Example 2.7.1*  **Some finite-dimensional simplexes**. A zero-dimensional simplex is a point; a one-dimensional simplex is a straight-line segment. A two-dimensional simplex is a triangle (including the plane region which it bounds), and a three-dimensional simplex is a tetrahedron. Finite-dimensional simplexes are used in the development of homology theory (see Chap. 3).

**Definition 2.7.2**  Any $(m + 1)$ of the $(n + 1)$ vertices of a given $n$-dimensional simplex $s_n$ determine an $m$-dimensional simplex, called an $m$-**dimensional face** of the simplex $s_n$ for every $m$ such that $0 \leq m \leq n$.

*Example 2.7.3*  The 0-dimensional simplexes of a simplex are its vertices.

**Definition 2.7.4**  Let $\sigma_p$ be a $p$-dimensional simplex. If $f$ is a simplex such that the vertices of $f$ form a subset of the vertices of $\sigma_p$, then $f$ is said to be a **face** of $\sigma_p$. A face $f$ of $\sigma_p$ is said to be proper if $f$ is neither $\emptyset$ nor the whole of $\sigma_p$.

**Definition 2.7.5**  A $k$-**dimensional polyhedron** $P$ in $\mathbf{R}^n$ is a point set in $\mathbf{R}^n$ endowed with the subspace topology inherited from the Euclidean topology on $\mathbf{R}^n$, which can be decomposed into simplexes of dimensions less than or equal to $k$, but there is at least one $k$-simplex such that the simplexes satisfy the property

(i)  Two simplexes of this decomposition have either a common face as their intersection
(ii)  Or no point in common.

The collection of all of the simplexes which belong to simplicial decomposition of a polyhedron is called a  **geometric complex**. A polyhedron is a geometric object such that the boundary of its two faces is an edge and its two edges meet at a vertex.

*Example 2.7.6*  A tetrahedron is polyhedron having 4 faces, 6 sides (edges) and 4 vertices.

**Definition 2.7.7**  Let $X$ be a subspace of $\mathbf{R}^3$ such that it is homeomorphic to a polyhedron $P$. The **Euler characteristic** of $X$, denoted by $\kappa(X)$, is given by

$$\kappa(X) = \mathbf{V} - \mathbf{E} + \mathbf{F},$$

where
    $\mathbf{V}$ denotes the number of vertices in $P$,
    $\mathbf{E}$ denotes the number of edges in $P$ and
    $\mathbf{F}$ denotes the number of faces in $P$.

***Example 2.7.8*** (i) If X is one-point, then $\kappa(X) = 1$.

(ii) If X is a line segment, then $\kappa(X) = 1$.

(iii) If X a triangle (i.e., made of 3 edges of a triangle), then $\kappa(X) = 0$.

(iv) If X is a triangular region, called triangular lamina, which includes the plane region the triangle bounds (i.e., made of 3 vertices, 3 edges and one face of a triangle), then $\kappa(X) = 1$.

(v) If X is the circle $S^1$, then $\kappa(X) = 0$, since X is homeomorphic to the sides of a triangle.

***Example 2.7.9*** (i) If X is the tetrahedron, then $\kappa(X) = 2$.

(ii) If X is the cube, then $\kappa(X) = 2$.

(iii) If X is the 2-sphere $S^2$, then $\kappa(X) = 2$, since X is homeomorphic to the surface of a tetrahedron.

***Example 2.7.10*** (i) Let **R** be the real line space. If X is a one-point set and Y is a line segment with topology induced from **R**, then $\kappa(X) = 1 = \kappa(Y)$, but they are not homeomorphic (though they are homotopy equivalent spaces).

(ii) The circle $S^1$ and the sphere $S^2$ are not homeomorphic, since $\kappa(S^1) = 0$ but $\kappa(S^2) = 2$.

***Remark 2.7.11*** Two homeomorphic compact polyhedra have the same Euler characteristic; i.e., if X and Y are two compact polyhedra, which are homeomorphic, then their Euler characteristics $\kappa(X) = \kappa(Y)$. Its proof follows from Euler characteristic theorem for simplicial homology (see Chapter 3).

### 2.7.2   Homotopy and Euler Characteristic of a Finite Graph

This subsection studies finite graphs from the viewpoint of homotopy theory and their Euler characteristics. The **'Seven Bridge Problem of Königsberg'** posed by Euler in 1752 initiated the concept of a new geometry, now called topology, without the concept of distance, which may be considered as the starting point of graph theory. Several problems of combinatorial nature may be solved by converting them in the language of graph theory. The 'Seven Bridge Problem of Königsberg' is an outstanding example. The diagrams commonly used by electrical engineers are practical examples of graphs. Jordan curve Theorem 2.25.29 has wide applications in graph theory (see Chap. 6).

The precise definition of a graph from the topological viewpoint is given in Definition 2.7.12.

**Definition 2.7.12  A (topological) graph** G is a topological space having a collection of points, called vertices of G together with a collection of edges such that every edge is either homeomorphic to the closed interval $\mathbf{I} = [0, 1]$ with subspace topology inherited from the real line space **R** with usual topology and joins two distinct vertices of G or homeomorphic to a circle $S^1$ and joins a given vertex of G to itself.

**Fig. 2.7** Graph $G_1$ is a tree

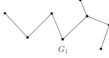

**Fig. 2.8** Graph $G_2$ is not a tree

***Remark 2.7.13*** If $G$ is a geometric graph, then its underlying space $|G|$ is a topologi-cal space satisfying the conditions of Definition 2.7.12. On the other hand, Definition 2.7.12 has an alternative form in the language of simplicial complex given in Defi-nition 2.7.14. Detailed description of a simplicial complex is available in Chapter 3.

**Definition 2.7.14** A simplicial complex of dimension $\leq 2$ is called a **graph**. A graph which does not contain any loop is called a **tree**. An end of a graph is the vertex of at most one-simplex.

***Example 2.7.15*** The graph $G_1$ shown in Fig. 2.7 is a tree. On the other hand, the graph $G_2$ shown in Fig. 2.8 is not a tree.

**Definition 2.7.16** A graph $X$ is said to be **contractible** if its underlying space $|X|$ is contractible.

**Theorem 2.7.17** *Let $X$ be any tree. Then, it is contractible.*

***Proof*** **Case I:** Let $X$ be a finite tree. The theorem is proved by induction on the number of vertices of $X$. If $X$ has only one vertex or exactly one edge, then the the-orem holds trivially. Assume that the theorem is true for all trees with $n$ vertices. If the tree $X$ has n+1 vertices with $v_{n+1}$ an end vertex, then there is a unique 1-simplex $s \in X$ with $v_{n+1}$ as its vertex. If $S = X - \{v_{n+1}, (s)\}$, then $S$ is a simplicial complex such that its underlying space $|S| = |X| - (s) \cup v_{n+1}$. Hence, $S$ is also a tree, since if $s^1$ is 1-simplex in $S$ such that $|S| - (s^1)$ is connected, then $|X| - (s^1)$ is to be connected. Since $S$ has only $n$-vertices, by induction hypothesis, $S$ is contractible. Consider the maps $f : |X| \to |S|$ such that $f$ maps $(s) \cup v_{n+1}$ into the other vertex of $s$ and $S$ onto itself. Let $g : |S| \to |X|$ be the inclusion map. Then, $g \circ f \simeq 1d$ and $f \circ g \simeq 1d$ assert that $|S|$ and $|X|$ are of the same homotopy type. This implies that $|X|$ is contractible.

**Case II** : Let $X$ be an arbitrary tree. Select a vertex $v_0 \in X$. Now construct a homotopy

$H : |X| \times \mathbf{I} \to X$, *such that* $H(x, 0) = x$, $H(x, 1) = v_0$ *and* $H(v_0, t) = v_0$, $\forall x \in |X|$, $\forall t \in \mathbf{I}$.

The construction of $H$ is done in several successive steps. The homotopy $H$ implies any arbitrary tree $X$ is also contractible.                                         □

**Definition 2.7.18** Let $G$ be a finite graph with **V** vertices and **E** edges (number of 1-simplexes), then the Euler characteristic $\kappa(G)$ is defined to be the integer

$$\kappa(G) = \mathbf{V} - \mathbf{E}.$$

The integer $\kappa(G)$ of any graph $G$ is invariant under subdivisions, since inserting an additional vertex into $G$ splits some 1-simplex into two 1-simplexes, and hence this process enhances both the numbers **V** and **E** by one keeping the number $\kappa(G) = \mathbf{V} - \mathbf{E}$, unchanged.

**Proposition 2.7.19** *For any tree $G$, its Euler characteristic $\kappa(G) = 1$.*

**Proof** The proposition is proved by induction on the number $n = \mathbf{V}$ (the number of vertices of $G$). For $n = 1$, the proposition is trivial. Assume that the proposition is true for trees with $n$ vertices. If the tree $G$ has n+1 vertices with $v_{n+1}$ an end vertex of $G$, then there is a unique 1-simplex $(s)$ with $v_{n+1}$ as its vertex. If $S = G - \{v_{n+1}, (s)\}$, then $S$ is a simplicial complex such that $|S| = |G| - (s) \cup v_{n+1}$. $S$ is also a tree, since if $s^1$ is an 1-simplex in $S$ such that $|S| - (s^1)$ is connected, then $|G| - (s^1)$ is to be connected. Since $S$ has only $n$-vertices, by induction hypothesis, $\kappa(S) = 1$. But the number of vertices of $G$ = number of vertices of $S$ plus 1 and also the number of edges of $G$ = number of edges of $S$ plus 1. This implies that $\kappa(G) = \kappa(S) = 1$.   □

**Definition 2.7.20** A topological space $X$ is said to be arcwise connected or pathwise connected if any two points of $X$ can be joined by a path in $X$.

**Definition 2.7.21** Let $G$ be an arcwise connected graph. If $m$ is the maximum number of open 1-simplexes that can be deleted from $G$ without disconnecting the space, then the number $m$ is said to be the number of **basic circuits** in $G$.

**Theorem 2.7.22** *Let $G$ be an arcwise connected graph. If $m$ is the number of basic circuits in $G$, then $m = 1 - \kappa(G)$.*

**Proof** Case I : If $G$ is a tree, then $m = 0$ and hence the theorem follows from Theorem 2.7.19.
Case II : If $G$ is not a tree, let $(s_1)$ be an open 1-simplex such that $X_1 = |G| - (s_1)$ is connected. If $X_1$ is a tree, then the theorem follows from Case I. Again if $X_1$ is not a tree, let $(s_2)$ be an open 1-simplex such that $X_2 = |G| - (s_1) \cup (s_2)$ is connected. Continue the process, which will stop after only a finite number of steps, since in $G$, there are only finite number of 1-simplexes. Hence, there exists some $m \in \mathbf{N}$ such that

$$X_m = |G| - (s_1) \cup (s_2) \cup \cdots \cup (s_m)$$

is a tree. Hence, it follows that

$$\kappa(G) = \kappa(X_m) - m = 1 - m \implies m = 1 - \kappa(G).$$

☐

**Remark 2.7.23** For the fundamental group of a connected graph which is not a tree, see Exercise 22 of Sect. 2.28.

### 2.7.3 Euler Characteristic of a Polyhedron

This subsection studies Euler characteristic of a polyhedron in $\mathbf{R}^3$, which is a topological (integral) invariant. Euler's theorem formulated in Corollary 2.7.26 is considered the first theorem on polyhedra which conveys the geometric properties of a polyhedron without using the concept of distance. Euler characteristic is an integral invariant. As $\mathbf{Z}$ is an algebraic object, the concept of Euler characteristic establishes an interplay between geometry and algebra, giving the birth of combinatorial topology and algebraic topology. Euler sent a letter to C. Goldbach (1690–1764) in 1750 giving his formula for a connected graph $G$ on a 2-dimensional sphere $S^2$ :

**V** (number of vertices of $G$) - **E** (number of edges of $G$) + **F** (number of regions of the sphere divided by the graph $G$) = 2, **notationally,** it is expressed as

$$\mathbf{V} - \mathbf{E} + \mathbf{F} = \mathbf{2}.$$

**Theorem 2.7.24** *Let S be a closed surface divided into* **F** *regions, called faces with the help of* **E** *arcs, joining in pairs,* **V** *vertices with the property that at least two arcs meet at a vertex. Then, the integer*

$$\mathbf{V} - \mathbf{E} + \mathbf{F}$$

*is independent of the choice of dividing up the surface S.*

**Proof** Let $T_1$ and $T_2$ be two given divisions of the surface $S$ into faces, arcs and vertices. Construct a third division $T_3$ containing all the regions, arcs and vertices of both divisions $T_1$ and $T_2$. This construction is possible, by adding new vertices at points of intersection of the arcs of $T_1$ and $T_2$. This process gives rise to new arcs and new faces. To prove the theorem, it is sufficient to show that for any division $T$ obtained from a given division of $S$ by adding new vertices, arcs and faces, the integral value of $\mathbf{V} - \mathbf{E} + \mathbf{F}$ remains the same. To show it, let one new vertex be inserted and this new vertex be joined to some of the existing vertices by $m$ new arcs. By this process, the number of faces is increased by $m - 1$, since one of the original faces is replaced by $m$ new faces. This asserts that the integral value of $\mathbf{V} - \mathbf{E} + \mathbf{F}$ is unaltered by this process of construction. This value remains also unaltered if new arcs are added without adding new vertices, since each new arc gives one new face. This asserts that the integral value of $V - E + F$ remains the same for any division obtained from the existing division by adding new vertices, arcs and faces. It concludes that $V - E + F$ is the same for all divisions of the given surface $S$. □

**Definition 2.7.25** The integral value of $V - E + F$ for any closed surface $S$ is well defined and is called the **Euler characteristic** of $S$, abbreviated by $\kappa(S)$.

**Corollary 2.7.26** (Euler theorem on polyhedron) Let $V$ be the number of vertices, $E$ the number of edges and $F$ the number of faces of a polyhedron $P$ homeomorphic to the 2-sphere $S^2$. Then its Euler characteristic

$$\kappa(P) = V - E + F = 2.$$

**Proof** It follows from Theorem 2.7.24 by using $\kappa(S^2) = 2$. □

**Remark 2.7.27** Euler characteristic of a topological space $X$ is independent of a polyhedron $P$ as long as $X$ is homeomorphic to $P$ by Poincaré–Alexander theorem; see Exercise 2.28.1 of Sect. 2.28. Euler characteristic of a compact surface is a **topological invariant** by Exercise 48 of Sect. 2.28. Thus, if $X$ and $Y$ are two compact homeomorphic surfaces in $\mathbf{R}^3$, then $\kappa(X) = \kappa(Y)$. This implies that if $\kappa(X) \neq \kappa(Y)$, then $X$ and $Y$ cannot be homeomorphic. But its converse is not always true in the sense that $\kappa(X) = \kappa(Y)$ does not guarantee that the spaces $X$ and $Y$ are homeomorphic. In support, see Example 2.7.30.

**Example 2.7.28** (i) If $S$ is the torus $T$, then $\kappa(S) = 0$, since for the torus $S$,

$$V = 1, \ E = 2, \ F = 1.$$

(ii) If $S$ is the real projective plane $\mathbf{R}P^2$, then $\kappa(S) = 1$.

**Proposition 2.7.29** $S^1$ *is not homeomorphic to* $S^2$.

**Proof** Since $\kappa(S^1) = 0$ and $\kappa(S^2) = 2$, they cannot be homeomorphic, since Euler characteristic of a compact surface is a topological invariant. □

**Example 2.7.30** Two nonhomeomorphic spaces may have the same Euler characteristic. For example, if $X = [a, b]$ and $Y = \{x_0\}$, a one-point space in $\mathbf{R}$, then $\kappa(X) = 1 = \kappa(\{x_0\})$. They are not homeomorphic but they are homotopy equivalent.

**Remark 2.7.31** Example 2.7.30 motivates to study Euler characteristic from the viewpoint of algebraic topology. More precisely, a study on Euler characteristic such as Euler–Poincaré theorem, the topological and homotopy invariance of the Euler characteristic of compact polyhedra by using simplicial homology groups of a compact polyhedra is available in Chapter 3. But a study of Euler characteristic for a surface is available in Exercises 44–48 of Sect. 2.28.

## 2.8  Exponential Maps and Homotopy Classification of Complex-Valued Functions

This section studies exponential map and index function from the viewpoint of homotopy. More precisely, **the homotopy classification** of complex-valued continuous functions from compact metric spaces is given in Theorem 2.8.4. In particular, the homotopy of complex-valued continuous maps from $S^1$ is characterized with the help of index numbers (see Exercise 53 of Sect. 2.28 ). For the study of continuous maps from a topological space to the punctured complex plane $\mathbf{C}^* = \mathbf{C} - \{0\}$ from the viewpoint of homotopy theory, the concept of exponential map is necessary, which is given in Definition 2.8.1.

**Definition 2.8.1**  Given any topological space $X$, a continuous map $f : X \to \mathbf{C} - \{0\}$ is said to be **exponential** if the map $f$ can be expressed as $f = e^k$ for some map $k : X \to \mathbf{C}$.

***Example 2.8.2***  (i)  Every continuous function $f : X \to \mathbf{R}^+$ is exponential, since $f$ can be expressed as $k = \log f$ (usual logarithm function) for some map $k : X \to \mathbf{C}$.

(ii)  Let $f : X \to \mathbf{C} - \{0\}$ be a continuous map such that $f(X) \cap (-\infty, 0) = \emptyset$; i.e., $f$ escapes the part of negative real axis. Then, $f$ is exponential.

**Proposition 2.8.3**  *Let* $G = \mathcal{C}(X, \mathbf{C} - \{0\})$ *be the group of all continuous maps from* $X$ *to* $\mathbf{C} - \{0\}$ *under pointwise multiplication*

$$(fg)(x) = f(x)g(x)$$

*and*

$$H = \{f \in G : f \text{ is exponential}\}.$$

*Then, $H$ is a subgroup of $G$.*

***Proof***  It follows from Definition 2.8.1.  □

Theorem 2.8.4 characterizes homotopy classes of continuous maps from a compact metric space to the space $\mathbf{C}^* = \mathbf{C} - \{0\}$ by exponential maps.

**Theorem 2.8.4**  *Let $X$ be a compact metric space and $f, g : X \to \mathbf{C} - \{0\}$ be two continuous maps. Then, $f \simeq g$ iff the map $f/g$ is exponential.*

***Proof***  Let the map $f/g$ be exponential. Then, there exists some map $k : X \to \mathbf{C}$ such that $f/g = e^k$. Define a homotopy

$$H : f \simeq g : X \times \mathbf{I} \to \mathbf{C} - \{0\}, (x, t) \mapsto g(x)e^{(1-t)k(x)}.$$

Conversely, let $H : f \simeq g : X \times \mathbf{I} \to \mathbf{C} - \{0\}$. Then, $H(x, 0) = f(x)$ and $H(x, 1) = g(x)$ for all $x \in X$. Since $X$ is compact by hypothesis, and $I$ is compact, their product space $X \times \mathbf{I}$ is also compact. Then, the continuous positive-valued function $|H| : X \times \mathbf{I} \to \mathbf{R}^+$ attains its minimum value $m$ and hence

$$0 < m = \inf \{|H(x, t)| : x \in X, t \in I\}.$$

Since $H$ is uniformly continuous on $X \times \mathbf{I}$, there exists a real number $\delta > 0$, such that whenever $|t - s| < \delta$, then

$$|H(x, t) - H(x, s)| < m, \ \forall x \in X.$$

For an integer $n > 1/\delta$, consider the maps

$$f_m : X \to \mathbf{C} - \{0\}, x \mapsto H(x, m/n), \ \forall x \in X, 1 \le m \le n.$$

Since each of the maps $f_{m-1}/f_m$ is exponential, $H(x, 0) = f(x) = f_0(x)$, say, and $H(x, 1) = g(x)$ for all $x \in X$, it follows that

$$f/g = (f_0/f_1)(f_1/f_2)(f_2/f_3) \cdots (f_{n-1}/f_n)$$

is exponential, because $f_0(x) = H(x, 0) = f(x)$ and $f_n(x) = H(x, 1) = g(x)$ for all $x \in X$. □

**Corollary 2.8.5** *Given a compact metric space $X$, a map $f : X \to \mathbf{C} - \{0\}$ is exponential iff $f \simeq c$, where $c : X \to \mathbf{C} - \{0\}$, $x \mapsto z_0 \in \mathbf{C} - \{0\}$ is a constant map.*

**Proof** It follows from Theorem 2.8.4 as a particular case. □

**Corollary 2.8.6** *Given a compact contractible space $X$, every map $f : X \to \mathbf{C} - \{0\}$ is exponential.*

**Proof** By hypothesis, $X$ is contractible. Then, there exists a homotopy $H : X \times \mathbf{I} \to X$ such that $H : I_X \simeq c$, where $c : X \to \mathbf{C} - \{0\}$, $x \mapsto z_0 \in \mathbf{C} - \{0\}$ is a constant map. Then, $f \circ H : f \simeq c$. Hence, it follows by Corollary 2.8.5 that $f$ is exponential. □

Example 2.8.7 leads to define the index number of a continuous map $f : S^1 \to \mathbf{C} - \{0\}$ formulated in Definition 2.8.8.

**Example 2.8.7** Given a continuous map $f : S^1 \to \mathbf{C} - \{0\}$, define

$$\psi_f : [0, 2\pi] \to \mathbf{C} - \{0\}, t \mapsto f(e^{it}).$$

Since $[0, 2\pi]$ is compact and contractible, it follows by Corollary 2.8.6 that there exists an exponential map

$$g : [0, 2\pi] \to \mathbf{C} - \{0\}$$

such that

$$f(e^{it}) = e^{g(t)}, \quad \forall t \in [0, 2\pi].$$

If $h : [0, 2\pi] \to \mathbf{C} - \{0\}$ is another map such that $f(e^{it}) = e^{h(t)}, \forall t \in [0, 2\pi]$. Then, $e^{g(t)-h(t)} = 1, \forall t \in [0, 2\pi]$ implies that $g - h$ assumes only values which are integral multiples of $2\pi i$. Again since the range of the map $g - h$ is discrete, it follows that $g - h$ is constant. This shows that the number $g(2\pi) - g(0)$ is independent of the choice of $g$ satisfying the condition $f(e^{it}) = e^{g(t)}, \forall t \in [0, 2\pi]$. Hence, corresponding to each $f$, there exists a number, denoted by $Ind\ f$ defined by

$$Ind\ f = [g(2\pi) - g(0)]/2\pi i.$$

**Definition 2.8.8** The number Ind  f defined in Example 2.8.7 is called the **index of the continuous map** $f : S^1 \to \mathbf{C} - \{0\}$, and it is denoted by $Ind\ f$.

**Example 2.8.9** For the map $f : S^1 \to \mathbf{C} - \{0\}$, $z \mapsto z^m$, its index $Ind\ f$ is m.

**Remark 2.8.10** For more study of exponential maps and index number of a map $f :$ $S^1 \to \mathbf{C} - \{0\}$ and their relations with homotopy, see Exercises 49–54 of Sect. 2.28.

## 2.9 Fundamental Groups

This section starts with the concept of fundamental group and its motivation. The fundamental group is an algebraic object which is assigned to a geometric space. This group provides useful tools to measure the number of holes in the geometric space. For example, the fundamental group of

  (i) The Euclidean line is 0, which indicates that it has no hole.
 (ii) The Euclidean plane is 0, which indicates that it has no hole.
(iii) The circle is $\mathbf{Z}$, which indicates that it has one hole.
(iv) The torus is $\mathbf{Z} \oplus \mathbf{Z}$ which indicates that it has two holes.

On the other hand, in more general, the fundamental group $\pi_1(X, x_0)$ of an arbitrary pointed topological space $(X, x_0)$ is defined by using the homotopy classes of loops in $X$ based at the point $x_0$ as its elements. For this construction, we start with the set $\Omega(X, x_0)$ of all loops in the space $X$ based at a point $x_0$. Two loops in $X$ based at $x_0$ are said to be **equivalent** if one loop can be continuously deformed to the other. This defines an equivalence relation $\sim$ on $\Omega(X, x_0)$ providing the quotient set $\Omega(X, x_0)/\sim$, denoted $\pi_1(X, x_0)$, which admits a group structure by Theorem 2.9.10, called the **fundamental group** of $X$ based at $x_0$.

This group is a powerful topological invariant and characterizes the connectivity properties of topological spaces related to properties of loops in these spaces. Its basic motivation is to detect a hole in the plane by letting loops in the plane shrunk to a point. This facilitates to attack some topological problems. One of the main problems in topology is the classification of topological spaces up to homeomorphism. To solve such a problem, either we have to find a homeomorphism between two given topological spaces or we have to show that no such homeomorphism exists. In the latter case, a special property or a characteristic is searched which is shared by homeomorphic spaces. This search led to the invention of homotopy, fundamental group and homology groups in algebraic topology.

**Historically**, Euler characteristic invented by L. Euler (1703–1783) in 1752, an integral invariant, is the first invented topological invariant which distinguishes nonhomeomorphic spaces. The search of other invariants establishes a connection between topology and modern algebra in such a way that homeomorphic spaces have isomorphic algebraic structures. More precisely, fundamental group and homology groups invented by Poincaré in 1895 are the first powerful topological (algebraic) invariants which came from such a search.

**Fundamental group** is the first of a sequence of functors $\pi_n$ designed in Sect. 2.20, called homotopy group functors from the category of pointed topological spaces and their continuous maps to the category of groups and their homomorphisms. Such functors occupy a vast territory in algebraic topology and are still the subject of intensive study. More precisely, given a pointed topological space $(X, x_0)$, the set $\pi_1(X, x_0)$ is defined to be the set of homotopy classes of paths $\alpha : \mathbf{I} \to X$ that send 0 and 1 to $x_0$. Each such path is called a loop in $X$ based at $x_0$. It is shown that $\pi_1(X, x_0)$ admits a group structure. The group $\pi_1(X, x_0)$ depends on $X$ as well as on $x_0 \in X$ and is called the fundamental group or **Poincare group** of the space $X$ based at $x_0$. It is a homotopy-type invariant in the sense that homotopy equivalent spaces $(X, x_0)$ and $(Y, y_0)$ have the isomorphic fundamental groups $\pi_1(X, x_0)$ and $\pi_1(Y, y_0)$. The algebraic properties of the fundamental group reflect the topological properties of some specified spaces. The study of this group is easier than the study of the topological space $X$ directly.

**Higher homotopy groups** $\pi_n(X, x_0)$ of pointed topological spaces which are natural generalization of $\pi_1(X, x_0)$ and **the sequence of higher homotopy functors** $\pi_n$ in the homotopy category of pointed topological spaces are studied in Sect. 2.20. On the other hand, the sequence of homology functors $H_n$ given by Poincaré is studied in Chap. 3. Further generalization of $\pi_n(X, x_0)$ by defining $\pi_n(X, A, x_0)$ of any triplet $(X, A, x_0)$ of topological spaces and a detailed study of $\pi_n(X, A, x_0)$ are available in Chap. 5.

## 2.9.1  Basic Motivation of Fundamental Group

The concept of the fundamental group arose to distinguish two geometrical objects such as a disk $\mathbf{D}^*$ with a hole and a disk $\mathbf{D}$ without a hole in the Euclidean plane $\mathbf{R}^2$ as

**Fig. 2.9**  A disk **D**\* with a
hole

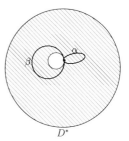

**Fig. 2.10**  A disk **D** without
a hole

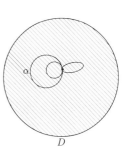

displayed in Figs. 2.9 and 2.10. While considering homotopy, the essential property
of a circle is the existence of an inside hole. Topological spaces homeomorphic to a
circle have the same connectivity properties, and they are homotopy equivalent but
its converse is not true.

All loops in the punctured disk **D**\*, as shown in Fig. 2.9, cannot be continuously
shrunk to a point, i.e., cannot be continuously deformed to a point in **D**\*; on the other
hand, all loops in the disk **D**, see Fig. 2.10, can be continuously shrunk to a point in
**D**. For example, the loop $\beta$ in Fig. 2.9 cannot be continuously deformed to a point
as there is a hole in **D**\*. Some loops in **D**\* such as $\alpha$ may be continuously deformed
to a point but not all loops. This characterizes the difference between the spaces **D**
and **D**\* and leads to the concept of fundamental group.

### 2.9.2  Construction of Fundamental Group

This subsection constructs the fundamental group $\pi_1(X, x_0)$ of a pointed topological
space $(X, x_0)$ by using the homotopy classes of loops in $X$ based at the point $x_0$ as
its elements. For this construction, we start with the set $\Omega(X, x_0)$ of all loops in the
space $X$ based at a point $x_0$ and define a group structure on the set of homotopy
equivalence classes of loops in $\Omega(X, x_0)$.

**Definition 2.9.1**  Let $\mathcal{L}(X, x_0)$ be the set of all loops in the topological $X$ based at
the point $x_0 \in X$ and the boundary points of **I** be $\dot{\mathbf{I}} = \{0, 1\} \subset \mathbf{R}$. Then, it follows

from Theorem 2.6.12 that the relation of homotopy $\alpha \simeq \beta$ rel $\dot{\mathbf{I}}$ on $\mathcal{L}(X, x_0)$ is an equivalence relation. Hence, it generates the set of homotopy classes $\mathcal{L}(X, x_0)/ \simeq$ of loops relative to $\dot{\mathbf{I}}$, denoted by $\pi_1(X, x_0)$.

A suitable composition on the set $\pi_1(X, x_0)$ is defined to make it a group.

**Definition 2.9.2** Given $\alpha, \beta \in \mathcal{L}(X, x_0)$, their product $\alpha * \beta : (\mathbf{I}, \dot{\mathbf{I}}) \to (X, x_0)$ is defined by

$$(\alpha * \beta)(t) = \begin{cases} \alpha(2t), & 0 \le t \le 1/2 \\ \beta(2t - 1), & 1/2 \le t \le 1. \end{cases} \qquad (2.1)$$

Then, $\alpha * \beta$ is well defined and it is continuous by pasting lemma. It is a loop in $X$ based at $x_0$ and hence $\alpha * \beta \in \mathcal{L}(X, x_0)$.

Similarly, given three loops $\alpha, \beta, \gamma \in \mathcal{L}(X, x_0)$, their product $\alpha * \beta * \gamma$ is defined by

$$(\alpha * \beta * \gamma)(t) = \begin{cases} \alpha(3t), & 0 \le t \le 1/3 \\ \beta(3t - 1), & 1/3 \le t \le 2/3 \\ \gamma(3t - 2), & 2/3 \le t \le 1. \end{cases}$$

Then, $\alpha * \beta * \gamma \in \mathcal{L}(X, x_0)$.

**Definition 2.9.3** If $\alpha \in \mathcal{L}(X, x_0)$, then its inverse $\alpha^{-1} : (\mathbf{I}, \dot{\mathbf{I}}) \to (X, x_0)$ defined by $\alpha^{-1}(t) = \alpha(1 - t), \forall\ t \in \mathbf{I}$ asserts that $\alpha^{-1} \in \mathcal{L}(X, x_0)$.

**Remark 2.9.4** The two paths $\alpha$ and $\alpha^{-1}$ have the same set of points of $X$, but they are of opposite directions.

**Proposition 2.9.5** *If $\alpha_1, \alpha_2, \beta_1, \beta_2 \in \mathcal{L}(X, x_0)$ are loops such that $\alpha_1 \simeq \alpha_2$, rel $\dot{\mathbf{I}}\beta_1 \simeq \beta_2$ rel $\dot{\mathbf{I}}$, then $\alpha_1 * \beta_1 \simeq \alpha_2 * \beta_2$ rel $\dot{\mathbf{I}}$.*

**Proof** Let $H : \alpha_1 \simeq \alpha_2$ *rel* $\dot{\mathbf{I}}$ and $K : \beta_1 \simeq \beta_2$ *rel* $\dot{\mathbf{I}}$. Then,

$$H(t, 0) = \alpha_1(t), H(t, 1) = \alpha_2(t), \forall t \in \mathbf{I}, H(0, s) = x_0 = H(1, s), \forall\ s \in \mathbf{I}$$

and

$$K(t, 0) = \beta_1(t), \ K(t, 1) = \beta_2(t), \forall\ t \in \mathbf{I}, \ K(0, s) = x_0 = K(1, s), \forall\ s \in \mathbf{I}.$$

Define a map $F : \mathbf{I} \times \mathbf{I} \to X$ by

$$F(t, s) = \begin{cases} H(2t, s), & 0 \le t \le 1/2 \\ K(2t - 1, s), & 1/2 \le t \le 1. \end{cases}$$

Then, $F$ is well defined. Moreover, it is continuous by pasting lemma. Again,

$$F(t, 0) = \begin{cases} H(2t, 0), & 0 \le t \le 1/2 \\ K(2t - 1, 0), & 1/2 \le t \le 1 \end{cases}$$
$$= \begin{cases} \alpha_1(2t), & 0 \le t \le 1/2 \\ \beta_1(2t - 1), & 1/2 \le t \le 1 \end{cases}$$
$$= (\alpha_1 * \beta_1)(t), \ \forall \ t \in \mathbf{I}.$$

Similarly, $F(t, 1) = (\alpha_2 * \beta_2)(t), \ \forall \ t \in \mathbf{I}, F(0, s) = x_0 = H(0, s), \ \forall \ s \in I$ and $F(1, s) = x_0 = K(1, s), \ \forall \ s \in \mathbf{I}$. It asserts that $F : \alpha_1 * \beta_1 \simeq \alpha_2 * \beta_2$ rel $\dot{\mathbf{I}}$.     □

**Proposition 2.9.6** *If $\alpha, \beta \in \mathcal{L}(X, x_0)$ and $\alpha \simeq \beta$ rel $\dot{\mathbf{I}}$, then $\alpha^{-1} \simeq \beta^{-1}$ rel $\dot{\mathbf{I}}$.*

**Proof** Let $H : \alpha \simeq \beta$ rel $\dot{\mathbf{I}}$. Then,

$$H(t, 0) = \alpha(t), H(t, 1) = \beta(t), \ \forall \ t \in \mathbf{I} \text{ and } H(0, s) = x_0 = H(1, s), \ \forall s \in \mathbf{I}.$$

Define

$$F : \mathbf{I} \times \mathbf{I} \to X, (t, s) \mapsto H(1 - t, s).$$

Then, $F$ is a continuous function such that

$$F(t, 0) = H(1 - t, 0) = \alpha(1 - t) = \alpha^{-1}(t), \ \forall \ t \in I,$$
$$F(t, 1) = H(1 - t, 1) = \beta(1 - t) = \beta^{-1}(t), \ \forall \ t \in I$$
$$\text{and } F(0, s) = H(1, s) = x_0, \ F(1, s) = H(0, s) = x_0.$$

It asserts that $F : \alpha^{-1} \simeq \beta^{-1}$ rel $\dot{\mathbf{I}}$.     □

**Proposition 2.9.7** *If $\alpha, \beta, \gamma \in \mathcal{L}(X, x_0)$, then $\alpha * (\beta * \gamma) \simeq (\alpha * \beta) * \gamma$ rel $\dot{\mathbf{I}}$.*

**Proof** Define $\alpha * (\beta * \gamma) : (\mathbf{I}, \dot{\mathbf{I}}) \to (X, x_0)$ by the rule

$$(\alpha * (\beta * \gamma))(t) = \begin{cases} \alpha(2t), & 0 \le t \le 1/2 \\ (\beta * \gamma)(2t - 1), & 1/2 \le t \le 1 \end{cases}$$
$$= \begin{cases} \alpha(2t), & 0 \le t \le 1/2 \\ \beta(4t - 2), & 1/2 \le t \le 3/4 \\ \gamma(4t - 3), & 3/4 \le t \le 1. \end{cases}$$

Then, $\alpha * (\beta * \gamma)$ is well defined and it is continuous by pasting lemma. It is a loop in $X$ based at $x_0$ and therefore $\alpha * (\beta * \gamma) \in \mathcal{L}(X, x_0)$. On the other hand,

$$((\alpha * \beta) * \gamma)(t) = \begin{cases} (\alpha * \beta)(2t), & 0 \le t \le 1/2 \\ \gamma(2t - 1), & 1/2 \le t \le 1 \end{cases}$$
$$= \begin{cases} \alpha(4t), & 0 \le t \le 1/4 \\ \beta(4t - 1), & 1/4 \le t \le 1/2 \\ \gamma(2t - 1), & 1/2 \le t \le 1. \end{cases}$$

It shows that $(\alpha * \beta) * \gamma \in \mathcal{L}(X, x_0)$.
Define a map $F : \mathbf{I} \times \mathbf{I} \to X$ by the rule

$$F(t, s) = \begin{cases} \alpha(4t/(1+s)), & 0 \leq t \leq (1+s)/4 \\ \beta(4t-1-s), & (1+s)/4 \leq t \leq (2+s)/4 \\ \gamma(1-(4(1-t)/(2-s))), & (2+s)/4 \leq t \leq 1. \end{cases}$$

Then, $F$ is well defined and continuous by pasting lemma. Clearly,

$$F(t, 0) = \begin{cases} \alpha(4t), & 0 \leq t \leq 1/4 \\ \beta(4t-1), & 1/4 \leq t \leq 1/2 \\ \gamma(2t-1), & 1/2 \leq t \leq 1 \end{cases}$$
$$= ((\alpha * \beta) * \gamma)(t), \ \forall \ t \in \mathbf{I},$$

$$F(t, 1) = \begin{cases} \alpha(2t), & 0 \leq t \leq 1/2 \\ \beta(4t-2), & 1/2 \leq t \leq 3/4 \\ \gamma(4t-3), & 3/4 \leq t \leq 1 \end{cases}$$
$$= (\alpha * (\beta * \gamma))(t), \ \forall \ t \in \mathbf{I},$$

Hence, it follows that

$$F(0, s) = \alpha(0) = x_0, \ \text{ and } F(1, s) = \gamma(1) = x_0.$$

It proves that $(\alpha * \beta) * \gamma \simeq \alpha * (\beta * \gamma)$ rel $\dot{\mathbf{I}}$. □

**Proposition 2.9.8** *If $\alpha \in \mathcal{L}(X, x_0)$ and $\delta : \mathbf{I} \to X$ is the constant loop at $x_0$ defined by $\delta(t) = x_0, \ \forall \ t \in \mathbf{I}$, then $\alpha * \delta \simeq \alpha$ rel $\dot{\mathbf{I}}$ and $\delta * \alpha \simeq \alpha$ rel $\dot{\mathbf{I}}$.*

**Proof** $\alpha * \delta : \mathbf{I} \to X$ is defined by the rule

$$(\alpha * \delta)(t) = \begin{cases} \alpha(2t), & 0 \leq t \leq 1/2 \\ \delta(2t-1), & 1/2 \leq t \leq 1 \end{cases}$$
$$= \begin{cases} \alpha(2t), & 0 \leq t \leq 1/2 \\ x_0, & 1/2 \leq t \leq 1. \end{cases}$$

This asserts that $\alpha * \delta \in \mathcal{L}(X, x_0)$. Define a map $F : \mathbf{I} \times \mathbf{I} \to X$ by the rule

$$F(t, s) = \begin{cases} \alpha(2t/(1+s)), & 0 \leq t \leq (1+s)/2 \\ x_0, & (1+s)/2 \leq t \leq 1. \end{cases}$$

This asserts that $F : \alpha * \delta \simeq \alpha$ rel $\dot{\mathbf{I}}$. Similarly, $\delta * \alpha \simeq \alpha$ rel $\dot{\mathbf{I}}$. □

**Proposition 2.9.9** *If $\alpha \in \mathcal{L}(X, x_0)$, then $\alpha * \alpha^{-1} \simeq \delta$ rel $\dot{\mathbf{I}}$ and $\alpha^{-1} * \alpha \simeq \delta$ rel $\dot{\mathbf{I}}$.*

**Proof** $\alpha * \alpha^{-1} : \mathbf{I} \to X$ is defined by the rule

$$(\alpha * \alpha^{-1})(t) = \begin{cases} \alpha(2t), & 0 \leq t \leq 1/2 \\ \alpha^{-1}(2t-1), & 1/2 \leq t \leq 1 \end{cases}$$
$$= \begin{cases} \alpha(2t), & 0 \leq t \leq 1/2 \\ \alpha(1-\overline{2t-1}), & 1/2 \leq t \leq 1 \end{cases}$$
$$= \begin{cases} \alpha(2t), & 0 \leq t \leq 1/2 \\ \alpha(2-2t), & 1/2 \leq t \leq 1. \end{cases}$$

This asserts that $\alpha * \alpha^{-1} \in \mathcal{L}(X, x_0)$. Define $H : \mathbf{I} \times \mathbf{I} \to X$ by

$$H(t, s) = \begin{cases} \alpha(2t(1 - s)), & 0 \le t \le 1/2 \\ \alpha((2 - 2t)(1 - s)), & 1/2 \le t \le 1. \end{cases}$$

This shows that $F : \alpha * \alpha^{-1} \simeq \delta$ rel $\dot{\mathbf{I}}$. Similarly, $\alpha^{-1} * \alpha \simeq c$ rel $\dot{\mathbf{I}}$. $\square$

**Theorem 2.9.10** $\pi_1(X, x_0)$ *is a group under the usual composition of homotopy classes of loops in $\mathcal{L}(X, x_0)$.*

**Proof** Let $[\alpha], [\beta] \in \pi_1(X, x_0)$. Then, $\alpha, \beta \in \mathcal{L}(X, x_0)$ and $\alpha * \beta$ given in Definition 2.9.2 is in $\mathcal{L}(X, x_0)$. This law of composition '$*$' is carried over $\pi_1(X, x_0)$ to define the composition '$\circ$' by the rule $[\alpha] \circ [\beta] = [\alpha * \beta]$. The composition '$\circ$' is well defined by Proposition 2.9.5, because it is independent of the choice of the representatives of the classes. This composition is associative by Proposition 2.9.7 with $[\delta]$ as the identity element by Proposition 2.9.8 and for any element $[\alpha] \in \pi_1(X, x_0)$, it has an inverse $[\alpha^{-1}] \in \pi_1(X, x_0)$ by Proposition 2.9.9. Consequently, $\pi_1(X, x_0)$ is a group under the composition '$\circ$'. This proves that $\pi_1(X, x_0)$ is a group under the usual composition of homotopy classes of loops in $\mathcal{L}(X, x_0)$. $\square$

**Remark 2.9.11** The inverse of an element in the group $\pi_1(X, x_0)$ is represented by a loop traveling the same loop in the reverse direction. Thus, traveling a loop in two opposite directions determines elements of the group $\pi_1(X, x_0)$ which are inverse to each other.

**Definition 2.9.12** The group $\pi_1(X, x_0)$ defined in Theorem 2.9.10 is said to be the **fundamental group or Poincaré group** of the topological space $X$ based at the point $x_0 \in X$.

**Remark 2.9.13** For an equivalent definition of $\pi_1(X, x_0)$ and an alternative proof of its group structure, see Sect. 2.16. **The index '1'** in $\pi_1(X, x_0)$ used now appeared later than the notation $\pi(X, x_0)$ used by Poincaré in 1895. It is also known as the first or one-dimensional homotopy group. There is an infinite sequence of groups $\pi_n(X, x_0)$ with $n = 1, 2, 3, \ldots$, called higher-dimensional homotopy groups of the pointed space $(X, x_0)$, and the first one is the fundamental group. The higher-dimensional homotopy groups (see Sect. 2.20) were introduced by W. Hurewicz in 1935 in his paper [Hurewicz, 1935]. For $n = 0$, $\pi_0(X, x_0)$ taken to be the set of path-connected components of $X$ is not necessarily a group.

**Example 2.9.14** (i) The fundamental group of $\mathbf{I} = [0, 1]$ at the base point $1 \in \mathbf{I}$ is the trivial group, because if $\alpha : \mathbf{I} \to \mathbf{I}$ is a loop, then $H : \mathbf{I} \times \mathbf{I} \to \mathbf{I}, (x, t) \mapsto t + (1 - t)\alpha(x)$ is a homotopy between $\alpha$ and the constant loop at the base point $1 \in \mathbf{I}$. This asserts that $\pi_1(\mathbf{I}, 1)$ consists of only element, denoted by 0.

(ii) For a contractible space $X$ and a point $x_0 \in X$, $\pi_1(X, x_0) = 0$. It is also true that $\pi_1(X, x) = 0$ for any $x \in X$. Hence for any tree $X$ (topological) and a vertex $v_0$ of $X$, the fundamental group $\pi_1(X, v_o) = 0$.

**Fig. 2.11** Geometrical
description of $\psi_\alpha$

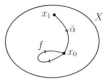

(iii) For the $n$-dimensional Euclidean space $\mathbf{R}^n$, its fundamental group $\pi_1(\mathbf{R}^n, x) = 0$
for any $x \in \mathbf{R}^n$.
(iv) For the $n$-dimensional Euclidean disk $\mathbf{D}^n$, its fundamental group $\pi_1(D^n, x) = 0$
for any $x \in D^n$.
(v) For any convex set $X \subset \mathbf{R}^n$, its fundamental group $\pi_1(X, x_0) = 0$ for any $x_0 \in X$.

**Remark 2.9.15** It is a natural question : Does the fundamental group of a topological
space depend on its base points $x_0, x_1 \in X$? If it depends on base points, how are
the groups $\pi_1(X, x_0)$ and $\pi_1(X, x_1)$ related to arbitrary topological spaces? Theorem
2.9.16 gives its answer.

**Theorem 2.9.16** *Let $X$ be a path-connected space and $x_0, x_1 \in X$ be any two distinct
points. Then, $\pi_1(X, x_0)$ and $\pi_1(X, x_1)$ are isomorphic groups.*

**Proof** By hypothesis, $X$ is a path-connected space and $x_0, x_1 \in X$ are two distinct
points. Then, there exists a path $\alpha : \mathbf{I} \to X, \alpha(0) = x_0, \alpha(1) = x_1$, with its inverse
path $\bar{\alpha} : \mathbf{I} \to X, t \mapsto \alpha(1 - t)$, shown in Fig. 2.11. Define a map

$$\psi_\alpha : \pi_1(X, x_0) \to \pi_1(X, x_1), [f] \mapsto [\bar{\alpha} * f * \alpha].$$

It is geometrically represented in Fig. 2.11. Then, $\psi_\alpha$ is well defined and a group
isomorphism with its inverse $\psi_{\bar{\alpha}} : \pi_1(X, x_1) \to \pi_1(X, x_0), [f] \mapsto [\alpha * f * \bar{\alpha}]$. $\square$

**Remark 2.9.17** For any path-connected space $X$, the fundamental group $\pi_1(X, x_0)$
is independent of its base point $x_0$ and hence it is unique up to isomorphism by
Theorem 2.9.16. This group is abbreviated to $\pi_1(X)$, without specifying the base
point of the path-connected space $X$.

**Corollary 2.9.18** *Let $X$ be a topological space and $x_0, x_1 \in X$. If $\alpha$ is a path in $X$
from $x_0$ to $x_1$, then $\alpha$ induces an isomorphism*

$$\psi_\alpha : \pi_1(X, x_0) \to \pi_1(X, x_1), [f] \mapsto [\bar{\alpha} * f * \alpha].$$

**Proof** It follows from the proof of Theorem 2.9.16. $\square$

Proposition 2.9.19 relates path homotopic maps to isomorphism of fundamental
groups.

**Proposition 2.9.19** *If $\alpha$ and $\beta$ are two path homotopic maps in $X$ joining the points
$x_0$ to $x_1$, then their induced isomorphisms $\psi_\alpha = \psi_\beta : \pi_1(X, x_0) \to \pi_1(X, x_1)$, i.e.,*

**Fig. 2.12**  Path homotopic
maps

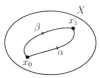

(i)  *the isomorphism* $\psi_\alpha : \pi_1(X, x_0) \to \pi_1(X, x_1)$ *and*
(ii)  *the isomorphism* $\psi_\beta : \pi_1(X, x_0) \to \pi_1(X, x_1)$

*defined in Corollary 2.9.18 are the same.*

**Proof**  If $\alpha$ and $\beta$ are two homotopic paths as shown in Fig. 2.12, then $\bar{\alpha}$ and $\bar{\beta}$ are
also path homotopic. Therefore, it follows that for any loop $f$ in $X$ based at $x_0$, the
path $\bar{\alpha} * f * \alpha$ is a path homotopic to the path $\bar{\beta} * f * \beta$. Consequently, $\psi_\alpha([f]) = \psi_\beta([f]) \; \forall \; [f] \in \pi_1(X, x_0)$. This asserts that $\psi_\alpha = \psi_\beta : \pi_1(X, x_0) \to \pi_1(X, x_1)$.  $\square$

## 2.10  Homotopy Property of Connectedness

This section establishes an interesting relation between homotopy and connectedness properties of topological spaces by showing that **connectedness is a homotopy
property** of the spaces in the sense that this property is preserved by every homotopy equivalence, which is in addition to its topological property proved in Chap. 5
of **Basic Topology, Volume 1** of the present series of books.

**Definition 2.10.1**  Given a topological space $X$, let $\mathcal{C}(X)$ denote the set of all connected components or simply called components of $X$. If $f : X \to Y$ is an arbitrary
continuous map, then for each component $C$ of $X$, $f(C) \subset Y$ is connected and hence
it is contained in a component $D$ of $Y$. This defines a map

$$\psi(f) : \mathcal{C}(X) \to \mathcal{C}(Y), \; C \to D$$

which is well defined, since components of a topological space are mutually disjoint.

**Proposition 2.10.2**  *If $f, g : X \to Y$ are homotopic maps, then $\psi(f) = \psi(g)$.*

**Proof**  Let $f, g : X \to Y$ be homotopic maps. Then, there exists a homotopy

$$F : X \times \mathbf{I} \to Y$$

such that $F(x, 0) = f(x)$ and $F(x, 1) = g(x)$ for all $x \in X$. Since $\mathbf{I}$ is connected, the
product space $C \times \mathbf{I}$ of two connected sets is connected for every component $C \in \mathcal{C}(X)$. Again since, $F$ is continuous, it follows that $F(C \times \mathbf{I}) \subset Y$ is also connected.

It shows that $F(C \times \mathbf{I})$ is contained in a component $D$ in $\mathcal{C}(Y)$. Again since both $f(C)$ and $g(C)$ are contained in $H(C \times \mathbf{I})$, it follows that

$$[\psi(f)](C) = D = [\psi(g)](C)$$

for every $C \in \mathcal{C}(X)$. This proves that $\psi(f) = \psi(g)$.     □

**Proposition 2.10.3** *Given a homotopy equivalence $f : X \to Y$, the map*

$$\psi(f) : \mathcal{C}(X) \to \mathcal{C}(Y)$$

*is a bijection.*

**Proof** As $f : X \to Y$ is a homotopy equivalence by hypothesis, there exists a continuous map $g : Y \to X$ such that $f \circ g \simeq 1_Y$ and $g \circ f \simeq 1_X$, where $1_X$ and $1_Y$ are identity maps on $X$ and $Y$, respectively. Then by Proposition 2.10.2, it follows that

$$\psi(f) \circ \psi(g) = \psi(f \circ g) = \psi(1_Y)$$

and

$$\psi(g) \circ \psi(f) = \psi(g \circ f) = \psi(1_X),$$

where $\psi(1_X)$ and $\psi(1_Y)$ are identity maps on the sets $\mathcal{C}(X)$ and $\mathcal{C}(Y)$, respectively. This asserts that $\psi(f)$ is a bijection with $\psi(g)$ its inverse.     □

**Definition 2.10.4** A property of a topological space is said to be a **homotopy property** if this property is preserved by every homotopy equivalence; i.e., a homotopy property of a topological $X$ is such a property that every topological space homotopy equivalent to $X$ has also the same property.

**Corollary 2.10.5** *Connectedness is a homotopy property.*

**Proof** It follows from Proposition 2.10.3.     □


## 2.11 Functorial Property of $\pi_1$

Functors play a key role in algebraic topology. This section proves the functorial property of $\pi_1$, where $\pi_1(X, x_0)$ denotes the fundamental group of the pointed space $(X, x_0)$. In the language of category theory, $\pi_1$ is a covariant functor from the homotopy category $\mathcal{H}tp_*$ of pointed topological spaces and homotopy classes of continuous maps to the category $\mathcal{G}rp$ of groups and homomorphisms. This functor transfers topological problems into algebraic ones to have a better chance for solution. The fundamental group of a pointed topological space is a topological (algebraic) invariant as well as it is a homotopy invariant. The aim of this section is to study $\pi_1$ through the category theory and establish some important properties of fundamental group.

**Theorem 2.11.1** *Let $f : (X, x_0) \to (Y, y_0)$ be a base point preserving continuous map. Then, it induces a group homomorphism*

$$f_* : \pi_1(X, x_0) \to \pi_1(Y, y_0), \quad [\alpha] \mapsto [f \circ \alpha]$$

*such that*

(i) *For the identity map $1_X : (X, x_0) \to (X, x_0)$, its induced homomorphism $1_{X*}$ is the identity automorphism on $\pi_1(X, x_0)$.*

(ii) *For the base point preserving continuous maps $f : (X, x_0) \to (Y, y_0)$ and $g : (Y, y_0) \to (Z, z_0)$, their composite map $g \circ f$ induces homomorphism with the property*

$$(g \circ f)_* = g_* \circ f_* : \pi_1(X, x_0) \to \pi_1(Z, z_0).$$

(iii) *For homotopic maps $f \simeq g : (X, x_0) \to (Y, y_0)$ rel $\{x_0\}$, their induced homomorphisms*

$$f_* = g_* : \pi_1(X, x_0) \to \pi_1(Y, y_0).$$

(iv) *For a homeomorphism $f : (X, x_0) \to (Y, y_0)$, its induced homomorphism*

$$f_* : \pi_1(X, x_0) \to \pi_1(Y, y_0)$$

*is an isomorphism of groups.*

(v) *For a homotopy equivalence $f : (X, x_0) \to (Y, y_0)$ its induced homomorphism*

$$f_* : \pi_1(X, x_0) \to \pi_1(Y, y_0)$$

*is an isomorphism of groups.*

***Proof*** Let $\alpha$, $\beta$ be two homotopic loops in $X$ based at $x_0$ and $H : \alpha \simeq \beta$ rel $\dot{\mathbf{I}}$ be a homotopy. Since $f \circ \alpha$ is a loop in $Y$ based at $y_0$, for every loop $\alpha$ in $X$ based at $x_0$,

$$f \circ H : f \circ \alpha \simeq f \circ \beta \text{ rel } \dot{\mathbf{I}}.$$

It shows that $f_*$ is well defined, and moreover, it is a homomorphism.

(i) If $1_X : (X, x_0) \to (X, x_0)$ is the identity map, then $1_X \circ \alpha = \alpha$ for every loop $\alpha$ in $X$ based at $x_0$ asserts that $1_{X*}([\alpha]) = [1_X \circ \alpha] = [\alpha], \ \forall \ [\alpha] \in \pi_1(X, x_0)$. It shows that $1_{X*} : \pi_1(X, x_0) \to \pi_1(X, x_0)$ is the identity automorphism.

(ii) Given any element $[\alpha] \in \pi_1(X, x_0)$, it follows that

$$(g \circ f)_*([\alpha]) = [(g \circ f) \circ \alpha] = [g \circ (f \circ \alpha)] = g_*([f \circ \alpha]) = g_*(f_*([\alpha])) = (g_* \circ f_*)[\alpha].$$

It asserts that $(g \circ f)_* = g_* \circ f_*$.

(iii) If $f \simeq g$ rel $\{x_0\}$, then $\forall \ [\alpha] \in \pi_1(X, x_0), f \circ \alpha \simeq g \circ \alpha$ rel $\{y_0\}$. It asserts that

$$f_*([\alpha]) = [f \circ \alpha] = [g \circ u] = g_*([\alpha]).$$

Hence, it follows that $f_* = g_*$.

(iv) If $f : (X, x_0) \to (Y, y_0)$ is homeomorphism, there exists a continuous map $g :$ $(Y, y_0) \to (X, x_0)$ such that $f \circ g = 1_Y, g \circ f = 1_X$. Hence, it follows from **(i)** **and (ii)** that $f_* : \pi_1(X, x_0) \to \pi_1(Y, y_0)$ is an isomorphism with $g_*$ as its inverse.

(v) If $f : (X, x_0) \to (Y, y_0)$ is a homotopy equivalence, there exists a continuous map $g : (Y, y_0) \to (X, x_0)$ such that $f \circ g \simeq 1_Y, g \circ f \simeq 1_X$. Hence, it from **(iii)** that

$$f_* : \pi_1(X, x_0) \to \pi_1(Y, y_0)$$

is an isomorphism with $g_*$ as its inverse.                    □

**Corollary 2.11.2** (i) *The fundamental group is a **topological invariant**.*
(ii) *The fundamental group is a **homotopy invariant**.*

**Proof** (i) The fundamental group of a pointed topological space is invariant under homeomorphisms by Theorem 2.11.1(iv) in the sense that homeomorphic pointed topological spaces have isomorphic fundamental groups. This implies that it is a topological invariant.

(ii) The fundamental group of a pointed topological space is invariant under homotopy equivalence by Theorem 2.11.1(v) in the sense that homotopy equivalent pointed topological spaces have isomorphic fundamental groups. This implies that it is a homotopy invariant.                    □

Theorem 2.11.3 expresses $\pi_1$ in the language of category theory by using Theorem 2.11.1.

**Theorem 2.11.3** *Let $\mathcal{T}op_*$ be the category of pointed topological spaces and their base point preserving continuous maps, $\mathcal{G}rp$ be the category of groups and their homomorphisms and $\mathcal{H}tp_*$ be the homotopy category of pointed topological spaces and their homotopy classes of maps. Then,*

(i) *$\pi_1 : \mathcal{T}op_* \to \mathcal{G}rp$ is a covariant functor from the category $\mathcal{T}op_*$ to the category $\mathcal{G}rp$ such that if $f, g : (X, x_0) \to (Y, y_0)$ are continuous and $f \simeq g$ rel $\{x_0\}$, then $\pi_1(f) = f_* = g_* = \pi_1(g)$.*

(ii) *$\pi_1 : \mathcal{H}tp_* \to \mathcal{G}rp$ is also a covariant functor from the category $\mathcal{H}tp_*$ to the category $\mathcal{G}rp$.*

**Proof** (i) The assignment

$$\pi_1 : \mathcal{T}op^* \to \mathcal{G}rp, (X, x_0) \mapsto \pi_1(X, x_0)$$

defines an object function, and the assignment

$$\pi_1 : \mathcal{T}op^* \to \mathcal{G}rp, f \mapsto \pi_1(f) = f_*$$

defines a morphism function. Hence, **(i)** follows from Theorem 2.11.1.

(ii) It follows from **(i)** and Theorem 2.11.1.                    □

## 2.12 Link between Fundamental Groups and Retractions

This section studies the concepts of retraction and strong deformation retractions by using fundamental group. It is proved in Proposition 2.12.2 that every strong deformation retract $A$ of $X$ has the isomorphic fundamental groups.

**Proposition 2.12.1** *Let $X$ be a topological space and $A$ be a subspace of $X$. Then,*

*(i)  Every retraction $r : X \to A$ induces an epimorphism*

$$r_* : \pi_1(X, a) \to \pi_1(A, a), \ \forall\, a \in A.$$

*(ii)  The inclusion map $i : A \hookrightarrow X$ induces a monomorphism*

$$i_* : \pi_1(A, a) \to \pi_1(X, a), \ \forall\, a \in A.$$

**Proof** Let $r : X \to A$ be a retraction and $i : A \hookrightarrow X$ be the inclusion map. Then given a point $a \in A$, the map $r : (X, a) \to (A, a)$ is a continuous map such that $r(a) = a$ and the induced maps

$$r_* : \pi_1(X, a) \to \pi_1(A, a)$$

and

$$i_* : \pi_1(A, a) \to \pi_1(X, a)$$

are both homomorphisms. Since the composite map $r \circ i : (A, a) \xrightarrow{i} (X, a) \xrightarrow{r} (A, a)$ is the identity map on $(A, a)$, the composite of induced homomorphisms

$$r_* \circ i_* : \pi_1(A, a) \xrightarrow{i_*} \pi_1(X, a) \xrightarrow{r_*} \pi_1(A, a)$$

is the identity automorphism on $\pi_1(A, a)$ by Theorem 2.11.1 by using the functorial property of $\pi_1$. This asserts from group theory that $r_*$ is an epimorphism and $i_*$ is a monomorphism for every $a \in A$. $\qquad\square$

**Proposition 2.12.2** *Let $X$ be a topological space and $A \subset X$ be a strong deformation retract of $X$. Then, the fundamental groups $\pi_1(A, a)$ and $\pi_1(X, a)$ are isomorphic for each $a \in A$.*

**Proof** Let $A \subset X$ be a strong deformation retract of $X$. Hence, there exists a retraction $r : (X, a) \to (A, a)$ for every $a \in A$ with the property that $1_X \simeq i \circ r$ rel $A$. This implies that the composite homomorphism $i_* \circ r_*$ is the identity automorphism. Hence, $i_*$ is an epimorphism. Again, $r \circ i = \mathrm{id}$ asserts that the composite homomorphism $r_* \circ i_*$ is an automorphism. Hence, $i_*$ is a monomorphism. This implies that

$$i_* : \pi_1(A, a) \to \pi_1(X, a)$$

is an isomorphism for every $a \in A$. $\qquad\square$

***Remark 2.12.3*** Let $X$ be a given topological space and $x_0 \in X$. If $C_{x_0}$ is the path component of $X$ containing $x_0$, then $\pi_1(C_{x_0}, x_0) = \pi_1(X, x_0)$. Since all the loops based at the point $x_0$ and their homotopies in $X$ relative to $\{x_0\}$ lie entirely in the subspace $C_{x_0}$, it asserts that the fundamental group $\pi_1(X, x_0)$ depends only on the path component $C_{x_0}$. Unfortunately, this does not provide any information on the complement set $X - C_{x_0}$.

## 2.13  Simply Connectedness

This section formalizes the intuitive concept of geometrical objects $X$ having no hole to prevent any closed path in $X$ from shrinking to a point in $X$. For example, every ordinary closed curve on the sphere can be deformed continuously into a point without leaving the surface. On the other hand, there exist circles $C$ and $C'$ on the surface of the torus as shown in Fig. 2.14 which cannot be deformed continuously into a point without leaving its surface. The former example leads to the concept of simply connected spaces, and the latter example leads to concept of nonsimply connected spaces. Moreover, simply connected spaces are characterized in Theorem 2.13.9 with the help of homotopy theory.

**Definition 2.13.1**  A topological space $X$ is said to be **simply connected** if it is path connected and its fundamental group is trivial for some base point $x_0 \in X$ and hence for every base point of $X$. In other words, a path-connected space $X$ is simply connected if $\pi_1(X, x_0) = 0$ for some $x_0 \in X$ (hence for every base point $x_0 \in X$).

***Remark 2.13.2***  **Geometrically,** a path-connected topological space $X$ is said to be simply connected if there is no hole in $X$ to prevent any closed path in $X$ from shrinking to a point in $X$. More precisely, a path-connected topological space is simply connected if on the topological space every closed curve can be continuously deformed into a point without leaving the space. For example, it appears geometrically that the $n$-sphere $S^n$ is simply connected for $n > 1$ and hence, $\pi_1(S^n) = 0$, $\forall n > 1$.

***Example 2.13.3***  The real line space $\mathbf{R}$ and the Euclidean $n$-space $\mathbf{R}^n$ are simply connected spaces by Corollary 2.13.6. The circle $S^1$ is not simply connected, because $\pi_1(S^1) \neq 0$ by Corollary 2.19.2, though it is path connected.

**Theorem 2.13.4**  *Let $\alpha, \beta : \mathbf{I} \to X$ be two paths in a path-connected space $X$ with the same initial point and the same terminal point. Then, the paths $\alpha$ and $\beta$ are homotopic, $X$ is simply connected.*

**Fig. 2.13** Paths $\alpha, \beta$ in $X$

***Proof*** Let $\alpha, \beta : \mathbf{I} \to X$ be two paths in a path-connected space $X$ from the point $x_0$ to the point $x_1$, as shown in Fig. 2.13. Then, $\alpha * \bar{\beta}$ is a loop in $X$ based at $x_0$, where $\bar{\beta}$ is the inverse path of $\beta$ in $X$.

By hypothesis, $X$ is simply connected. Hence, it follows that $\alpha * \bar{\beta} \simeq \delta_{x_0}$ (constant loop at $x_0$). It shows that $\alpha * \bar{\beta} * \beta \simeq \delta_{x_0} * \beta$. Consequently, $[(\alpha * \bar{\beta}) * \beta] = [\delta_{x_0} * \beta = [\beta]$. Again, $[(\alpha * \bar{\beta}) * \beta] = [\alpha * (\bar{\beta} * \beta)] = [\alpha * \delta_{x_1}] = [\alpha]$. Hence, $[\alpha] = [\beta]$ shows that $\alpha \simeq \beta$.                                            □

**Theorem 2.13.5** *Let $X$ be a contractible space. Then, it is simply connected.*

***Proof*** Since $X$ is a contractible space by hypothesis, the identity map $1_X \simeq \delta$ for some constant map $\delta$. Hence, there is a point $x_0 \in X$ and a homotopy $F : X \times \mathbf{I} \to X$ such that $F(x, 0) = x$ and $F(x, 1) = x_0$, $\forall \; x \in X$. Define a path

$$\alpha_x = F(x, -) : \mathbf{I} \to X, t \mapsto \alpha_x(t) = F(x, t)$$

in $X$ from $\alpha_x(0) = F(x, 0) = x$ to $\alpha_x(1) = F(x, 1) = x_0$. Again, for any $y \in X$, $\alpha_y$ is a path from $y$ to $x_0$ and hence $\bar{\alpha}_y$ (the inverse path of $\alpha_y$) is a path in $X$ from $x_0$ to $y$. Thus, any two points $x$ and $y$ can be joined by the path $\alpha_x * \bar{\alpha}_y$ in $X$. This asserts that the space $X$ is path-connected space. Since $\pi_1(X, x_0) = 0$ (see Example 2.9.14), it follows that the space $X$ is simply connected.                                            □

**Corollary 2.13.6** *The Euclidean n space $\mathbf{R}^n$ and any convex subspace of $\mathbf{R}^n$ are simply connected.*

**Definition 2.13.7** A subspace $X \subset \mathbf{R}^n$ is called **star convex** if for some $x_0 \in X$,

$$(1 - t)x + tx_0 \in X, \; \forall t \in \mathbf{I}$$

for any other point $x \in X$; i.e., geometrically, it means that all the line segments joining the point $x_0$ to any other point $x \in X$ completely lie in $X$.

**Proposition 2.13.8** *Every star convex subspace of $\mathbf{R}^n$ is simply connected.*

***Proof*** Let $X \subset \mathbf{R}^n$ be a star convex space and $x_0 \in X$. Then, there exists a continuous map

$$F : X \times \mathbf{I} \to X, \; (x, t) \mapsto (1 - t)x + tx_0.$$

This shows that $F : I_X \simeq c_{x_0}$ (constant map at $x_0$) and hence the space is contractible. This asserts by Theorem 2.13.5 that $X$ is simply connected.                                            □

### 2.13.1   A Characterization of Simply Connected Spaces

Theorem 2.13.9 gives a characterization of simply connected spaces with the help of homotopy of paths.

**Theorem 2.13.9** *Let $X$ be a path-connected space. Then, it is simply connected iff every pair of paths in $X$ with the same initial point and same terminal point is homotopic.*

**Proof** First suppose that the path-connected space $X$ is simply connected. Given two points $x_0, x_1 \in X$, let $\alpha$ and $\beta$ be two paths in $X$ from $x_0$ to $x_1$. Then, it follows by Theorem 2.13.4 that $\alpha \simeq \beta$. Next suppose that the space $X$ is path connected and $[\alpha] \in \pi_1(X, x_0)$. Then by hypothesis, $\alpha \simeq c_{x_0}$ (constant path at $x_0$). It shows that $[\alpha] = [c_{x_0}]$ and hence $\pi_1(X, x_0) = 0$. As $X$ is path connected, it follows that the space $X$ is simply connected.                                                                 □

### 2.13.2   Simply Connected Surfaces

This subsection continues the study of connected and disconnected spaces $X$, when $X$ is in particular a surface and also provides concrete examples of simply connected surfaces. A connected surface $S$ is said to be **simply connected** if every closed path on $S$ can be deformed continuously into a point without leaving the surface $S$.

**Definition 2.13.10** A surface is said to be **arc connected (or path connected)** in $\mathbf{R}^n$ if any two points on the surface can be joined by a continuous path.

**Example 2.13.11** (i)   The sphere $S^2$ is arc connected.
(ii)   The hyperboloid of two sheets is not arc connected.
(iii)   Every arc-connected surface is connected. A surface which is not connected consists of two or more distinct surfaces.

**Definition 2.13.12** A connected surface is said to be **simply connected** if every closed curve on the surface can be continuously deformed to a point without leaving the surface.

**Example 2.13.13** (i)   The sphere $S^2$ is a simply connected surface.
(ii)   The torus $T = S^1 \times S^1$ is not a simply connected surface, since every closed curve on the surface of the torus $T$ cannot be continuously deformed to a point without leaving the surface of the torus $T$. For example, the closed curve $C''$ on $T$ can be continuously deformed to a point without leaving the torus, but the closed curves $C$ and $C'$ as shown in Fig. 2.14 cannot be continuously deformed to a point.

**Fig. 2.14** Circles $C, C', C''$
on the surface of the torus

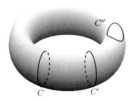

## 2.14   Fundamental Groups of Product Spaces

This section gives a relation between the fundamental group of a product space and the fundamental group of its factor spaces.

From abstract group theory, it is well known that

(i) Given two groups $G$ and $H$, their Cartesian product $G \times H$ admits a group structure with pointwise multiplication given by

$$(g, h)(g', h') = (gg', hh'), \ \forall \, g, g' \in G, h, h' \in H.$$

(ii) If $f : K \to G$ and $s : K \to H$ are any two homomorphisms from any group $K$, then the map

$$\psi : K \to G \times H, \ k \mapsto \ (f(k), s(k))$$

is also a group homomorphism.

Moreover, from general topology, it is known that

(i) Given two canonical projections $p_1 : (X \times Y) \to X$, $p_2 : X \times Y \to Y$ and a given a pair of continuous maps $f : \mathbf{I} \to X, g : \mathbf{I} \to Y$, there is a continuous map

$$(f, g) : \mathbf{I} \to X \times Y, \ t \mapsto \ (f(t), g(t)).$$

(ii) Conversely, any continuous map $h : \mathbf{I} \to X \times Y$ defines a pair of continuous maps

$$p_1 \circ h : \mathbf{I} \to X \ \text{ and } p_2 \circ h : \mathbf{I} \to Y.$$

Proof of Theorem 2.14.1 is based on the above algebraic and topological results.

**Theorem 2.14.1** *If $X$ and $Y$ are two pointed topological spaces having base points $x_0 \in X$ and $y_0 \in Y$, then the fundamental groups $\pi_1(X, x_0) \oplus \pi_1(Y, y_0)$ and $\pi_1(X \times Y, (x_0, y_0))$ are isomorphic.*

***Proof*** If $\alpha : (\mathbf{I}, \dot{\mathbf{I}}) \to (X \times Y, (x_0, y_0))$ is a loop in $X \times Y$ at $(x_0, y_0)$, then the canonical projections $p : X \times Y \to X$ and $q : X \times Y \to Y$ are continuous maps. Hence, they induce homomorphisms

(i) $p_* : \pi_1(X \times Y, (x_0, y_0)) \to \pi_1(X, x_0), [\alpha] \mapsto [p \circ \alpha]$ and
(ii) $q_* : \pi_1(X \times Y, (x_0, y_0)) \to \pi_1(Y, y_0), [\alpha] \mapsto [q \circ \alpha]$.

Define a map

$$f = (p_*, q_*) : \pi_1(X \times Y, (x_0, y_0)) \to \pi_1(X, x_0) \oplus \pi_1(Y, y_0), [\gamma] \mapsto (p * ([\gamma]),$$
$$q_*([\gamma])) = ([p \circ \gamma], [q \circ \gamma]).$$

Then, $f = (p_*, q_*)$ is a group isomorphism with its inverse isomorphism

$$g : \pi_1(X, x_0) \oplus \pi_1(Y, y_0) \to \pi_1(X \times Y, (x_0, y_0)), ([\alpha], [\beta]) \mapsto [(\alpha, \beta)],$$

where $(\alpha, \beta)$ is defined by

$$(\alpha, \beta) : \mathbf{I} \to X \times Y, t \mapsto (\alpha(t), \beta(t)). \qquad \square$$

**Corollary 2.14.2** *If $X = S^1 \times \mathbf{R}$ is a cylinder, then $\pi_1(X) \cong \mathbf{Z}$.*

***Proof*** It follows from Theorem 2.14.1, since both of $S^1$ and $\mathbf{R}$ are arcwise-connected spaces with $\pi_1(S^1) \cong \mathbf{Z}$ and $\pi_1(\mathbf{R}) \cong \mathbf{0}$. $\qquad \square$

## 2.15 Fundamental Groups of Hopf Spaces and Topological Groups

This section discusses Hopf spaces, named after H. Hopf (1894–1971), by generalizing topological groups, and studies their fundamental groups. Fundamental groups of arbitrary topological spaces may be abelian or nonabelian. For example, the fundamental group of the circle is abelian but that of figure-eight is nonabelian (see Theorems 2.19.1 and 2.19.16). On the other hand, **the fundamental group of every H-space is always abelian** and hence fundamental group of every topological group is abelian, which is an interesting result proved in this section.

### 2.15.1 H-spaces and their Fundamental Groups

This subsection defines Hopf spaces and proves that their fundamental groups are always abelian. Every topological group is an $H$-space, but its converse is not necessarily true (see Example 2.15.4 ). For the study of Hopf group formulated in Definition 2.15.3 from the viewpoint of homotopy theory, the book [Adhikari, 2016] is referred.

**Definition 2.15.1** Let $X$ be a pointed topological space with base point $x_0 \in X$. Then, $X$ is said to be an **Hopf space**, in brief, $H$-**space** if there exists a continuous multiplication

$$\mu : X \times X \to X,$$

whose image is abbreviated, $\mu(x, y) = xy$ such that the constant map

$$c : X \to X, \; x \mapsto x_0$$

is a homotopy identity in the sense that the maps

$$\alpha, \beta : X \to X, \; x \mapsto x_0 x, \; x x_0$$

are homotopic.

**Definition 2.15.2** An $H$-space $X$ with associative continuous multiplication

$$\mu : X \times X \mapsto X, (x, y) \to xy$$

and homotopy identity

$$c : X \to X, \; x \mapsto x_0$$

is said to be an **Hopf group (H-group)** if there exists a continuous map

$$\psi : X \to X$$

such that each of the maps

$$X \to X, \; x \mapsto \psi(x)x, \; x\psi(x)$$

is homotopic to the map $c$ rel $\{x_0\}$. The map $\psi$ is said to be **homotopy inverse** for $X$ and $\mu$.

**Definition 2.15.3** An associative Hopf space with homotopy inverse is said to be an **Hopf group or $H$-group**. It is sometimes called a **generalized topological group.**

*Example 2.15.4* Every topological group is an Hopf space. Let $(X, x_0)$ be a topological group with the identity element $x_0$. Then, the maps $\alpha, \beta : X \to X, \; x \mapsto x_0 x, \; x x_0$ are equal and hence they are homotopic. But its converse is not necessarily true. For example,

(i)  The infinite real projective space $\mathbf{R}P^{\infty} = \bigcup_{n \geq 0} \mathbf{R}P^n$ endowed with weak topology is an Hopf space but it is not a topological group.

(ii)  The infinite complex projective space $CP^\infty = \bigcup_{n \geq 0} CP^n$ endowed with weak topology is an Hopf space but it is not a topological group.

**Example 2.15.5** Every Lie group is an $H$-space. Hence, the general linear group $GL(n, \mathbf{R})$ and the orthogonal groups $O(n, \mathbf{R})$ form an important family of $H$-spaces.

**Remark 2.15.6** Example 2.15.4 asserts that every topological group $X = G$ is an Hopf space, where $x_0 = e$ is taken to be the identity element of the topological group.

**Theorem 2.15.7** *Let $X$ be an $H$-space with continuous multiplication $\mu : X \times X \to X$ and homotopy identity $c : X \to X$, $x \mapsto x_0$. Then, $\pi_1(X, x_0)$ is abelian.*

**Proof** The isomorphism $f$ and $g$ defined in Theorem 2.14.1 are now applied to prove the theorem by taking, in particular, $X = Y$. Then,

$$f : \pi_1(X \times X, (x_0, x_0)) \to \pi_1(X, x_0) \oplus \pi_1(X, x_0)$$

is an isomorphism with its inverse isomorphism $g : \pi_1(X, x_0) \times \pi_1(X, x_0) \to \pi_1(X \oplus X, (x_0, x_0))$ and $1_X : X \to X$, $x \mapsto x$ is the identity map. Now, for $[\alpha], [\beta] \in \pi_1(X, x_0)$,

$$[\beta] = (\mu \circ (c, 1_X))_*[\beta] = \mu_*([(c, 1_X) \circ \beta]) = \mu_*([(c \circ \beta, \beta)]) = (\mu_* \circ g)([c \circ \beta],$$
$$[\beta]) = (\mu_* \circ g)(e, [\beta]),$$

where $e$ is the identity element of $\pi_1(X, x_0)$. Similarly, $[\alpha] = (\mu_* \circ g)([\alpha], e)$. Clearly,

$$\mu_* \circ g : \pi_1(X, x_0) \oplus \pi_1(X, x_0) \to \pi_1(X, x_0)$$

is a homomorphism such that

$$(\mu_* \circ g)(([\alpha], [\beta])) = (\mu_* \circ g)(([\alpha], e), (e, [\beta])) = (\mu_* \circ g)(([\alpha], e)) \cdot (\mu_* \circ g)$$
$$(e, [\beta]) = [\alpha] \circ [\beta]$$

and also

$$(\mu_* \circ g)(([\alpha] \cdot [\beta])) = (\mu_* \circ g)((e, [\beta]), ([\alpha], e)) = (\mu_* \circ g)$$
$$(e, [\beta]) \cdot (\mu_* \circ g)([\alpha], e) = [\beta] \circ [\alpha].$$

Hence,

$$[\alpha] \circ [\beta] = [\beta] \circ [\alpha], \ \forall \ [\alpha], [\beta] \in \pi_1(X, x_0)$$

asserts that the group $\pi_1(X, x_0)$ is abelian. $\qquad\qquad\qquad\Box$

## 2.15.2  Fundamental Groups of Topological Groups

This subsection proves that the fundamental group of every topological group is abelian. As topological groups occupy a vast territory in topology, geometry and algebra, it is necessary to study their fundamental groups.

**Proposition 2.15.8** *Let G be a topological group with identity element e. Then, the fundamental $\pi_1(G, e)$ is abelian.*

**Proof** By hypothesis, $G$ is a topological group with identity element $e$. Hence, it is an H-space by Remark 2.15.6. Consequently, the proposition is proved by using Theorem 2.15.7.  □

Example 2.15.9 provides an extensive class of topological spaces having abelian fundamental groups.

**Example 2.15.9** (Some topological spaces having abelian fundamental groups) There are plenty of such examples in matrix algebra. To show it, consider the set $M(n, \mathbf{R})$ of all square matrices of order $n$ over the field $\mathbf{R}$ identified with the Euclidean $n^2$-space $\mathbf{R}^{n^2}$. The fundamental groups of $GL(n, \mathbf{R})$, $SL(n, \mathbf{R})$, $O(n, \mathbf{R})$ and $SO(n, \mathbf{R})$ and their complex analogues $GL(n, \mathbf{C})$, $SL(n, \mathbf{C})$, $U(n, \mathbf{C})$, $SU(n, \mathbf{C})$, and symplectic group $Sp(n, \mathbf{H})$ based at identity matrix are all abelian by Proposition 2.15.8, where these classical topological groups of matrices are defined as follows.

(i) Let $GL(n, \mathbf{R})$ be the set of all $n \times n$ nonsingular matrices over $\mathbf{R}$. It forms a topological group under usual multiplication of matrices, called general linear group over $\mathbf{R}$.

(ii) Let $SL(n, \mathbf{R})$ be defined by $SL(n, \mathbf{R}) = \{A \in GL(n, \mathbf{R}) : \det A = 1\}$. It is a subgroup of $GL(n, \mathbf{R})$, called special linear group.

(iii) Let $O(n, \mathbf{R})$ be defined by $O(n, \mathbf{R}) = \{A \in GL(n, \mathbf{R}) : AA^t = I\}$ is a subgroup of $GL(n, \mathbf{R})$ called orthogonal group.

(iv) Let $SO(n, \mathbf{R})$ be defined by $SO(n, \mathbf{R}) = \{A \in O(n, \mathbf{R}) : \det A = 1\}$. It forms topological group, called special orthogonal group.

(v) The topological groups $GL(n, \mathbf{C})$, $SL(n, \mathbf{C})$, $U(n, \mathbf{C})$, $SU(n, \mathbf{C})$ are defined in analogous ways.

(vi) The quaternionic analogue of orthogonal analogue is the symplectic group $Sp(n, \mathbf{H})$ defined by

$$Sp(n, \mathbf{H}) = \{A \in GL(n, \mathbf{H}) : AA^* = I\},$$

where $A^*$ is the quaternionic conjugate transpose of $A$, and conjugation is in the sense of reversal of all three imaginary components. It is a subgroup of the topological group $GL(n, \mathbf{H})$.

**Remark 2.15.10** Given a pointed topological space $(X, x_0)$, its fundamental group $\pi_1(X, x_0)$ may be abelian or nonabelian. But for a pointed topological space $(X, x_0)$, if its fundamental group $\pi_1(X, x_0)$ is nonabelian, then there exists no multiplication on $X$ making it a topological group and even such a space cannot be equipped with the structure of an $H$-space because of Theorem 2.15.7 and Proposition 2.15.8.

## 2.16   Alternative Approach to Fundamental Group

This section presents an alternative approach to the concept of fundamental groups defined in Sect. 2.9, by considering a loop on a pointed topological space $(X, x_0)$ as a continuous map from a circle into $X$, and a chosen point of the circle is being sent to $x_0$ by this map. The basic objective of the concept of fundamental groups is to classify all loops in a topological space based at a point up to homotopy equivalence. This leads to define an alternative definition of fundamental groups. This section gives a convenient approach to define fundamental group initiated by Hurewicz (1904–1956) in his paper [Hurewicz, 1935], which is equivalent to this concept inaugurated by Poincaré in 1895, described in Theorem 2.9.10. For this alternative approach, an element $[f] \in \pi_1(X, x_0)$ is defined to be the homotopy class of the continuous map $f : (S^1, 1) \to (X, x_0)$, where $S^1 = \{e^{2\pi i t} : 0 \leq t \leq 1\}$ is the unit circle in the complex plane $\mathbf{C}$. The motivation of this approach was born through the observation that every loop $\alpha : (I, \dot{I}) \to (X, x_0)$ gives rise to a pointed continuous map

$$\tilde{\alpha} : (S^1, 1) \to (X, x_0), \ e^{2\pi i t} \mapsto \alpha(t),$$

and conversely, every continuous map $\beta : (S^1, 1) \to (X, x_0)$ gives a loop

$$\tilde{\beta} : (\mathbf{I}, \dot{\mathbf{I}}) \to (X, x_0), \ t \mapsto \beta(e^{2\pi i t}).$$

**Proposition 2.16.1** *For a pointed topological space* $(X, x_0)$, *the map*

$$f : \pi_1(X, x_0) \to [(S^1, 1), (X, x_0)], \ [\alpha] \mapsto [\tilde{\alpha}]$$

*is bijective.*

**Proof** Since every loop $\alpha : (\mathbf{I}, \dot{\mathbf{I}}) \to (X, x_0)$ determines a continuous map

$$\tilde{\alpha} : (S^1, 1) \to (X, x_0), \ e^{2\pi i t} \mapsto \alpha(t)$$

and every continuous map $\beta : (S^1, 1) \to (X, x_0)$ determines a loop

$$\tilde{\beta} : (\mathbf{I}, \dot{\mathbf{I}}) \to (X, x_0), \ t \mapsto \beta(e^{2\pi i t}),$$

the proposition follows, since the maps $\tilde{\alpha}$ and $\tilde{\beta}$ are determined uniquely up to homotopy. $\qquad\square$

Let $\Omega(X, x_0)$ be the set of homotopy classes $[(S^1, 1), (X, x_0)]$ of the continuous maps $\psi : (S^1, 1) \to (X, x_0)$. The identification map $f$ described in Proposition 2.16.1 identifies $\Omega(X, x_0)$ with the group $\pi_1(X, x_0)$.

**Theorem 2.16.2**  $\Omega(X, x_0)$ is a group isomorphic to the fundamental group $\pi_1(X, x_0)$.

**Proof**  Let $\alpha, \beta : (\mathbf{I}, \dot{\mathbf{I}}) \to (X, x_0)$ be two given loops. Then by Proposition 2.16.1, there exist two continuous maps

$$\tilde{\alpha} : (S^1, 1) \to (X, x_0), \; e^{2\pi it} \mapsto \alpha(t)$$

and

$$\tilde{\beta} : (S^1, 1) \to (X, x_0), \; e^{2\pi it} \mapsto \beta(t).$$

Define a composition 'o' on the $\Omega(X, x_0) = [(S^1, 1) \to (X, x_0)]$ by the rule

$$[\tilde{\alpha}] \circ [\tilde{\beta}] = [\widetilde{(\alpha * \beta)}],$$

where $\alpha * \beta$ is given in Definition 2.9.2. Then, $\Omega(X, x_0)$ is a group under the composition 'o' with homotopy class of the constant map $\delta : S^1 \to \{x_0\}$ as identity element and the inverse of the homotopy class $[\tilde{\alpha}]$ is represented by the map

$$\tilde{\alpha}' : (S^1, 1) \to (X, x_0), \; e^{i\theta} \mapsto \tilde{\alpha}(e^{-i\theta}), 0 \leq \theta \leq 2\pi.$$

Hence, the bijection

$$f : \pi_1(X, x_0) \to [(S^1, 1), (X, x_0)], \; [\alpha] \mapsto [\tilde{\alpha}]$$

defines an isomorphism of groups. This establishes the equivalence of the two groups $\pi_1(X, x_0)$ and $\Omega(X, x_0)$ defined in two different ways. $\qquad\square$

**Remark 2.16.3**  At many situations, it is convenient to consider an element $[\alpha] \in \pi_1(X, x_0)$ as the homotopy class of the map

$$\tilde{\alpha} : (S^1, 1) \to (X, x_0), \; e^{2\pi it} \mapsto \alpha(t).$$

Again, the circle $S^1$ defined in the complex plane is a group under usual multiplication of complex numbers. This multiplication rule can also be used to define the composition law on $\pi_1(S^1, 1)$.

## 2.17  Degree and Winding Number of a Loop on the Circle

This section communicates the concept of degree of a loop $f : (\mathbf{I}, \dot{\mathbf{I}}) \to (S^1, 1)$ on the circle $S^1$ by using homotopy theory with an eye to compute the fundamental

**Fig. 2.15**  Exponential map
$p : \mathbf{R} \to S^1, \; t \mapsto e^{2\pi i t}$

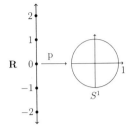

**Fig. 2.16**  Lifting $\tilde{f}$ of $f$ in $\mathbf{R}$

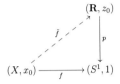

group of the circle $S^1$. On the other hand, its generalization for the degree of spherical maps $f : S^n \to S^n$ ($n \geq 1$) using homotopy theory is studied in Sect. 2.22 and by
using homology theory the same concept is available in Chapter 3. This section also
studies winding number of a closed curve from the viewpoint of homotopy theory.

The exponential map $p : \mathbf{R} \to S^1, \; t \mapsto e^{2\pi i t}$ is a continuous map onto $S^1$, and
geometrically, it wraps the real line $\mathbf{R}$ rounding the circle infinite number of times
as represented in Fig. 2.15.

**Definition 2.17.1**  The exponential map $p : \mathbf{R} \to S^1, t \mapsto e^{2\pi i t}$ defines a group homomorphism $p : (\mathbf{R}, +) \to (S^1, \cdot)$ from the additive group of reals under usual addition
to the multiplicative group $S^1$ under usual multiplication of complex numbers with
its kernel $ker \; p = \mathbf{Z}$.

**Definition 2.17.2**  Let $f : (X, x_0) \to (S^1, 1)$ be a continuous map and $z_0 \in ker \; p$.
The map $f$ is said to have a **lifting** if there is a continuous map $\tilde{f} : (X, x_0) \to (\mathbf{R}, z_0)$
making diagram in Fig. 2.16 commutative, i.e., satisfying the property $p \circ \tilde{f} = f$.
Then, $\tilde{f}$ is said to be lifting of $f$.

**Remark 2.17.3**  It is a natural question: Given a continuous map $f : (X, x_0) \to$
$(S^1, 1)$, if its lifting $\tilde{f}$ exists, is it unique? The answer is positive under suitable
situations given in Exercise 26 and Exercise 29 of Sect. 2.28.

**Proposition 2.17.4**  (*Path Lifting Property*) *Given a loop* $\alpha : (\mathbf{I}, \dot{\mathbf{I}}) \to (S^1, 1)$ *on*
$S^1$, *there exists a unique path* $\tilde{\alpha} : \mathbf{I} \to \mathbf{R}$ *such that* $p \circ \tilde{\alpha} = \alpha$ *and* $\tilde{\alpha}(0) = 0$ *as shown*
*in Fig. 2.17.*

**Proof**  Since the subspace $\mathbf{I} \subset \mathbf{R}$ is convex and compact, corresponding to the given
$\alpha$, there exists a unique lifting $\tilde{\alpha} : \mathbf{I} \to \mathbf{R}$ such that $\alpha(0) = 0$ by using Exercise 29
of Sect. 2.28.                                                                                                   □

**Fig. 2.17** Path lifting $\tilde{\alpha}$ of $\alpha$ in $\mathbf{R}$

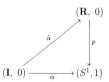

**Definition 2.17.5** Given a loop $\alpha : (\mathbf{I}, \dot{\mathbf{I}}) \to (S^1, 1)$, **the winding number** $w(\alpha)$ of $\alpha$ is given in complex analysis by the integral

$$w(\alpha) = \frac{1}{2\pi i} \int_{\alpha} \frac{dz}{z}.$$

We now study the winding number from the viewpoint of homotopy theory.

**Definition 2.17.6** **The winding number of a closed curve** $\alpha$ in the Euclidean plane $\mathbf{R}^2$ around a given point $a \in \mathbf{R}^2$ is the total number of times that the curve $\alpha$ travels anticlockwise around the point $a$ and hence it is an integer and depends on orientation of the curve. This number may be positive, 0 or negative. The winding number of a loop $\alpha$ is the same as its degree by Definition 2.17.9.

***Example 2.17.7*** Given point $a \in \mathbf{R}^2$, the space $\mathbf{R}^2 - \{a\}$ is homotopy equivalent to the circle $S^1$. The set $[f]$ of homotopy classes of continuous maps $f : S^1 \to X$ based at a point $x_0$ of the topological space $X$ is its fundamental group $\pi_1(X, x_0)$. The fundamental group $\pi_1(S^1)$ is isomorphic to $\mathbf{Z}$ (see Sect. 2.19.1), and the homotopy class of a complex closed curve is precisely represented by its winding number. Given a loop $\alpha : (\mathbf{I}, \dot{\mathbf{I}}) \to (S^1, 1)$, its winding number $w(\alpha)$ is given in complex analysis by the integral

$$w(\alpha) = \frac{1}{2\pi i} \int_{\alpha} \frac{dz}{z}.$$

If $p : \mathbf{R} \to S^1, t \mapsto e^{2\pi it}$, then there exists a real-valued function $\tilde{\alpha}$ such that $\alpha = p \circ \tilde{\alpha}$ as shown in Fig. 2.17. Taking $z = \alpha(t) = p \circ \tilde{\alpha}(t) = e^{2\pi i\tilde{\alpha}(t)}$, it follows that

$$dz = 2\pi i e^{2\pi i\tilde{\alpha}(t)} \tilde{\alpha}'(t) dt = 2\pi i z \tilde{\alpha}'(t) dt.$$

This shows that

$$w(\alpha) = \frac{1}{2\pi i} \int_{\alpha} \frac{dz}{z} = \int_0^1 \tilde{\alpha}'(t) dt = \tilde{\alpha}(1) - \tilde{\alpha}(0).$$

***Example 2.17.8*** If $\alpha : \mathbf{I} \to S^1, \ t \mapsto e^{2\pi int}$, i.e., if $\alpha(t) = e^{2\pi int}$, then

$$\tilde{\alpha}(t) = nt \text{ and } w(\alpha) = \tilde{\alpha}(1) - \tilde{\alpha}(0) = n.$$

**Geometrically,** the map $\alpha : \mathbf{I} \to S^1$ wraps the unit interval $\mathbf{I}$ around the circle $S^1$ anticlockwise $n$ times if $n \geq 0$ and clockwise $|n|$ times if $n < 0$.

**Definition 2.17.9  (Degree of a loop)** Given a loop $\alpha : (\mathbf{I}, \dot{\mathbf{I}}) \to (S^1, 1)$ on $S^1$, there exists a unique path $\tilde{\alpha} : \mathbf{I} \to \mathbf{R}$ such that $p \circ \tilde{\alpha} = \alpha$ and $\tilde{\alpha}(0) = 0$. The winding number $w(\alpha)$ defined by

$$w(\alpha) = \tilde{\alpha}(1) - \tilde{\alpha}(0) = \tilde{\alpha}(1)$$

is called the degree of $\alpha$, denoted by $deg\ \alpha$. It is an integer and is homotopy invariant by Proposition 2.17.11.

***Example 2.17.10***  For the loop $\alpha : (\mathbf{I}, \dot{\mathbf{I}}) \to (S^1, 1)$, $t \mapsto e^{2\pi nit}$, $\tilde{\alpha}(1) = n$. Thus for the map $\alpha : S^1 \to S^1$, $z \mapsto z^n$, $deg\ \alpha = n$ justifies term ' degree.'

**Proposition 2.17.11**  *The degree of any loop $\alpha : (\mathbf{I}, \dot{\mathbf{I}}) \to (S^1, 1)$ is an integer, and it is homotopy invariant.*

***Proof*** Let $p : \mathbf{R} \to S^1, t \mapsto e^{2\pi it}$ be the exponential map. For the loop $\alpha : (\mathbf{I}, \dot{\mathbf{I}}) \to (S^1, 1)$ if $\tilde{\alpha} : \mathbf{I} \to \mathbf{R}$ is its unique lifting, then $p \circ \tilde{\alpha} = \alpha$. Hence, $(p \circ \tilde{\alpha})(1) = \alpha(1) = 1$ implies that $deg\ \alpha = \tilde{\alpha}(1) \in ker\ p = \mathbf{Z}$. Again homotopy lifting property (HLP) asserts that if $\alpha, \beta : (\mathbf{I}, \dot{\mathbf{I}}) \to (S^1, 1)$ are two loops such that $\alpha \simeq \beta$ rel $\dot{\mathbf{I}}$, then their corresponding unique liftings $\tilde{\alpha}, \tilde{\beta} : \mathbf{I} \to \mathbf{R}$ with $\tilde{\alpha}(0) = \tilde{\beta}(0)$ are such that

$$\tilde{\alpha} \simeq \tilde{\beta} \text{ rel } \dot{\mathbf{I}} \text{ and } \tilde{\alpha}(1) = \tilde{\beta}(1).$$

Since $deg\ \alpha = \tilde{\alpha}(1)$, it follows by the above homotopy lifting property (see Exercise 30 of Sect. 2.28) that the degree of any loop $\alpha : (\mathbf{I}, \dot{\mathbf{I}}) \to (S^1, 1)$ is homotopy invariant.                                                                                    □

**Corollary 2.17.12**  *If two differentiable functions*

$$\alpha,\ \beta : (\mathbf{I}, \dot{\mathbf{I}}) \to (S^1, 1)$$

*are such that $\alpha \simeq \beta$ rel $\dot{\mathbf{I}}$, then they have the same winding numbers.*

***Proof*** It follows from Proposition 2.17.11.                                                        □

Proposition 2.17.13 characterizes homotopy of loops in terms of their winding numbers.

**Proposition 2.17.13**  *Let $\alpha, \beta : (\mathbf{I}, \dot{\mathbf{I}}) \to (S^1, 1)$ be any two loops. Then, the winding numbers $w(\alpha) = w(\beta)$ iff $\alpha \simeq \beta$ rel $\dot{\mathbf{I}}$.*

***Proof*** Consider the exponential map

$$p : \mathbf{R} \to S^1, t \mapsto e^{2\pi it}.$$

If $\alpha \simeq \beta$ rel $\dot{\mathbf{I}}$, then it follows by Proposition 2.17.11 that $w(\alpha) = w(\beta)$. Conversely, let $w(\alpha) = w(\beta)$ and $\tilde{\alpha}$, $\tilde{\beta} : \mathbf{I} \to \mathbf{R}$ be two liftings of $\alpha$ and $\beta$. Since $\mathbf{I}$ is convex, the map

$$\tilde{H} : \mathbf{I} \times \mathbf{I} \to \mathbf{R}, \ (t, s) \mapsto (1 - s)\tilde{\alpha}(t) + s\tilde{\beta}(t)$$

is a well-defined and continuous map such that

$$\tilde{H} : \tilde{\alpha} \simeq \tilde{\beta} \ \text{rel} \ \dot{\mathbf{I}} \quad \text{and} \quad \text{hence} \ H = p \circ \tilde{H} : \alpha \simeq \beta \ \text{rel} \ \dot{\mathbf{I}}.$$

$\square$

**Corollary 2.17.14** *Any loop* $(\mathbf{I}, \dot{\mathbf{I}}) \to (S^1, 1)$ *is homotopic to the unique loop*

$$\alpha_n : (\mathbf{I}, \dot{\mathbf{I}}) \to (S^1, 1), \ t \mapsto e^{2\pi i n t}.$$

**Proof** Let $\alpha : (\mathbf{I}, \dot{\mathbf{I}}) \to (S^1, 1)$ be an arbitrary loop and $w(\alpha) = n$ be its winding number around the origin $(0, 0)$. Again, $w(\alpha_n)$ also moves $n$ times around the origin $(0, 0)$. Hence, it follows from Proposition 2.17.13 that $\alpha \simeq \alpha_n$ rel $\dot{\mathbf{I}}$ and for any integer $m \neq n$, the loops $\alpha$ and $\alpha_m$ cannot be homotopic rel $\dot{\mathbf{I}}$.                      $\square$

## 2.18   Vector Fields and their Applications

This section is devoted to the study of nonvanishing continuous vector fields on spheres $S^n$ from the viewpoint of algebraic topology. On the other hand, **Basic Topology, Volume 2** studies vector fields on smooth manifolds from the viewpoint of differential topology. Theorem 2.25.23 provides a necessary and sufficient condition for existence of a nonvanishing vector field on $S^n$. Vector fields establish a close connection among geometry, analysis and topology.

### 2.18.1   Basic Concepts of Vector Fields

This subsection starts with basic concepts of vector fields on Euclidean space $\mathbf{R}^n$.

**Definition 2.18.1** A **vector field** on the Euclidean space $\mathbf{R}^n$ is a continuous vector-valued function

$$v : \mathbf{R}^n \to \mathbf{R}^m.$$

If $x_1, x_2, \ldots, x_n$ are coordinates of any point $x \in \mathbf{R}^n$, then the vector field $v$ is represented by $m$ continuous real-valued functions

$$f : \mathbf{R}^n \to \mathbf{R}^m, \ (x_1, x_2, \ldots, x_n) \mapsto (f_1(x_1, x_2, \ldots, x_n), \ldots, f_m(x_1, x_2, \ldots, x_n)),$$

in brief,

$$f : \mathbf{R}^n \to \mathbf{R}^m, \ x \mapsto (f_1(x), \dots, f_m(x)).$$

There are two natural problems:

**(a)** Does there exist a vector field tangent to a given surface?
**(b)** Has every continuous map $f : X \to X$ a fixed point for an arbitrary topological space $X$ ?

**Proposition 2.18.2** *Let $A \subset \mathbf{R}^2$ be a subset of the Euclidean plane $\mathbf{R}^2$. A **vector field** $V$ (or $V_A$ ) on A is a continuous function which assigns to each point $x \in A$, a vector $v(x)$ in $\mathbf{R}^2$ with its initial point $a \in A$.*

**Remark 2.18.3** A vector field $V$ on $A \subset \mathbf{R}^2$ may be intuitively thought as the velocity of a particle on $A$ under certain situations. As the essential components of a vector $v(x)$ are its length and direction, without loss of generality it is assumed that the vector $v(x)$ starts from the origin.

### 2.18.2   Vector Fields on $S^n$

This subsection conveys the concept of vector field on $S^n$ with its geometrical interpretation.

**Definition 2.18.4**  A **vector field** on $S^n$ is a continuous map $v : S^n \to \mathbf{R}^{n+1}$ $(n \geq 1)$ such that the inner product $< x, v(x) > = 0$, $\forall x \in S^n$; i.e., the vector $v(x)$ is orthogonal to the vector $v(x)$ in $\mathbf{R}^{n+1}$ for every $x \in S^n$. Moreover, if $v(x) \neq 0$ for all $x \in S^n$, we say that the **vector field is nonvanishing**.

**Remark 2.18.5** (Geometrical Interpretation) Definition 2.18.4 implies that a vector field $v$ on $S^n$ is a continuous map which assigns to every vector $x$ of unit length in $\mathbf{R}^{n+1}$, a unit vector $v(x)$ in $\mathbf{R}^{n+1}$ such that $x$ and $v(x)$ are orthogonal, i.e., $x \perp v(x)$, $\forall x \in S^n$. If we consider the vector $v(x)$ starting from the point $x \in S^n$, then this vector $v(x)$ must be tangent to $S^n$ at each point $x$ of $S^n$. If $x$ moves in $S^n$, then end point of the vector $v(x)$ varies continuously in $\mathbf{R}^{n+1}$. Geometrically, Fig. 2.18 represents it for the particular case, when $n = 1$.

**Example 2.18.6** A continuous tangent vector field (or simply a vector field ) on $S^2$ is an ordered pair $(x, f(x))$, where $f : S^2 \to \mathbf{R}^3$ is a continuous map such that $f(x)$ is tangent to $S^2$ at the point $x$ for every $x \in S^2$ and it is written simply by $f$. This vector field $f$ on $S^2$ is nonvanishing if $f(x) \neq 0$ for every $x \in S^2$; hence, a nonvanishing (nonzero ) vector field $f$ on $S^2$ is considered as continuous mapping $f : S^2 \to \mathbf{R}^3 - \{0\}$.

**Fig. 2.18** Geometrical interpretation of a vector field on $S^1$

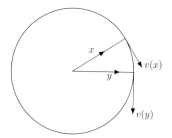

**Definition 2.18.7** A continuous map $A_n : S^n \to S^n$ is said to be **antipodal** map if $A_n(x) = -x$, $\forall x \in S^n$, where for any $x = (x_1, x_2, \cdots x_n, x_{n+1}) \in S^n$, its antipode $-x$ is given by

$$-x = (-x_1, -x_2, \cdots, -x_n, -x_{n+1}) \in S^n.$$

Theorem 2.18.8 provides a necessary and sufficient condition for existence of a nonvanishing vector field on $S^n$ in terms of homotopy.

**Theorem 2.18.8** *Let $A_n : S^n \to S^n$, $x \mapsto -x$ be the antipodal map and $1_{S^n} : S^n \to S^n$ be the identity map. Then, there exists a continuous nonvanishing tangent vector field on $S^n$ iff $1_{S^n} \simeq A_n$.*

**Proof** First suppose that there exists a continuous nonvanishing tangent vector field

$$v : S^n \to \mathbf{R}^{n+1}$$

on $S^n$. Define a continuous map

$$F : S^n \times \mathbf{I} \to S^n, \quad (x, t) \mapsto (1 - 2t)x + 2\sqrt{t - t^2}\, v(x)/||v(x)||.$$

Hence, it follows that $F : 1_{S^n} \simeq A_n$. For the converse part, let a homotopy $F : 1_{S^n} \simeq A$ can be approximated by a differentiable homotopy. This produces tangent curve elements to $S^n$. Since the tangent line to the curve $\beta_x : \mathbf{I} \to S^n$, $t \mapsto F(x, t)$ at $t = 0$ contains a unit vector in the direction of increasing $t \in \mathbf{I}$, it follows that this tangent line is tangent to $S^n$ at $x \in S^n$ and it is nonvanishing. $\square$

**Proposition 2.18.9** *There does not exist antipodal map $f : S^2 \to S^2$ with the property that $x$ and $f(x)$ are orthogonal for every $x \in S^2$.*

**Proof** Let $f : S^2 \to S^2$ be a continuous map; for example, $x$ and $f(x)$ are orthogonal vectors for every $x \in S^2$. Suppose there is some point $y \in S^2$ for which $f(y) = -y$. Then, the inner product $< y, f(y) >= -1$, and hence the vectors $y$ and $f(y)$ cannot be orthogonal, which gives a contradiction, because by hypothesis $< y, f(y) >= 0$. This contradiction proves that there does not exist antipodal map $f : S^2 \to S^2$ with the property that $x$ and $f(x)$ are orthogonal for every $x \in S^2$. $\square$

**Proposition 2.18.10** *For all odd integers $n \geq 1$, there is a nonvanishing vector field* $f : S^n \to S^n$.

**Proof** Let $n$ be an odd integer. Define a continuous map

$$f : S^n \to S^n, \quad x = (x_1, x_2, \ldots, x_{n+1}) \mapsto (x_2, -x_1, x_4, -x_3, \ldots, x_{n+1}, -x_n).$$

Then, the inner product

$$< x, f(x) >= (x_1 x_2 - x_1 x_2) + (x_3 x_4 - x_3 x_4) + \cdots + (x_n x_{n+1} - x_n x_{n+1}) = 0, \quad \forall x \in S^n.$$

This shows that $f : S^n \to S^n$ is a continuous map such that the vectors $x$ and $f(x)$ are orthogonal and hence it is a vector field on $S^n$.  $\square$

**Remark 2.18.11** For converse of Proposition 2.18.10, see Brouwer–Poincaré Theorem 2.25.23.

**Proposition 2.18.12** *If $v : S^{2n+1} \to \mathbf{R}^{2n+2}$ is a nonvanishing tangent vector field on* $S^{2n+1}$, *then the map*

$$f : S^{2n+1} \to S^{2n+1}, \quad x \mapsto \frac{v(x)}{||v(x)||}$$

*is homotopic to the identity map.*

**Proof** Consider the map

$$H : S^{2n+1} \times I \to S^{2n+1}, \quad (x, t) \mapsto x \cos\frac{\pi t}{2} + \frac{v(x)}{||v(x)||} \sin\frac{\pi t}{2}.$$

Then, $H$ is a homotopy between the identity maps on $S^{2n+1}$ and $f$.

$\square$

**Remark 2.18.13** Let $f_n : S^n \to S^n$ be the antipodal map. If there exists a tangent vector field on $S^n$, then $f_n \simeq 1_d$, where $1_d$ is the identity map on $S^n$.

**Proof** Construct a continuous map

$$H : S^n \times I \to S^n, \quad (x, t) \mapsto H(x, t) = \alpha(t)x + \beta(t)f_n(x) : ||H(x, t)||^2 = 1,$$

where $\alpha$ and $\beta$ are real-valued functions of $t$. This determines the equation

$$\alpha(t)^2 + \beta(t)^2 \langle f_n(x), f(x) \rangle = 1.$$

Choose $\alpha(t) = 1 - 2t$. Then, $\beta(t)^2 = 4(t - t^2)/||f_n(x)||^2$, since $f_n(x) \neq 0$, gives $\beta(t) = 2\sqrt{t - t^2}/||f_n(x)||$.    Consequently,    $H(x, t) = (1 - 2t)x + 2\sqrt{t - t^2}$ $f_n(x)/||f_n(x)||$ shows that $H : 1_d \simeq f_n$, which implies that $f_n \simeq 1_d$.  $\square$

**Remark 2.18.14** The following basic results on vector fields on $n$-sphere $S^n$ are also proved in this chapter.

(i) **Brouwer–Poincaré Theorem** 2.25.23 asserts that the $n$-sphere $S^n$ admits a continuous nonvanishing vector field iff $n$ is odd.

(ii) For all even integers $n > 1$, there is no vector field $f : S^n \to S^n$ proved in Corollary 2.25.24.

### 2.18.3  Applications of Vector Fields

This subsection proves hairy ball and Brouwer fixed-point theorems by using the concept of vector fields. A hair is said to be combed if it is kept flat, which means mathematically that it is tangent to the sphere. To establish a criterion that a continuous map from a topological space to itself must have a fixed point is one of the most important applications of topology in mathematics and other sciences.

**Historically**, L. E. J. Brouwer (1881–1967) first proved during 1910–1912 that every continuous map $f : \mathbf{D}^n \to \mathbf{D}^n, n \geq 1$ has a fixed point by using the concept of degree of spherical maps $f : S^n \to S^n$. This theorem is subsequently proved by using either the **homology or homotopy groups**. But Brouwer used neither of them, which had not been invented at that time. This theorem named after L. E. J. Brouwer is applied in Corollary 2.23.8 to determine the nature of eigenvalues of a special class of matrices.

**Remark 2.18.15** The absence of any nonvanishing vector field on $S^2$ presents a popular result called hairy ball theorem 2.18.16 saying that the hair on $S^2$ cannot be combed flat . So, this theorem is called 'hairy ball theorem.' It is proved by the concept of nonzero vector field on $S^2$.

**Theorem 2.18.16** (Hairy Ball Theorem) There is no nonzero vector field on $S^2$.

**Proof** Suppose that there is a nonzero vector field $v : S^2 \to \mathbf{R}^3$ on $S^2$. Then, $v$ is a continuous map such that $v(x) \neq 0, \ \forall x \in S^2$ and the vector $v(x)$ is tangent to $S^2$ at $x$ for every $x \in S^2$, since by hypothesis $v$ is a nonzero vector field. Consider the map

$$f : S^2 \to S^2 \subset \mathbf{R}^3, \ x \mapsto v(x)/||v(x)||.$$

Clearly, the map $f$ is continuous. Since $f(x)$ is tangent to $S^2$ at the point $x$ for every $x \in S^2$, this produces a contradiction by Proposition 2.18.9. This contradiction proves the theorem.

    **Geometrical interpretation of the proof** : If one imagines that he has a hair growing out from each point on the surface of a ball, then it is not possible to brush them flat. Otherwise, the tangent vectors to the hairs would show that $S^2$ would admit a continuous nonvanishing vector field. As it would contradict Proposition 2.18.9, we conclude that a hairy ball cannot be combed flat.                               □

**Definition 2.18.17** A continuous map $v : \mathbf{D}^2 \to \mathbf{R}^2$ is said to be a vector field on $\mathbf{D}^2$. It is said to be nonvanishing if $v(x) \neq 0, \ \forall x \in \mathbf{D}^2$.

**Proposition 2.18.18** *Let v be a nonvanishing vector field on 2-dimensional disk* $\mathbf{D}^2$. *Then, there exist points* $x, y \in \partial \mathbf{D}^2$ *such that*

$$v(x) = rx, \quad v(y) = -sy \text{ for some real numbers } r, s > 0.$$

**Proof** Let $v : \mathbf{D}^2 \to \mathbf{R}^2$ be a vector field such that there exists no point $x \in \partial \mathbf{D}^2$ with $v(x) = -sx, s > 0$, i.e., $v(x)$ points directly inward. Let $f = v|_{\partial \mathbf{D}^2}$ be the restriction map of $v$ to $\partial \mathbf{D}^2$. Then, $f$ has a continuous extension

$$\tilde{f} : \mathbf{D}^2 \to \mathbf{R}^2 - \{0\}.$$

Hence, it follows that $f$ is nullhomotopic. Again, $f$ is homotopic to the inclusion map

$$i : S^1 = \partial \mathbf{D}^2 \hookrightarrow \mathbf{R}^2 - \{0\}$$

by a homotopy $H : \partial \mathbf{D}^2 \times \mathbf{I} \to \mathbf{R}^2 - \{0\} : (x, t) \mapsto tx + (1 - t)f(x)$. Clearly, $H(x, t) \neq 0$ for $t = 0, 1$. If $H(x, t) = 0$ for some $t \in (0, 1)$, then the equality $f(x) = -\frac{t}{1-t}x$ would hold. This would imply that the point $f(x)$ points outward at $x$, which contradicts our supposition. For the second part, consider $-v$ for $v$.  □

Proposition 2.18.18 is now used to prove Brouwer fixed-point theorem 2.18.19 for dimension 2. It is named after L. E. J. Brouwer (1881–1967). Its alternative proof is given in Theorem 2.23.6 by showing that $S^1 \subset \mathbf{D}^2$ not a retract of $\mathbf{D}^2$.

**Theorem 2.18.19** (Brouwer Fixed-Point Theorem for Dimension 2) *Every continuous map* $f : \mathbf{D}^2 \to \mathbf{D}^2$ *has a fixed point.*

**Proof** If possible, $f : \mathbf{D}^2 \to \mathbf{D}^2$ has no fixed point. Then, $f(x) \neq x, \forall x \in \mathbf{D}^2$. Consider the map

$$v : \mathbf{D}^2 \to \mathbf{R}, x \mapsto f(x) - x.$$

Then, $v$ defines a nonvanishing vector field on $\mathbf{D}^2$. Clearly, $v$ cannot point directly outward at any point $x \in \partial \mathbf{D}^2$, because this would assert that there exists some $r > 0$ such that $f(x) - x = rx$, and hence $f(x) = (r + 1)x \notin \mathbf{D}^2$. But this contradicts Proposition 2.18.18. This contradiction proves the theorem.  □

**Remark 2.18.20** For more study on vector fields, and Brouwer fixed-point theorem, see Chap. 3.

## 2.18.4   Link Between Homotopy and Flow

This subsection continues the study of vector fields by introducing the concept of flow, which is a generalization of the concept of vector field. It establishes an interplay between flow and homotopy. Let $X$ be a topological space and $\psi_t : X \to X$ ($t \in \mathbf{R}$)

be a family of continuous maps. By continuity of $\psi_t$, it is meant jointly continuous in $t$ and $x$.

**Definition 2.18.21** A continuous family $\psi_t : X \to X$ ($t \in \mathbf{R}$) of maps is said to be a **flow** if

(i) $\psi_0 = 1$, i.e., $\psi_0(x) = x$, $\forall x \in X$.
(ii) $\psi_t$ is a homeomorphism for all $t \in \mathbf{R}$.
(iii) $\psi_t \circ \psi_s = \psi_{t+s}$, i.e., $\psi_t(\psi_s(x)) = \psi_{t+s}(x)$, $\forall x \in X$ and $\forall t, s \in \mathbf{R}$.

A flow is also known as a **one parameter group of homeomorphisms.**

***Remark 2.18.22*** The first condition asserts that the map $\psi_0$ is the identity map. The third condition defining property of a flow on $X$ implies that the family $\{\psi_t\}$ of flows on $X$ forms a group isomorphic to the additive group of real numbers. It asserts that if a ball rolls for s seconds and then it rolls for $t$ seconds, the result says that the ball rolls for $s + t$ seconds. The concept of flows plays a key role in topology and differential equations [Hirsch et al., 1974].

A flow $\psi_t$ on a topological space $X$ is sometimes considered as a continuous map

$$\psi : X \times \mathbf{R} \to X, \ (x, t) \mapsto \psi_t(x).$$

This motivates to link the concept of flow with homotopy in Proposition 2.18.23.

**Proposition 2.18.23** *Given a topological space $X$, every flow $\psi_t : X \to X$ is homotopic to $I_X$.*

***Proof*** Define the map

$$F : X \times \mathbf{I} \to X, \ (x, s) \mapsto \psi(x, (1 - s)t).$$

This map is continuous and is such that

$$F(x, 0) = \psi(x, t) = \psi_t(x)$$

and

$$F(x, 1) = \psi(x, 0) = \psi_0(x) = 1_X(x), \ \forall x \in X, \ \forall t \in \mathbf{R}.$$

This asserts that $\psi_t$ is homotopic to $1_X$ under $F$. $\qquad\square$

## 2.18.5 Topological Dynamics

This subsection initiates a study of topological dynamics by using the concept of a flow, which establishes a close connection between topology and analysis. The concept of flow on a given topological space $X$ is an extension of the flows in $\mathbf{R}^2$ arising from vector fields. The tools of vector fields and differential equations can be used as tools in topology, facilitating topological ideas to penetrate in the premises of differential equations. **Topological dynamics** is the study of flows and gives an abstract form of differential equations. The study of vector field is closely related to the study of differential equations

$$\frac{dx}{dt} = f(x, y)$$

$$\frac{dy}{dt} = g(x, y).$$

**Remark 2.18.24**  The three basic concepts in topological dynamics such as continuous functions, vector fields and system of differential equations are closely interlinked, because vector fields and differential equations are used as tools in topology enroading their entry into the premises of topology. For example, vector fields are used to prove Brouwer fixed-point theorem, which is a classical result in topology (see Theorem 2.18.19), and in more general, fixed-point theorem on compact manifold (see Exercise 64 of Section 2.28) is also an important result of fixed-point theory. Fixed-point theorems facilitate to study a given system of equations in several variables.

There is a natural question: Whether a given system of equations of the form $f(x) = 0$ has a solution or not, where $f(x) = 0$ is a polynomial equation or a complicated equation of several variables. If $g(x) = f(x) + x$ and $g(x)$ has a fixed point, then this fixed point will be a solution of the given equation $f(x) = 0$.

**Definition 2.18.25**  For a given $p = (x, y) \in A \subset \mathbf{R}^2$, and a map $v : A \to \mathbf{R}^2$, its image $v(p)$ can be expressed as

$$v(x, y) = (U(x, y), V(x, y)),$$

where $U$ and $V$ are real-valued continuous functions of $x$ and $y$.
A vector field $v$ on a subset $A \subset \mathbf{R}^2$ is said to be continuous if the function

$$v : A \to \mathbf{R}^2, \ (x, y) \mapsto (U(x, y), V(x, y))$$

is continuous.

Example 2.18.26 implies that the study of vector fields $v$ on a set $A \subset \mathbf{R}^2$ is precisely the study of continuous transformations $f : A \to \mathbf{R}^2$.

**Example 2.18.26**  Let $A \subset \mathbf{R}^2$ and $v : A \to \mathbf{R}^2$ be a map such that $f : A \to \mathbf{R}^2$ and $f(x, y)$ is the end point of the vector $v(x, y)$. Then,

$$f(x, y) = (x, y) + v(x, y) = (x + U(x, y), y + V(x, y)). \qquad (2.2)$$

Hence, it follows that $f$ is continuous, if $v$ is a continuous vector field. Conversely, given $A \subset \mathbf{R}^2$ and a transformation $f : A \to \mathbf{R}^2$, the vector field $v$ is defined by taking $v(a)$ to be the vector from the point $a \in A$ to the point $f(a)$. Then, $v$ is continuous if $f$ is continuous.

**Example 2.18.27**   Some important examples of vector fields.

  (i) The force field arising from gravitation is a vector field.
  (ii) Electromagnetism is a vector field.
(iii) Velocity vector of a moving fluid is a vector field.
(iv) Pressure gradient on a weather map is a vector field.

A vector field $v(x, y) = (U(x, y), V(x, y))$ gives **a system of differential equations** in two unknowns $x$ and $y$ representing the position of a moving particle in $\mathbf{R}^2$, which depends on time $t$. This produces a system of differential equations

$$\mathrm{d}x/\mathrm{d}t = U(x, y), \quad \mathrm{d}y/\mathrm{d}t = V(x, y). \tag{2.3}$$

The system of Eq. (2.3) is said to be **autonomous,** if the functions $U(x, y)$ and $V(x, y)$ do not depend on $t$. A solution of the system of Eq. (2.3) involves two functions in $x$ and $y$ expressed in terms of $t$. These are the parametric equations of a path in $\mathbf{R}^2$. This idea may be applied to study the orbit space of a planet in Newtonian mechanics or the orbit of a molecule or the path of a molecule of a fluid. The original vector field $v(x, y)$ conveys the tangent vector to the path of the particle at the point $(x, y) \in \mathbf{R}^2$.

**Example 2.18.28**   (i) The vector field $v(x, y) = (1, y)$ has the exponential curves given by $y = ce^x$ as the solution curves, since here

$$\mathrm{d}x/\mathrm{d}t = U(x, y) = 1, \mathrm{d}y/\mathrm{d}t = V(x, y) = y \implies x = t + c_1, y = c_2 e^t.$$

 (ii) The vector field $v(x, y) = (x, -y)$ has the rectangular hyperbolas given by $xy = c$ as the solution curves, since here $\mathrm{d}x/\mathrm{d}t = U(x, y) = x$, $\mathrm{d}y/\mathrm{d}t = V(x, y) = -y$ and hence $x = c_1 e^t$, $y = c_2 e^{-t}$.
(iii) The vector field $v(x, y) = (y, x)$ has the hyperbolas given by $x^2 - y^2 = c$ as the solution curves, since here $\mathrm{d}x/\mathrm{d}t = U(x, y) = y$, $\mathrm{d}y/\mathrm{d}t = V(x, y) = x$ and hence $\mathrm{d}^2x/\mathrm{d}t^2 = \mathrm{d}y/\mathrm{d}t = x$, $\mathrm{d}^2y/\mathrm{d}t^2 = \mathrm{d}x/\mathrm{d}t = y$. Consequently, $x = c_1 e^t + c_2 e^{-t}$, $y = c_1 e^t + c_2 e^{-t}$.

## 2.19   Computation of the Fundamental Groups

This section computes the fundamental groups of the circle $S^1$, punctured Euclidean plane $\mathbf{R}^2 - \{(0, 0)\}$ and figure-eight $F_8$ which are used in our subsequent study.

## *2.19.1  Computation of the Fundamental Group of the Circle*

This subsection computes the fundamental group $\pi_1(S^1, 1)$ of the circle by utilizing the degree function. As the circle is a path-connected space, its fundamental group does not depend on its base point. So without loss of generality, $1 \in S^1$ is taken as its base point. Its fundamental group $\pi_1(S^1)$ is isomorphic to $\mathbf{Z}$, the infinite cyclic group by Corollary 2.19.2.

**Theorem 2.19.1**  *The degree function* $\psi : \pi_1(S^1, 1) \to \mathbf{Z}$, $[\alpha] \mapsto \deg \alpha$ *is a group isomorphism.*

**Proof**  The map

$$\psi : \pi_1(S^1, 1) \to \mathbf{Z}, \ [\alpha] \mapsto \deg \alpha$$

is independent of the choice of the representative of the class $[\alpha] \in \pi_1(S^1, 1)$ by Proposition 2.17.11, and hence, the map $\psi$ is well defined. Moreover, it clearly follows that $\psi$ is a homomorphism and also a monomorphism. Finally, we claim that $\psi$ is an epimorphism. To prove it, given an integer $n$, define a loop $\alpha : \mathbf{I} \to S^1$, $t \mapsto e^{2\pi i nt}$. Then, the path $\tilde{\alpha} : \mathbf{I} \to \mathbf{R}$, $t \mapsto nt$ is such that $\tilde{\alpha}(0) = 0$ and lifts the path $\alpha$. Hence, the path $\tilde{\alpha}$ is unique and $\tilde{\alpha}(1) = n = \deg \alpha$. Consequently, it follows that the map $\psi$ is a group isomorphism.  □

**Corollary 2.19.2**  $\pi_1(S^1) \cong \mathbf{Z}$.

**Proof**  Since the circle $S^1$ is path connected, the fundamental group $\pi_1(S^1)$ does not depend on the choice of the base point of $S^1$. This implies by Theorem 2.19.1 that $\pi_1(S^1) = \pi_1(S^1, 1) \cong \mathbf{Z}$.  □

**Corollary 2.19.3**  *The circle* $S^1$ *is not simply connected.*

**Proof**  It follows from Corollary 2.19.2.  □

**Remark 2.19.4  Geometrical interpretation** of the result $\pi_1(S^1, 1) \cong \mathbf{Z}$: Let $\alpha : \mathbf{I} \to S^1$ be a loop based at the point $1 \in S^1$. Then, if $t$ travels from 0 to 1, its image point $\alpha(t)$ travels on the circle and at $t = 1$, it returns to this base point. This shows that the total number of times for its complete winding the circle is an integer. If the loops $\alpha, \beta$ on $S^1$ based at the point 1 are such that $\alpha$ describes $S^1$ $m$ times and $\beta$ describes it $n$ times. With $m > n$, then the loops $\alpha, \beta$ cannot be homotopic; otherwise, $\alpha * \beta$ would be a path describing $(m - n)$ times the circle $S^1$. But such a path cannot be a null path. This implies that assigning to each loop the number of times it winds around $S^1$ establishes a bijective correspondence between the set of homotopy classes of loops in $S^1$ with the set of integers $\mathbf{Z}$, which is also an isomorphism between the groups $\pi_1(S^1, 1)$ and $\mathbf{Z}$.

**Theorem 2.19.5**  *The circle* $S^1 \subset \mathbf{D}^2$ *is not a retract of* $\mathbf{D}^2$.

**Proof** Let $1_{S^1} : S^1 \to S^1$ be the identity map and $i : S^1 \hookrightarrow D^2$ be the inclusion map. If $S^1$ is a retract of $D^2$, there is a retraction $r : D^2 \to S^1$ such that the equality $r \circ i = 1_{S^1}$ holds. This asserts that the composite homomorphism $\pi_1(S^1, 1) \xrightarrow{i_*} \pi_1(D^2, 1) \xrightarrow{r_*} \pi_1(S^1, 1)$ is the identity automorphism of $\pi_1(S^1, 1) = \mathbf{Z}$. This is impossible, since $D^2$ is contractible and hence $\pi_1(D^2, 1) = 0$. □

**Corollary 2.19.6** *There exists no retraction $r : D^2 \to S^1$.*

**Proof** If possible, there exists a retraction $r : D^2 \to S^1$. Then, $S^1$ is a retract of $D^2$. But it contradicts Theorem 2.19.5. □

**Corollary 2.19.7** *The identity map $1_{S^1} : S^1 \to S^1$ has no continuous extension over $D^2$.*

**Proof** If $1_{S^1}$ has a continuous extension $r : D^2 \to S^1$, then its restriction to $S^1$ is the identity map $1_{S^1}$. This implies that $r \circ i = 1_{S^1}$ and hence $S^1$ is a retract of $D^2$. But this contradicts Theorem 2.19.5 that asserts $S^1$ is not a retract of $D^2$. □

**Remark 2.19.8** The homotopy class of a loop $\alpha : (\mathbf{I}, \dot{\mathbf{I}}) \to (S^1, 1)$ is completely determined by *deg* $\alpha$ in Theorem 2.19.9.

**Theorem 2.19.9** *Let $\alpha, \beta : (\mathbf{I}, \dot{\mathbf{I}}) \to (S^1, 1)$ be two loops. Then, $\alpha \simeq \beta$ rel $\dot{\mathbf{I}}$ if and only if* deg $\alpha$ = deg $\beta$.

**Proof** For the loops $\alpha \simeq \beta$ rel $\dot{\mathbf{I}}$, it follows that $\tilde{\alpha} \simeq \tilde{\beta}$ *rel* $\dot{\mathbf{I}}$ with $\tilde{\alpha}(1) = \tilde{\beta}(1)$. This shows that deg $\alpha$ = deg $\beta$. Again, since the degree function $\psi$ is injective, it follows that if deg $\alpha$ = deg $\beta$, then $[\alpha] = [\beta]$, and hence $\alpha \simeq \beta$ rel $\dot{\mathbf{I}}$. □

## 2.19.2 Computation of the Fundamental Group of the Punctured Plane

This subsection computes the fundamental group of the punctured Euclidean plane by utilizing its special homotopic properties. It is an important geometrical object for the study of topology.

**Theorem 2.19.10** *Let $X = \mathbf{R}^2 - \{\mathbf{0}\}$ be the punctured Euclidean plane, where $\mathbf{0}$ stands for $(0, 0) \in \mathbf{R}^2$ and $x_0 \in S^1 \subset X$ be an arbitrary point. Then, the homomorphism*

$$i_* : \pi_1(S^1, x_0) \to \pi_1(X, x_0)$$

*induced by the inclusion map*

$$i : (S^1, x_0) \hookrightarrow (X, x_0)$$

*is an isomorphism.*

**Proof** Define a continuous map

$$r : X \to S^1, \ x \mapsto \frac{x}{||x||}.$$

Geometrically, the map $r$ collapses each radial ray in $X$ onto the point where the ray intersects $S^1$, and in particular, it maps every point $x \in S^1$ onto itself. As the composite map $r \circ i$

$$(S^1, x_0) \xrightarrow{i} (X, x_0) \xrightarrow{r} (S^1, x_0)$$

is the identity map $1_{S^1}$ on $S^1$, its induced homomorphism

$$(r \circ i)_* = r_* \circ i_* = 1^*_{S^1} : \pi_1(S^1, x_0) \to \pi_1(S^1, x_0)$$

is the identity automorphism of $\pi_1(S^1, x_0)$ by the functorial property of $\pi_1$ and hence $i_*$ is a monomorphism. Again, given a loop $\alpha$ in $X$ based at the point $x_0$,

$$i \circ r \circ \alpha = \beta : \mathbf{I} \to X, \ t \mapsto (i \circ r)(\alpha(t)) = i(\frac{\alpha(t)}{||\alpha(t)||}) = \frac{\alpha(t)}{||\alpha(t)||}$$

is also a loop in $X$ based at the point $x_0$. Define a map

$$H : \mathbf{I} \times \mathbf{I} \to X, \ (t, s) \mapsto \frac{\alpha(t)}{||\alpha(t)||} s + (1 - s)\alpha(t).$$

Clearly, $H(t, s) \neq 0$ for every $(t, s) \in \mathbf{I} \times \mathbf{I}$ and hence it is well defined. Moreover, it is a continuous map such that $H : \alpha \simeq \beta$ rel $\dot{\mathbf{I}}$. This asserts that

$$(i_* \circ r_*)([\alpha]) = [i \circ r \circ \alpha] = [\beta] = [\alpha], \ \forall \ [\alpha] \in \pi_1(X, x_0).$$

It implies that $i_* \circ r_*$ is the identity automorphism of $\pi_1(X, x_0)$ and hence $i_*$ is an epimorphism. Consequently, $i_*$ is an isomorphism with $r_*$ as its inverse.  $\square$

**Theorem 2.19.11** $\pi_1(\mathbf{R}^2 - \{0\}) \cong \mathbf{Z}.$

**Proof** Let $X = \mathbf{R}^2 - \{0\}$ be the punctured Euclidean plane and $x_0 \in S^1 \subset X$ be an arbitrary point. Then, the inclusion map

$$i : (S^1, x_0) \hookrightarrow (X, x_0)$$

induces an isomorphism

$$i_* : \pi_1(S^1, x_0) \to \pi_1(X, x_0)$$

by Theorem 2.19.10. Define a map

$$H : X \times \mathbf{I} \to X, (x, t) \mapsto (1 - t)x + t\frac{x}{||x||}.$$

Then, $H(x, t) \neq 0$ for every $(x, t) \in X \times \mathbf{I}$ and hence it is well defined. Moreover, $H$ is a continuous map satisfying the properties:

$$H(x, 0) = x, \ \forall \ x \in X, \ H(x, 1) = \frac{x}{||x||} \in S^1, \ \forall \ x \in X, \ H(a, t) = a, \ \forall \ a \in S^1.$$

This asserts that $S^1$ is a strong deformation retract of $\mathbf{R}^2 - \{\mathbf{0}\}$ and hence by Proposition 2.12.2 $\pi_1(S^1, x_0) \cong \pi_1(\mathbf{R}^2 - \{\mathbf{0}\}, x_0) \cong \mathbf{Z}$. As $\mathbf{R}^2 - \{\mathbf{0}\}$ is path connected, it follows that its fundamental group is independent of the choice of its base point $x_0$ and hence it follows that $\pi_1(\mathbf{R}^2 - \{\mathbf{0}\}) \cong \mathbf{Z}$.                                □

**Corollary 2.19.12** *The punctured Euclidean plane* $\mathbf{R}^2 - \{\mathbf{0}\}$ *is not simply connected.*

**Proof** It follows from Theorem 2.19.11.                                          □

**Example 2.19.13** The circle $S^1$ is not simply connected by Corollary 2.19.3. On the other hand, the $n$-sphere $S^n$ is simply connected for all $n > 1$ (see Theorem 2.26.7).

### 2.19.3   Computation of Fundamental Group of Figure-Eight $F_8$ and a Wedge of Circles

This subsection computes the fundamental group of figure-eight $F_8$, which is the union of two circles in the plane with one point in common. Its fundamental group is nonabelian, which is different from fundamental groups computed in previous subsections, because they are abelian. Its computation by another method using the concept of covering spaces is given in Chapter 5.

**Remark 2.19.14** Theorem 2.19.15 proves that the fundamental group of figure-eight $F_8$ is not abelian and Corollary 2.19.18 asserts that this fundamental group is a free group on two generators.

**Definition 2.19.15** Let $F_8 \subset S^1 \times S^1$ be the subspace of the product space $S^1 \times S^1$ defined by

$$F_8 = \{(u, v) \in S^1 \times S^1 : u = (1, 0) \ or \ v = (1, 0)\},$$

which is the union of two circles in the plane touching each of other at one point. $F_8$ is known as **figure-eight**.

Theorem 2.19.16 proves by using Van Kampen theorem that the fundamental group of figure-eight is nonabelian. The computation of this group by two other methods is also available in Chapter 5.

**Theorem 2.19.16** *The fundamental group of figure-eight is nonabelian.*

***Proof*** For the figure-eight $F_8$, let $x_0$ be the common point of the two circles and $x_1, x_2$ be two points taking one point from each circle such that $x_1 \neq x_0 \neq x_2$. Then, the spaces

$$(F_8 - x_1, x_0) \simeq (S^1, s_0) \; and \; (F_8 - x_2, x_0) \simeq (S^1, s_0),$$

for an arbitrary point $s_0 \in S^1$. Moreover,

$$((F_8 - x_1, x_0) \cap (F_8 - x_2, x_0), x_0) \simeq (s_0, s_0).$$

Hence by using Van Kampen theorem (see Exercise 23 of Sect. 2.28 and see Chapter 6 for its detailed proof in another form by using graph theory), it follows that

$$\pi_1(F_8, x_0) \cong \mathbf{Z} *_{\{1\}} \mathbf{Z},$$

which is the amalgamated product of the groups $\mathbf{Z}$ and $\mathbf{Z}$ over the trivial group $\{1\}$ (see Chapter 1). Hence, the group $\pi_1(F_8, x_0)$ is a free group having two generators. If we take the closed paths $\alpha$ at $x_0$ going once around the one circle $C_1$ and another closed path $\beta$ based at $x_0$ going once around the other circle $C_2$, then their homotopy classes can be taken as generators. As $\alpha$ and $\beta$ encircle different holes, their product closed path $\alpha * \beta * \alpha^{-1}$ cannot be deformed into $\beta$. This asserts that

$$[\alpha] * [\beta] * [\alpha^{-1}] \neq [\beta] \implies [\alpha] * [\beta] \neq [\beta] * [\alpha].$$

This proves that the group $\pi_1(F_8, x_0)$ is not abelian. □

**Theorem 2.19.17** *Let $X$ be a Hausdorff space such that $X = X_1 \cup X_2$, $X_1 \cap X_2 = \{x_0\}$ and $X_1$ and $X_2$ are both homeomorphic to the circle $S^1$. Then, $\pi_1(X, x_0)$ is a free group on two generators and it is not abelian.*

***Proof*** Let $x_1 \in X_1$ and $x_2 \in X_2$ be two distinct points. Then, both $\pi_1(X_1, x_1)$ and $\pi_1(X_2, x_2)$ are infinite cyclic groups. Hence by Theorem 2.19.16, it follows $\pi_1(X, x_0)$ is a free group on two generators by taking one generator the homotopy class of the closed path $\alpha$ which goes once around the circle homeomorphic to $X_1$ and the other generator is the homotopy class of the closed path $\beta$ which goes once around the circle homeomorphic to $X_2$. □

**Corollary 2.19.18** *The fundamental group of the figure-eight is a free group on two generators.*

***Proof*** It follows from Theorem 2.19.17 that the fundamental group of the figure-eight $F_8$ is a free group on two generators with one generator the homotopy class of the closed path $\alpha$ which goes once around the one circle $S_1^1$ and the other generator is the homotopy class of the closed path $\beta$ which goes once around the other circle $S_2^1$. $\square$

**Remark 2.19.19** Theorem 2.19.20 gives a generalization of Theorem 2.19.17 by induction. It computes the fundamental group of one-point union of $n$ circles $\bigvee\limits_{i=1}^{n} S_i^1$ (wedge ). On the other hand, the homology groups of $\bigvee\limits_{i=1}^{n} S_i^1$ (wedge) are computed in Chapter 6.

**Theorem 2.19.20** *Let $X$ be a Hausdorff space such that*

$$X = X_1 \cup X_2 \cup \cdots \cup X_n : \quad for \ \ i \neq j, \ X_i \cap X_j = \{x_0\},$$

*where each $X_i$ with subspace topology is homeomorphic to the circle $S^1$ ( $X$ is* ***called the wedge of the $n$ circles $X_i$ denoted by*** $\bigvee\limits_{i=1}^{n} X_i$ ***or*** $\bigvee\limits_{i=1}^{n} S_i^1$). *Then, $\pi_1(X, x_0)$ is a free group on $n$ generators and it is not abelian.*

***Proof*** Use induction on $n$. For $n = 2$, it is proved in Corollary 2.19.18. Take a point $x_i \in X_i$ such that $x_i \neq x_0$. Then, $\pi_1(X, x_0)$ is a free group on $n$ generators and it is not abelian. Moreover, the homotopy class of each loop $\alpha_i$ is a generator and the loops $\{\alpha_i : i \in A\}$ form a representative system of the free generators for the fundamental group $\pi_1(X, x_0)$.

$$U = X - \{x_n\} \quad \text{and} \quad V = X - \{x_1, x_2, \ldots, x_n\}$$

are both open sets such that

(i) $X_1 \cup X_2 \cup \cdots \cup X_{n-1}$ is a deformation retract of $U$.
(ii) $X_n$ is a deformation retract of $V$.
(iii) $U \cap V$ is contractible.

Hence, the corollary follows from Theorem 2.19.16. $\square$

Theorem 2.19.21 gives a further a generalization of Theorem 2.19.20 by taking $X$ as the wedge of the circles $\{X_i : i \in A\}$ with $X_i$ having the subspace topology of $X$.

**Theorem 2.19.21** *Let $X$ be a Hausdorff space such that*

(i)

$$X = \bigcup_{i \in A} X_i$$

*(ii)*

$$\bigcap_{i \in \mathbf{A}} X_i = \{x_0\}$$

*where each $X_i$ with subspace topology is homeomorphic to the circle $S^1$ (X is **called the wedge of the circles** $X_i$). Then, $\pi_1(X, x_0)$ is a free group. Moreover, if $\alpha_i$ is a loop in $X_i$, then its homotopy class is a generator of $\pi_1(X, x_0)$ and $\{\alpha_i : i \in \mathbf{A}\}$ forms a representative system of free generators for the fundamental group $\pi_1(X, x_0)$.*

**Proof** Proceed as in Theorem 2.19.17.                                                    □

**Remark 2.19.22** If $(X, x_0)$ is a pointed topological space such that its fundamental group $\pi_1(X, x_0)$ is nonabelian, then there exists no multiplication on $X$ making the space $X$ a topological group and even such a space cannot be equipped with the structure of an $H$-space structure or an $H$-group, because of Theorem 2.15.7 and Proposition 2.15.8. This implies by Theorem 2.19.16 that there does not exist any continuous multiplication on the topological space figure-eight $F_8$ admitting it a topological group structure or an $H$-space or an $H$-group structure.


## 2.20  Higher Homotopy Groups

The fundamental group of a topological space studies the properties of loops under continuous deformation. It characterizes the topological properties of a space in which the loops are defined. The study of these properties exhibited by loops in dimension higher than two fails. This failure motivates to define higher-dimensional analog of the loop (one-dimensional), leading to the concept of higher homotopy groups. This section studies higher homotopy groups $\pi_n(X, x_0)$, for integers $n > 1$, of a pointed topological space $(X, x_0)$. The concept of higher homotopy groups of a pointed space is a natural generalization of the concept of fundamental group of the pointed space defined in Sect. 2.16. H. Hurewicz in his paper [Hurewicz, 1935] chose spheres directly (in place of loops) to define higher homotopy groups formulated in Definition 2.20.8 and classified continuous maps $f$ from the $n$-sphere $S^n$:

$$f : S^n \to X$$

to an arbitrary topological space $X$ up to homeomorphism. There is an alternative approach to define higher homotopy groups starting with $n$-cubes $\mathbf{I}^n$. It is formulated in Proposition 2.20.11. In topology, homotopy groups are applied to classify topological spaces; on the other hand, they are used in physics to classify continuous maps. The relative homotopy groups $\pi_n(X, A, x_0)$ are natural generalization of $\pi_n(X, x_0)$ (see Chap. 5). Higher homotopy groups carry special importance in the study of bundles (see Chap. 5).

## 2.20.1  Introductory Concepts in Higher Homotopy Groups $\pi_n(X, x_0)$

This subsection introduces the basic concepts in higher homotopy groups $\pi_n(X, x_0)$, for integers $n > 1$, needed for our further study of algebraic topology.

**Definition 2.20.1** Let $X$ be a topological space and $A$ be a subspace of $X$. Then, the pair $(X, A)$ is said to be a **topological pair of spaces**. If $A = \emptyset$, the pair $(X, \emptyset)$ is identified with the space $X$. If $A = \{x_0\}$, the pair $(X, x_0)$ is identified with the pointed topological space $X$ having base point $x_0$.

**Example 2.20.2** If $s_0 \in S^n$, then $(S^n, s_0)$ is a pointed topological space. On the other hand, $(\mathbf{D}^{n+1}, S^n)$ is a pair of spaces, where the $n$-sphere $S^n$ is the boundary of the $(n + 1)$-disk $\mathbf{D}^{n+1}$ in $\mathbf{R}^{n+1}$.

**Definition 2.20.3** Let $(X, A)$ and $(Y, B)$ be two pairs of topological spaces. A map $f : (X, A) \to (Y, B)$ is said to be continuous if $f : X \to Y$ is continuous and $f(A) \subset B$.

**Remark 2.20.4** The homotopy of continuous maps between two topological spaces is now extended to pairs of topological spaces by using the usual convention of writing $(X \times \mathbf{I}, A \times \mathbf{I})$ to represent the product space $(X, A) \times \mathbf{I}$.

**Definition 2.20.5** Let $(X, A)$ and $(Y, B)$ be two pairs of topological spaces and $f, g : (X, A) \to (Y, B)$ be two continuous maps such that for a subset $X' \subset X, f(x') = g(x')$, $\forall x' \in X'$. Then, they are homotopic relative to $X'$, symbolized, $f \simeq g$ rel $X'$, if $\exists$ a continuous map

$$H : (X \times \mathbf{I}, A \times \mathbf{I}) \to (Y, B)$$

such that $H(x, 0) = f(x)$ and $H(x, 1) = g(x)$, $\forall x \in X$ and $H(x', t) = f(x') = g(x')$ for every $x' \in X'$ and for every $t \in \mathbf{I}$. It is expressed by the symbol $H : f \simeq g$ rel $X'$. In particular, if $X' = \emptyset$, then the term relative to $X'$ is not used.

**Definition 2.20.6** Given a pointed topological space $(X, x_0)$, the topological space $F^n(X, x_0)$ is defined to be the space of all continuous maps $f : (\mathbf{I}^n, \partial \mathbf{I}^n) \to (X, x_0)$ endowed with compact open topology.

**Remark 2.20.7** **Motivation of Hurewicz original** Definition 2.20.8 of the absolute group $\pi_n(X, x_0)$.
This definition was given by Hurewicz (1904–1956) in 1935. He studied the groups $\pi_n(S^n, s_0)$ during 1935–36 by considering elements of $\pi_n(S^n, s_0)$ as the homotopy classes of continuous maps $f : (S^n, s_0) \to (X, x_0)$, where $S^n$ is a fixed $n$-sphere. Let $S^n$ be oriented by a choice of a generator $\tau_n$ of the homology group $H_n(S^n)$ (see Chap. 3) and take a continuous map $g : (\mathbf{I}^n, \partial \mathbf{I}^n) \to (S^n, s_0)$ such that $g_*(u_n) = \tau_n$,

where $u_n$ is a generator of $\pi_n(X, x_0)$ and $g$ maps $\mathbf{I}^n - \partial \mathbf{I}^n$ topologically onto $S^n - s_0$. Then, the composite map

$$f \circ g : (\mathbf{I}^n, \partial \mathbf{I}^n) \to (X, x_0) \in F^n(X, x_0).$$

If $h$ is the second map having the properties of $g$, then $g_*(u_n) = h_*(u_n)$. It implies that $g \simeq h$, since they represent the same generator of $\pi_n(X, x_0)$. This implies that the homotopy class of the composite map $f \circ g$ depends only on $f$ and hence $f$ represents a unique element $c(f)$ in $\pi_n(X, x_0)$. Conversely, given $k \in F^n(X, x_0)$ the continuous map $f = k \circ g^{-1}$ is single-valued and hence any element of $\pi_n(X, x_0)$ is a $c(f)$. This assignment sets up a 1-1 correspondence between the elements of $\pi_n(X, x_0)$ and the homotopy classes of continuous map $f : (S^n, s_0) \to (X, x_0)$. It is pointed out that $c(f)$ depends on orientation of $S^n$ : Its other orientation reverses the sign of $c(f)$. Its generalization for higher homotopy groups is given in Chapter 5.

**Definition 2.20.8** Let $(X, x_0)$ be a pointed space and $(S^n, s_0)$ be the $n$- sphere with base point $s_0$. Then, the set homotopy classes of continuous maps $f : (S^n, s_0) \to (X, x_0)$, abbreviated by the set $\pi_n(X, x_0) = [(S^n, s_0); (X, x_0)]$, admit an abelian group structure for all $n > 1$, called the $n$-**th homotopy group** of the pointed space $(X, x_0)$, denoted by $\pi_n(X, x_0)$.

The group structure of $\pi_n(X, x_0)$ can be equally well defined in a way analogous to that of the fundamental group. Let $\mathbf{I}^n$ be the unit $n$-cube and $\partial \mathbf{I}^n = \dot{\mathbf{I}}^n$ be its boundary defined as follows

$$\mathbf{I}^n = \{t = (t_1, t_2, \ldots, t_n) : 0 \le t_i \le 1, \ i = 1, 2, \ldots, n\};$$

$$\dot{\mathbf{I}}^n = \partial \mathbf{I}^n = \{(t_1, t_2, \ldots, t_n) \in \mathbf{I}^n : t_i = 0 \text{ or } t_i = 1 \text{ for } \text{ atleast } \text{ one } i\}.$$

Consider the set $\Omega_n(X, x_0)$ of all continuous maps $f : (\mathbf{I}^n, \partial \mathbf{I}^n) \to (X, x_0)$ endowed with the compact open topology. Then, $\Omega_n(X, x_0)$ consists of all continuous maps

$$f : \mathbf{I}^n \to X : f(t) = x_0, \ \forall t \in \partial \mathbf{I}^n.$$

$f(\mathbf{I}^n)$ is called an $n$-loop in $X$ based at the point $x_0 \in X$. In view of identification of the points on the boundary of $\mathbf{I}^n$, these $n$-loops become topologically equivalent to $n$-spheres $S^n$.

**Definition 2.20.9** Two continuous maps $f, g \in \Omega_n(X, x_0)$ are said to be homotopic relative to the subspace $\partial \mathbf{I}^n$, written as $f \simeq g \ rel \ \partial \mathbf{I}^n$ if there exists a continuous map

$$G : \mathbf{I}^n \times \mathbf{I} \to X : G(t, 0) = f(t), \ G(t, 1) = g(t), \ \forall t \in \mathbf{I}^n \ and \ G(t, s) = x_0, \ \forall t \in \partial \mathbf{I}^n, \ \forall s \in \mathbf{I}.$$

This homotopy is an equivalence relation $\sim$ between $n$-loops on $\Omega_n(X, x_0)$, the class corresponding to the $n$-loop $f$ is as usual denoted by $[f]$, and the corresponding quotient set $\Omega_n(X, x_0)/ \sim$ is denoted by $\pi_n(X, x_0)$. It can be endowed with a group structure by defining a product $f * g$ of $f$ and $g$ connecting along a common part of boundaries as formulated in Definition 2.20.10.

**Definition 2.20.10**  A composition $*$ is defined on the set $\Omega_n(X, x_0)$ by the rule:

$$(f * g)(t) = \begin{cases} f(2t_1, t_2, \ldots, t_n), & 0 \le t_1 \le 1/2 \\ g(2t_1 - 1, t_2, \ldots, t_n), & 1/2 \le t_1 \le 1 \end{cases} \forall f, g \in \Omega_n(X, x_0), t = (t_1, t_2, \ldots, t_n) \in \mathbf{I}^n$$

$$(2.4)$$

$f * g$ is continuous and hence $f * g \in \Omega_n(X, x_0)$. Consequently, the composition $*$ is well defined.

**Proposition 2.20.11**  $\pi_n(X, x_0)$ *is a group under the composition $\circ$ defined by* $[f] \circ [g] = [f * g]$, *called the* **higher homotopy group** *of the pointed space* $(X, x_0)$ *for* $n > 1$.

**Proof**  The composition $\circ$ defined by $[f] \circ [g] = [f * g]$ is clearly well defined. The unit element $e$ is defined by the homotopy class of the constant map

$$e(t) = e(t_1, t_2, \ldots, t_n) = x_0, \quad \forall t \in \mathbf{I}^n.$$

The inverse element of $[f]$ is defined by the homotopy class of

$$f^{-1}(t_1, t_2, \ldots, t_n) = f(1 - t_1, t_2, \ldots, t_n).$$

Proceed as in the case of $\pi_1(X, x_0)$ to complete the proof of the group structure of $\pi_n(X, x_0)$ for $n > 1$.  $\square$

**Remark 2.20.12**  The higher homotopy group $\pi_n(X, x_0)$ of all pointed spaces $(X, x_0)$ is abelian for all $n > 1$, because there exists a rotation of $S^n$ interchanging the two hemispheres of $S^n$ and keeping its base point $s_0$ unchanged. Its analytical proof is available in Theorem 2.20.14. On the other hand, the fundamental group $\pi_1(X, x_0)$ is not abelian for all pointed spaces $(X, x_0)$. For example, the fundamental group of figure-eight $F_8$ is not abelian by Theorem 2.19.16.

## 2.20.2   Abelian Group Structure of $\pi_n(X, x_0)$ for $n > 1$

This subsection proves some properties of $\pi_n(X, x_0)$. For example, Theorem 2.20.14 proves that $\pi_n(X, x_0)$ is abelian for all $n > 1$, and for all pointed topological spaces $(X, x_0)$ which is not true for fundamental groups for all $(X, x_0)$. This provides a difference between $\pi_n(X, x_0)$ for all $n \ge 2$ and fundamental group $\pi_1(X, x_0)$.

**Definition 2.20.13** Given a pointed topological space $(X, x_0)$, its **loop space** denoted by $\Omega(X, x_0)$ is defined to be the space of all loops in $X$ based at the point $x_0$ endowed with compact open topology. It is a pointed topological space with the constant map $c : \mathbf{I} \to x_0$ its base point.

**Theorem 2.20.14** *Let* $(X, x_0)$ *be a pointed topological space. Then, the higher homotopy group* $\pi_n(X, x_0)$ *is abelian for all* $n > 1$.

**Proof** Since the loop space $\Omega(X, x_0)$ endowed with compact open topology is an $H$-space, with the constant loop $c$ as its homotopy identity, it follows that the group $\pi_2(X, x_0) = \pi_1(\Omega(X, x_0), c)$ is abelian by Theorem 2.15.7. Using this result, the theorem is proved by induction on $n$.                                                                $\square$

**Proposition 2.20.15** *Let* $X = \{x_0\}$ *be a topological space consisting of a single element. Then,* $\pi_n(\{x_0\}) = 0, \forall n \geq 0$.

**Proof** By the given condition, there exists only one map $f : S^n \to \{x_0\}$, which is the constant map and hence the proposition follows.                                                $\square$

## 2.20.3  Functorial Property of $\pi_n$

This subsection proves that $\pi_n$ is a covariant functor from the homotopy category $\mathcal{T}op_*$ of pointed topological spaces to the category $\mathcal{A}b$ of abelian groups and homomorphisms for every $n \geq 2$. For the case, when $n = 1$, the functorial property of $\pi_1$ has been proved in Theorem 2.11.3, which says that

$$\pi_1 : \mathcal{H}tp_* \to \mathcal{G}rp$$

is a covariant functor from the category $\mathcal{H}tp_*$ to the category $\mathcal{G}rp$.

**Theorem 2.20.16** *Let* $f : (X, x_0) \to (Y, y_0)$ *be a base point preserving continuous map. Then,* $f$ *induces a homomorphism*

$$f_* : \pi_n(X, x_0) \to \pi_n(Y, y_0), [\sigma] \mapsto [f \circ \sigma], \forall n \geq 1.$$

   *such that*

(i) *For the identity map* $1_X : (X, x_0) \to (X, x_0)$, *its induced homomorphism* $1_{X*}$ *is the identity automorphism on* $\pi_n(X, x_0)$.

(iii) *For the base point preserving continuous maps* $f : (X, x_0) \to (Y, y_0)$ *and* $g : (Y, y_0) \to (Z, z_0)$, *their composite map* $g \circ f : (X, x_0) \to (Z, z_0)$ *induces a homomorphism* $(g \circ f)_*$ *with the property*

$$(g \circ f)_* = g_* \circ f_* : \pi_n(X, x_0) \to \pi_n(Z, z_0).$$

(iii)  *For homotopic maps $f \simeq g : (X, x_0) \to (Y, y_0)$ rel $\{x_0\}$, their induced homo-morphisms*

$$f_* = g_* : \pi_n(X, x_0) \to \pi_n(Y, y_0).$$

(iv)  *For a homeomorphism $f : (X, x_0) \to (Y, y_0)$, its induced homomorphism*

$$f_* : \pi_n(X, x_0) \to \pi_n(Y, y_0)$$

*is an isomorphism.*

(v)  *For a homotopy equivalence $f : (X, x_0) \to (Y, y_0)$, its induced homomorphism*

$$f_* : \pi_n(X, x_0) \to \pi_n(Y, y_0)$$

*is an isomorphism.*

**Proof** **Case $n = 1$** : For fundamental group, it has been proved in Theorem 2.11.3.

For **Case $n > 1$**, take any element $\sigma \in \Omega_n(X, x_0)$ and consider the assignment $\sigma \to f \circ \sigma$. This assignment induces a transformation

$$f_* : \pi_n(X, x_0) \to \pi_n(Y, y_0), [\sigma] \mapsto [f \circ \sigma], \ \forall n > 1.$$

Then, $f_*$ carries the zero element of $\pi_n(X, x_0)$ to the zero element of $\pi_n(Y, y_0)$ and for any two elements $\sigma, \tau \in \Omega_n(X, x_0)$,

$$f(\sigma * \tau) = (f \circ \sigma) * (f \circ \tau)$$

asserts that the induced transformation

$$f_* : \pi_n(X, x_0) \to \pi_n(Y, y_0), \ [\sigma * \tau] \mapsto f_*[\sigma] \circ f_*[\tau], \ \forall n > 1$$

is a homomorphism of groups, called the **induced homomorphism of $f$**. The prop-erties $(i) - (v)$ follow from the definition of induced homomorphism, which are similar to the corresponding properties for fundamental groups proved in Theorem 2.11.3.

□

**Corollary 2.20.17** (Homotopy invariance) *Let $(X, x_0)$ and $(Y, y_0)$ be two homotopy equivalent spaces. Then, the groups $\pi_n(X, x_0)$ and $\pi_n(Y, y_0)$ are isomorphic for every $n \geq 1$.*

**Proof** By hypothesis, $(X, x_0)$ and $(Y, y_0)$ are homotopy equivalent spaces. Then, there exists a homotopy equivalence

$$f : (X, x_0) \to (Y, y_0).$$

Hence, it follows by Theorem 2.20.16(v) that

$$f_* : \pi_n(X, x_0) \to \pi_n(Y, y_0)$$

is an isomorphism for every $n \geq 1$. ☐

In Theorem 2.20.18 and thereafter, the following notations are used:

(i) $Top_*$ stands for the category of pointed topological spaces and their base point preserving continuous maps.

(ii) $Ab$ stands for the category of abelian groups and their homomorphisms.

(iii) $Htp_*$ stands for the homotopy category of pointed topological spaces and their homotopy classes of maps.

**Theorem 2.20.18** (Functorial property of $\pi_n$) *For every $n \geq 2$,*

(i) *$\pi_n : Top_* \to Ab$ is a covariant functor such that if $f, g : (X, x_0) \to (Y, y_0)$ are continuous and $f \simeq g$ rel $\{\partial \, I^n\}$, then $\pi_n(f) = f_* = g_* = \pi_n(g)$.*

(ii) *$\pi_n : Htp_* \to Ab$ is also a covariant functor.*

**Proof** (i) The assignment

$$\pi_n : Top^* \to Ab, (X, x_0) \mapsto \pi_n(X, x_0)$$

gives an object function, and the assignment

$$\pi_n : Top^* \to Ab, f \mapsto \pi_n(f) = f_*$$

is well defined and gives a morphism function. This proves the first part of the theorem by using Theorem 2.20.16.

(ii) The assignment

$$\pi_n : Htp^* \to Ab, (X, x_0) \mapsto \pi_n(X, x_0)$$

defines an object function, and the assignment

$$\pi_n : Htp^* \to Ab, [f] \mapsto \pi_n(f) = f_*$$

is well defined and gives a morphism function. This proves the last part of the theorem by using Theorem 2.20.16. ☐

## 2.20.4   *Role of Base Point x in Groups $\pi_n(X, x)$ for $n \geq 1$*

This subsection studies the role of base point of a homotopy group and proves Theorem 2.20.19 saying that for a path-connected space $X$, the groups $\pi_n(X, x)$ do not depend on the choice of base points in the sense that all the groups $\pi_n(X, x)$

are isomorphic with various base points $x \in X$ for every $n \geq 1$. Its detailed study is available in Chapter 5.

**Theorem 2.20.19** *Let $X$ be a path-connected space and $x_0, x_1 \in X$ be two arbitrary points. Then there exists an isomorphism*

$$\alpha_n : \pi_n(X, x_1) \to \pi_n(X, x_0), \ \forall n \geq 1.$$

*Proof* **Proof I (Geometric proof)**: By hypothesis, $X$ is a path-connected space. Hence there exists a path $\alpha : \mathbf{I} \to X$ from the point $x_0$ to the point $x_1$. Pull the image of the boundary $\partial \mathbf{I}^n$ of $\mathbf{I}^n$ along the path $\alpha$ back to the point $x_0$ with the image of $\mathbf{I}^n$ being pulled in an arbitrary way. The map thus obtained represents the homotopy class of an element $\theta \in \pi_n(X, x_0)$. It depends only on $\sigma$ and the homotopy class of $\alpha$. This correspondence $\sigma \to \theta$ constructs the map $\alpha_n$

$$\alpha_n : \pi_n(X, x_1) \to \pi_n(X, x_0), \ \sigma \mapsto \theta, \ \forall n \geq 1.$$

This yields the required isomorphism $\alpha_n$ of Theorem 2.20.19.

 **Proof II (Analytical proof)**: An analytical proof formulating the geometric construction of $\alpha_n$ described in Proof I, is available in Chap. 5.    □

**Corollary 2.20.20** *Let $X$ and $Y$ be two homotopy equivalent path-connected spaces. Then the groups $\pi_n(X)$ and $\pi_n(Y)$ are isomorphic and independent of base points for every $n \geq 1$.*

*Proof* By hypothesis, $X$ and $Y$ are two homotopy equivalent path-connected spaces. Let $x_0 \in X$ and $y_0 \in Y$. Then there exists a homotopy equivalence

$$f : X \to Y : f(x_0) = y_0.$$

This $f$ induces an isomorphism

$$f_* : \pi_n(X, x_0) \to \pi_n(Y, y_0), \ \forall n \geq 1.$$

Hence the corollary follows from Theorem 2.20.19.    □

## 2.21 Freudenthal Suspension Theorem and Its Applications

This section defines Freudenthal suspension homomorphism, which is an isomorphism under some specified condition prescribed in Theorem 2.21.3. Hans Freudenthal proved this theorem in 1937 showing the stable range for homotopy groups. This theorem is known as Freudenthal suspension theorem named after Hans Freudenthal. The most interesting feature of the theorem is that the suspension homomorphism lowers the dimension of homotopy groups by one and it facilitates the study of the problem of computing the homotopy groups $\pi_{m+n}(S^n)$, one of the deepest problem in homotopy theory. The other aspect of this theorem is to introduce the concept of stabilization of homotopy groups leading to stable homotopy theory (see Chap. 5).

### 2.21.1   Freudenthal Suspension Theorem

This subsection describes a geometric construction of Freudenthal suspension homomorphism

$$\sigma_n : \pi_n(S^n) \to \pi_{n+1}(S^{n+1}).$$

H. Freudenthal proved in 1937 a basic theorem in homotopy theory saying that

$$\sigma_n : \pi_n(S^n) \to \pi_{m+1}(S^{n+1}), \ [\alpha] \mapsto [\tilde{\alpha}]$$

is an isomorphism for $m < 2n - 1$ and is surjective for $m \leq 2n - 1$, where $\tilde{\alpha} : S^{n+1} \to S^{n+1}$ is an continuous extension of the continuous map $\alpha : S^n \to S^n$ determined uniquely up to homotopy. This theorem is known as **Freudenthal Suspension Theorem**.

**Definition 2.21.1** (*Geometric Construction*) To define Freudenthal suspension homomorphism geometrically, consider

$$\pi_m(S^n) = \{[\alpha] \ such \ that \ \alpha : (S^m, 1) \to (S^n, 1) \ is \ a \ continuous \ map \}$$

and $S^n$ as the equator of $S^{n+1}$ defined as the subspace of $S^{n+1}$ consisting of all points of $S^{n+1}$ with last coordinate 0. Let $N$ and $S$ be the north and south points of $S^{n+1}$ defined by

$$N = (0, 0, \ldots, 1) \in S^{n+1}, \ and \ S = (0, 0, \ldots, -1) \in S^{n+1}.$$

Let $[\alpha] \in \pi_m(S^n)$. Then $\alpha : S^m \to S^n$ is a continuous map. Extend $\alpha$ to a continuous function $\tilde{\alpha} : S^{n+1} \to S^{n+1}$ as follows: $\tilde{\alpha}|_{S^m} = \alpha$ and it maps the equator of $S^{m+1}$ to the equator of $S^{n+1}$. The map is then extended radially as shown in Fig. 2.19.

**Fig. 2.19** Radially extended map

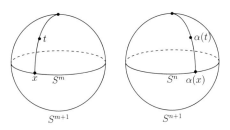

The arc from the north pole to a point $x \in S^m$ is mapped linearly onto the arc from the north pole of $S^{n+1}$ to $\alpha(x)$. This defines the map $\tilde{\alpha}$ on the northern hemisphere. For the southern hemisphere it is similarly defined. The extended map $\tilde{\alpha}$ is called the suspension of $\alpha$.

We are now in a position to define Freudenthal suspension homomorphism $\sigma_n$.

**Definition 2.21.2** (*Freudenthal suspension homomorphism*) The natural map

$$\sigma_n : \pi_n(S^n) \to \pi_{n+1}(S^{n+1}), \ [\alpha] \mapsto [\tilde{\alpha}]$$

is a homomorphism, called the Freudenthal suspension homomorphism.

**Theorem 2.21.3** (*Freudenthal suspension theorem*) *Freudenthal suspension homomorphism*

$$\sigma_n : \pi_m(S^n) \to \pi_{m+1}(S^{n+1}), \ [\alpha] \mapsto [\tilde{\alpha}]$$

*is an isomorphism for $m < 2n - 1$ and is surjective for $m \leq 2n - 1$, where $\tilde{\alpha} : S^{n+1} \to S^{n+1}$ is an continuous extension of the continuous map $\alpha : S^n \to S^n$ determined uniquely up to homotopy.*

**Proof** The paper [Freudenthal, 1937] is referred.                    □

## 2.21.2   Hurewicz and Hopf Theorems on $S^n$

This subsection proves Hurewicz and Hopf Theorems on $S^n$, which are two basic results in homotopy theory. The proof is based on Freudenthal Suspension Theorem 2.21.3.

**Theorem 2.21.4** (Hurewicz) $\pi_m(S^n) = 0, \ \forall m, n$ satsfying $0 < m < n$.

**Proof** Let $k$ be any positive integer such that $k < m$. Then $k + m + 1 < 2n \implies m - k < 2(n - k) - 1$ and hence by Freudenthal suspension theorem 2.21.3, it follows that

$$\pi_m(S^n) \cong \pi_{m-1}(S^{n-1}) \cong \pi_{m-2}(S^{n-2}) \cong \pi_{m-1}(S^{n-1}) \cong \cdots \cong \pi_1(S^{n-m+1}) = 0$$

by Theorem 2.26.7, since $n - m + 1 > 1$.                    □

**Theorem 2.21.5** (Hopf) $\pi_n(S^n) \cong \mathbf{Z}$ for every integer $n \geq 1$.

**Proof** Since $\pi_1(S^1) \cong \mathbf{Z}$, it follows from the above discussion that $\mathbf{Z} \cong \pi_1(S^1) \cong \pi_2(S^2)$. For $n > 1, n < 2n - 1$ asserts by Freudenthal suspension theorem that

$$\pi_2(S^2) \cong \pi_3(S^3) \cong \pi_4(S^4) \cong \cdots \cong \pi_n(S^n) \cong \mathbf{Z}.$$                    □

**Table 2.1** Table of $\pi_i(S^n)$ for $1 \le i \le 8$, $1 \le n \le 8$

|         | $i = 1$ | 2 | 3 | 4 | 5 | 6 | 7 | 8 |
|---------|---------|---|---|---|---|---|---|---|
| $n = 1$ | $\mathbf{Z}$ | 0 | 0 | 0 | 0 | 0 | 0 | 0 |
| 2 | 0 | $\mathbf{Z}$ | $\mathbf{Z}$ | $\mathbf{Z}_2$ | $\mathbf{Z}_2$ | $\mathbf{Z}_{12}$ | $\mathbf{Z}_2$ | $\mathbf{Z}_2$ |
| 3 | 0 | 0 | $\mathbf{Z}$ | $\mathbf{Z}_2$ | $\mathbf{Z}_2$ | $\mathbf{Z}_{12}$ | $\mathbf{Z}_2$ | $\mathbf{Z}_2$ |
| 4 | 0 | 0 | 0 | $\mathbf{Z}$ | $\mathbf{Z}_2$ | $\mathbf{Z}_2$ | $\mathbf{Z} \times \mathbf{Z}_{12}$ | $\mathbf{Z}_2 \times \mathbf{Z}_2$ |
| 5 | 0 | 0 | 0 | 0 | $\mathbf{Z}$ | $\mathbf{Z}_2$ | $\mathbf{Z}_2$ | $\mathbf{Z}_{24}$ |
| 6 | 0 | 0 | 0 | 0 | 0 | $\mathbf{Z}$ | $\mathbf{Z}_2$ | $\mathbf{Z}_2$ |
| 7 | 0 | 0 | 0 | 0 | 0 | 0 | $\mathbf{Z}$ | $\mathbf{Z}_2$ |
| 8 | 0 | 0 | 0 | 0 | 0 | 0 | 0 | $\mathbf{Z}$ |

*Remark 2.21.6* Theorem 2.21.5 proved by H. Hopf is called **Hopf degree theorem**. The justification towards naming it degree theorem is given in Sect. 2.22.

## 2.21.3    Table of $\pi_i(S^n)$ for $1 \le i, n \le 8$

This subsection gives the Table 2.1 displaying a small sample of the values of the groups $\pi_i(S^n)$, the results are extracted from the paper [Toda, 1962]. It is an immediate consequence of Freudenthal Suspension Theorem 2.21.3. An analogue table is also given in Chapter 5 for readers interested in topology of fiber bundles.

## 2.22    Degree of a Spherical Map on $S^n$ from Viewpoint of Homotopy Theory

This section gives the concept of **degree of a spherical map by using homotopy theory** in Definition 2.22.2 by establishing a close relation between the group $\pi_n(S^n) \cong \mathbf{Z}$ and degree of a continuous spherical map $f : S^n \to S^n$. On the other hand, **degree of a spherical map by using homology theory** is equally well-defined in Chapter 3. The concept of degree of such maps play a key role in analysis and topology and it generalizes the concept of degree of a loop on the circle, studied in Sect. 2.17.

**Proposition 2.22.1** *Let* $f : S^n \to S^n$ *be a continuous map and* $\alpha_n$ *be a generator of the group* $\pi_n(S^n) \cong \mathbf{Z}$. *If* $f_* : \pi_n(S^n) \to \pi_n(S^n)$ *is the homomorphism induced by* $f$, *then* $f_*(\alpha_n)$ *determines an integer* $d$ *such that* $f_*(\alpha_n) = d\alpha_n$.

*(i)  The integer* $d$ *is independent of the choice of the generator* $\alpha_n$ *of* $\pi_n(S^n)$.

(ii) *If $f, g : S^n \to S^n$ are two continuous maps such that $f \simeq g$. Then the homomorphisms $f_* = g_* : \pi_n(S^n) \to \pi_n(S^n), \ \forall n \geq 1$.*

**Proof** (i) Let $x \in \pi_n(S^n)$ be an arbitrary element. There is some integer $p$ such that $x = p\alpha_n$. Now,

$$f_*(x) = f_*(p\alpha_n) = pf_*(\alpha) = pd\alpha_n = d(p\alpha_n) = dx, \ \forall x \in \pi_n(S^n).$$

It also follows in particular that

$$f_*(-\alpha_n) = d(-\alpha_n).$$

Consequently, it asserts that the degree $d$ corresponding to $f : S^n \to S^n$ does not depend on a particular choice of the generator of $\pi_n(S^n)$.

(ii) It follows from the functorial property of $\pi_n$ given in Corollary 2.20.17.          $\square$

**Definition 2.22.2** Let $f : S^n \to S^n$ be a continuous map and $\alpha_n$ be a generator of the group $\pi_n(S^n) = \mathbf{Z}$. If

$$f_* : \pi_n(S^n) \to \pi_n(S^n), \ [\beta] \mapsto [f \circ \beta]$$

is the homomorphism induced by $f$, then $f_*(\alpha_n)$ determines an integer $d$ such that $f_*(\alpha_n) = d\alpha_n$. The integer $d$ is independent of the choice of the generator and hence it is well-defined and is called the **degree of the spherical map** $f : S^n \to S^n$, denoted by $deg\ f = d$.

The above discussion is summarized in a basic result embodied in Proposition 2.22.3.

**Proposition 2.22.3** *Let $f : S^n \to S^n$ be a continuous map. Then $f$ has degree $d$ iff its induced homomorphism*

$$f_* : \pi_n(S^n) \to \pi_n(S^n)$$

*has the property*
$$f_*(x) = dx, \ \forall x \in \pi_n(S^n).$$

**Example 2.22.4** The continuous map $f : S^1 \to S^1, \ z \mapsto z^m$ has degree $m$. Geometrically, $f$ rounds the circle $S^1$ completely $m$ times.

**Remark 2.22.5** Given a continuous spherical map $f : S^n \to S^n$, its degree $deg\ f = d$ is the number of the homotopy classes $[f] \in \pi_n(S^n)$. Because, for the continuous map $f : S^n \to S^n$,

$$f_*(\alpha_n) = f_*([1_d]) = [f].$$

Since $\alpha_n$ is the homotopy class of the identity map $1_d : S^n \to S^n$, it follows that $d\alpha_n$ is the homotopy class of $f$.

## 2.23 Applications

This section presents some interesting applications of homotopy and fundamental groups to algebra, matrix theory, atmospheric science, vector field and extension problems.

### 2.23.1 Fundamental Theorem of Algebra and Algebraic Completeness of the Field C

This subsection proves the fundamental theorem of algebra through homotopy theory. The proof of this theorem in an algebraic way is complicated, but it is less complicated by using complex analysis. One role of algebraic topology is mainly 'algebra serving topology,' but in this subsection the role is reversed to prove 'Fundamental Theorem of Algebra' by using the concept of homotopy followed by proving the algebraic completeness of the field C of complex numbers.

**Theorem 2.23.1** (Fundamental Theorem of Algebra) Let C be the field of complex numbers. Every nonconstant polynomial over C has a root in C.

*Proof* **Proof I** : Let $f(z) = a_0 + a_1 z + \cdots + a_{n-1} z^{n-1} + a_n z^n$, $a_n \neq 0$, $n \geq 1$ be an arbitrary polynomial over C. If the theorem is not true, the restriction of the map

$$\psi : C \to C - \{0\}, z \mapsto f(z)$$

to different circles $C_r : |z| = r, r \geq 0$ are loops in $C - \{0\}$. Define the two maps

$$F : I \times [0, \infty) \to S^1, (t, r) \mapsto \frac{f(re^{2\pi it})/f(r)}{|f(re^{2\pi it})/|f(r)|}$$

and

$$H : I \times I \to S^1 \subset C, (t, s) \mapsto \begin{cases} F(t, s/(1-s)), & 0 \leq t \leq 1, 0 \leq s < 1 \\ e^{2\pi int}, & 0 \leq t \leq 1, s = 1. \end{cases}$$

The continuity of $F$ asserts that

$$\lim_{s \to 1} H(t, s) = \lim_{s \to 1} F(t, s/(1-s)) = \lim_{r \to \infty} F(t, r) = e^{2\pi int}.$$

This proves the continuity of $H$. If $H(t, 0) = g_0(t)$ and $H(t, 1) = g_1(t)$, then

$$H : g_0 \simeq g_1 \text{ rel } \dot{I}.$$

This shows that deg $g_0$ = deg $g_1$ by Theorem 2.24.1. Again, by Definition 2.17.9, it follows that deg $g_0$ = 0 and deg $g_1$ = $n \geq 1$. This contradiction implies that $f(z)$ has a root in $\mathbf{C}$.

**Proof II** : Let $f(z) = a_0 + a_1 z + \cdots + a_{n-1} z^{n-1} + a_n z^n$, $a_n \neq 0$, $n \geq 1$ be an arbitrary polynomial over $\mathbf{C}$. If the theorem is not true, then $f(z) \neq 0$, $\forall z \in \mathbf{C}$. Define a map

$$H_t : S^1 \to S^1, z \mapsto \frac{f(tz)}{|f(tz)|}$$

for every nonnegative real number $t$. Then $H_0$ is a constant map at $a_0$. Taking $t$ large enough, it follows that $H_t$ is homotopic to the map

$$g : S^1 \to S^1, z \mapsto z^n.$$

Since any two of the maps $H_t$ are homotopic, $H_0$ and $g$ are homotopic. Again two closed paths in $S^1$ are homotopic iff they have the same degree. This implies that $deg\ H_0 = deg\ g$, which is not possible, since $deg\ H_0 = 0$ and $deg\ g = n$. □

**Definition 2.23.2** A field $F$ is called **algebraically complete (or closed)** if every polynomial over $F$ of degree $\geq 1$, has a root in $F$.

**Corollary 2.23.3** *The complex field* $\mathbf{C}$ *is algebraically complete.*

**Proof** Theorem 2.23.1 asserts that the field $\mathbf{C}$ is algebraically complete. □

**Example 2.23.4** The field $\mathbf{C}$ of complex numbers is algebraically complete but the real field $\mathbf{R}$ is not complete. Because there are some polynomials over $\mathbf{R}$ of degree $\geq 1$ having no root in $\mathbf{R}$. For example, $x^2 + 2$ has no root in $\mathbf{R}$. But it has a root in $\mathbf{C}$.

**Corollary 2.23.5** *The real field* $\mathbf{R}$ *is embedded in the algebraically complete field* $\mathbf{C}$.

**Proof** The corollary follows from Corollary 2.23.3. □

## 2.23.2   Fixed Point and Homotopy: Brouwer Fixed-Point Theorem for Dimension 2

The aim of this subsection is to study links between fixed-point theory and homotopy. For example, this subsection proves Brouwer fixed-point theorem (BFT) for dimension 2 as an application of Theorem 2.19.5 asserting that $S^1 \subset \mathbf{D}^2$ not a retract of $\mathbf{D}^2$. Moreover, Proposition 2.23.11 establishes a close relation between fixed point theorem and homotopy. There is a natural question: if $X = \mathbf{D}^2 - \{0\}$ is the punctured disk, does there exist a continuous map $f : X \to X$ without fixed points? For existence of such an example see Exercise 43 of Sect. 2.28.

Theorem 2.23.6 proves Brouwer fixed-point theorem for dimension 2. Its generalization for an arbitrary finite dimension $n$ is proved in Theorem 2.25.58.

**Theorem 2.23.6** (Brouwer Fixed-Point Theorem for Dimension 2) Every continuous map $h : \mathbf{D}^2 \to \mathbf{D}^2$ has a fixed point.

**Proof** Suppose $h$ has a fixed point. Then there exists a point $x \in \mathbf{D}^2$ such that $h(x) = x$. If no such point $x \in \mathbf{D}^2$ exists, then $h(x) \neq x$, $\forall x \in \mathbf{D}^2$. Define a map

$$r : \mathbf{D}^2 \to S^1, x \mapsto r(x) \in S^1,$$

where $r(x)$ is the point of intersection of the circle $S^1$ with the ray in $\mathbf{R}^2$ starting from the point $h(x)$ and passing through the point $x$. Since this ray intersects $S^1$ at exactly one point, the map $r$ is well-defined. This implies that

$$x = (1 - t)h(x) + tr(x)$$

for some $t > 0$. Hence the point $r(x)$ is determined by the rule

$$r(x) = \frac{(x - (1 - t)h(x))}{t}.$$

This shows that $r$ is continuous. Moreover, $r(x) = x$, $\forall~x \in S^1$ shows that $r : \mathbf{D}^2 \to S^1$ is a retraction which implies that $S^1$ is a retract of $\mathbf{D}^2$. But it contradicts Theorem 2.19.5. □

**Remark 2.23.7** For an alternative proof of Theorem 2.23.6 by using the concept of vector field is already given in Theorem 2.18.19. Brouwer fixed-point theorem for dimension $n = 1$ is also proved in Theorem 2.26.6. Its generalization for an arbitrary finite dimension $n$ is given in Theorem 2.25.58.

**Corollary 2.23.8** *Every real $3 \times 3$ matrix with positive real entries has a positive real eigenvalue.*

**Proof** Let $A$ be a real $3 \times 3$ matrix with positive real entries and associated linear transformation $L : \mathbf{R}^3 \to \mathbf{R}^3$ with respect to the standard basis of the real vector space $\mathbf{R}^3$ and $X_1 = \{(x, y, z) \in \mathbf{R}^3 : x, y, z \geq 0\}$ be the first octant of $\mathbf{R}^3$. If $X = S^2 \cap X_1 \subset \mathbf{R}^3$, then the subspace $X$ is homeomorphic to the disk $\mathbf{D}^2$. This implies that Brouwer fixed-point theorem is valid for every continuous map $f : X \to X$. For every $w = (x, y, z) \in X$, all the components $x, y, z$ of the point $w$ must be nonnegative and at least one of them is positive. Since by hypothesis, all the entries of the matrix $A$ are positive by hypothesis, the vector $L(x)$ is a vector having its all components positive. Hence the map

$$f : X \to X, w \mapsto \frac{L(w)}{||L(w)||}$$

is continuous. This asserts that $f$ has a fixed point $x_0$ (say). Therefore, $x_0 = \frac{L(x_0)}{||L(x_0)||}$ shows that $L(x_0) = ||L(x_0)||x_0$. This implies that the linear transformation $L$ and hence the matrix $A$ has the positive real eigenvalue $||L(x_0)||$. It proves that the matrix $A$ has a positive real eigenvalue. □

**Remark 2.23.9** A generalization of the Corollary 2.23.8 is given in the **Perron–Frobenius theorem** in $\mathbf{R}^n$. It asserts that any square matrix with positive entries has a unique eigenvector with positive entries (up to a multiplication by a positive constant) and the corresponding eigenvalue has multiplicity one and is strictly greater than the absolute value of any other eigenvalue.

**Definition 2.23.10** For a point $x \in S^n$, its antipode is the point $-x \in S^n$ and the pair $\{x, -x\}$ of points is said to be a pair of **antipode points**. A continuous map $f : S^n \to S^m$ is called **antipodal** if $f(-x) = -f(x)$, $\forall x \in S^n$.

Proposition 2.23.11 establishes a close relation between fixed point theorem and homotopy.

**Proposition 2.23.11** *Let the continuous map* $h : S^1 \to S^1$ *be nullhomotopic, i.e., $h$ is homotopic to a constant map. Then*

(i) *$h$ has a fixed point.*
(ii) *There exists some point $x_0 \in S^1$ such that $h(x_0) = -x_0$ (i.e., $h$ maps the point $x_0$ to its antipode $-x_0$).*

**Proof** (i) Let $h : S^1 \to S^1$ be nullhomotopic. Then the map $h$ has a continuous extension

$$\tilde{h} : \mathbf{D}^2 \to S^1 \subset \mathbf{D}^2$$

(see Theorem 2.23.22). Hence by Brouwer fixed-point theorem, there is a point $x_0 \in \mathbf{D}^2$ such that $\tilde{h}(x_0) = x_0$. But $\tilde{h}(x_0) \in S^1$, since Im $\tilde{h} \subset S^1$. Hence $\tilde{h}(x_0) = x_0$ shows that $x_0 \in S^1$. This concludes that $x_0$ is a fixed point of $h$.

(ii) Consider the map $g : S^1 \to S^1, x \mapsto -x$. Then the composite map $g \circ h : S^1 \to S^1$ is also nullhomotopic. This shows that $g \circ h$ has a fixed point $x_0$. Thus $x_0 = (g \circ h)(x_0) = g(h(x_0)) = -h(x_0)$ shows that $h(x_0) = -x_0$. □

**Corollary 2.23.12** *There is no antipodal continuous map $h : S^2 \to S^1$.*

### 2.23.3 Borsuk–Ulam Theorem for Dimension 2

This subsection proves the Borsuk–Ulam theorem for dimension 2 and applies this theorem to solve the following two problems.

**Problem 1**: Can the standard sphere $S^2$ be put in the Euclidean plane $\mathbf{R}^2$ ? Its negative answer is available in Remark 2.23.17.

**Problem 2** : Does there exist a pair of antipode points $\{x, -x\}$ on the earth $S^2$ (assumed spherical) having the same temperature and same barometric pressure at any point of time? Its affirmative answer is available in Remark 2.23.17.

***Remark 2.23.13*** Borsuk–Ulam Theorem 2.23.14 for dimension 2 solves the problems posed as above. A generalization of Borsuk–Ulam Theorem 2.23.14 for higher dimension asserts that given any continuous map $f : S^n \to \mathbf{R}^n$ $(n \geq 2)$, there exists a pair of antipodal points $\{x, -x\}$ on $S^n$ with the property that $f(x) = f(-x)$. Its proof is available in Chap. 6.

**Theorem 2.23.14** (Borsuk–Ulam Theorem for dimension 2) *Given any continuous map $f : S^2 \to \mathbf{R}^2$, there exists a pair of antipodal points $\{x, -x\}$ on $S^2$ with the property that $f(x) = f(-x)$.*

***Proof*** If possible, $f(x) \neq f(-x)$ for all $x \in S^2$. Then there exists a continuous map

$$h : S^2 \to S^1, \ x \mapsto \frac{f(x) - f(-x)}{\|f(x) - f(-x)\|}.$$

This asserts that $h(-x) = -h(x)$, $\forall x \in S^2$ and hence it contradicts the Corollary 2.23.12.                                                                                                            □

***Remark 2.23.15*** A generalization of Theorem 2.23.14 asserting that given two integers $m, n$ with $m > n \geq 0$, there does not exist a continuous map $f : S^m \to S^n$ preserving the antipodal points is proved by using homology theory in Chapter 6.

**Corollary 2.23.16** *The sphere $S^2$ cannot be embedded in the Euclidean plane $\mathbf{R}^2$.*

***Proof*** Suppose $f : S^2 \to \mathbf{R}^2$ is an embedding. Then it is continuous and injective. By applying Borsuk–Ulam Theorem 2.23.14, it follows that there exists a pair $\{x, -x\}$ of antipode points on $S^2$ with the property that $f(x) = f(-x)$. This gives a contradiction, since the map $f$ is injective.                                                                                □

***Remark 2.23.17*** **Corollary** 2.23.16 asserts there exists no homeomorphism from the sphere $S^2$ on the plane $\mathbf{R}^2$ and hence 2-sphere $S^2$ cannot be put in the plane $\mathbf{R}^2$. This implies that **no map of the earth can be drawn** (up to homeomorphism) on a page of an atlas. This gives a negative answer of the Problem 1.

***Remark 2.23.18*** **Atmospherically,** Borsuk–Ulam Theorem 2.23.14 asserts that there exist two different places on the earth represented by a pair of points $x$ and $-x$ having the same temperature and same barometric pressure at any point of time. To show it, consider the earth as 2-sphere $S^2$. Let $T, P$ be functions on the earth defining temperature and barometric pressure at any point of time and at a place on the earth. These are continuous functions and define a map

$$f : S^2 \to \mathbf{R}^2, \ x \mapsto (T(x), P(x)).$$

Then by Borsuk–Ulam Theorem 2.23.14, at any point of time, there exists a pair of points $x$ and $-x$ on the earth $S^2$ such that the temperature and barometric pressure both are identical at $x$ and $-x$. This gives an affirmative answer of the Problem 2.

## 2.23.4   Lusternik–Schnirelmann Theorem for $S^2$

This subsection proves Lusternik–Schnirelmann theorem by applying Borsuk–Ulam Theorem 2.23.14.

**Theorem 2.23.19** (Lusternik–Schnirelmann theorem for $S^2$) *If* $S^2 = G_1 \cup G_2 \cup G_3$, *where each* $G_i$ *is a closed subset of* $S^2$, *(i.e., if* $S^2$ *is covered by its three closed subsets* $G_1, G_2$ *and* $G_3$), *then one of the sets contains a pair of antipodal points.*

**Proof** Define a function

$$f : S^2 \to \mathbf{R}^2, \ x \mapsto (d(x, G_1), d(x, G_2)).$$

where $d(x, G_i)$ denotes the Euclidean distance of the point $x$ from $G_i$. Since $f$ is continuous, it identifies a pair of opposite points by Theorem 2.23.14. This asserts that there is a point $x_0 \in S^2$ such that

$$d(x_0, G_i) = d(-x_0, G_i) \ for \ 1 \leq i \leq 2.$$

If $d(x_0, G_i) > 0$, *for* $1 \leq i \leq 2$, then $x_0, \ -x_0 \in G_3$, since $S^2 = G_1 \cup G_2 \cup G_3$. Again, if $d(x, G_i) = 0$ for some $i$, then both $x_0$ and $-x_0$ lie in $G_i$, since each $G_i$ is a closed subset of $S^2$.                                                                                      $\square$

**Remark 2.23.20**  A generalization of Lusternik–Schnirelmann Theorem 2.23.19 for $S^n$ asserts that if $S^n = G_1 \cup G_2 \cup \cdots \cup G_n$, where each $G_i$ is a closed subset of $S^n$, then one of the sets contains a pair of antipodal points. Its proof is similar to that of Theorem 2.23.19. Its detailed proof is available in Chapter 6.


## 2.23.5   Cauchy's Integral Theorem of Complex Analysis

Cauchy's integral theorem is one of the basic theorems of complex analysis. This subsection proves this theorem from the viewpoint of homotopy by utilizing the concept of winding number of a differentiable loop in the complex plane and the exponential map $p : \mathbf{R} \to S^1, \ t \mapsto e^{2\pi i t}$.


**Theorem 2.23.21** (Cauchy's Integral Theorem) Let $X \subset \mathbf{C}$ be an open subset and $f : X \to \mathbf{C}$ be an analytic function. If $\beta$ is a simple closed piecewise differentiable loop in $X$ with the property that $\beta$ is nullhomotopic, then $\int_{\beta} f(z)dz = 0$.

**Proof** As $\beta$ is nullhomotopic, there exists a constant curve $\alpha$ in $X$ such that $\beta \simeq \alpha$. If $H : \beta \simeq \alpha$, define a map

$$g : \mathbf{I} \to \mathbf{Z}, \ t \mapsto w(\beta_t; z_0).$$

That is, $g(t) = w(\beta_t; z_0)$, where $\beta_t(s) = H(s, t)$ for $0 \leq s, t \leq 1$ and $z_0$ is a fixed point in $\mathbf{C} - X$ and $w(\beta; z_0)$ denotes the winding number of $\beta$ about the point $z_0$. Clearly, the map $g$ is continuous on $\mathbf{I}$. Again since $g$ is an integral valued function and $g(0) = 0$, it follows that $g(t) \equiv 0$. Consequently, the winding number $w(\beta; z_0) = 0$ for all $z_0 \in \mathbf{C} - X$. This proves the theorem by using Exercise 31 of Sect. 2.28. $\square$

### 2.23.6   Extension Problem in Homotopy Theory

Extension problem in topology is a basic problem. This subsection solves an extension problem by using homotopy of maps but it is different from Tietze extension theorem of real-valued continuous functions on normal spaces (see Chap. 6).

Extension problem formulated in Theorem 2.23.22 is solved by homotopy theory.

**Theorem 2.23.22** (Extension Problem)   Let $S^n \subset \mathbf{D}^{n+1}$ be the $n$-sphere in the Euclidean $(n + 1)$-space $\mathbf{R}^{n+1}$ and $X$ be an arbitrary topological space. Then a continuous map $f : S^n \to X$ has a continuous extension over the $(n + 1)$-disk $\mathbf{D}^{n+1}$ iff $f$ is homotopic to some constant map $c : S^n \to X$ (i.e., iff $f$ is nullhomotopic).

**Proof**   Suppose $c : S^n \to X$, $y \mapsto x_0 \in X$ is a constant map such that $f \simeq c$. Then there exists a homotopy

$$H : S^n \times I \to X, \quad (x, 0) \mapsto f(x), \quad and \quad (x, 1) \mapsto x_0$$

such that $H : f \simeq c$. Define a map

$$\tilde{f} : \mathbf{D}^{n+1} \to X, \; y \mapsto \begin{cases} x_0, & 0 \leq ||y|| \leq 1/2 \\ H(\frac{y}{||y||}, 2 - 2||y||), & 1/2 \leq ||y|| \leq 1. \end{cases}$$

Let $A = \{y \in \mathbf{D}^{n+1} : 0 \leq ||y|| \leq 1/2\}$ and $B = \{y \in \mathbf{D}^{n+1} : 1/2 \leq ||y|| \leq 1\}$. Then $\tilde{f}$ agrees on their intersection $A \cap B$, since at $||y|| = \frac{1}{2}$, $H(\frac{y}{||y||}, 1) = x_0$. Again $A, B \subset \mathbf{D}^{n+1}$ are closed sets such that $\mathbf{D}^{n+1} = A \cup B$. Since the map $\tilde{f}$ is well defined and is continuous on each of closed sets $A$ and $B$, by Pasting Lemma, $\tilde{f}$ is continuous. Again, since $\forall \; y \in S^n$, $||y|| = 1$ and $\tilde{f}(y) = H(y, 0) = f(x)$, it follows that $f$ has a continuous extension $\tilde{f}$ over $\mathbf{D}^{n+1}$. Conversely, if $\tilde{f} : \mathbf{D}^{n+1} \to X$ is a continuous extension of $f : S^n \to X$, then $\tilde{f}(y) = f(y) \forall \; y \in S^n$. Let $s_0 \in S^n$ and $f(s_0) = x_0 \in X$. Define a map

$$H : S^n \times I \to X, \quad (y, t) \mapsto \tilde{f}((1 - t)y + ts_0).$$

Again, $((1 - t)y + ts_0) \in \mathbf{D}^{n+1} \forall y \in S^n$, since $\mathbf{D}^{n+1}$ is a convex set. Clearly, $H$ is a continuous map such that $H : f \simeq c$. $\square$

Another extension problem is solved in Corollary 2.23.23 by the value of $c(f)$ formulated in Remark 2.20.7.

**Corollary 2.23.23** $c(f)$ *formulated in Remark 2.20.7 vanishes iff $f$ has a continuous extension over* $\mathbf{D}^{n+1}$.

**Proof** It follows from definition of $c(f)$ by using Theorem 2.23.22.                                □

**Remark 2.23.24** For a study of extension problem by using the degree function, see Theorem 2.25.1.

## 2.24 Hopf Classification Theorem by Using Homotopy Theory

Hopf classification theorem 2.24.2 characterizes the homotopy of continuous spherical maps by their degrees formulated in Definition 2.22.2 by **homotopy theory**, which are integers and hence it gives a complete classification of spherical maps by integers. On the other hand, the same homotopy classification theorem of continuous spherical maps by their degrees formulated by **homology theory** is available in Chapter 3.

The main tool used here is the concept degree of spherical maps given in Definition 2.22.2 and their basic properties are given in Theorem 2.24.1.

**Theorem 2.24.1** *(i) If $1_d : S^n \to S^n$ is the identity map, then its degree is 1.*
*(ii) If $f \simeq c : S^n \to S^n$, where $c$ is a constant map, then its $deg\, f$ is 0.*
*(iii) If $f, g : S^n \to S^n$ are continuous maps, then $deg\, (f \circ g) = deg\, f\ deg\, g$.*
*(iv) If $f \simeq g : S^n \to S^n$, then $deg\, f = deg\, g$.*
*(v) If $f, g : S^n \to S^n$ are continuous maps such that $deg\, f = deg\, g$, then $f \simeq g$.*

**Proof** It follows from the definition of degree of a continuous spherical map $f : S^n \to S^n$.                                □

**Theorem 2.24.2** (Hopf classification theorem) *Two continuous maps $f, g : S^n \to S^n$ are homotopic iff $deg\, f = deg\, g$.*

**Proof** It follows from Theorem 2.24.1. An alternative proof is given in Theorem 2.25.20.                                □

## 2.25  Application of Degree of Spherical Maps on $S^n$ and Hopf Classification Theorem

This section applies the degree of continuous map to solve some classical problems such as extension problem given in Theorem 2.25.1. Moreover, Hopf classification theorem 2.25.20 completely classifies continuous spherical maps $f : S^n \to S^n$ by degree function $deg : f \mapsto deg\, f$.

Given a continuous spherical map $f : S^n \to S^n$, its degree $deg\, f = d$ is the number of the homotopy classes $[f] \in \pi_n(S^n)$. Because, for the continuous map $f : S^n \to S^n$, $f_*(\alpha_n) = f_*[1_d] = [f]$, since $\alpha_n$ is the homotopy class of the identity map $1_d : S^n \to S^n$ and hence $d\alpha$ is the homotopy class of $f$. For any continuous map $f : S^n \to S^n$, the $deg\, f$ is equally well defined by **homology theory** and is studied in Chap. 3. Let $f : S^n \to \mathbf{R}^{n+1} - \{0\}$ be a continuous map. It is a natural question: whether $f$ has a continuous extension $\tilde{f}$ over the closed $(n + 1)$- disk $D^{n+1}$? Since the punctured Euclidean space $\mathbf{R}^{n+1} - \{0\}$ is homotopy equivalent to $S^n$ by a homotopy equivalence $g$, it follows that the groups $\pi_n(S^n)$ and $\pi_n(\mathbf{R}^{n+1} - \{0\})$ are isomorphic by Corollary 2.20.17 by the isomorphism $g_*$. Hence the $deg(g \circ f)$ of the map $g \circ f : S^n \to S^n$ leads to define the degree of a map $f : S^n \to \mathbf{R}^{n+1} - \{0\}$, called the **characteristic of the vector field** $f$, **usually denoted by** $\kappa_{S^n}(f)$. It is used to solve an extension problem in Theorem 2.25.1.

**Theorem 2.25.1**  *A continuous map $f : S^n \to \mathbf{R}^{n+1} - \{0\}$ has a continuous extension $\tilde{f}$ over the closed $(n + 1)$-disk $\mathbf{D}^{n+1}$ iff $\kappa_{S^n}(f) = 0$.*

**Proof**  If $f$ is homotopic to a constant map $c$, i.e., if $f$ is nullhomotopic, then $\kappa_{S^n}(f) = 0$. Moreover, the continuous extension $\tilde{f}$ over the closed $(n + 1)$- disk $\mathbf{D}^{n+1}$ determines a homotopy

$$H : S^n \times \mathbf{I}, \ (x, t) \mapsto \tilde{f}(tx).$$

$H$ is a continuous map such that $H : f \simeq c$. Hence the theorem follows from Theorem 2.23.22.    □

**Corollary 2.25.2**  *Let a continuous map $f : S^n \to \mathbf{R}^{n+1} - \{0\}$ has a continuous extension $\tilde{f}$ over the closed $(n + 1)$-disk $\mathbf{D}^{n+1}$. If $\kappa_{S^n}(f) \neq 0$, then $\tilde{f}$ has a fixed point.*

**Remark 2.25.3**  Given a continuous extension $\tilde{f} : \mathbf{D}^{n+1} \to \mathbf{R}^{n+1}$ of the continuous map $f : S^n \to \mathbf{R}^{n+1} - \{0\}$, Corollary 2.25.2 is applied to solve the existence problem of a solution to the equation $\tilde{f}(x) = 0$.

**Theorem 2.25.4**  *Let $f : S^n \to S^n$ be a continuous map of degree $d$. Then its suspension map*

$$\Sigma f = f_{n+1} : S^{n+1} \to S^{n+1}, \ [x, t] \mapsto [f(x), t]$$

*has also the same degree $d$.*

**Fig. 2.20** Freudenthal
suspension isomorphism $\sigma_n$
diagram

$$\begin{array}{ccc}
\pi_n(S^n) & \xrightarrow{\ \sigma_n\ } & \pi_{n+1}(S^{n+1}) \\
{\scriptstyle f_{n*}}\downarrow & & \downarrow{\scriptstyle f_{(n+1)*}} \\
\pi_n(S^n) & \xrightarrow{\ \sigma_n\ } & \pi_{n+1}(S^{n+1})
\end{array}$$

**Proof** To prove the theorem, consider the commutative diagram in Fig. 2.20 of groups and homomorphisms where $\sigma_n$ is the Freudenthal suspension isomorphism and

$$f_{n*} : \pi_n(S^n) \to \pi_n(S^n)$$

is the homomorphism of a continuous map $f_n : S^n \to S^n$ of degree $d$. Since $\sigma_n$ is an isomorphism and the diagram in Fig 2.20 is commutative, it follows that

$$\sigma_n \circ f_{n*} \circ \sigma_n^{-1} = f_{(n+1)}^*.$$

Hence it follows that

$$f_{(n+1)*}(x) = \sigma_n \circ f_{n*} \circ \sigma_n^{-1}(x) = d\sigma_n[\sigma_n^{-1}(x)] = dx, \ \forall x \in \pi_{n+1}(S^{n+1}).$$

This proves the theorem by Proposition 2.22.3. □

**Corollary 2.25.5** *Given any positive integer $n$, there is a continuous map $f : S^n \to S^n$ such that $\deg f = d$.*

**Proof** For $n = 1$, it follows from Theorem 2.19.1. Suppose it is true for some $n \geq 1$ and $f : S^n \to S^n$ is a continuous map such that $\deg f = d$. Let $f_n : S^n \to S^n$ be a continuous map of degree $d$. Then its suspension map

$$\Sigma f = f_{n+1} : S^{n+1} \to S^{n+1}$$

has also the same degree $d$ by Theorem 2.25.4. By induction on $n$ the proof of the corollary is completed. □

**Theorem 2.25.6** *Let $f, g : S^n \to S^n$ be any two continuous maps. Then*

*(i)  $\deg (f \circ g) = \deg f \deg g$.*
*(ii)  If $f = 1_{S^n}$ is identity map, then $\deg f = 1$.*
*(iii)  If $f$ is a homeomorphism, then $\deg f = 1$ or $-1$.*
*(iv)  If $f \simeq g$, then $\deg f = \deg g$.*
*(v)  If $f \simeq c : S^n \to S^n$, where $c$ is a constant map, then its $\deg f$ is 0.*

**Proof** (i) $(f \circ g)_*(x) = f_*(g_*(x)) = f_*(\deg g \, x) = \deg f \, \deg g, \ \forall x \in S^n \implies \deg(f \circ g) = \deg f \deg g$;
(ii) $f = 1_{S^n}, f_*(x) = x = 1x, \forall x \in S^n \implies \deg f = 1$.

(iii) For a homeomorphism $f$ with its inverse $f^{-1}$, their composite map

$$f \circ f^{-1} = 1_{S^n} \implies \deg f \deg f^{-1} = 1 \implies \deg f = 1 \text{ or } -1,$$

since $\deg f$ and $\deg f^{-1}$ are both integers.

(iv) It follows from Proposition 2.22.1.

(v) It follows trivially.                                                                  □

**Definition 2.25.7** The map

$$r_m : S^n \to S^n, \; x = (x_1, x_2, \ldots, x_{m-1}, x_m, \ldots, x_{n+1}) \mapsto (x_1, x_2, \ldots, x_{m-1} - x_m, \ldots, x_{n+1})$$

is called the **reflection map** of $S^n$ about the $x_m$ axis for $m = 1, 2, \ldots, n+1$ and the map

$$A : S^n \to S^n, \; x = (x_1, x_2, \ldots, x_{m-1}, x_m, \ldots, x_{n+1}) \mapsto -x = (-x_1, -x_2, \ldots, -x_{m-1} - x_m, \ldots, -x_{n+1}),$$

i.e., $A(x) = -x$, $\forall x \in S^n$, is called the **antipodal map.**

**Proposition 2.25.8** *The degree of the refection map*

$$r_m : S^n \to S^n, \; x = (x_1, x_2, \ldots, x_{m-1}, x_m, \ldots, x_{n+1}) \mapsto (x_1, x_2, \ldots, x_{m-1}, -x_m, \ldots, x_{n+1}),$$

*is* $-1$, *i.e.,* $\deg r_m = -1$ *for all* $m = 1, 2, \ldots n+1$

**Proof** For each reflection map

$$r_m : S^n \to S^n, \; x = (x_1, x_2, \ldots, x_{m-1}, x_m, \ldots, x_{n+1}) \mapsto (x_1, x_2, \ldots, x_{m-1}, -x_m, \ldots, x_{n+1}),$$

$\deg r_m = -1$, because $r_m \neq 1_{S^n}$ but $r_m \circ r_m = 1_{S^n}$ and hence $\deg r_m \deg r_m = 1$ and $\deg r_m \neq 1$. This proves that the only possibility is that $\deg r_m = -1$. This is true for every $m = 1, 2, \ldots, n+1$.                                                 □

**Proposition 2.25.9** *The degree of the antipodal map* $A : S^n \to S^n$, $x \mapsto -x$ *is* $(-1)^{n+1}$ *for all* $n \geq 1$.

**Proof** The map $A$ is the composite map

$$A = r_1 \circ r_2 \circ \cdots \circ r_{n+1}.$$

Hence it implies that
$$\deg A = (-1)^{n+1},$$

since $\deg r_m = -1$, $\forall m = 1, 2, \ldots, n+1$.                                       □

Theorem 2.25.10 relates the nonexistence of fixed points of spherical maps to their degrees.

**Theorem 2.25.10** *Let* $f : S^n \to S^n$ *be a continuous map such that* $f$ *has no fixed point for all integers* $n > 0$, *then* $\deg f = (-1)^{n+1}$.

**Proof** By hypothesis $f$ has no fixed point. It asserts that the line segment $(1 - t)f(x) - tx \neq 0$, $\forall x \in S^n$, $\forall t \in \mathbf{I}$. Otherwise, if $(1 - t)f(x) - tx = 0$ for some $x \in S^n$ and any $t \in \mathbf{I}$, then

$$1 - t = ||(1 - t)f(x)|| = ||tx|| \implies 1 - t = t,$$

since $||f(x)|| = 1 = ||x||$, $\forall x \in S^n$ and hence from above it follows that $f(x) = x$, which is not possible, since $f(x) \neq x$ by hypothesis. Hence the map $H : S^n \times \mathbf{I} \rightarrow S^n$ defined by

$$H(x, t) = \frac{(1 - t)f(x) - tx}{||(1 - t)f(x) - tx||}$$

is well-defined and continuous. Then $H(x, 0) = f(x)$ and $H(x, 1) = -x = A(x)$, $\forall x \in S^n$ show that $H : f \simeq A$ and hence it follows that $deg f = (-1)^{n+1}$ from Proposition 2.25.9. $\qquad\square$

Theorem 2.25.10 is now applied to solve the following fixed-point problems by homotopy.

**Corollary 2.25.11** *Let $f : S^{2n} \rightarrow S^{2n}$ be a continuous map such that $f \simeq 1_{S^n}$. Then $f$ has a fixed point.*

**Proof** If $f \simeq 1_{S^n}$, then $deg f = 1$. Suppose $f$ has no fixed point. Then by $deg f = (-1)^{2n+1} = -1$ by using Theorem 2.25.10, a contradiction. This contradiction shows that $f$ has a fixed point. $\qquad\square$

**Corollary 2.25.12** *Let $f : S^n \rightarrow S^n$ be a continuous map such that*

$$deg f \neq (-1)^{n+1}.$$

*Then $f$ has a fixed point.*

**Proof** By hypothesis,

$$deg f \neq (-1)^{n+1}.$$

If possible, $f$ has no fixed point. Then it follows by Theorem 2.25.10 that

$$deg f = (-1)^{n+1}.$$

This contradiction proves the corollary. $\qquad\square$

**Proposition 2.25.13** *If $f, g : S^n \rightarrow S^n$ be two continuous maps such that $f(x) \neq g(x)$, $\forall x \in S^n$. Then*

$$deg f + (-1)^n deg\ g = 0.$$

**Proof** Proceed as in Theorem 2.25.10 by showing first that

$$(1 - t)f(x) - tg(x) \neq 0, \quad \forall x \in S^n, \forall t \in \mathbf{I}. \qquad \square$$

**Definition 2.25.14** A topological space $X$ is said to have the **fixed-point property** if every continuous map

$$f : X \to X$$

has a fixed point.

**Example 2.25.15** **Examples of topological spaces having fixed-point property.**

(i) Let $X = [a, b]$ be a subspace of the real line space $\mathbf{R}$. Then every continuous map $f : X \to X$ has a fixed point by Brouwer fixed-point theorem 2.26.6 of dimension 1. This implies that $[a, b]$ has the fixed-point property.

(ii) Let $X = \mathbf{D}^2$ be the unit disk in $\mathbf{R}^2$. Then every continuous map $f : X \to X$ has a fixed point by Brouwer fixed-point theorem 2.23.6 of dimension 2. This implies that $\mathbf{D}^2$ has the fixed-point property.

(iii) Let $X = \mathbf{D}^n$ be the unit disk in $\mathbf{R}^n$. Then every continuous map $f : X \to X$ has a fixed point by Brouwer fixed-point Theorem 2.25.58 of a finite dimension $n \geq 0$. This implies that $\mathbf{D}^n$ has the fixed-point property.

(iv) On the other hand, the sphere $S^n$ has **no fixed-point property** by Proposition 2.25.16.

All topological spaces do not enjoy the fixed-point property. For example, by Proposition 2.25.16, the sphere $S^n$ has no fixed-point property.

**Proposition 2.25.16** *For any positive integer $n$, the sphere $S^n$ has no fixed-point property.*

**Proof** Let $f : S^n \to S^n$ be an arbitrary continuous map. Since the antipodal map $A : S^n \to S^n$, $x \mapsto -x$ has no fixed point, it follows that $S^n$ has no fixed-point property. $\qquad \square$

Proposition 2.25.17 proves the nonexistence of continuous nonvanishing vector field on $S^n$ for every even integer $n$.

**Proposition 2.25.17** *If $n$ is an even integer, there is no continuous nonvanishing vector field $f : S^n \to S^n$.*

**Proof** Let $v$ be a continuous nonvanishing vector field on $S^n$. The map

$$f : S^n \to S^n, \quad x \mapsto \frac{v(x)}{||v(x)||}$$

is a continuous map homotopic to the identity map on $S^n$. As by hypothesis, $n$ is an even integer, it follows from Corollary 2.25.11 that $f$ has a fixed point $x_0$ such that $< x_0, f(x_0) > = 1$. This implies that the vector field must vanish at some point of $S^n$. $\qquad \square$

**Remark 2.25.18** An alternative proof of Proposition 2.25.17 is given in Corollary 2.25.24.

## 2.25.1 Brouwer Degree Theorem and Hopf Classification Theorem by Using Homotopy Theory

Hopf classification theorem 2.24.2 characterizes the homotopy of continuous spherical maps by their degrees given in Definition 2.22.2 by homotopy theory, which are integers and hence it gives a complete classification of spherical maps by integers. The same homotopy classification theorem of continuous spherical maps by their degrees defined by homology theory is available in Chapter 3.

**Theorem 2.25.19** (Brouwer degree theorem) Let $f, g : S^n \to S^n$ be two continuous maps such that $f \simeq g$, then $deg\, f = deg\, g$.

**Proof** Let $f \simeq g$. Then their induced homomorphisms $f_*$ and $g_*$ are such that

$$f_* = g_* : \pi_n(S^n) \to \pi_n(S^n) \text{ for all integers } n \geq 1.$$

This asserts that $deg\, f = deg\, g$.                    □

**Theorem 2.25.20** (Hopf Classification Theorem) Let $f, g : S^n \to S^n$ be two continuous maps. Then $f \simeq g$, iff $deg\, f = deg\, g$.

**Proof** First. let $f \simeq g$. Then it follows by Theorem 2.25.19 that $deg\, f = deg\, g$. Conversely, let $deg\, f = deg\, g$. If $f, g : S^1 \to S^1$ are continuous maps such that $deg\, f = deg\, g$, then by Theorem 2.19.9, it follows that $f \simeq g$. This shows that the theorem is true for $n = 1$. Use induction on $n$ and apply Theorem 2.25.4 to complete the proof.                    □

**Corollary 2.25.21** Let $I_d : S^n \to S^n$ be the identity map and n be an even integer. Then $I_d$ is not homotopic to any continuous map $f : S^n \to S^n$ free from fixed points.

**Proof** Let $n$ be an even integer and $f : S^n \to S^n$ be a continuous map having no fixed point. Then by Theorem 2.25.10, the map

$$deg\, f = (-1)^{n+1} = -1.$$

Since $deg\, 1_d = 1$ and $deg\, f = -1$, the maps $f$ and $1_d$ cannot be homotopic by Hopf classification theorem 2.25.20.                    □

**Remark 2.25.22** **Hopf classification theorem** 2.25.20 completely classifies continuous spherical maps $f : S^n \to S^n$ by degree function $deg : f \mapsto deg\, f$. The same classification is also given in Chapter 3 by degree function through **homology theory.**

### 2.25.2  Brouwer-Poincaré Theorem

Brouwer- Poincaré Theorem 2.25.23 asserts that there is a continuous nonvanishing vector field $f : S^n \to S^n$ ($n \geq 1$), iff $n$ is odd. On the other hand, Corollary 2.25.24 shows that for all even integers $n \geq 1$, there is no vector field $f : S^n \to S^n$.

**Theorem 2.25.23** (Brouwer-Poincaré)  The $n$-sphere $S^n$ admits a continuous non-vanishing vector field iff $n$ is odd.

**Proof**  Suppose the integer $n \geq 1$ is odd. Then it follows by Proposition 2.18.10 that there is a nonvanishing vector field $f : S^n \to S^n$. For the converse, let $v$ be a nonvanishing vector field on $S^n$. Define

$$H : S^n \times I \to S^n, \ (x, t) \mapsto x \cos(\pi t) + v(x) \sin(\pi t),$$

then $H$ is a homotopy between the identity map on $S^n$ and the antipodal map $A : S^n \to S^n$. Hence the antipodal map $A$ has degree 1. But $deg\, A = (-1)^{n+1}$ by Proposition 2.25.9. It proves that $n$ is odd.  □

**Corollary 2.25.24**  *The n-sphere $S^n$ admits no continuous nonvanishing vector field if n is even.*

**Proof**  It follows by method of contradiction. If $v$ is a nonvanishing vector field, then the map

$$H : S^{2n} \times I \to S^{2n}, \ (x, t) \mapsto x \cos \pi t + \frac{v(x)}{||v(x)||} \sin \pi t$$

defines a homotopy between the identity map and the antipodal map. This asserts a contradiction that the degree of the identity map on $S^{2n}$ is $(-1)^{2n+1} = -1$. This contradiction proves the corollary.  □

**Remark 2.25.25**  If $n$ is odd, the difficult problem of determining **the maximum number of linearly independent nowhere vanishing vector fields on $S^n$ was solved by J.F Adams in 1962 by using $K$-theory** [Adams, 1962].

### 2.25.3  Separation of the Euclidean Plane and Jordan Curve Theorem

This subsection studies separation of Euclidean plane $\mathbf{R}^2$ by proving some theorems including Jordan Curve Theorem. Many questions naturally arise while studying the topology of $\mathbf{R}^2$ as a continuation of study of analysis. Jordan curve theorem is one them. An alternative proof of Theorem 2.25.31 with a homological proof of the theorem is given in Chapter 6.

**Fig. 2.21**  Jordan curve
theorem

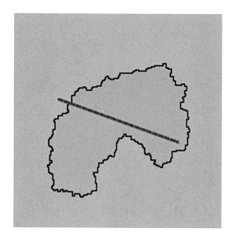

Recall that a surface $S$ is said to be **simply connected,** if every closed curve on $S$ can be continuously deformed into a point without leaving the surface $S$.

***Example 2.25.26***  (i)  The sphere $S^2$ is simply connected.
(ii)  The torus $T$ is not simply connected.

**Definition 2.25.27**  A subset $X \subset \mathbf{R}^2$ is said to **separate the Euclidean space $\mathbf{R}^2$** if $\mathbf{R}^2 - X$ has more than one component.

**Definition 2.25.28**  A subspace $J \subset \mathbf{R}^2$ homeomorphic to the circle $S^1$ is said to be a **Jordan curve or a simply closed curve** in $\mathbf{R}^2$. It is geometrically represented in Fig. 2.21.

**Theorem 2.25.29**  (Jordan Curve Theorem) If $J$ is a Jordan curve in $\mathbf{R}^2$, then $\mathbf{R}^2 - J$ has exactly two components, one is bounded and the other one is unbounded. The curve $J$ is their common boundary.

***Remark 2.25.30***  **Jordan Curve Theorem**  2.25.29 appears to be trivial geometrically but it is difficult to prove. It was a conjecture posed by Jordan in 1892 and its first correct proof was given by Oswald Veblen (1880–1960) in 1905. Its general statement known as known as Jordan–Brouwer theorem (see Chap. 6) was later proved by Brouwer. Original form of Jordan curve theorem 2.25.31 is proved first, from which its another form is given in Corollary 2.25.32. This proof is based on homotopy theory. The proof of Jordan curve theorem by homology theory is available in Chapter 6.

**Theorem 2.25.31**  (Jordan Curve Theorem) Let $J$ be a subspace of the Euclidean plane $\mathbf{R}^2$, which is homeomorphic to the circle $S^1$. Then $J$ separates the Euclidean plane $\mathbf{R}^2$.

**Proof** Let $p : S^2 - \{N\} \to \mathbf{R}^2$ be the stereographic projection with $f : \mathbf{R}^2 \to S^2 - \{N\}$ its inverse homeomorphism, where $N = (0, 0, 1)$ is the north pole of $S^2$. Given a point $q \in f(J)$, let $g : \mathbf{R}^2 \to S^2 - \{q\}$ be a homeomorphism. Let $X \subset \mathbf{R}^2$ be defined by

$$X = g^{-1}(f(J) - \{q\}).$$

Then $X$ is homeomorphic to the real line $\mathbf{R}$. So, $X$ may be considered as a line extended up to infinity on both sides in $\mathbf{R}^2$. Hence the spaces $\mathbf{R}^2 - J, S^2 - f(J)$ and $\mathbf{R}^2 - X$ have the same number of components. To prove the theorem, it is sufficient to prove that $\mathbf{R}^2 - X$ is not connected. If possible, let $\mathbf{R}^2 - X$ be connected. By construction of $X$, it follows that $\mathbf{R}^2 - X$ is path connected, since every connected open subset is path connected. Let $\mathbf{R}_+^3$ and $\mathbf{R}_-^3$ be the upper and lower open half-spaces of $\mathbf{R}^3$ given by $z > 0$ and $z < 0$. Define the sets

$$U = \mathbf{R}_+^3 \cup \{(x, y, z) : (x, y) \in \mathbf{R}^2 - X, -1 < z \le 0\}$$

and

$$V = \mathbf{R}_-^3 \cup \{(x, y, z) : (x, y) \in \mathbf{R}^2 - X, 0 \le z < 1\}.$$

Hence it follows that $U \cup V = \mathbf{R}^3 - X$, where $U \cap V$ is homeomorphic to the product space $(\mathbf{R}^2 - X) \times (-1, 1)$, which is path connected. Again since $U$ and $V$ are simply connected, because, any loop on them can be moved vertically until it stays in either $\mathbf{R}_+^3$ or $\mathbf{R}_-^3$ followed by deforming it to a point and hence the space $\mathbf{R}^3 - X$ is simply connected by using Exercise 60 of Section 2.28 and hence its fundamental group $\pi_1(\mathbf{R}^3 - X) \cong \{0\}$. Moreover, there is a homeomorphism $h : \mathbf{R}^3 \to \mathbf{R}^3$ with the image $h(X)$ of $X$ is the $z$-axis by using Exercise 63 of Section 2.28. Then it follows that the spaces $\mathbf{R}^3 - X$ and $\mathbf{R}^3 - (z \ axis)$ are homeomorphic. But the space $\mathbf{R}^3 - (z \ axis)$ is homotopy equivalent to space $\mathbf{R}^2 - \{0\}$. This asserts that the fundamental groups $\pi_1(\mathbf{R}^3 - X)$ and $\pi_1(\mathbf{R}^2 - \{0\})$ are isomorphic. But $\pi_1(\mathbf{R}^2 - \{0\}) \cong \pi_1(S^1) \cong \mathbf{Z}$. On the other hand, $\pi_1(\mathbf{R}^3 - X) \cong \{0\}$. This contradiction implies that the space $\mathbf{R}^2 - X$ is not connected. This proves that the space $J$ separates $\mathbf{R}^2$.  □

To prove Jordan curve theorem it is sufficient to prove its equivalent statement in Corollary 2.25.32.

**Corollary 2.25.32** *(An Alternative Form of Jordan Curve Theorem) If $J$ is Jordan curve in $\mathbf{R}^2$, then $\mathbf{R}^2 - J$ has exactly two components, one bounded and the other one is unbounded. The curve $J$ is their common boundary as shown in Fig. 2.21.*

**Proof** The corollary follows from Theorem 2.25.31.  □

**Remark 2.25.33** **The Jordan curve theorem** 2.25.29 **also asserts** that every Jordan curve divides the surface of the sphere into two disjoint regions, for which the given Jordan curve is a common boundary. Because, $S^2 - J$ is locally path connected, its path components and connected components coincide.

### 2.25.4　Separation of the Euclidean Plane Problem: Tietze Extension Theorem

This subsection proves that analogous of Jordan Curve theorem 2.25.31 fails if $S^1$ is replaced by $\mathbf{I} = [0, 1]$ in Definition 2.25.28. The proof is based on Tietze extension theorem, which asserts that if $A$ is a closed subspace of a normal space $X$, then every continuous map $g : A \to \mathbf{R}$ has a continuous extension over $X$ (see Chap. 6, **Basic Topology, Volume 1** of the present series of books).

**Theorem 2.25.34** Let $X \subset \mathbf{R}^2$ be a subspace which is homeomorphic to the closed interval $[0, 1]$. Then $X$ does not separate the Euclidean plane $\mathbf{R}^2$.

**Proof** If possible, let $\mathbf{R}^2 - X$ have more than one component. By hypothesis, $X$ is homeomorphic to the closed set $\mathbf{I} = [0, 1]$. Then $X$ is compact and hence it is bounded and $\mathbf{R}^2 - X$ has a unique unbounded component. Let $B$ be a bounded component of $\mathbf{R}^2 - X$. Construct a disk $\mathbf{D}^2$ with center at the origin having sufficiently large radius so that $X \cup B$ is in its interior. For $b \in B$, let $r : \mathbf{D}^2 - \{b\} \to S^1$ be the retraction along the straight lines joining the point $b$ to the points of the boundary $S^1$ of $\mathbf{D}^2$. Consider its restriction map

$$f = r|_{\mathbf{D}^2 - B} : \mathbf{D}^2 - B \to S^1.$$

By hypothesis, $X$ is homeomorphic to $[0, 1]$ hence it follows that the map $g = r|_X : X \to S^1$ has a lifting $\tilde{g} : X \to \mathbf{R}$ such $p \circ \tilde{g} = g$, where $p : \mathbf{R} \to S^1, t \mapsto e^{2\pi i t}$. Hence by **Tietze extension theorem** $\tilde{g}$ extends to a map $\tilde{h} : X \cup B \to \mathbf{R}$. If $h = p \circ \tilde{h} : X \cup B \to S^1$, then the map $f \cup h : \mathbf{D}^2 \to S^1$ obtained by gluing the maps $f$ and $h$ together is continuous by the pasting lemma. But $(f \cup h)(x) = f(x), \forall x \in S^1$. This implies that the map $f \cup h$ is a retraction, which contradicts the Corollary 2.19.6 saying that there exists no retraction of $\mathbf{D}^2$ onto its boundary $\partial \mathbf{D}^2 = S^1$. This contradiction proves that $X$ does not separate the Euclidean plane $\mathbf{R}^2$. □

### 2.25.5　Applications of the Euler Characteristic in the Theory of Convex Polyhedra

This subsection proves a classical theorem asserting that there are only five different types of platonic solids in Theorem 2.25.38. It is an interesting application of the Euler characteristic in the theory of convex polyhedra. Its proof is based on considering the surface of a convex polyhedron as glued together a finite number of convex polygons with respect to identity map on edges glued.

**Definition 2.25.35** For a given polyhedron $P$, if $n$ edges meet at each vertex and each face is a convex $m$-gon, then the polyhedron $P$ is said to be of **type** $[m, n]$. In particular, $P$ is said to regular if every $m$-gon is regular.

**Remark 2.25.36** If the type $[m, n]$ of a polygon $P$ is known, then the number $\mathbf{V}$ of the vertices, the numbers $\mathbf{E}$ of the edges and the number $\mathbf{F}$ of the faces of $P$ can be calculated.

**Definition 2.25.37** A **platonic solid** is a polyhedron such that its faces are congruent regular polygons and each vertex lies in the same number of edges. It is also called a **regular simple polyhedron.**

**Theorem 2.25.38** *There are only five platonic solids which are precisely of types:* $[3, 3], [4, 3], [3, 4], [5, 3],$ *and* $[3, 5]$.

**Proof** Let $P$ be a platonic solid with $\mathbf{V}$ number of vertices, $\mathbf{E}$ number of edges, and $\mathbf{F}$ number of faces. If $n$ is the number of edges meeting in each vertex and m is the number of edges in each face, then from the geometric point of view, it is assumed that $n, m, \mathbf{V} \geq \mathbf{3}$ and it follows that

$$n\mathbf{V} = \mathbf{2E} \text{ and } \mathbf{m\,F} = \mathbf{2E},$$

since $n$ edges meet in each vertex and each edge joins two vertices and two faces. This asserts by Euler formula $\mathbf{V} - \mathbf{E} + \mathbf{F} = \mathbf{2}$ for P such that

$$\frac{\mathbf{V}}{n^{-1}} = \frac{\mathbf{E}}{2^{-1}} = \frac{\mathbf{F}}{m^{-1}} = \frac{\mathbf{V} - \mathbf{E} + \mathbf{F}}{n^{-1} - 2^{-1} + m^{-1}} = \frac{2}{n^{-1} - 2^{-1} + m^{-1}} = \frac{4mn}{2m - mn + 2n}.$$

Then the values $\mathbf{V}, \mathbf{E}$ and $\mathbf{F}$ which are all positive integers, are determined. Hence it follows that for $m, n \in \mathbf{N}$, there are only five solutions of $[m, n]$ which are precisely, $[3, 3], [4, 3], [3, 4], [5, 3],$ *and* $[3, 5]$. They represent geometrically precisely the five regular polyhedra which are tetrahedron ($[3, 3]$), cube ($[4,3]$), octahedron ($[3, 4]$), dodecahedron ($[5, 3]$), and icosahedron ($[3, 5]$). $\qquad\square$

*Example 2.25.39* Possible values of $\mathbf{V}, \mathbf{E}, \mathbf{F}$ are given for

  (i)   **tetrahedron** ($[3, 3]$): $\mathbf{V} = \mathbf{4}, \mathbf{E} = \mathbf{6}, \mathbf{F} = \mathbf{4}$.
 (ii)   **cube** ($[4,3]$): $\mathbf{V} = \mathbf{8}, \mathbf{E} = \mathbf{12}, \mathbf{F} = \mathbf{6}$.
(iii)   **octahedron** ($[3, 4]$): $\mathbf{V} = \mathbf{6}, \mathbf{E} = \mathbf{12}, \mathbf{F} = \mathbf{8}$.
 (iv)   **dodecahedron** ($[5, 3]$): $\mathbf{V} = \mathbf{12}, \mathbf{E} = \mathbf{30}, \mathbf{F} = \mathbf{20}$.
  (v)   **icosahedron** ($[3, 5]$): $\mathbf{V} = \mathbf{20}, \mathbf{E} = \mathbf{30}, \mathbf{F} = \mathbf{12}$.

### 2.25.6   *Interior and Boundary Point Problem of a Surface*

This subsection applies homotopy and fundamental group to prove in Theorem 2.25.40 that a point cannot lie both in the interior and the boundary of a surface.

A **surface** is Hausdorff space in which every point has a nbd homeomorphic either to $\mathbf{R}^2$ or to the closed half space $\mathbf{R}_+^2 = \{(x, y) \in \mathbf{R}^2 : y \geq 0\}$. The interior points of S, denoted by *Int S*, are precisely, the points of S, each having a nbd homeomorphic to $\mathbf{R}^2$. On the other hand, the set of points x of S for each of which, there is a nbd $U_x$ and a homeomorphism $f : \mathbf{R}_+^@ \to U_x$ such that $f(x) = 0$, form the boundary of S, denoted by $\partial S$.

**Theorem 2.25.40** *Given a surface S, let Int S be its interior and $\partial S$ be its boundary. Then are disjoint.*

**Proof** Suppose there is a point $x \in Int\ S \cap \partial S$. Then there exist nbds $U_1$, $U_2$ of the point x in S and homeomorphisms

$$h_1 : \mathbf{R}_+^2 \to U_1 \quad \text{and} \quad h_2 : \mathbf{R}^2 \to U_2$$

with the property that $h_1(0) = x = h_2(0)$. Select a half disk $\mathbf{D}_+ \subset \mathbf{R}_+^2$ with center at origin and radius small enough such that $h_1(\mathbf{D}_+) \subset U_2$. Consider the map

$$h = h_2^{-1} \circ h_1 : \mathbf{D}_+ \to \mathbf{R}^2.$$

Since, the maps $h_1$, $h_2$ are homeomorphisms, the subset $h(\mathbf{D}_+) \subset \mathbf{R}^2$ is a nbd of the point 0 in $\mathbf{R}^2$. Take a disk $\mathbf{D}$ in $\mathbf{R}^2$ with center at the origin and radius sufficiently small so that $\mathbf{D} \subset h(\mathbf{D}_+)$. Let $\partial D$ denote the boundary circle of radius $\rho$ of $\mathbf{D}$ and

$$p : \mathbf{R}^2 - \{0\} \to \partial D, \ x \mapsto \rho(\frac{x}{||x||})$$

denote the radial projection. Then the restriction map

$$r|_{h(\mathbf{D}_+ - \{0\})} : h(\mathbf{D}_+ - \{0\}) \to \partial D$$

is a retraction. Hence this map induces an epimorphism

$$r_* : \pi_1(h(\mathbf{D}_+ - \{0\}) \to \pi_1(\partial(\mathbf{D})),$$

which is not possible since $\pi_1(h(\mathbf{D}_+ - \{0\})) = 0$ and $\pi_1(\partial(\mathbf{D})) = \mathbf{Z}$. This contradiction proves that the sets *Int S* and $\partial S$ are disjoint. $\square$

**Theorem 2.25.41** *Given two homeomorphic surfaces $S_1$ and $S_2$, every homeomorphism $f : S_1 \to S_2$*

(i) *sends the interior Int $S_1$ of $S_1$ to the interior Int $S_2$ of $S_2$; and*
(ii) *sends the boundary $\partial S_1$ of $S_1$ to the boundary $\partial S_2$ of $S_2$.*

**Proof** (i) Suppose $x \in Int\ S_1$. Then there exists a nbd $U_x$ of x in the surface $S_1$ and a homeomorphism $h : \mathbf{R}^2 \to U_x$. As by hypothesis, f is a homeomorphism, $f(U_x)$

is a nbd of $f(x)$ in the surface $S_2$ and $f \circ h : \mathbf{R}^2 \to f(U_x)$ is a homeomorphism. This shows that the point $f(x)$ is in *Int* $S_2$ and hence the first part is proved.

(iii) Again since, $f^{-1} : S_2 \to S_1$ is also a homeomorphism, it follows by the first part that $f^{-1}$ sends the interior *Int* $S_2$ of $S_2$ to the interior *Int* $S_1$ of $S_1$. Thus $f$ maps *Int* $S_1$ of $S_1$ onto *Int* $S_2$ of $S_2$. Then the second part follows under the given homeomorphism $f$, since *Int* $S$ and $\partial S$ for any surface $S$ are disjoint by Theorem 2.25.40. $\qquad\square$

**Corollary 2.25.42** *Homeomorphic surfaces contain homeomorphic boundaries.*

**Proof** It follows from Theorem 2.25.41. $\qquad\square$

### 2.25.7  Homotopy Property of Infinite Dimensional Sphere $S^\infty$

This subsection studies homotopy property of the infinite dimensional sphere $S^\infty$ and proves that this space is contractible.

For construction of $S^\infty$, we need the concepts of infinite dimensional Euclidean space $\mathbf{R}^\infty$ and infinite dimensional unitary space $\mathbf{C}^\infty$.

**Definition 2.25.43** (i) **(Infinite dimensional Euclidean space $\mathbf{R}^\infty$)** It consists of all sequences $x = (x_1, x_2, \ldots, x_n, \ldots)$ of real numbers such that $\sum_{1}^{\infty} |x_n|^2$ converges, i.e.,

$$\mathbf{R}^\infty = \{x = (x_1, x_2, \ldots, x_n, \ldots) : x_n \in \mathbf{R} \text{ and } \sum_{1}^{\infty} |x_n|^2 < \infty\}.$$

$\mathbf{R}^\infty$ is a vector space over $\mathbf{R}$ under pointwise addition and scalar multiplication. $\mathbf{R}^\infty$ endowed with a norm function

$$\|x\| = (\sum_{1}^{\infty} |x_n|^2)^{1/2},$$

is a real Banach space. The space $\mathbf{R}^\infty$ is called infinite dimensional Euclidean space.

(ii) **(Infinite dimensional unitary space $\mathbf{C}^\infty$)** It is defined in a way analogous to infinite dimensional Euclidean space $\mathbf{R}^\infty$. The space $\mathbf{C}^\infty$ is a complex Banach space and as a topological space $\mathbf{C}^n$ is homeomorphic to $\mathbf{R}^{2n}$ and $\mathbf{C}^\infty$ is homeomorphic to $\mathbf{R}^\infty$.

The concept of $S^\infty$ is formulated in Definition 2.25.44.

**Fig. 2.22** Commutative
diagram involving $S^{2n+1}$
and $\mathbf{C}^{n+1}$

**Definition 2.25.44** (*Infinite dimensional sphere* $S^\infty$) It is the subspace of $\mathbf{R}^\infty$ consisting of all real sequences $x = (x_1, x_2, x_3, \cdots)$ such that $x_1^2 + x_2^2 + x_3^2 + \cdots = 1$, i.e.,

$$S^\infty = \{x = (x_1, x_2, x_3, \cdots) \in \mathbf{R}^\infty : x_1^2 + x_2^2 + x_3^2 + \cdots = 1\}$$

and endowed with weak topology in the sense that a subset $A \subset S^\infty$ is closed iff $A \cap S^n$ is closed for each $n \geq 1$, where the chain of inclusions are represented in the commutative diagram as shown in Fig.2.22.
$S^\infty$ as the subspace of $\mathbf{C}^\infty$ consists of the sequences $z = (z_1, z_2, \dots)$ over $\mathbf{C}$ such that $|z_1|^2 + |z_2|^2 + \cdots = 1$.

Proposition 2.25.45 proves the contractibility of the topological space $S^\infty$, though $S^n$ is not contractible for any finite value of $n$.

**Proposition 2.25.45** *The infinite dimensional sphere* $S^\infty$ *is contractible.*

**Proof** To prove the proposition, define the map

$$H : S^\infty \times \mathbf{I} \to S^\infty, (x_1, x_2, x_3, \dots, t) \mapsto ((1-t)x_1, tx_1 + (1-t)x_2, tx_2 + (1-t)x_3, \dots)/N_t,$$

where   the   denominator   $N_t = [((1-t)x_1)^2 + (tx_1 + (1-t)x_2)^2 + (tx_2 + (1-t)x_3)^2 + \dots]^{1/2}$, is the norm of the nonzero vector of the numerator. Parameterize $H$ as

$$H_t(x_1, x_2, x_3, \dots) = H(x_1, x_2, x_3, \dots, t), \forall t \in \mathbf{I}.$$

Then $H_0(x_1, x_2, x_3, \dots) = (x_1, x_2, x_3, \dots)$, since $N_0 = 1$ and $H_1(x_1, x_2, x_3, \dots) = (0, x_1, x_2, x_3, \dots)$, since $N_1 = 1$. This asserts that $H_0$ is the identity map $1_d : S^\infty \to S^\infty$, the image of $H_1$ is given by $X = \{x \in S^\infty : x_1 = 0\}$ and $H : H_0 \simeq H_1$.
Define another homotopy

$$G : X \times \mathbf{I} \to S^\infty : G(x_1 = 0, x_2, x_3, \dots, t) \mapsto (t, (1-t)x_2, (1-t)x_3, \dots)/N_t',$$

where $N_t' = [t^2 + ((1-t)x_2)^2 + ((1-t)x_3)^2 + \cdots]^{1/2}$.
Let $i : X \hookrightarrow S^\infty$ be the inclusion map and $c : X \hookrightarrow S^\infty$ be a constant map. Then $G : i \simeq c$.
Define the map

$$G * H : X \times \mathbf{I} \to S^\infty, (x, t) \mapsto \begin{cases} H(x, 2t), 0 \leq t \leq 1/2 \\ G(x, 2t - 1), 1/2 \leq t \leq 1. \end{cases}$$

The map $G * H$ is clearly a contraction. Hence it is proved that the space $S^\infty$ is contractible. □

**Corollary 2.25.46** *The inclusion map* $i : S^{n-1} \hookrightarrow S^n$ *is nullhomotopic in the sense that it is homotopic to a constant map.*

**Remark 2.25.47** The infinite dimensional sphere $S^\infty$ is contractible. On the other hand, the $n$-sphere $S^n$ is not contractible for any finite integer $n \geq 0$, because $\pi_n(S^n) \cong \mathbf{Z} \neq \{0\}$ by Hopf Theorem 2.21.5 (see also Remark 2.25.56).

### 2.25.8  Homotopy Property of Infinite Symmetric Product Space $SP^\infty X$

This subsection studies homotopy property of the infinite symmetric product space $SP^\infty X$ for any pointed topological space $X$ and proves that this space is contractible.

**Construction of** $SP^\infty X$: Given a pointed topological space $X$ with base point $x_0$, let $X^n = X \times \times \cdots \times X$ denote its $n$th Cartesian product for every $n \geq 1$. Let $S_n$ denote the symmetric groups on the set $\{1, 2, \ldots, n\}$ of $n$ elements. Consider a right action

$$\sigma : X^n \times S^n \to X^n, \ (x_1, x_2, \ldots, x_n) \cdot \sigma = (x_{\sigma(1)}, x_{\sigma(2)}, \ldots, x_{\sigma(n)}).$$

Then its orbit space denoted by $X^n$ mod $S_n$ is called the $n$th **symmetric product** of $X$ and the equivalence class of $(x_1, x_2, \ldots, x_n)$ is denoted by $[x_1, x_2, \ldots, x_n]$. Consider the inclusion maps

$$i : SP^n X \hookrightarrow SP^{n+1} X, \ [x_1, x_2, \ldots, x_n] \mapsto [x_0, x_1, \ldots, x_n], \ \forall n \geq 1$$

and construct the union $SP^\infty X = \bigcup_n SP^n X$ endowed with the weak topology in the sense that a subset $A \subset SP^\infty X$ is closed iff $A \cap SP^n X$ is closed for each $n \geq 1$, where the chain of inclusions are represented in the diagram as shown in Fig.2.23. The topological space $SP^\infty X$ is called the **infinite symmetric product space** of $X$. The elements of $SP^n X$ are thus unordered $n$-tuples $[x_1, x_2, \ldots, x_n]$, $\forall n \geq 1$ and $SP^\infty X$ is a pointed topological space with the base point $0 = [x_0]$. Clearly, there is a natural inclusion $i : X \hookrightarrow SP^\infty X$, where $X = SP^1 X$ is taken. Every continuous map $f : X \to Y$ in $\mathcal{T}op_*$ induces a sequence of continuous maps

$$f^n : X^n \to Y^n, \ (x_1, x_2, \ldots, x_n) \mapsto (f(x_1), f(x_2), \ldots, f(x_n)).$$

Since these maps are compatible with the action, they induce another sequence of continuous maps $SP^n(f) : SP^n X \to SP^n Y$ and hence they induce a continuous map

$$SP^\infty(f) = f_* : SP^\infty X \to SP^\infty Y$$

for every continuous map $f : X \to Y$ in $\mathcal{T}op_*$.

**Fig. 2.23** Commutative
diagram involving infinite
symmetric product spaces
and their maps

$$
\begin{array}{ccc}
\cdots \longrightarrow SP^n X \longrightarrow SP^{n+1} X \cdots \longrightarrow \\
f_*^{(n)} \downarrow \qquad\qquad f_*^{(n+1)} \downarrow \\
\cdots \longrightarrow SP^n Y \longrightarrow SP^{n+1} Y \cdots \longrightarrow
\end{array}
$$

**Remark 2.25.48** Every base point preserving continuous map $f : X \to Y$ induces a map $f_*^{(n)} : SP^n X \to SP^n Y$, compatible with the action of the group $S_n$ making the diagram in Fig. 2.23 commutative.
This induces a continuous map $f_* : SP^\infty X \to SP^\infty Y$ in $Top_*$. Moreover,

(i)   Proposition 2.25.49 proves the functorial properties of $SP^\infty$ in category $Top_*$.
(ii)  Proposition 2.25.50 proves that $SP^\infty : Top_* \to Top_*$ is a covariant functor.
(iii) Theorem 2.25.54 proves that $SP^\infty : Htp_* \to Htp_*$ is a covariant functor.

**Proposition 2.25.49** *Let $Top_*$ be the category of pointed topological spaces.*

(i) *If $f = 1_X : X \to X$ is the identity map on $X$, then its induced map*

$f_* = 1_{SP^\infty X} : SP^\infty X \to SP^\infty X$ *is the identity map on $SP^\infty X$.*
(ii) *If $f : X \to Y$ and $g : Y \to Z$ are in $Top_*$, then $(g \circ f)_* = g_* \circ f_* : SP^\infty X \to SP^\infty Z$.*

**Proof** It follows that from the definition of the induced map $f_*$. □

**Proposition 2.25.50** *$SP^\infty : Top_* \to Top_*$ is a covariant functor.*

**Proof** The objective function assigns to every topological $X$ in $Top_*$, the topological space $SP^\infty X$ in $Top_*$ and the morphism function assigns to every continuous map $f : X \to Y$ in $Top_*$, the continuous map $SP^\infty(f) = f_* : SP^\infty X \to SP^\infty Y$ in $Top_*$. By using Proposition 2.25.49, it follows that $SP^\infty : Top_* \to Top_*$ is a covariant functor. □

Theorem 2.25.51 proves that for any map $f : X \to Y$ in $Top_*$, the map

$$SP^\infty(f) : SP^\infty X \to SP^\infty Y$$

is homotopy preserving.

**Theorem 2.25.51** *If $f \simeq g : X \to Y$ in $Top_*$, then $SP^\infty(f) \simeq SP^\infty(g)$.*

**Proof** By hypothesis, $f \simeq g : X \to Y$ in $Top_*$. Then there exists a homotopy $H : X \times I \to Y$ such that

$$H(x, 0) = f(x) \text{ and } H(x, 1) = g(x), \ \forall x \in X.$$

Define a sequence of maps

$$H^{(n)} : X^n \times I \to Y^n, (x_1, x_2, \ldots, x_n, t) \mapsto (H(x_1, t), H(x_2, t), \ldots, H(x_n, t)), \ \forall n \geq 1.$$

Then every map $H^{(n)}$ is continuous, since its projection onto each coordinate function is continuous. Again, since the symmetric group $S_n$ acts on $X^n \times \mathbf{I}$ by permuting the coordinate of $X^n$ and fixing $\mathbf{I}$, it follows that every continuous map $H^{(n)}$ respects this action. Hence $H^{(n)}$ induces maps $SP^n(H) : SP^n X \to SP^n Y$. Now passing to the limit, it induces a continuous map

$$SP^\infty(H) : SP^\infty X \times \mathbf{I} \to SP^\infty Y.$$

Parametrize $H$ as $H_t : X \to Y, x \mapsto H(x, t), \forall t \in \mathbf{I}$ and consider the map

$$SP^\infty(H) : SP^\infty X \times \mathbf{I} \to SP^\infty Y, (x, t) \mapsto SP^\infty(H_t)(x).$$

This implies that
$$SP^\infty(H) : SP^\infty(f) \simeq SP^\infty(g). \qquad \square$$

**Corollary 2.25.52** *Let $X$ and $Y$ be two topological spaces in $\mathcal{T}op_*$ such that $X \simeq Y$. Then the corresponding infinite symmetric spaces satisfy the same relation in the sense that $SP^\infty X \simeq SP^\infty Y$.*

**Proof** By hypothesis, $X \simeq Y$. Then the exists a homotopy equivalence $f : X \to Y$ with homotopy inverse $g : Y \to X$. This implies that $SP^\infty(g)$ is a homotopy inverse of $SP^\infty(f)$. This proves that $SP^\infty X \simeq SP^\infty Y$. $\qquad \square$

**Corollary 2.25.53** *For every contractible space $X \in \mathcal{T}op_*$, the space $SP^\infty X$ is also contractible.*

**Proof** By hypothesis, the space $X$ is contractible. Then $X \simeq \{*\}$. This asserts that $SP^\infty X \simeq SP^\infty \{*\} = \{*\}$. It proves that the space $SP^\infty X$ is also contractible. $\qquad \square$

Proposition 2.25.50 says that $SP^\infty : \mathcal{T}op_* \to \mathcal{T}op_*$ is a covariant functor on the category $\mathcal{T}op_*$. It is also a covariant functor on the category $\mathcal{H}tp_*$ as proved in Theorem 2.25.54.

**Theorem 2.25.54** $SP^\infty : \mathcal{H}tp_* \to \mathcal{H}tp_*$ *is a covariant functor.*

**Proof** Using Theorem 2.25.51 and Proposition 2.25.50, the theorem is readily proved. $\qquad \square$

## 2.25.9  Brouwer Fixed-Point Theorem for Dimension n

L. E. J. Brouwer (1881–1967) took the first step toward connecting homotopy and homology by demonstrating in 1912 that two continuous mappings of a two-dimensional sphere into itself can be continuously deformed into each other if and only if they have the same degree (that is, if they are equivalent from the point of view of homology theory). The papers of H. Poincaré (1854–1912) during 1895-1904 can be considered as blue prints for theorems to come. The results of Brouwer during 1910–1912 may be considered the first one of the proofs in algebraic topology. He proved the celebrated theorem 'Brouwer fixed-point theorem 2.25.58' by using the concept of degree of a continuous spherical map defined by Brouwer himself. This section proves Brouwer fixed-point theorem for an arbitrary dimension $n \geq 0$. In particular, for $n = 2$, i.e., Brouwer fixed-point theorem for dimension 2 is also proved in Theorem 2.23.6 and that of for dimension $n = 1$ is also proved in Theorem 2.26.6.

**Proposition 2.25.55** *The n-sphere $S^n$ is not contractible for every finite $n \geq 0$.*

**Proof** If possible, suppose $S^n$ is contractible. Then the identity map $1_{S^n} : S^n \to S^n$ is homotopic to some constant map. But the identity map $1_{S^n} : S^n \to S^n$ has degree 1 for $n \geq 1$, and any constant map $f : S^n \to S^n$ has degree 0. This contradicts Hopf classification theorem 2.25.20 and hence $S^n$ is not contractible for $n \geq 1$. For $n = 0$, $S^0 = \{-1, 1\}$ is a discrete space. Hence $S^0$ cannot be contractible.                    □

**Remark 2.25.56** The finite dimensional $n$-sphere $S^n$ is not contractible. On the other hand, the infinite dimensional sphere $S^\infty$ is contractible (see also Remark 2.25.47 )

**Theorem 2.25.57** (Brouwer no retraction theorem) There does not exist any continuous onto map $f : \mathbf{D}^{n+1} \to S^n$ which leaves each point of $S^n$ fixed for every finite $n \geq 0$.

**Proof** Suppose for every integer $n \geq 0$, there exists a continuous map $f : \mathbf{D}^{n+1} \to S^n$ such that $f(x) = x$ for all $x \in S^n$. Define a homotopy

$$H : S^n \times \mathbf{I} \to S^n, (x, t) \mapsto f((1 - t)x).$$

This shows that $S^n$ is contractible. But this contradicts Proposition 2.25.55.                    □

Theorem 2.25.58 proves Brouwer fixed-point theorem for dimension $n \geq 0$. Separate proofs of Brouwer fixed-point theorem for dimension $n = 1$ and $n = 2$ are given in Theorem 2.26.6 and Theorem 2.23.6 respectively. This theorem provides a sufficient condition such that every continuous map from a particular topological space to itself must have a fixed point. This result is one of the most important results having applications of topology in mathematics and other sciences.

**Theorem 2.25.58** (Brouwer fixed-point theorem for dimension $n$) Every continuous map $f : \mathbf{D}^{n+1} \to \mathbf{D}^{n+1}$ has a fixed point for every finite $n \geq 0$.

**Proof** If possible, $f$ has no fixed point. This implies that $f(x) \neq x$ for each $x \in \mathbf{D}^n$. For $n = 0$, as it an immediate contradiction, it is well assumed from now that $n \geq 1$. By assumption, for each $x \in \mathbf{D}^n$, the points $x$ and $f(x)$ are distinct. For any $x \in \mathbf{D}^n$ we now consider the half-line in the direction from $f(x)$ to $x$. Let $h(x)$ denote the point of intersection of this ray with $S^n$. Then the map $h : \mathbf{D}^{n+1} \to S^n$ is continuous. Moreover, $h(x) = x$ for every $x \in S^n$. This contradicts the Brouwer no retraction theorem 2.25.57. This proves that $f(x)$ has a fixed point.                    □

## 2.26   More Applications

This section conveys more applications of homotopy and fundamental groups.

**Theorem 2.26.1** *Let $f : S^2 \to S^2$ be a continuous map.*

(i) *If $f$ has no fixed point, then $f$ is homotopic to the*
    *antipodal map $g : S^2 \to S^2, x \mapsto -x$.*
(ii) *If $f$ is nullhomotopic, then it has a fixed point.*

**Proof** (i) Let $f : S^2 \to S^2$ have no fixed point. Consider the map

$$H : S^2 \times I \to S^2, : (x, t) \mapsto \frac{(1 - t)g(x) + tf(x)}{||(1 - t)g(x) + tf(x)||} \in S^2,$$

provided $(1 - t)g(x) + tf(x) \neq 0$. But $(1 - t)g(x) + tf(x)$ cannot be 0, otherwise we have a contradiction. Because if $(1 - t)g(x) + tf(x) = 0$, then $f(x) = -(1 - t)/t)g(x)$ implies that $||f(x)|| = 1 = || - (1 - t)/t)g(x)||$ and hence $(1-t)/t = 1$ would imply that $f(x) = -g(x) = x$. This shows that $f$ has a fixed point, which is a contradiction.

(ii) Let $f : S^2 \to S^2$ be nullhomotopic. Suppose the continuous map $f : S^2 \to S^2$ has no fixed point. Then by (i), $f$ is homotopic to the antipodal map $g : S^2 \to S^2, x \mapsto -x$ and hence *deg f* $=$ *deg g*. But *deg f* $= 0$, since $f$ is nullhomotopic and *deg g* $\neq 0$. Hence it produces a contradiction, which asserts that $f$ has a fixed point.                                                     □

**Proposition 2.26.2** *If $v : S^{2n+1} \to \mathbf{R}^{2n+2}$ is a nowhere vanishing tangent vector field on $S^{2n+1}$, then the map*

$$f : S^{2n+1} \to S^{2n+1}, \quad x \mapsto \frac{v(x)}{||v(x)||}$$

*is homotopic to the identity map.*

**Proof** Consider the map

$$H : S^{2n+1} \times I \to S^{2n+1}, \quad (x, t) \mapsto x \cos\frac{\pi t}{2} + \frac{v(x)}{||v(x)||} \sin\frac{\pi t}{2}.$$

Then $H$ is a homotopy between the identity map on $S^{2n+1}$ and $f$.                                                     □

Torus (2-torus) is an important geometrical object, which is homeomorphic to the product space $S^1 \times S^1$. It can be studied through homotopy theory.

**Theorem 2.26.3** *For the 2-torus $T^2 = S^1 \times S^1$, its fundamental group $\pi_1(S^1 \times S^1) \cong \mathbf{Z} \oplus \mathbf{Z}$.*

***Proof*** Using the Theorem 2.14.1, it follows that the fundamental groups $\pi_1(S^1, 1) \times \pi_1(S^1, 1)$ and $\pi_1(S^1 \times S^1, (1, 1)))$ are isomorphic. Again $\pi_1(S^1, 1) \cong \mathbf{Z}$ by Theorem 2.19.1. Since both the spaces $S^1$ and $S^1 \times S^1$ are path connected, hence it follows that $\pi_1(S^1 \times S^1) \cong \mathbf{Z} \times \mathbf{Z} \cong \mathbf{Z} \oplus \mathbf{Z}$.                                    □

**Corollary 2.26.4**  *For the n-torus* $T^n = \overbrace{S^1 \times S^1 \times \cdots \times S^1}^{n}$ *(n components),*

$$\pi_1(T_n) \cong \overbrace{\mathbf{Z} \oplus \mathbf{Z} \oplus \cdots \oplus \mathbf{Z}}^{n}.$$

**Proposition 2.26.5**  *Let* $X = [a, b]$ *be a subspace of the real line space* $\mathbf{R}$ *and* $A = \{a, b\}$ *be the two-point subspace of* $X$. *Then there is no retraction* $r : X \to A$.

***Proof*** If possible, there is a retraction $r : X \to A$. Then its restricted map $r|_A = 1_A$ (identity map on A) and hence $r \circ i = 1_A$ asserts that the map $r$ is surjective. $X = [a, b]$ is a connected space, but $A$ is not connected, because, the open sets $\{a\}$ and $\{b\}$ of the discrete space $A$ constitute a nontrivial separation of the space $A$. This gives a contradiction. This proves that there is no retraction $r : X \to A$.                □

**Theorem 2.26.6**  (Brouwer fixed-point theorem of dimension 1) Let $X = [a, b]$ be a subspace of the real line space $\mathbf{R}$. Then every continuous map $f : X \to X$ has a fixed point.

***Proof*** Without loss of generality, assume that $X = [-1, 1]$. If possible, $f : X \to X$ has no fixed point. Then for every point $x \in X$, $f(x) \neq x$. Define a map

$$r : X \to X, \ x \mapsto \frac{x - f(x)}{|x - f(x)|}.$$

This map $r$ is well defined and continuous, because, by assumption, for every point $x \in X$, $f(x) \neq x$ and $f$ is continuous. If $A = \{-1, 1\}$, then $r : X \to A$ is a retraction. This contradicts Proposition 2.26.5.                    □

**Theorem 2.26.7**  *For* $n > 1$, *the n- sphere* $S^n$ *is simply connected.*

***Proof*** If $N = (0, 0, \ldots, 0, 1) \in \mathbf{R}^{n+1}$ is the north pole of $S^n$ and $S = (0, 0, \ldots, 0, -1) \in \mathbf{R}^{n+1}$ is the south pole of $S^n$, then $U_1 = S^n - \{N\}$ and $U_2 = S^n - \{S\}$ are both open sets homeomorphic to $\mathbf{R}^n$, which is a simply connected space. Consequently, $U_1$ and $U_2$ are both simply connected. Again, since $U_1 \cap U_2 = S^n - \{N\} - \{S\} \approx \mathbf{R}^n - \{0\}$ and the space $\mathbf{R}^n - \{0\}$ is path connected, by using Exercise 25 of Section 2.28. it follows that for $n > 1$, the $n$-sphere $S^n$ is simply connected.                □

**Corollary 2.26.8**  *The Euclidean plane* $\mathbf{R}^2$ *cannot be homeomorphic to the Euclidean space* $\mathbf{R}^n$ *for* $n > 2$.

**Proof** Consider the punctures $n$-dimensional Euclidean space $\mathbf{R}^n - \{0\}$ for $n > 2$ and the punctured plane $\mathbf{R}^2 - \{0\}$. The former space $\mathbf{R}^n - \{0\}$ is homotopy equivalent to $S^{n-1}$; on the other hand $\mathbf{R}^2 - \{0\}$ is homotopy equivalent to $S^1$. Since $S^{n-1}$ is simply connected for $n > 2$ by Theorem 2.26.7 but $S^1$ is not so by Corollary 2.19.12, hence the corollary follows.                                                                              □

## 2.27 Invariance of Dimensions of Spheres and Euclidean Spaces

This section solves a **homeomorphism problem and a homotopy equivalence problems on spheres and Euclidean spaces** of different dimensions by proving the topological and homotopy invariance of dimensions of spheres and Euclidean spaces using homotopy theory. It is proved by using homotopy theory in this section that the dimension of the spheres and Euclidean spaces are both topological invariants. The same properties by another method using the homology theory is also proved in Chapter 5.

**Proposition 2.27.1** *Let $m$ and $n$ be two distinct nonnegative integers. Then the spheres $S^m$ and $S^n$ cannot be homotopically equivalent.*

**Proof** Without loss of generality, assume that $0 \leq m < n$. Then it follows from Theorem 2.21.4 and Theorem 2.21.5 that

$$\pi_n(S^n) \cong \mathbf{Z}, \ and \ \pi_n(S^m) = \{0\},$$

(though the usual convention in algebraic topology to use 0 in place of the trivial group $\{0\}$). This asserts that the spheres $S^m$ and $S^n$ cannot be homotopically equivalent.                                                                              □

**Corollary 2.27.2** *Let $m$ and $n$ be two distinct nonnegative integers. Then the spheres $S^m$ and $S^n$ cannot be homeomorphic.*

**Proof** It follows from Proposition 2.27.1.                                                                              □

**Corollary 2.27.3** *The dimension of the sphere is a topological invariant.*

**Proof** It follows from Corollary 2.27.2.                                                                              □

**Remark 2.27.4** (i) Proposition 2.27.1 solves homotopy equivalence problem on spheres of different dimensions.

(ii) Corollary 2.27.2 solves the homeomorphism problem on spheres of different dimensions.

**Proposition 2.27.5** *Let $n$ be a nonnegative integer and $0 = (0, 0, \ldots, 0)$ be the origin of the Euclidean $(n + 1)$-space $\mathbf{R}^{n+1}$. Then the inclusion map*

$$i : S^n \hookrightarrow \mathbf{R}^{n+1} - \{0\}$$

*is a homotopy equivalence.*

**Proof** Let $X = \mathbf{R}^{n+1} - \{0\}$ and $i : S^n \hookrightarrow X$ be the inclusion map. Define maps

$$r : X \rightarrow S^n, \ x \mapsto \frac{x}{||x||}$$

and

$$H : X \times \mathbf{I} \rightarrow X, \ (x, t) \mapsto (1 - t)x + \frac{tx}{||x||}.$$

Then $H$ and $r$ are continuous maps such that

$$H(x, 0) = x \text{ and } H(x, 1) = \frac{x}{||x||} = (i \circ r)(x), \ \forall x \in X.$$

This shows that $H : 1_X \simeq i \circ r$ and hence $i \circ r \simeq 1_X$. Moreover, $r \circ i = 1_{S^n}$. This asserts that $i$ is a homotopy equivalence and hence $S^n \simeq \mathbf{R}^{n+1} - \{0\}$.  $\square$

**Corollary 2.27.6** *Let $X = \mathbf{R}^{n+1} - \{0\}$ be the punctured Euclidean $(n + 1)$-space and $S^n$ be the n-sphere in $\mathbf{R}^{n+1}$. Then $S^n$ is a deformation retract of $X$.*

**Proof** It follows from Proposition 2.27.5 that $S^n$ is a deformation retract of $X$ with $r$ as a deformation retraction.  $\square$

**Theorem 2.27.7** *Let m and n be two distinct positive integers. Then the Euclidean m-space $\mathbf{R}^m$ and the Euclidean n-space $\mathbf{R}^n$ cannot be homotopically equivalent.*

**Proof** Suppose that $\mathbf{R}^m$ and $\mathbf{R}^n$ are homeomorphic. Then there exists a homeomorphism

$$f : \mathbf{R}^m \rightarrow \mathbf{R}^n.$$

If $f(0) = y \in \mathbf{R}^n$, and if

$$t : \mathbf{R}^n \rightarrow \mathbf{R}^n, \ x \mapsto x - y$$

is the translation of $\mathbf{R}^n$, then the map

$$h = t \circ f : \mathbf{R}^m \rightarrow \mathbf{R}^n$$

is a homeomorphism such that $h$ sends the origin $0 \in \mathbf{R}^m$ to the origin $0 \in \mathbf{R}^n$. This determines a homeomorphism

$$h' : h|_{\mathbf{R}^m - \{0\}} : \mathbf{R}^m - \{0\} \rightarrow \mathbf{R}^n - \{0\}.$$

Since $S^{m-1}$ is homotopy equivalent to $\mathbf{R}^m - \{0\}$ and $S^{n-1}$ is homotopy equivalent to $\mathbf{R}^n - \{0\}$ by Proposition 2.27.5, it follows that $S^{m-1}$ and $S^{n-1}$ are homotopically equivalent. By hypothesis, $m$ and $n$ be two distinct positive integers and hence $m - 1$ and $n - 1$ are two distinct nonnegative integers such that $S^{m-1}$ and $S^{n-1}$ are homotopically equivalent. But it contradicts the Proposition 2.27.2. □

**Corollary 2.27.8** *Let $m$ and $n$ be two distinct positive integers. Then the Euclidean m-space $\mathbf{R}^m$ and the Euclidean n-space $\mathbf{R}^n$ cannot be homeomorphic.*

**Proof** It follows from Theorem 2.27.7.                                         □

**Corollary 2.27.9** *The dimension of the Euclidean space is a topological invariant.*

**Proof** It follows from Corollary 2.27.8.                                       □

**Remark 2.27.10** (i) Theorem 2.27.7 solves a homotopy equivalence problem on Euclidean spaces of different dimensions.
(ii) Corollary 2.27.8 solves a homeomorphism problem on Euclidean spaces of different dimensions.

## 2.28 Exercises and Multiple Choice Exercises

As solving exercises plays an essential role of learning mathematics, various types of exercises and multiple choice exercises are given in this section. They form an integral part of the book series.

### 2.28.1 Exercises

1. Let $X, Y, Z$ be topological spaces and $f_1, g_1 \in C(X, Y)$ and $f_2, g_2 \in \mathbf{C}(Y, Z)$ be maps such that $f_1 \simeq g_1$ and $f_2 \simeq g_2$. Show that $f_2 \circ f_1 \simeq g_2 \circ g_1 : X \to Z$. Hence prove that the composites of homotopic maps are homotopic.
2. Given a matrix $M$ in the topological general linear group $GL(2, \mathbf{R})$, show that there is a path $\alpha : \mathbf{I} = [0, 1] \to GL(2, \mathbf{R})$ such that

    (i) if $\det(M) > 0$, then

    $$\alpha(0) = M$$

    and

    $$\alpha(1) = \begin{pmatrix} 1 & 0 \\ 0 & 1 \end{pmatrix}.$$

    (ii) if $\det(M) < 0$, then

    $$\alpha(0) = M$$

and

$$\alpha(1) = \begin{pmatrix} 1 & 0 \\ 0 & -1 \end{pmatrix}.$$

(iii) Hence, show that $GL(2, \mathbf{R})$ has two connected components.

[ Hint: Construct a family of paths, each taking the matrix $M$ to a simpler form such as $M = \begin{pmatrix} 1 & 1 \\ 0 & 1 \end{pmatrix}$, when $det\ M > 0$ and $M = \begin{pmatrix} 1 & 1 \\ 0 & -1 \end{pmatrix}$, when $det\ M < 0$. Then

for the case (i), the path

$$\alpha : \mathbf{I} = [0, 1] \to GL(2, \mathbf{R}), \ t \mapsto \begin{pmatrix} 1 & 1 - t \\ 0 & 1 \end{pmatrix},$$

and

for the case (ii), the path

$$\alpha : \mathbf{I} = [0, 1] \to GL(2, \mathbf{R}), \ t \mapsto \begin{pmatrix} 1 & 1 - t \\ 0 & -1 \end{pmatrix}$$

solve the problem.]

3. Show that

   (i) Every contractible space is path connected.
   (ii) If $X$ is a contractible space and $Y$ is a path-connected space, then the set of homotopy classes $[X, Y]$ of continuous maps from $X$ to $Y$ consists of single element.

4. Show that the pair $(X, A)$ of topological spaces with $A$ a closed subset of $X$ has the homotopy extension property iff the space $(X \times 0) \cup (A \times \mathbf{I})$ is a retract of $X \times \mathbf{I}$.

5. (**Homotopy invariance of mapping cones and mapping cylinders**) Let $f, g : X \to Y$ be two homotopic maps. Show that the corresponding

   (i) Mapping cones $C_f$ and $C_g$ are homotopy equivalent.
   (ii) Mapping cylinders $M_f$ and $M_g$ are homotopy equivalent.

6. Let $X$ be a topological space and $A$ be a closed subspace of $X$. If the pair $(X \times \mathbf{I}, X \times \{0\} \cup (A \times \mathbf{I}) \cup (X \times \{1\}))$ of spaces has the homotopy extension property (HEP) with respect to X, show that $A$ is a deformation retract of $X$ iff $A$ is a strong deformation retract of $X$.

7. Show that a topological space is contractible iff it is deformable into one of its points.

8. Let $X$ and $Y$ be two topological spaces having the same homotopy type. Show that $X$ is path connected iff $Y$ is also so.

9. Show that

   (i) The cylinder has the same homotopy type of the circle.
   (ii) The Möbius strip has the same homotopy type of the circle.

10. Given a path-connected space $X$ and two points $x_0, x_1 \in X$, show that the fundamental group $\pi_1(X, x_0)$ is abelian if and only if for each pair of paths $\alpha, \beta$ in $X$ from $x_0$ to $x_1$, the homomorphisms

$$\psi_\alpha : \pi_1(X, x_0) \to \pi_1(X, x_1), \ [f] \mapsto [\overline{\alpha} * f * \alpha].$$

   and

$$\psi_\beta : \pi_1(X, x_0) \to \pi_1(X, x_1), \ [f] \mapsto [\overline{\beta} * f * \beta].$$

   are the same, where $\overline{\alpha}$ and $\overline{\beta}$ are the inverse paths of $\alpha$ and $\beta$, respectively.

11. Given a loop $\alpha : (\mathbf{I}, \dot{\mathbf{I}}) \to (S^1, 1)$ on $S^1$, show that

$$deg(\alpha^n) = n \, deg \, \alpha.$$

12. Show that the map $f : S^1 \to S^1, z \mapsto z^n$ induces a monomorphism

$$f_* : \pi_1(S^1, 1) \to \pi_1(S^1, 1), \ [\alpha] \mapsto [f \circ \alpha].$$

13. Show that $S^1$ and $S^n$ $(n \geq 2)$ are not of the same homotopy type.

   [ Hint: $\pi_1(S^1) \cong \mathbf{Z}$. On the other hand, $\pi_1(S^n) \cong 0$, $\forall n \geq 2$. ]

14. Show that given any topological space $X$, the homotopy set $[X, \mathbf{I}]$ consists of a single element.

15. Let $X_\beta, Y_\beta$ $(\beta \in \mathbf{A})$ be pointed topological spaces and $f_\beta \simeq g_\beta : X_\beta \to Y_\beta : \beta \in \mathbf{A}$ be base point preserving homotopic maps. Show that $\Pi_{\beta \in \mathbf{A}} f_\beta \simeq \Pi_{\beta \in \mathbf{A}} g_\beta$ relative to the base point.

   [ Hint : Let $H_\beta : X_\beta \times \mathbf{I} \to Y_\beta$ be a homotopy between $f_\beta$ and $g_\beta$. Define

$$H : \Pi_{\beta \in \mathbf{A}} X_\beta \times \mathbf{I} \to \times Y_\beta, \ (x_\beta, t) \mapsto (H_\beta(x_\beta, t))$$

   is continuous and is the required homotopy relative to base point.]

16. Given a pointed topological space $A$, let $[A, X]$ denote the set of homotopy classes of base point preserving continuous maps from $A$ to $X$.
   Prove that every base point preserving continuous map $f : X \to Y$ induces a function

$$f_* : [A, X] \to [A, Y]$$

such that

(i) If $f \simeq g$, then $f_* = g_*$.

(ii) If $1_X : X \to X$ is the identity map, then $1_{X^*} : [A, X] \to [A, X]$ is the identity function.

(iii) If $g : Y \to Z$ is another continuous map, then $(g \circ f)_* = g_* \circ f_*$.

Hence show that if $X \simeq Y$, then there exists a bijective correspondence between the sets $[A, X]$ and $[A, Y]$.

What are the corresponding results for the sets $[X, A]$ for a fixed pointed topological space $A$ ?

[ Hint Use Theorems 2.3.2 and 2.3.6 and their corollaries. ]

17. Consider the topological group $S^1 = \{z \in \mathbf{C} : |z| = 1\}$ under usual multiplication of complex numbers. Show that

   (i) Given any pointed topological space $X$, the pointwise multiplication makes the set of base point preserving continuous maps $X \to S^1$ an abelian group.

   (ii) The set $[X, S^1]$ admits the structure of an abelian group.

   (iii) If $f : Y \to X$ is a base point preserving continuous, then its induced map

   $$f^* : [X, S^1] \to [Y, S^1]$$

   is a group homomorphism.

18. Given any topological space $X$, show that the cardinal number card $(\mathbf{I}, X])$ of the homotopy set $[I, X]$ is the number path-connected components of $X$.

19. Show that a path-connected space is simply connected iff for any two paths in the space with the same initial point and the same terminal point are homotopic.

20. m Let $X$ be a topological space and $U, V$ be two simply connected open set in $X$ such

   (i) $X = U \cup V$.

   (ii) $U \cap V$ is path connected.
      Show that $X$ is also simply connected.

21. Let $X$ be a topological group with identity element $x_0$ and $\Omega(X, x_0)$ be the set of all loops in $X$ based at $x_0$. For any two loops $f, g \in \Omega(X, x_0)$, define a loop $f \# g$ by

   $$(f \# g)(t) = f(t)g(t).$$

Show that

   (i) The set $\Omega(X, x_0)$ endowed with the binary operation '#' forms a group.

(ii) The operation $'\#'$ on $\Omega(X, x_0)$ induces a group $'\#'$ on $\pi_1(X, x_0)$.
(iii) The two group operations $\circ$ (usual) and $'\#'$ on $\pi_1(X, x_0)$ are identical.
(iv) The group $\pi_1(X, x_0)$ is abelian.
   [ Hint: Compute $(f \circ x_0)\#(x_0 \circ g)$ and use $(i)$, $(ii)$, $(iii)$. ]

22. Let $X$ be a connected graph which is not a tree. Show that $\pi_1(X)$ is a nontrivial free group.

23. **(Van Kampen theorem)** Let $(X, \tau)$ be a topological space and $U, V$ be two open sets in $(X, \tau)$ such that

(i) $X = U \cup V$.
(ii) $U, V$ *and* $U \cap V$ are nonempty path connected open subsets in $(X, \tau)$.

Then for any point $x_0 \in U \cap V$, show that

$$\pi_1(X, x_0) \cong \pi_1(U, x_0) *_{\pi_1(U \cap V, x_0)} \pi_1(V, x_0)$$

where the right hand side denotes the **amalgamated product** of the groups $\pi_1(U, x_0)$ and $\pi_1(V, x_0)$ over the group $\pi_1(U \cap V, x_0)$ through the group homomorphisms $i_*$ and $j_*$ defined by the inclusion maps

$$i : (U, x_0) \hookrightarrow (X, x_0) \ and$$

$$j : (V, x_0) \hookrightarrow (X, x_0).$$

Hence show that

(i) The fundamental group of the wedge of $n$ circles is a free group having $n$ generators.
(ii) If $U \cap V$ is simply connected, then $\pi_1(X, x_0)$ is the free product of the groups $\pi_1(U, x_0)$ and $\pi_1(V, x_0)$.
(iii) $\pi_1(S^n, s) = 0$, $\forall n > 1$ *and* $\forall s \in S^n$.

   [ Hint: See Chapter 6 for the proof of Van Kampen theorem in an alternative form. For the other part, consider the open sets $U$ and $V$ in $S^n$ defined by

$$U = \{(x_1, x_2, \ldots, x_n, x_{n+1}) \in S^n : x_{n+1} < 1\}$$

and
$$V = \{(x_1, x_2, \ldots, x_n, x_{n+1}) \in S^n : x_{n+1} > -1\},$$

which are homeomorphic to $\mathbf{R}^n$. This implies that $\pi_1(U) = 0 = \pi_1(V)$. Then use Van Kampen theorem. ]

24. A subspace $Y \subset \mathbf{R}^n$ is called **star-shaped** if there exists a point $y \in Y$ such that for any point $x \in Y$, the line segment $[y, x]$ joining the points $y$ and $x$ entirely

lies in $Y$. Show that any continuous map from an arbitrary topological space $X$ to a star-shaped subspace $Y$ of $\mathbf{R}^n$ is nullhomotopic.

25. If $\{U_a : a \in \mathbf{A}\}$ be an open covering of a space $X$, where each $U_a$ is simply connected, then prove that $X$ is itself simply connected if

   (i) $\cap U_a \neq \emptyset$.
   (ii) For $a \neq b \in \mathbf{A}$, $U_a \cap U_b$ is path connected.

26. Let $X$ be a connected space and $\tilde{f}, \tilde{g}$ be two liftings of $f : (X, x_0) \to (S^1, 1)$. If $\tilde{f}, \tilde{g} : (X, x_0) \to (\mathbf{R}, z_0)$ be two lifting of $f$ such that $\tilde{f}(x_0) = \tilde{g}(x_0)$, then show that $\tilde{f} = \tilde{g}$.

27. Show that if $n$ is odd, then the antipodal map $A : S^n \to S^n, x \mapsto -x$ is homotopic to the identity map $1_d : S^n \to S^n$.

28. Let $f : S^n \to S^n$ be a continuous map such that $f(x) \neq x, \forall x \in X$. Show that $f$ is homotopic to the map $g :: S^n \to S^n, x \mapsto -x$.

29. Let $X \subset \mathbf{R}^n$ be compact and convex. If $f : (X, x_0) \to (S^1, 1)$ is continuous and $z_0 \in \mathbf{Z}$, show that its lifting

$$\tilde{f} : (X, x_0) \to (\mathbf{R}, z_0)$$

with $p \circ \tilde{f} = f$ is unique, where $p : \mathbf{R} \to S^1, t \mapsto e^{2\pi i t}$ is the exponential map.

30. **(Homotopy Lifting Property)** .Let $\alpha, \beta : (\mathbf{I}, \dot{\mathbf{I}}) \to (S^1, 1)$ be two loops such that $\alpha \simeq \beta$ rel $\dot{\mathbf{I}}$, and $p \circ \tilde{\alpha} = \alpha, \ p \circ \tilde{\beta} = \beta$ and $\tilde{\alpha}(0) = \tilde{\beta}(0) = 0$, where $p : \mathbf{R} \to S^1, t \mapsto e^{2\pi i t}$ is the exponential map. Show that

$$\tilde{\alpha} \simeq \tilde{\beta} \ \text{rel} \ \dot{\mathbf{I}} \ \text{and} \ \tilde{\alpha}(1) = \tilde{\beta}(1).$$

31. If $f$ is a piecewise differentiable loop in the complex plane $\mathbf{C}$ and $z_0$ is a point in $\mathbf{C}$ but not in $\text{Im} f$, show that the winding number $w(f; z_0)$ is given by the integral

$$w(f; z_0) = \frac{1}{2\pi i} \int_f \frac{dz}{z - z_0}.$$

32. Show that the circle $S^1$ is path connected but not simply connected.

33. Given a topological space $X$, let $s_0 \in S^n \subset \mathbf{D}^{n+1}$ be an arbitrary point and $f : S^n \to X$ be a continuous map. Show the following three statements are equivalent.

   (i) The map $f$ is nullhomotopic (i.e., $f$ is homotopic to a constant map).
   (ii) The map $f$ has a continuous extension over $\mathbf{D}^{n+1}$.
   (iii) The map $f$ is nullhomotopic relative to $\{s_0\}$.

34. Show that the map $f : S^1 \to \mathbf{R}^2 - \{(0, 0)\}, z \mapsto z^n$ is not nullhomotopic.

35. Let $f : S^1 \to S^1$ be an antipode preserving map. Show that $f$ is not nullhomotopic.

36. Show that there is no antipode preserving continuous map $f : S^2 \to S^2$.

37. Show that there exists a continuous map $\psi : \mathbf{D}^n \to S^{n-1}$ with the property that for the inclusion map $i : S^{n-1} \hookrightarrow D^n$, the composite

$$\psi \circ i = 1_d : S^{n-1} \to S^{n-1}$$

is the identity map iff the identity map $1_d$ is nullhomotopic.

38. Let $f : S^1 \to \mathbf{R}^2$ be a continuous map. Show that there exists a point $x_0 \in S^1$ such that $f(x_0) = f(-x_0)$.

39. Let $f : S^2 \to \mathbf{R}^2$ be a continuous map such that $f(x) = -f(x)$, $\forall x \in S^2$. Show that exists a point $x_0 \in S^2$ such that $f(x_0) = 0$.

40. Let $f : S^2 \to S^2$ be a continuous map. Show that

   (i) Either $f$ has a fixed point or
   (ii) There is a point $x_0 \in S^2$ such that $f(x_0) = -x_0$.

41. **(Hopf)** Show that a continuous map $f : S^n \to S^n$ of degree 0 is homotopic to a constant map.

42. Show that there does not exist any homeomorphic copy of the ordinary sphere $S^2$ in the Euclidean plane $\mathbf{R}^2$.

[ Hint: Use Borsuk–Ulam Theorem 2.23.14].

43. Let $X$ be the disk with $g$ holes. Construct (if possible) a continuous map $f : X \to X$ without fixed points.

[ Hint: Let 0 be an arbitrary point inside one hole of the disk. Let $f : X \to X$ be a continuous map mapping each point $X$ to the point where the ray $OX$ meets the boundary circle of the disk. Then under $f$ only the points of its boundary point are fixed. If you turn the circle by a nonzero angle, then a continuous map is obtained without fixed points.]

44. Let $S_1$ and $S_2$ be two surfaces with boundaries $\partial S_1$ and $\partial S_2$, which are homeomorphic to $S^1$. Then they can be glued boundarywise by the homeomorphism $f : \partial S_1 \to \partial S_2$. If $S_1 \cup_f S_2$ is the corresponding factor space, show that the Euler characteristic satisfies the relation

$$\kappa(S_1 \cup_f S_2) = \kappa(S_1) + \kappa(S_2).$$

45. Show that the Euler characteristic of

   (i) The sphere $S^2$ is 2.
   (ii) The disk is 1.
   (iii) The 2-torus $T$ is 0.
   (iv) The Möbius strip is 0.
   (v) The sphere $S_n^2$ with n holes is 2 -n.

Hence show that the sphere $S^2$ and the torus 2- torus $T$ are homotopically distinct in the sense that one cannot be deformed into the other by a continuous
deformation without tearing or cutting.
46. Show that the cylinder and the Möbius strip are not homeomorphic.

   [ Hint: Use Theorem Th 10.21.39. ]
47. **(Poincaré - Alexander theorem)** Let $X$ be any topological space homeomorphic
    to a polyhedron $P$. Show that its Euler characteristic $\kappa(X)$ is independent of the
    polyhedron $P$ provided $P$ is homeomorphic to $X$.
48. Let $S_1$ and $S_2$ be two compact surfaces. Show that they are homeomorphic iff

   (i) $\kappa(S_1) = \kappa(S_2)$.
   (ii) $S_1$ and $S_2$ are both orientable or both nonorientable.

49. Let $f, g : X \to \mathbf{C}$ be two continuous maps such that

$$|f(x) - g(x)| < |f(x) + |g(x)|, \quad \forall x \in X.$$

   Show that

   (i) Both the maps $f/g$ and $g/f$ are exponential.
   (ii) $f$ is exponential if $g$ is also exponential.

50. Let $f, g : S^1 \to \mathbf{C} - \{0\}$ be two continuous maps. Let Ind f denotes the index of
    $f$ given in Definition 2.8.8.

   Show that

   (i) $Ind(fg) = Ind\ f + Ind\ g$, where $fg : S^1 \to \mathbf{C} - \{0\}$, $x \mapsto f(x) \cdot g(x)$.
   (ii) Ind  f $=0$, iff $f$ is an exponential map.
   (iii) Ind  f $=$ Ind $\left(\frac{f}{|f|}\right)$.

51. Let $f : S^1 \to S^1$ be a continuous map such that $deg\ f = 1$. Show that $f \simeq 1_{S^1}$.

   [Hint: Consider the exponential map $p : \mathbf{R} \to S^1$ and the lift of the composite
   map $f \circ p$ to a map $\mathbf{R} \to \mathbf{R}$. ]
52. If $f : S^1 \to S^1$ is a continuous map such that $f(1) = 1$. Show that Ind f and the
    index of the loop $\alpha : (\mathbf{I}, \dot{\mathbf{I}}) \to (S^1, 1)$, $t \mapsto f(e^{2\pi it})$ are the same.
53. Show that for the maps $f, g : S^1 \to \mathbf{C} - \{0\}$, the following statements are equiv-
    alent:

   (i) $f \simeq g$.
   (ii) Ind  f $=$ Ind  g.
   (iii) The map $f/g$ is exponential.

54. If $f : S^1 \to \mathbf{C} - \{0\}$ is a continuous map such that its index $Ind\ f = n$, show that $f$ is homotopic to the exactly one map defined by

$$g : S^1 \to \mathbf{C} - \{0\}, z \mapsto z^n.$$

55. Let $f, g : (\mathbf{I}^n, \partial\ \mathbf{I}^n) \to (X, x_0)$ be two homotopic maps relative to $\partial\ \mathbf{I}^n$ and $\psi, \phi : \mathbf{I} \to X$ are two homotopic paths relative to end points. If $F_t, G_t : \mathbf{I}^n \to X : t \in \mathbf{I}$, are homotopies of $f$ along the path $\psi$ and that of $g$ along the path $\phi$ respectively, show that $F_1$ and $G_1$ are homotopic relative to $\partial\ \mathbf{I}^n$.

56. Let the set $M(n, \mathbf{R})$ of all $n \times n$ real matrices be identified with the Euclidean $\mathbf{R}^{n^2}$-space and $M \in M(n, \mathbf{R})$. Then its associated linear transformation $T_M : \mathbf{R}^n \to \mathbf{R}^n$ has a continuous extension $f : S^n \to S^n$. Show that deg f = det M.

57. Show that every rotation of $S^{2n}$ has a fixed point.

58. Show that there exists a continuous tangent vector field on $S^n$ for every odd integer n.

59. Show that for every even integer $n > 0$, there exists no continuous tangent vector field on $S^n$.

60. Let $X$ be a topological space such that it can be expressed as the union of two simply connected open sets $X_1$ and $X_2$ with their path-connected intersection $X_1 \cap X_2$. Show that $X$ is simply connected.

61. Show that there exists no continuous map $f : \mathbf{D}^2 \to S^1$ such that its restriction map $f|_{S^1} : S^1 \to S^1$ is the identity map.
    [ Use the results that $\pi(S^1, 1) = \mathbf{Z}$ and $\pi_1(\mathbf{D}^2, 1) = \{0\}$. ]

62. If $X$ and $Y$ are two arcwise connected topological spaces, show that

$$\pi_n(X \times Y) \cong \pi_n(X) \oplus \pi_n(Y), \quad \forall\, n \geq 1.$$

63. Let $X$ be the closed subset of $\mathbf{R}^3$ defined in Theorem 2.25.31, which is homeomorphic to $\mathbf{R}$. Show that there exists a homeomorphism $\psi : \mathbf{R}^3 \to \mathbf{R}^3$ such that its image $\psi(X)$ is the $z$-axis.

64. Let $M$ be compact manifold of dimension $n$ and $v$ be a nonzero vector field in $M$. Show that there is a continuous map $f : M \to M$ without a fixed point such that $f$ is homotopic to the identity map $1_M : M \to M$.

65. Let $M$ be a compact differentiable manifold and $m_0 \in M$. Show that its fundamental group $\pi_1(M, m_0)$ is finitely generated.

66. Given any two connected $n$-manifolds $M$ and $N$ $(n > 2)$, show that the fundamental group $\pi_1(M \# N)$ of their connected sum, is the free product of the fundamental groups $\pi_1(M)$ and $\pi_1(N)$.

### 2.28.2  Multiple Choice Exercises

Identify the correct alternative (s) (there may be none or more than one ) from the following list of exercises:

1. Let $X$ be a subspace of $\mathbf{R}^3$ and $\kappa(X)$ be its Euler characteristic.

   (i) If $X$ is the tetrahedron, then $\kappa(X) = 2$.
   (ii) If $X$ is the cube, then $\kappa(X) = 3$.
   (iii) If $X$ is the 2-sphere $S^2$, then $\kappa(X) = 2$.

2. Let $X$ be a subspace of $\mathbf{R}^3$ homeomorphic to a polyhedron and $\kappa(X)$ be its Euler characteristic.

   (i) $\kappa(X)$ is an integer.
   (ii) $\kappa(X)$ is a topological invariant.
   (iii) $\kappa(X)$ is a homotopy invariant.

3. Let $\pi_1(X, x_0)$ be the fundamental group of a topological space $X$ based at the point $x_0 \in X$.

   (i) If $X$ is any topological group with identity element $x_0$, then $\pi_1(X, x_0)$ is abelian.
   (ii) If $X$ is any Hopf group with homotopy identity element $x_0$, then $\pi_1(X, x_0)$ is abelian.
   (iii) If $X$ is the circle $S^1$ and $x \in X$ is an arbitrary point, then $\pi_1(X, x) \cong \mathbf{Z}$ for all $x \in X$.

4. Let $\mathbf{C}$ be the field of complex numbers.

   (i) Every polynomial with coefficients in $\mathbf{C}$ may not have a root in $\mathbf{C}$.
   (ii) The field $\mathbf{C}$ is algebraically closed.
   (iii) The field $\mathbf{R}$ of real numbers is algebraically embedded in the field $\mathbf{C}$.

5. Let $\mathbf{R}^n$ be the $n$-dimensional Euclidean space.

   (i) Every Jordan curve $J$ separates the plane $\mathbf{R}^2$.
   (ii) The punctured Euclidean 5-space $\mathbf{R}^5 - \{0\}$ is homotopy equivalent to the 4-sphere $S^4$.
   (iii) The Euclidean plane $\mathbf{R}^2$ can be continuously deformed into the Euclidean space $\mathbf{R}^3$.

6. Let $\mathbf{D}^n = \{x \in \mathbf{R}^n : ||x|| \leq 1\}$ be the unit disk in the Euclidean $n$ space $\mathbf{R}^n$.

   (i) A continuous map $f : \mathbf{D}^2 \to \mathbf{D}^2$ may not have a fixed point.
   (ii) Every continuous map $f : \mathbf{D}^2 - \{0\}$ from the punctured disk : $\mathbf{D}^2 - \{0\}$ to itself has a fixed point.
   (iii) Every continuous map $S^2 \to S^2$ has a continuous extension over $\mathbf{D}^3$ iff $f$ is homotopic to a constant map, where $S^2$ is the boundary of $\mathbf{D}^3$.

7. Let $S^n$ be the $n$-sphere in $\mathbf{R}^{n+1}$.

   (i) The odd dimensional sphere $S^{2n+1}$ admits a nowhere nonvanishing tangent vector field for every $n \geq 0$.

(ii) Every vector field on an even dimensional sphere $S^{2n}$ vanishes.

(iii) If $v : S^{2n+1} \to \mathbf{R}^{2n+2}$ is a nowhere vanishing tangent vector field on $S^{2n+1}$, then the map

$$f : S^{2n+1} \to S^{2n+1}, \ x \mapsto \frac{v(x)}{||v(x)||}$$

is homotopic to the identity map.

# References

Adams JF. Algebraic topology: a student's guide. Cambridge: Cambridge University Press; 1972.

Adams JF. Vector fields on spheres. Ann of Math. 1962;75:603–32.

Adhikari MR, Adhikari A. Groups. Rings and modules with applications. Hyderabad: Universities Press; 2003.

Adhikari MR, Adhikari A. Textbook of linear algebra: An introduction to modern algebra. New Delhi: Allied Publishers; 2006.

Adhikari A, Adhikari MR. Basic topology, vol. 1: metric spaces and general topology. India: Springer; 2022a.

Adhikari A, Adhikari MR, Basic topology, vol. 2: topological groups, topology of manifolds and lie groups. India: Springer; 2022b.

Adhikari MR. Basic algebraic topology and its applications. India: Springer; 2016.

Adhikari MR, Adhikari, A. Basic modern algebra with applications. India, New York, Heidelberg, Springer; 2014.

Martin A. Introduction to homotopy theory. New York: Springer; 2011.

Armstrong MA. Basic topology. New York: Springer; 1993.

Bredon GE. Topology and geometry. New York: Springer; 1983.

Brown R. Topology: a geometric account of general topology, homotopy types, and the fundamental groupoid. New York: Wiley; 1988.

Chatterjee BC, Ganguly S, Adhikari MR. Introduction to topology. New Delhi: Asian Books; 2002.

Freudenthal H. Über die Klassen von Sphärenabbildungen. Compositio Mathematica. 1937;5:299–314.

Gray B. Homotopy theory, an introduction to algebraic topology. New York: Academic Press; 1975.

Hatcher A. Algebraic topology. Cambridge University Press; 2002.

Henle M. A combinatorial introduction to topology. New York: Dover Publications; 1979.

Hirsch M, Smale S. Differential equations, dynamical systems and linear algebra. New York: Academic Press; 1974.

Hu ST. Homotopy Theory. New York: Academic Press; 1959.

Hurewicz W. Beitrage der Topologie deformation. Proc K Akad Wet Ser. 1935;38:112–119, 521-528.

Lefschetz S. Introduction to topology. Princeton; 1949.

Lie S. Theorie der Transformations gruppen. Math Annalen. 1880;16:441–528.

Massey WS. A basic course in algebraic topology. New York: Springer; 1991.

Maunder CRF. Algebraic topology. London: Van Nostrand Reinhhold; 1970.

Geometry Nakahara M. Topology and physics. Bristol: Institute of Physics Publishing, Taylor and Francis; 2003.

Patterson EM. Topology. Oliver and Boyd; 1959.

Prasolov VV. Intuitive topology. India: Universities Press; 1995.

Rotman JJ. An introduction to algebraic topology. New York: Springer; 1988.

Singer IM, Thorpe JA. Lecture notes on elementary topology and geometry. New York: Springer; 1967.

Spanier E. Algebraic topology. New York: McGraw-Hill Book Company; 1966.

Steenrod N. The topology of fibre bundles. Prentice: Prentice University Press; 1951.

Toda H. Composition methods in homotopy groups of spheres. Ann Math Stud. 49:19621

Whitehead GW. Elements of homotopy theory. New York: Springer; 1978.

# Chapter 3
# Homology and Cohomology Theories: An Axiomatic Approach with Consequences

As the classical constructions of homology and cohomology groups are available in almost every textbook of algebraic topology, this chapter develops homology and cohomology theories with an emphasis on the axiomatic approach of Eilenberg and Steenrod with various interesting applications for accessible presentation. It starts with a brief discussion on **simplicial homology theory with Euler-Poincaré theorem**, and then it defines **singular homology theory** which is a natural generalization of simplicial homology theory. Their dual theories are also discussed. This chapter also gives a presentation of an approach formulating axiomatization of homology and cohomology theories which makes the subject algebraic topology elegant and provides a quick access to further study. These axioms, now called **Eilenberg and Steenrod axioms for homology and cohomology theories**, were announced by S. Eilenberg (1915–1998) and N. Steenrod (1910–1971) in 1945, but they first appeared in their celebrated book 'The Foundations of Algebraic Topology' in 1952. This approach classifies and unifies different homology (cohomology) groups. Chapter 2 studies fundamental and higher homotopy groups, where the base point plays a key role. There are several advantages of homology groups over them.

**Advantages of homology groups over fundamental and higher homotopy groups**:

(i) All homology groups and higher homotopy groups (of dimension more than 1) are abelian but all fundamental groups are not abelian.
(ii) Every homotopy group needs a base point, but a homology group needs no base point.
(iii) From the intuitive viewpoint, homology groups are less intuitive than the fundamental group and the higher homotopy groups. Computation of fundamental group and the higher homotopy groups is more difficult than that of homology groups.

**Historically**, homology invented by Henri Poincaré in 1895 was studied by him during 1895–1904. This homology, now called the **simplicial homology**, is one of the

© The Author(s), under exclusive license to Springer Nature Singapore Pte Ltd. 2022  163
M. R. Adhikari, *Basic Topology 3*,
https://doi.org/10.1007/978-981-16-6550-9_3

most fundamental powerful inventions in mathematics. Poincaré in 1899 remarked on his homology theory which runs as follows:

' **Assume that one can find in** $V$ **a manifold of** $p + 1$ **dimension whose boundary consists of** $n$ **manifolds of** $p$ **dimension** $V_1, V_2, \ldots, V_n$ **I will express this fact with the relation** $V_1 + V_2 + \cdots + V_n \sim 0$, **that I will call it homology.**' Henri Poincaré, 1899.

For the study of geometric properties of a topological space, Henri Poincaré prescribed a method of constructing homology groups of the topological space in 1895 and contributed to the fundamental idea of converting the topological problems to algebraic ones for the first time in the history of topology. He started with a geometric object, which is a topological space and gave a combinatorial data providing a simplicial complex. These data were used to construct homology groups by utilizing the tools of linear algebra and boundary relations. **The classical simplicial homology theory** is involved of tedious discussion on the concepts of triangulability of the topological spaces, orientations of simplexes, incidence numbers, subdivisions, simplicial approximation and also the topological invariance of the simplicial homology groups. **Emmy Noether** (1882–1935) formulated an algebraic approach corresponding to the geometric approach of homology theory invented by Poincaré S. Lefschetz (1884–1972) extended simplicial homology theory to **singular homology theory** in 1931 in his paper [ Lefschetz, 1933] by considering continuous maps from the standard simplexes to a topological space $X$, which are called singular simplexes in $X$ and by using the algebraic properties of singular chain complexes. There are other important homology groups (topological invariants) such as **homology groups for compact metric spaces** constructed by L. Vietoris (1891-2002) in 1927 and **homology groups for compact Hausdorff spaces** constructed by E. Čech (1893–1960) in 1932.

Lefschetz remarked

'**Others (topological invariants) were discovered by Poincaré. They are all tied up with his homology theory which is perhaps the most profound and far reaching creation in all topology.**'

Historically, algebraic topologists began around 1940 to compare different definitions of homology and cohomology formulated in the earlier years. Eilenberg and Steenrod formulated a new approach in 1945 by considering a small number of their properties (without focusing the tools used for construction of homology and cohomology groups) as axioms to characterize a theory of homology and cohomology. This approach is the most surprising contribution to algebraic topology since the invention of the homology groups by Poincaré in 1895. Now, it is called the **axiomatic approach** formulated by a set of seven axioms announced by S. Eilenberg and N. Steenrod in 1945 and published in their book in 1952. This approach gives the subject **algebraic topology** conceptual coherence and elegance and provides quick techniques for computing homology and cohomology groups. **It unifies different homology groups (modules) on the category of compact triangulable spaces. Their dual theories are called cohomology theories**. Homology and cohomology theories are basic topics of study of algebraic topology and are used in different disciplines. Initially, no important result connecting these theories was found but the twentieth century witnessed their greatest development.

For this chapter, the books [Adhikari 2016], [Bredon 1993], [Eilenberg and Steenrod 1952], [Gray 1975], [Hatcher 2002], [Hilton and Wylie 1960], [Hu 1966], [Poincaré 2010] , [Rotman 1988], [Switzer 1975], [Adhikari and Adhikari 2003] and [Adhikari and Adhikari 2006] some others are referred in the bibliography.

# 3.1 Manifolds with Motivation

Before starting simplicial homology theory in Sect. 3.2, one needs an idea of a manifold. This section provides the concept of manifolds with its motivation. A manifold is a generalization of the familiar geometric objects like curves and surfaces which are considered locally homeomorphic to $\mathbf{R}$ and $\mathbf{R}^2$, respectively. An $n$-dimensional real manifold looks locally like $\mathbf{R}^n$ but not necessarily globally. A local Euclidean structure to manifold by introducing the concept of a chart is utilized to use the conventional calculus of several variables. Due to linear structure of vector spaces, for their applications in mathematics and in other areas, it needs generalization of metrizable vector spaces, maintaining only the local structure of the latter. On the other hand, every manifold can be considered as a (in general, nonlinear) subspace of some vector space. Both aspects are used to approach the theory of manifolds. Since dimension of a manifold is a locally defined property, a manifold has a dimension. Our main interest in this chapter is on finite dimensional topological manifold.

**Remark 3.1.1** A topological manifold is a Hausdorff space $M$ such that every point of $M$ has a nbd homeomorphic to the Euclidean $n$-space $\mathbf{R}^n$, i.e., every point of $M$ has a nbd homeomorphic to an open subset of $\mathbf{R}^n$. This means that every topological manifold is locally Euclidean. In particular, the $n$-dimensional Euclidean space $\mathbf{R}^n$ is a topological space as well as an $n$-dimensional real vector space. If $x = (x_1, x_2, \ldots, x_n) \in \mathbf{R}^n$, then the real numbers $x_1, x_2, \ldots, x_n$ are called the coordinates of the point $x$. **A manifold is a geometric object in topology which looks locally like a small piece of Euclidean space, where a local coordinate system can be introduced.** As calculus is developed on the local geometry of Euclidean space, tools of calculus can be used to develop the structure and properties of manifolds. For example, every point on the sphere lies on small curved disk which can be flattened into a disk in the Euclidean plane $\mathbf{R}^2$. On the other hand, the vertex of the cone has no nbd which looks like a small piece of $\mathbf{R}^2$. Curves are manifolds of dimension one, and surfaces are manifolds of dimension two.

Definition 3.1.2 formalizes the intuitive idea of an $n$-dimensional manifold $M$ such that each point of $M$ is in a nbd that looks like an open ball in $\mathbf{R}^n$.

**Definition 3.1.2** Let $M$ be a Hausdorff space. Then $M$ is said to be an $n$-dimensional manifold or simply $n$-**manifold,** if there is an open covering $\mathcal{C} = \{V_i : i \in \mathbf{A}\}$ of $M$ such that for every $i \in \mathbf{A}$, there is a map $\psi_i : V_i \to \mathbf{R}^n$ mapping $V_i$ homeomorphically onto an open subset $U_i$ of $\mathbf{R}^n$ homeomorphically. Each ordered pair $(\psi_i, V_i)$ is called a **local chart** (or simply **a chart or a coordinate system** on $V_i$), and the family $\psi = \{(\psi_i, V_i) : i \in \mathbf{A}\}$ is called an **atlas of the manifold** $M$.

***Example 3.1.3*** (i) The Euclidean $n$-space $\mathbf{R}^n$ is a manifold of dimension $n$. Here
for each $x \in \mathbf{R}^n$, the map $\psi_x$ is taken to be the identity map on $\mathbf{R}^n$.
(ii) The $n$-sphere $\mathbf{S}^n$ is a manifold of dimension $n$. Here for the points $x, y \in \mathbf{S}^n, x \neq y$, the map $\psi_x$ is taken to be the stereographic projection $\psi_x : S^n - \{y\} \to \mathbf{R}^n$, which is a homeomorphism.
(iii) The $n$-dimensional projective space $\mathbf{R}P^n$ (the space of all lines through the origin $\mathbf{0} \in \mathbf{R}^{n+1}$) is a manifold of dimension $n$.
(iv) Let $M(n, \mathbf{R})$ be the set of all $n \times n$ matrices over $\mathbf{R}$, identified with the Euclidean $\mathbf{R}^{n^2}$-space and $GL(n, \mathbf{R}) = \{M \in M(n, \mathbf{R}) : detM \neq 0\}$. The set $GL(n, \mathbf{R})$ of nonsingular matrices is a manifold of dimension $n^2$.
(v) The 2-torus $S^1 \times S^1$ with product topology is a compact and connected manifold, which is a closed manifold of dimension 2.

***Remark 3.1.4*** A detailed study of manifolds including differentiable manifolds is available in **Chap. 3, Basic Topology, Volume 2** of the present series of books.

## 3.2 Simplicial Homology Theory and Euler-Poincaré Theorem

This section begins with the homology theory born through the work of **Henri Poincaré** (1854–1912) published in his land-marking paper **'Analysis Situs' of 1895** [Poincaré 1895].

Poincaré himself defined homology in 1895 in the following way:

> Let us consider a manifold $V$ of $p$ dimensions; now let $W$ be a manifold of $q$ dimensions ( $q \leq p$) which is part of $V$. Assume the complete boundary of $W$ consists of $\lambda$ manifolds of $q - 1$ dimensions $V_1, V_2, \ldots, V_\lambda$ We express this fact with the notation $V_1 + V_2 + \cdots + V_\lambda \sim 0$. More generally, the notation $k_1 V_1 + k_2 V_2 - k_3 V_3 - k_4 V_4 \sim 0$, where the $k's$ are integers and $V's$ are manifolds of $q - 1$ dimensions, will denote that there exists a manifold $W$ of $q$ dimension in $V$ such that the complete boundary of $W$ consists of $k_1$ manifolds similar to $V_1$, $k_2$ manifolds similar to $V_2$, $k_3$ manifolds similar to $V_3$ but oppositely oriented, and $k_4$ manifolds similar to $V_4$, but oppositely oriented. Relations of this form are called homologies. Homologies can be considered like ordinary equations". [Poincaré 1895]

### The motivation leading to the concept of simplicial homology:
While studying two-dimensional manifolds, one can distinguish intuitively non-homeomorphic manifolds. But for a study of higher-dimensional manifolds, geometric intuition is less effective. So an alternative approach came through the work of Henri Poincaré: Let $M^n$ and $N^n$ be two $n$-dimensional manifolds, and $X \subset M^n$ and $Y \subset N^n$ be two compact submanifolds (subspaces which are $C^0$-manifolds).

If any $m$-dimensional submanifold in $M^n$ ($m \le n$) is the boundary of an $(m+1)$-submanifold of $M^n$, there is an $m$-dimensional submanifold in $N^n$, and then the manifolds $M^n$ and $N^n$ are nonhomeomorphic. For example, any one-dimensional compact submanifold of the 2-sphere $S^2$ is a boundary, but there is one-dimensional manifold such as the circle $S^1$ on the torus $T = S^1 \times S^1$ which is not the boundary of any two-dimensional submanifold of $T$. On the other hand, if there are submanifolds, which are not boundaries both in $M^n$ and $N^n$, we can find certain quantities associated with these manifolds.

## 3.2.1 Simplicial Complex

This subsection formally defines simplicial complex based on which simplicial homology groups are formulated. It is assumed that the readers are familiar with the concept of standard simplexes in Euclidean space $\mathbf{R}^n$.

**Definition 3.2.1** A **finite simplicial complex** $K$ is a finite collection of simplexes in some Euclidean space $\mathbf{R}^n$ such that

(i) If $\sigma_p$ is a simplex in $K$, then all of its faces are also in $K$.
(ii) If $\sigma_p$ and $\sigma_q$ are simplexes in $K$, then $\sigma_p \cap \sigma_q$ is either empty or is a common face of $\sigma_p$ and $\sigma_q$ in $K$.

The **dimension** of $K$ denoted by dim $K$ is defined to be $-1$ if $K = \emptyset$ and to be $m \ge 0$ if $m$ is the greatest integer such that $K$ has an $m$-simplex.

**Remark 3.2.2** A **simplicial complex $K$ is said to be finite** if it consists of a finite number of simplexes. Let $K$ be any simplicial complex and $|K| \subset \mathbf{R}^n$ be the set-theoretic union of all simplexes in $K$ endowed with a topology such that a subset $Y \subset |K|$ is closed if $Y \cap \sigma_p^i$ is closed in $\sigma_p^i$ for every $\sigma_p^i \in K$. If $K$ is finite, then this topology coincides with the subspace topology on $|K|$ inherited from the Euclidean topology on $\mathbf{R}^n$.

**Definition 3.2.3** Let $K$ be a finite dimensional simplicial complex. A topological space $X$ is said to be a **polyhedron** if $X$ is homeomorphic to $|K|$ and then $K$ is said to be a **triangulation** of X.

**Remark 3.2.4** A triangulation of a topological space $X$ consists of a simplicial complex $K$ together with a homeomorphism $h : |K| \to X$. Since a finite simplicial complex is made up of a finite number of simplexes lying in a Euclidean space, its polyhedron $K$ is compact, hence, every triangulable space is compact, and it is also a metric space.

**Example 3.2.5** Every closed surface is triangulable.

**Definition 3.2.6** *(Orientation of a simplicial complex)* An oriented $p$-simplex $\sigma_p$ for $p \geq 1$ is obtained from a $p$-simplex $\sigma_p = < v_0, v_1, \ldots, v_p >$ by selecting an ordering of its vertices. An oriented simplicial complex is obtained from a simplicial complex by choosing an orientation of each of its simplexes.

Proposition 3.2.7 follows from the above discussion.

**Proposition 3.2.7** *Let $K$ be a finite-dimensional-oriented simplicial complex. Then*

(i) *Every polyhedron is a normal Hausdorff space.*
(ii) *$|K|$ is compact.*
(iii) *$K$ is finite iff $|K|$ is compact.*

Let $K$ be an oriented simplicial complex of dimension $n$ and $\Omega(K)$ be the set of the linear combinations of oriented $p$-simplexes $\sigma_p$ over the field $\mathbf{Q}$ given by

$$\Omega(K) = \{q_1\sigma_1 + q_2\sigma_2 + \cdots + q_n\sigma_n : \text{each } q_p \in \mathbf{Q} \text{ and each } \sigma_p \text{ is an oriented}$$
$$p\text{-simplex of } K \text{ for } p = 0, 1, 2, \ldots, n\}.$$

Every element of $\Omega(K)$ is called a **rational $p$-chains of $K$.** Then $\Omega(K)$ forms a vector space over $\mathbf{Q}$ under usual compositions. Let $V(K)$ be the subspace of $\Omega(K)$, generated by the elements of the form $\sigma + \rho$, where $\sigma$ and $\rho$ are the same $p$-simplex of $K$ with opposite orientations. The quotient space $\Omega(K)/V(K)$ denoted by $C_p(K; \mathbf{Q})$ is called the vector space of rational $p$-chains of $K$. Its dimension is the number of $p$-simplexes in $K$. Consider a rational $p$-chain

$$c_p = q_1\sigma_p^1 + q_2\sigma_p^2 + \cdots + q_n\sigma_p^{\beta_i} = \Sigma_{i=1}^{\beta_i} q_i \, \sigma_p^i,$$

where $-\sigma_p$ is denoted by $(-1)\sigma_p$. The boundary of a $p$-simplex $\sigma_p^i$ $(p > 0)$ in $K$ is the rational $(p-1)$-chain

$$\partial\sigma_p = \Sigma_j[\sigma_p^i, \sigma_{p-1}^j] \, \sigma_{p-1}^j,$$

where $[\sigma_p^i, \sigma_{p-1}^j]$, called **incidence number**, is an integer defined as follows:

(i) If $\sigma_{p-1}^j$ is not a face of $\sigma_p^i$, then $[\sigma_p^i, \sigma_{p-1}^j]$ is taken to be 0.
(ii) If $\sigma_{p-1}^j$ is a face of $\sigma_p^i$, then $[\sigma_p^i, \sigma_{p-1}^j]$ is taken to be 1 or $-1$ according as the orientation of $\sigma_{p-1}^j$ induced by the orientation of $\sigma_p^i$ agrees or not.

**Proposition 3.2.8** *Let $K$ be an oriented simplicial complex and $p > 1$. Then*

$$\partial_{p-1} \circ \partial_p : C_p(K) \to C_{p-2}(K)$$

*is the trivial homomorphism, i.e., $\partial_{p-1} \circ \partial_p = 0$ for all $p > 1$.*

**Proof** It follows from the definition of the boundary homomorphism $\partial$. ☐

### 3.2.2 Simplicial Homology Group

This subsection prescribes a method of construction of homology groups, called simplicial homology groups invented by H. Poincaré in 1895 before giving formalization of the axioms of homology and cohomology theories. Homology group is a basic topological invariant different from the fundamental group and higher homotopy groups studied in Chap. 2. Instead of the closed loop at a base point in a topological space $X$, for defining the fundamental group, here the **sum of paths** $\sigma : \mathbf{I} \to X$ in a formal sense called 1-chains and in general $p$-chains are considered. For the path $\sigma : \mathbf{I} \to X$, its boundary $\partial \sigma = \sigma(1) - \sigma(0)$ is a formal sum of signed points. Hence for a chain $c = \Sigma \sigma_i$, its boundary $\partial c$ is defined by

$$\partial c = \Sigma \partial \sigma_i,$$

which is to be closed if $\partial c = 0$.

**Definition 3.2.9** Let $K$ be an oriented simplicial complex and $p \geq 0$ be an integer. Then the image of

$$\partial_{p+1} : C_{p+1}(K) \to C_p(K)$$

is a subgroup of $C_p(K)$, denoted by $B_p(K)$, which is called the $p$-**dimensional boundary group** of $K$, and each element $b_p \in B_p(K)$ is called a $p$-**dimensional boundary** of $K$.

**Definition 3.2.10** Let $K$ be an oriented simplicial complex and $p > 0$ be an integer. Then the kernel of

$$\partial_p : C_p(K) \to C_{p-1}(K)$$

is a subgroup of $C_p(K)$ denoted by $Z_p(K)$, which is called the $p$-**dimensional cycle group** of $K$, and each element $z_p \in Z_p(K)$ is called a $p$-**cycle** of $K$.

**Proposition 3.2.11** *Let $K$ be an oriented $n$-dimensional simplicial complex. Then $B_p(K)$ is a subgroup of $Z_p(K)$ for every integer $p$ with $0 \leq p \leq n$.*

**Proof** It follows from Proposition 3.2.8.                 □

**Definition 3.2.12** Let $K$ be an oriented simplicial complex and $p \geq 0$ be an integer. Then the quotient group

$$H_p(K; \mathbf{Z}) = Z_p(K; \mathbf{Z})/B_p(K; \mathbf{Z})$$

is called the $p$-**dimensional simplicial homology group** of $K$ in coefficient ring $\mathbf{Z}$.

***Example 3.2.13*** $K$ be the 2-simplex $<v_0, v_1, v_2>$ having orientation induced by the ordering $v_0 < v_1 < v_2$.

(i) 0-simplexes of $K$ are $<v_0>$, $<v_1>$ and $<v_2>$.

(ii)  Positively oriented 1- simplexes of $K$ are $<v_0, v_1>$,  $<v_1, v_2>$ and $<v_0, v_2>$.
(iii)  Positively oriented 2-simplex of $K$ is $<v_0, v_1, v_2>$.

Then

(i)  $H_0(K; \mathbf{Z}) \cong \mathbf{Z}$.
(ii)  $H_1(K; \mathbf{Z}) \cong \{0\}$.
(iii)  $H_2(K; \mathbf{Z}) \cong \{0\}$.

**Definition 3.2.14**  The quotient group $H_p(K; \mathbf{Q}) = Z_p(K; \mathbf{Q})/B_p(K; \mathbf{Q})$ is called the *p*th **simplicial homology group of** $K$ **with coefficients in Q**. It is a vector space over **Q**. Instead, the quotient group $H_p(K; \mathbf{Z}) = Z_p(K; \mathbf{Z})/B_p(K; \mathbf{Z})$, the *p*th simplicial homology group of $K$ with coefficients in **Z** is a module over **Z**.

There are several definitions of simplicial maps, but we use the Definition 3.2.15.

**Definition 3.2.15**  Let $K$ and $L$ be two simplicial complexes. **A simplicial map** between them is a continuous map $f : |K| \rightarrow |L|$ between their corresponding polyhedra which takes simplexes of $K$ linearly onto simplexes of $L$.

**Definition 3.2.16**  Let $f : |K| \rightarrow |L|$ be a simplicial map between oriented simplicial complexes $K$ and $L$. Then it induces a group homomorphism

$$f_* : H_n(K; \mathbf{Z}) \rightarrow H_n(L; \mathbf{Z}), \; [z_n] \mapsto [f(z_n)].$$

$f_*$ is called **the homomorphism induced by** $f$ on homology groups.

**Proposition 3.2.17**  *Let  Sim be the category of oriented simplicial complexes and their simplicial maps and* $\mathcal{A}b$ *be the category of abelian groups and their homomorphisms. Then* $H_n :$ *Sim* $\rightarrow \mathcal{A}b$ *is a covariant functor for every integer* $n \geq 0$.

***Proof***  Here the object function is defined by

$$H_n : Sim \rightarrow \mathcal{A}b, \; K \mapsto H_n(K; \mathbf{Z})$$

and for every simplicial map $f : |K| \rightarrow |L|$ in the category  *Sim*, the morphism function is defined by

$$f \mapsto f_* : H_n(K; \mathbf{Z}) \rightarrow H_n(L; \mathbf{Z}), \; [z_n] \mapsto [f(z_n)].$$

$\square$

**Corollary 3.2.18**  *Let $X$ and $Y$ be two homeomorphic compact polyhedral spaces. Then*
$$H_n(X; \mathbf{Z}) \cong H_n(Y; \mathbf{Z}), \; \forall n \geq 0.$$

***Proof***  It follows from the functorial property of $H_n$.

$\square$

**Corollary 3.2.19** *Let X and Y be two homotopy equivalent compact polyhedral spaces. Then*

$$H_n(X; \mathbf{Z}) \cong H_n(Y; \mathbf{Z}), \quad \forall\, n \geq 0.$$

**Proof** It follows from the functorial property of $H_n$. □

**Definition 3.2.20** A simplicial pair $(K, L)$ is a pair of the simplicial complex $K$ and its subcomplex $L$, and a simplicial map $f : (K, L) \rightarrow (X, A)$ of simplicial pairs is a simplicial map $f : K \rightarrow X$ such that $f(L) \subset A$.

### 3.2.3 Euler-Poincaré Theorem

Euler-Poincaré Theorem 3.2.24 establishes a close link among geometry, topology and algebra with the help of the Euler characteristic of compact polyhedra.

**Definition 3.2.21** Let $K$ be a simplicial complex of dimension $n$ and $\beta_p$ denotes the number of $p$-simplexes in $K$, for $p = 0, 1, 2, ..., n$. Then the alternative sum

$$\kappa(K) = \Sigma_{p=0}^{n}(-1)^p \beta_p \tag{3.1}$$

is called the **Euler characteristic** of $K$ with **Betti numbers** $\beta_p$.

**Remark 3.2.22** Betti numbers $\beta_p$ in Equation (3.1) coined by Poincaré (named after E. Betti) play an important role in algebraic topology to classify topological spaces based on the connectivity of a $p$-dimensional simplicial complex $K$. The number $\beta_p$ is the same as the rank of the $p$th homology group $H_p(K; \mathbf{Q})$ with rational coefficients.

**Example 3.2.23** By Theorem 3.12.9, the homology group of $S^n$ with coefficient group $\mathbf{Z}$ is

$$H_p(S^n; \mathbf{Z}) \cong \begin{cases} \mathbf{Z}, & \text{if } p = 0, \text{ or } n \\ \{0\}, & \text{otherwise.} \end{cases}$$

It shows that the Betti numbers $\beta_0 = 1$, $\beta_n = 1$, and all other Betti numbers are 0.

**Theorem 3.2.24** *(Euler-Poincaré theorem) The Euler characteristic of an oriented simplicial complex K of dimension n is given by*

$$\kappa(K) = \Sigma_{p=0}^{n}(-1)^p \beta_p = \Sigma_{p=0}^{n}(-1)^p \, rank(H_p(K; \mathbf{Q})). \tag{3.2}$$

**Proof** Consider the chain complex

$$C_* : \{0\} \xrightarrow{\partial_{n+1}} C_n \xrightarrow{\partial_n} C_{n-1} \rightarrow \cdots \rightarrow C_{p+1} \xrightarrow{\partial_{p+1}} C_p \rightarrow \cdots \xrightarrow{\partial_1} C_0 \xrightarrow{\partial_0} \{0\},$$

where $C_p$ stands for the vector space $C_p(K; \mathbf{Q})$, $\beta_p = dim_{\mathbf{Q}} C_p(K, \mathbf{Q})$ is the same as the number of $p$-simplexes in $K$, for $p = 0, 1, 2, \ldots, n$ and $\{0\}$ stands for the trivial group (though the usual convention in algebraic topology is to use $0$ in place of the trivial group $\{0\}$). Then

$$B_p = \partial_{p+1}(C_{p+1}) \cong C_{p+1}/ker\, \partial_{p+1} = C_{p+1}/Z_{p+1}, \quad \forall p \geq 0,$$

where $B_p$ stands for $B_p(K; \mathbf{Q})$ and $Z_p$ stands for $Z_p(K; \mathbf{Q})$. This asserts that

$$\dim B_p(K; \mathbf{Q}) = \dim C_{p+1}(K; \mathbf{Q}) - \dim Z_{p+1}(K; \mathbf{Q}), \quad \forall p \geq 0.$$

Again, $H_p(K; \mathbf{Q}) \cong Z_p(K; \mathbf{Q})/B_p(K, Q)$ implies $\dim H_p(K; \mathbf{Q}) = \dim Z_p(K; \mathbf{Q}) - \dim B_p(K; \mathbf{Q})$, $\forall p \geq 0$. These two relations assert that

$$\dim C_{p+1}(K; \mathbf{Q}) = \dim Z_p(K; \mathbf{Q}) - \dim H_p(K; \mathbf{Q}) + \dim Z_{p+1}(K; \mathbf{Q}), \quad \forall p \geq -1.$$

By taking the alternative sum for $p = -1, 0, 1, 2, \ldots, n$, it follows that the Euler characteristic is given by the formula

$$\kappa(K) = \Sigma_{p=0}^n (-1)^p \beta_p = \Sigma_{p=0}^n (-1)^p \, rankH_p(K; \mathbf{Q}),$$

where $\dim C_p(K; \mathbf{Q}) = \beta_p$ and $\dim H_p(K; \mathbf{Q}) = rank\, H_p(K; \mathbf{Q})$.     □

**Corollary 3.2.25** *The Euler characteristic of an oriented simplicial complex $K$ of dimension $n$ is also given by the formula*

$$\kappa(K) = \Sigma_{p=0}^n (-1)^p \beta_p = \Sigma_{p=0}^n (-1)^p \, rankH_p(K; \mathbf{Z}).$$

**Proof** It follows from Theorem 3.2.24.     □

**Corollary 3.2.26** *Let $X$ be a compact polyhedron. Then its Euler characteristic is given by the formula*

$$\kappa(X) = \Sigma_{p=0}^n (-1)^p \beta_p = \Sigma_{p=0}^n (-1)^p \, rankH_p(X; \mathbf{Z}).$$

**Proof** It follows from Theorem 3.2.24 by taking $K$ as a triangulation of $X$.     □

The above discussion is summarized in a basic result in Corollary 3.2.27.

**Corollary 3.2.27** *The rank of the free abelian part of $H_p(K; \mathbf{Q})$ of a finite oriented complex $K$ is the **Betti number** $\beta_p$ of $K$.*

### 3.2.4 Topological and Homotopy Invariance of Euler Characteristics

**Euler-Poincaré theorem** is a powerful result in topology. For example, this subsection applies this theorem to prove topological invariance of Euler characteristics in Theorem 3.2.28 in the sense that two homeomorphic compact polyhedra have the same Euler characteristic and homotopy invariance of compact polyhedra in Theorem 3.2.29 in the sense that two homotopy equivalent compact polyhedra have the same Euler characteristics.

**Theorem 3.2.28** *Two homeomorphic compact polyhedra have the same Euler characteristics.*

**Proof** Euler-Poincaré theorem of simplicial homology asserts that if $X$ is compact polyhedron, then its Euler characteristic

$$\kappa(X) = \Sigma_{p=0}^{n}(-1)^p \, \beta_p = \Sigma_{p=0}^{n}(-1)^p \, \text{rank} H_p(X; \mathbf{Z}), \tag{3.3}$$

which follows from Eq. (3.2). Homology group is a topological invariant in the sense that if $X$ and $Y$ are two homeomorphic compact polyhedra, then

$$H_p(X; \mathbf{Z}) \cong H_p(Y; \mathbf{Z}), \ \forall p \geq 0.$$

Consequently, it follows by using Eq. (3.3) that $\kappa(X) = \kappa(Y)$.     □

**Theorem 3.2.29** *Two homotopy equivalent compact polyhedra have the same Euler characteristics.*

**Proof** Since homology group is a homotopy invariant in the sense that, if $X$ and $Y$ are two homotopy equivalent compact polyhedra, $H_p(X; \mathbf{Z}) \cong H_p(Y; \mathbf{Z})$ for all $p \geq 0$. Consequently, it asserts by Eq. (3.3) that $\kappa(X) = \kappa(Y)$.     □

**Remark 3.2.30** Theorem 3.2.28 establishes a close link among geometry, topology and algebra by Euler characteristics of compact polyhedra, which are integers. For a generalization of theorem 3.2.29 asserting that if the compact polyhedra $X$ and $Y$ are two homotopy equivalent spaces, then $\kappa(X) = \kappa(Y)$. Moreover, $\kappa(X) \in \mathbf{Z}$, which is an algebraic object.

## 3.3 Topology of CW-Complexes

This section introduces the concept of CW-complexes by generalizing the concept of simplicial complexes. The process of this generalization is made by building up topological spaces by attaching cells successively, starting from a discrete set of points.

**Definition 3.3.1** Let $X$ be a Hausdorff space. It is said to be a *CW*-**complex** if $X$ is the union of disjoint subspaces $e_a$ $(a \in \mathbf{A})$, called **cells** such that it satisfies the axioms:

(i) **Axiom (i)**: Corresponding to each cell $e_a$, there is an integer $n \geq 0$, called its dimension. Then the cell $e_a$ is symbolized by $e_a^n$.

Let $X^n$ denote the union of all cells $e_a^k : k \leq n$, which is called the *n*-**skeleton** of $X$.

(ii) **Axiom (ii)**: There is a map $\kappa_a : (\mathbf{D}^n, S^{n-1}) \to (X, X^{n-1})$, called a **characteristic map** with the property that its restriction $\kappa_a \mid (\mathbf{D}^n - S^{n-1})$ is a homeomorphism from $\mathbf{D}^n - S^{n-1}$ onto $e_a^n$.

(iii) **Axiom (C)**: $X$ is **closure finite** in the sense that if for each cell $e_a^n$, the intersection $K(e_a^n)$ of all subcomplexes containing $e_a^n$ is a finite subcomplex. (A subset $A \subset X$ is said to be a **subcomplex** of $X$, if $A$ is a union of cells $e_a$ and $\overline{e_a} \subset A$ for $e_a \subset A$ in the sense of Definition 3.3.2).

(iv) **Axiom (W)**: $X$ has the **weak topology** in the sense that for every subset $B \subset X$, the subset $B$ is closed iff $B \cap \overline{e_a^k}$ is compact for every cell $e_a^k$.

$X$ satisfying the above conditions (i) and (ii) is called a **cell complex**. A cell complex satisfying the above additional conditions (iii) and (iv) is called a *CW*-**complex**, where $C$ stands for closure finite and $W$ stands for weak topology in the sense of Axioms (C) and (W).

**Definition 3.3.2** Let $X$ be a *CW*-complex and $A \subset X$. Then $A$ is said to be a **subcomplex** of $X$, if $A$ is a union of cells $e_a$ and $\overline{e_a} \subset A$, if $e_a \subset A$.

*Example 3.3.3* (i) If $K$ is a simplicial complex, then $|K|$ is a *CW*-complex. A polyhedron is usually representable as *CW*-complex with less number of cells than the original number of simplexes.

(ii) $S^n$ is a cell complex having two cells such as $e^0 = \{(1, 0, \ldots, 0\}$ and $e^n = S^n - e^0$

(see Proposition 3.3.4).

(iii) The real line space $\mathbf{R}$ is a *CW*-complex with 0-cells $e^0$ the integers and 1-cells $e^1$ the intervals $[n, n+1]$ for all integers $n$.

(iv) The 2-sphere $S^2$ considered as a cell complex with each point a 0-cell, is not a *CW*-complex, because, it is closure finite but it does not have the weak topology, and hence, it fails **Axiom (W)** of Definition 3.3.1.

(v) (**Wedge of the circles** $\bigvee\limits_{i=1}^{\infty} S_i^1$) Let $X = \bigvee\limits_{i=1}^{\infty} S_i^1$ be an infinite 1-point union of circles. Then $X$ is a *CW*-complex. In particular, the **figure-eight** defined in Chap. 2 is the 1-point union of two circles and is a *CW*-complex.

(vi) $\mathbf{D}^3$ with one 0-cell for every point of $S^2$ and the cell $e^3 = \mathbf{D}^3 - S^2$, is not a *CW*-complex, because it has the weak topology but it is not closure finite, and hence, it fails the **Axiom (C)** of Definition 3.3.1.

(vii) $RP^\infty = \bigcup_{n=1}^{\infty} RP^n$ is a closure finite cell complex, and it has the weak topology given by
$RP^n \subset RP^{n+1}$ as a subcomplex (see Exercise 3 of Sect. 3.16).

(viii) $CP^\infty = \bigcup_{n=1}^{\infty} CP^n$ is a closure finite cell complex, and it has the weak topology given by
$CP^n \subset CP^{n+1}$ as a subcomplex (see Exercise 3 of Sect. 3.16).

(ix) $HP^\infty = \bigcup_{n=1}^{\infty} HP^n$ is a closure finite cell complex, and it has the weak topology given by
$HP^n \subset HP^{n+1}$ as a subcomplex (see Exercise 3 of Sect. 3.16).

**Proposition 3.3.4** $S^n$ *is a CW-complex.*

**Proof** Let $e^0 = s_0 = \{(-1, 0, \ldots, 0)\} \in S^n$ and $e^n = S^n - e^0$.

$$\kappa_n : (\mathbf{D}^n, S^{n-1}) \to (S^n, s_0)$$

be the standard characteristic map. Then $\kappa_n|e^n$ is a homeomorphism onto $\kappa_n(e^n)$. Consider the characteristic map $\psi : \mathbf{D}^0 \to s_0$. Hence, it follows that $S^n$ is a CW-complex with one 0-cell and one $n$-cell with characteristic maps $\kappa_n$ and $\psi$. $\qquad\square$

**Definition 3.3.5** Let $X$ and $Y$ be two cell complexes. A map $f : X \to Y$ is said to be **cellular** if

$$f(X^k) \subset Y^k, \; \forall X^k,$$

where $X^k$ denotes a $k$-skeleton of $X$ and $Y^k$ denotes a $k$-skeleton of $Y$.

**Proposition 3.3.6** *(i) CW-complexes and their cellular maps forms a category, denoted $\mathcal{CW}$.*

*(ii) Pointed CW-complexes and their cellular maps form a category, denoted $\mathcal{CW}^*$.*

**Proof** Take all CW-complexes as objects and their cellular maps as morphisms for the category $\mathcal{CW}$ and take all pointed CW-complexes $X$, where $* \in X$ is a 0-cell as objects and their cellular maps as morphisms for the category $\mathcal{CW}^*$. $\qquad\square$

Theorem 3.3.7 proves a close link between a CW-complex and its path component.

**Theorem 3.3.7** *(i) Each path component of a CW-complex is also a CW-complex.*
*(ii) Each path component of a CW-complex is both open and closed.*
*(iii) A CW-complex is connected iff it is path connected.*

**Proof** Let $X$ be an arbitrary CW-complex.

(i) Since $X$ is a disjoint union of cells and each cell is path connected, it follows that every path component $P$ of $X$ is a union of cells, and hence, $P$ is a CW-complex.

(ii) Let $P$ be a path component of $X$. Then $P$ is closed in $X$. Let $U$ be the union of its other path components in $X$. Then $U$ is a CW-subcomplex of the complex $X$, and it is also closed. $P$ is also open in $X$, since it is the complement of the closed set $U$ in $X$.

(iii) Let $P$ be a path component of $X$. Since $P$ is both open and closed in $X$, it follows that it is connected. Then (iii) follows, since path components of $X$ are the components of $X$. $\qquad\qquad\square$

**Definition 3.3.8** A $CW$-complex pair $(X, A)$ is a pair consisting of a topological space $X$ and a closed subspace $A$ of $X$ together with a sequence of closed subspaces $(X, A)^n$, called the $n$-**skeleton** of $X$ relative to $A$, for all $n \geq 0$ such that

(i) $(X, A)^0$ is constructed from $A$ by adjoining 0-cells.
(ii) For every $n \geq 1$, the pair $(X, A)^n$ is constructed from $(X, A)^{n-1}$ by adjoining $n$-cells.
(iii) $X = \bigcup (X, A)^n$.
(iv) $X$ has the weak topology with respect to $\{(X, A)^n\}_n$.
     For $A = \emptyset$, the $CW$-complex pair $(X, \emptyset)$ coincides with the $CW$-complex $X$.

**Example 3.3.9** For the simplicial pair $(K, L)$, the pair $(|K|, |L|)$ is a $CW$-complex pair such that $(|K|, |L|)^n = |K^n \cup L^n|$.

**Remark 3.3.10** For more study of $CW$-complexes, see Exercises 1–4 of Section 3.16.

## 3.4 Singular Homology Theory

This section gives a brief discussion on **Singular Homology Theory introduced by S. Lefschetz** in his paper [ Lefschetz, 1933]. He extended simplicial homology theory to singular homology theory by considering continuous maps from the standard simplexes to a topological space $X$, which are called singular simplexes in $X$ and by using the algebraic properties of singular chain complexes.

**Definition 3.4.1** Let $e_0, e_1, e_2, \ldots, e_n$ be the standard basis of $\mathbf{R}^{n+1}$. Then the standard $n$-**simplex** $\Delta_n$ is defined by

$$\Delta_n = \{\textstyle\sum_{i=0}^{n} \lambda_i e_i : \Sigma \lambda_i = 1, \text{ where } 0 \leq \lambda_i \leq 1\} \subset \mathbf{R}^{n+1}.$$

It is also defined by

$$\Delta_n = \{(\lambda_0, \lambda_1, \ldots, \lambda_n) \in \mathbf{R}^{n+1} : 0 \leq \lambda_i \leq 1 \text{ and } \Sigma \lambda_i = 1\} \subset \mathbf{R}^{n+1}.$$

**Definition 3.4.2** A **singular $n$-simplex** $\sigma_n$ in a topological space $X$ is a continuous map

$$\sigma_n : \Delta_n \to X$$

where $\Delta_n$ is the standard $n$-simplex.

***Example 3.4.3*** (i) A singular 0-simplex $\sigma_0$ is identified to a point in $X$, since, $\Delta_0$ is a one-point set.

(ii) A singular **1**-simplex $\sigma_1$ is a path in $X$, since $\Delta_1$ is the closed interval **I** .

**Definition 3.4.4** Given a set of points $\{v_0, v_1, \ldots, v_n : v_i \in \mathbf{R}^m\}$, its **affine combination** is a point

$$x = \lambda_0 v_0 + \lambda_1 v_1 + \cdots + \lambda_n v_n : \lambda_i \in \mathbf{R} \text{ and } \Sigma_{i=0}^n \lambda_i = 1.$$

A convex combination is an affine combination for which $0 \leq \lambda_i \leq 1, \forall i$.

***Example 3.4.5*** A convex combination of $x, y \in \mathbf{R}^n$, is of the form

$$\lambda x + (1 - \lambda)y, \forall \lambda \in \mathbf{I}.$$

**Definition 3.4.6** Given a set of points $\{v_0, v_1, \ldots, v_n : v_i \in \mathbf{R}^m\}$, the set $[v_0, v_1, \ldots, v_i, \ldots, v_n]$ consisting of all convex combinations of $v_0, v_1, \ldots v_i, \ldots, v_n$ i.e., denotes the convex set spanned by the points $v_0, v_1, \ldots, v_i, \ldots, v_n$.

**Definition 3.4.7** An ordered set of points $\{v_0, v_1, \ldots, v_n\} \subset \mathbf{R}^m$, is said to be **affine independent** if

$$\{v_1 - v_0, v_2 - v_0, \ldots, v_n - v_0\}$$

is a linearly independent subset of the real vector space $\mathbf{R}^n$.

**Definition 3.4.8** If $[v_0, v_1, \ldots, v_n]$ denotes an $n$-simplex $\Delta_n$, then the map

$$T_n : \Delta_n \to \mathbf{R}^m, \ \Sigma_{i=0}^n \lambda_i e_i \mapsto \Sigma_{i=0}^n \lambda_i v_i$$

is called an **affine singular $n$-simplex**.

***Remark 3.4.9*** The image of $T_n$ is a convex subspace of $\mathbf{R}^m$, called the **convex space spanned by** $v_i's$ which may not be independent.

**Definition 3.4.10** Let $[v_0, v_1, \ldots \hat{v}_i, \ldots, v_n]$ denote the affine $(n - 1)$- simplex $\Delta_{n-1}$ obtained from the affine $n$-simplex $[v_0, v_1, \ldots v_i, \ldots, v_n]$ by deleting the $i$- th vertex $v_i$ (symolized by $\hat{v}_i$) counted from 0. Then the map

$$[v_0, v_1, \ldots \hat{v}_i, \ldots, v_n] : \Delta_{n-1} \to \Delta_n$$

is called the $n$**th face map**, denoted by $f_i^n$.

**Definition 3.4.11** Given a singular $n$-simplex in a topological space $X$, its $i$th face is defined

$$\sigma_n^i = \sigma_n \circ f_i^n,$$

and its boundary is defined by

$$\partial \sigma_n = \Sigma_{i=0}^n (-1)^i \sigma_n^i, \quad \forall n \geq 1.$$

For an $n$-chain $c = \Sigma n_\sigma \sigma$, define

$$\partial_n c = \partial \Sigma n_\sigma \sigma = \Sigma n_\sigma \partial_n \sigma \ \forall n \geq 1, \ and \ if \ n = 0, \partial_0(c) = 0,$$

which is extended to $\Delta_n(X)$, the free abelian group based on the singular $n$-simplexes, to obtain a homomorphism

$$\partial_n : \Delta_n(X) \to \Delta_{n-1}(X).$$

**Proposition 3.4.12** *The composite homomorphism* $\partial_n \circ \partial_{n+1} = 0, \ \forall n \geq 0.$

**Proof** It follows from definition of $\partial_n$.                                    □

The concepts of chains, cycles and boundaries defined for simplicial complex are extended in a natural way to define singular complex associated with a topological space $X$.

**Definition 3.4.13** For every integer $n \geq 0$, the **singular $n$-chain group** $C_n(X; \mathbf{Z})$ is defined as the free abelian group with basis on all singular $n$-simplexes in $X$ and $C_{-1}(X; \mathbf{Z}) = \{0\}$ is taken.

(i) An element of $C_n(X; \mathbf{Z})$ can be represented as a formal linear combination of singular $n$-simplices with integer coefficients.

(ii) The elements of the group $C_n(X; \mathbf{Z})$ are called singular $n$-chains in $X$ with coefficient group $\mathbf{Z}$.

The **group of singular $n$-cycles** in $X$, denoted by $Z_n(X; \mathbf{Z})$, is the *ker* $\partial_n$ and the **group of singular $n$-boundaries**, denoted by $B_n(X; \mathbf{Z})$ is the *Im* $\partial_{n+1}$. Hence, it follows that

(i) $B_n(X; \mathbf{Z})$ and $Z_n(X; \mathbf{Z})$ are both subgroups of $C_n(X; \mathbf{Z}) \ \forall n \geq 0$.

(ii) Moreover, the condition $\partial_n \circ \partial_{n+1} = 0$ asserts that $B_n(X; \mathbf{Z})$ is a subgroup of $Z_n(X; \mathbf{Z})$.

**Definition 3.4.14** Given a topological space $X$, for every integer $n \geq 0$, the quotient group

$$H_n(X; \mathbf{Z}) = Z_n(X; \mathbf{Z})/B_n(X; \mathbf{Z}) = ker \ \partial_n / Im \ \partial_{n+1}$$

is called the $n$th **singular homology group** of $X$ with coefficient group $\mathbf{Z}$. For an $n$-cycle $z_n \in Z_n(X; \mathbf{Z})$, the coset $z_n + B_n(X; \mathbf{Z})$ is called the **homology class** of $z_n$, denoted by $[z_n]$. The notion of the $n$th singular homology group $H_n(X; G)$ of $X$ with coefficient group $G$ (abelian) is analogous.

**Definition 3.4.15** Let $f : X \to Y$ be a continuous map of topological spaces. Then it induces a group homomorphism

$$f_* : H_n(X; \mathbf{Z}) \to H_n(Y; \mathbf{Z}), \; [z_n] \mapsto [f(z_n)].$$

**Proposition 3.4.16** $H_n : \mathcal{T}op \to \mathcal{A}b$ *is a covariant functor from the category of topological spaces and their continuous maps to the category of abelian groups and their homomorphisms, for every integer $n \geq 0$.*

**Proof** Here, the object function is defined by

$$H_n : \mathcal{T}op \to \mathcal{A}b, \; X \mapsto H_n(X; \mathbf{Z})$$

and for every continuous map $f : X \to Y$ in the category $\mathcal{T}op$, the morphism function is defined by

$$f \mapsto f_* : H_n(X; \mathbf{Z}) \to H_n(Y; \mathbf{Z}), \; [z_n] \mapsto [f(z_n)].$$

$\square$

**Corollary 3.4.17** *Let $X$ and $Y$ be two homeomorphic spaces. Then*

$$H_n(X; \mathbf{Z}) \cong H_n(Y; \mathbf{Z}), \; \forall\, n \geq 0.$$

**Proof** It follows from the functorial property of $H_n$. $\square$

**Proposition 3.4.18** *Let $X$ be a topological space and $\{X_j : j \in \mathbf{A}\}$ be the set of path components of $X$. Then every inclusion map $i_j : X_j \hookrightarrow X$ induces an 1-1 homomorphism*

$$i_j^* : H_n(X_j; \mathbf{Z}) \to H_n(X; \mathbf{Z}), \; \forall\, n \geq 0$$

*of groups such that for the natural homomorphism*

$$\psi : \oplus_{j \in \mathbf{A}} H_n(X_j; \mathbf{Z}) \to H_n(X; \mathbf{Z}), \; \forall\, n \geq 0,$$

$$\psi | H_n(X_j; \mathbf{Z}) = i_j^*, \; \forall\, j \in \mathbf{A}, \; \forall\, n \geq 0.$$

**Proof** Using the fact that the image of a singular simplex under a continuous map completely lies in one path component, it follows that every inclusion map $i_j : X_j \hookrightarrow X$ induces a homomorphism

$$i_j^+ : C_n(X_j; \mathbf{Z}) \to C_n(X; \mathbf{Z}).$$

Hence passing to the corresponding homology groups, the proposition is proved. $\square$

### 3.4.1 Advantage of Singular Homology over Simplicial Homology

There are several **advantages of singular homology theory over simplicial homology** theory as a generalization of simplicial homology theory. They include mainly as stated below:

(i) Singular homology theory is defined on all topological spaces, not on just polyhedra.

(ii) The induced homomorphism of a continuous map is easier to define in singular homology than in simplicial homology.

**Remark 3.4.19** Because of this advantage, **singular homology theory has developed extensively.** Instead of **Z**, called coefficient group of the singular homology theory, **any abelian group $G$ may be taken equally well as its coefficient group.** For detailed study of the singular homology theory, the book [Adhikari 2016] is referred.

### 3.4.2 The Singular Homology with Coefficient Group G

This subsection extends homology theory with integral coefficient formulated in Definition 3.4.14 to homology theory with an arbitrary abelian group $G$ as its coefficient group.

**Definition 3.4.20** Let $X$ be an arbitrary topological space and $G$ be an abelian group. Then $Z_0(X; G)$ of 0-cycles on $X$ is an abelian group, and $B_0(X; G)$ of 0-dimensional boundaries is a subgroup of $Z_0(X; G)$. Its quotient group

$$H_0(X; G) = \frac{Z_0(X; G)}{B_0(X; G)} \cong F,$$

where $F$ is the free abelian group on the set of path-connected components of the topological space $X$. The group $H_0(X; G)$ is called its 0-**dimensional homology group** of $X$ with $G$ as its coefficient group. In a analogous way, its 1-**dimensional homology group** is the quotient group

$$H_1(X; G) = \frac{Z_1(X; G)}{B_1(X; G)}$$

is called the first homology group of $X$ with coefficient in $G$, where $C_1(X; G)$ is the is the abelian group of 1-chains, $Z_1(X; G)$ is the subgroup of 1-cycles of $C_1(X; G)$ on $X$ and $B_1(X; G)$ of 1-dimensional boundaries is a subgroup of $Z_1(X; G)$. In general, the $n$-**dimensional homology group** of $X$ with coefficient group $G$, is the quotient group

$$H_n(X; G) = \frac{Z_n(X; G)}{B_n(X; G)}$$

## 3.5 Noether's Algebraic Approach to Homology and Cohomology Theories

This section gives an **algebraic approach formulated by Emmy Noether (1882–1935) corresponding to the geometric approach of homology theory** invented by Poincaré and described in Sect. 3.6.1. Inspired by her algebraic approach, P. Alexandroff (1896–1982) and H. Hopf (1894–1971) jointly published a detailed study of homology theory in 1935 for the first time. Its dual algebraic approach to cohomology theories is also given in Sect. 3.5.2.

### 3.5.1 Homology Groups of Chain Complexes

The geometric approach of homology theory invented by Poincaré is shifted to algebraic approach by Emmy Noether. In her algebraic approach, homology groups are defined for any topological space. This approach has created homological algebra, a new branch of mathematics. The classical method of construction consists of construction of homology groups of chain complexes associated with topological spaces.

**Definition 3.5.1** A sequence $C_*$ of abelian groups $\{C_n\}$ and their homomorphisms $\{\partial_n\}$

$$C_* : \cdots \xrightarrow{\partial_{n+1}} C_n \xrightarrow{\partial_n} C_{n-1} \xrightarrow{\partial_{n-1}} \cdots \xrightarrow{\partial_1} C_0 \xrightarrow{\partial_0} \{0\}$$

is said be a **chain complex** if

$$\partial_{n-1} \circ \partial_n = 0, \ \forall n \geq 1.$$

The groups $C_n$ are called **chain groups** and the homomorphisms $\partial_n : C_n \to C_{n-1}$ are called **boundary homomorphisms** of the chain complex $C_*$. This chain complex is sometimes written as $C_* = \{C_n, \partial_n\}$.

**Definition 3.5.2** Given a chain complex $C_* = \{C_n, \partial_n\}$, the condition

$$\partial_n \circ \partial_{n+1} = 0, \ \forall n \geq 1$$

asserts that Im $\partial_{n+1}$ is a subgroup of ker $\partial_n$. The quotient group ker $\partial_n$/ Im $\partial_{n+1}$ is called the $n$ th **homology group** of the chain complex $C_*$ denoted by $H_n(C_*)$. On the

**Fig. 3.1** Chain map
$f : C_* \to C'_*$

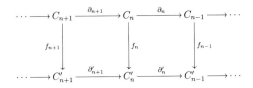

other hand, ker $\partial_n$ is called the group of $n$-**dimensional cycles,** denoted by $Z_n$ and Im $\partial_{n+1}$ is called the $n$-**dimensional boundaries,** denoted by $B_n$.

(i)   An element of $Z_n = ker\partial_n$ is called an $n$-dimensional **cycle.**
(ii)  An element of $B_n = Im\partial_{n+1}$ is called an $n$-dimensional **boundary.**
(iii) Elements of $H_n(C_*) = Z_n/B_n$ are called **homology classes**, denoted by $[z]$ for every $z \in Z_n$.

**Definition 3.5.3**   Let $M^n$ be an $n$-dimensional manifold and $\{C_i^m\}$ be the set of all $m$-dimensional cycles (which are $m$-dimensional submanifolds with boundary) of $M^n$ and $B^{m+1}$ be a submanifold of $M^n$ with a boundary consisting of connected manifolds $C_1^m, C_2^m, \ldots, C_i^m$, which are in $\{C_i^m\}$. Then the cycle $C_1^m + C_2^m + \cdots + C_i^m$ is said to be **homologous to 0.**

**Definition 3.5.4**   Two cycles in an $n$-dimensional manifold $M^n$ are said to be **homologous or equivalent** if they differ by a cycle homologous to 0. The set of equivalence classes of $m$-dimensional cycles is called the $m$-**dimensional homology group of** $M^n$.

**Definition 3.5.5**   Given two chain complexes $C_* = \{C_n, \partial_n\}$ and $C'_* = \{C'_n, \partial'_n\}$, a sequence $f = \{f_n : C_n \to C'_n\}$ of homomorphisms is said to be a **chain map** from $C_*$ to $C'_*$, if these homomorphisms commute with the boundary homomorphisms in the sense

$$f_n \circ \partial_{n+1} = \partial'_{n+1} \circ f_{n+1}, \ \forall n \geq 0,$$

i.e., if every square in Fig. 3.1 is commutative.

**Proposition 3.5.6**   *Given two chain complexes of abelian groups $C_* = \{C_n, \partial_n\}$ and $C'_* = \{C'_n, \partial'_n\}$, for a chain map $f = \{f_n : C_n \to C'_n\}$,*

(i)   *its images $f_n(Z_n) \subset Z'_n$ for every n, i.e., $f_n$ sends n-cycles of $C_*$ into n-cycles of $C'_*$. and*
(ii)  *$f_n(B_n) \subset B'_n$ for every n, i.e., $f_n$ sends n-boundaries of $C_*$ into n-boundaries of $C'_*$.*

**Proof**  Using the commutativity of each square in Fig. 3.1, the proposition follows. □

**Theorem 3.5.7**   *Given two chain complexes $C_* = \{C_n, \partial_n\}$ and $C'_* = \{C'_n, \partial'_n\}$, every chain map $f : C_* \to C'_*$ induces a homomorphism*

$$H_n(f_n) = f_{n*} : H_n(C_*) \to H_n(C'_*), [z] \mapsto [f_n(z)].$$

**Proof** It follows from Proposition 3.5.6                                   □

**Definition 3.5.8** The homomorphism $H_n(f_n) = f_{n*} : H_n(C_*) \to H_n(C'_*)$ defined in Theorem 3.5.7 is said to be the **homomorphism induced by $f_n$ in homology groups** for every integer $n$.

**Proposition 3.5.9** (a) *Given two chain maps $f : C_* \to C'_*$ and $g : C'_* \to C''_*$, their composite*

$$g \circ f : C_* \to C''_*$$

*is also a chain map such that*

$$(g \circ f)_* = g_* \circ f_* : H_n(C_*) \to H_n(C''_*).$$

(b) *If $1_d : C_* \to C_*$ is the identity chain map, then its induced map*

$$(1_d)_* : H_n(C_*) \to H_n(C_*)$$

*is the identity automorphism.*

**Proof** It follows from Definition 3.5.8 of induced homomorphism in the homology groups.                                                                  □

**Definition 3.5.10** Given two chain complexes $C_* = \{C_n, \partial_n\}$ and $C'_* = \{C'_n, \partial'_n\}$, $f, g : C_* \to C'_*$ their two chain maps $f, g : C_* \to C'_*$ are said to be **chain homotopic**, denoted by $f \simeq g$, if there is a sequence

$$\{H_n : C_n \to C'_{n+1}\}$$

of homomorphisms such that

$$\partial'_{n+1} H_n + H_{n-1} \partial_n = f_n - g_n : C_n \to C'_n, \ \forall n \geq 0.$$

On the other hand, a chain map

$$f : C_* \to C'_*$$

is said to be a **chain homotopy equivalence** if there exists a chain map

$$h : C'_* \to C_*$$

such that

$$h \circ f \simeq 1_d \text{ and } f \circ h \simeq 1_d.$$

**Proposition 3.5.11** *Let $\mathcal{S}(C_*, C'_*)$ be the set of all chain maps from $C_*$ to $C'_*$. Then the chain homotopy relation on $\mathcal{S}(C_*, C'_*)$ is an equivalence relation.*

**Proof** It follows from Definition 3.5.10.    □

**Theorem 3.5.12** *Any two chain homotopic maps induce the same homomorphisms in the homology.*

**Proof** Let $f, g : C_* \to C'_*$ be two chain maps such that $f \simeq g$ in the sense of Definition 3.5.10. Then there is a chain homotopy $\{H_n : C_n \to C'_{n+1}\}$ such that for $[z] \in H_n(C_*)$, the image $\partial_n([z]) = 0$. This implies that

$$f_n([z]) - g_n([z]) = \partial_{n+1} H_n([z])$$

is a boundary. Consequently,

$$[f_n[z]] = [g, [z]], \forall [z] \in H_n(C_*) \implies f_{n*}([z]) = g_{n*}([z]), \forall [z] \in H_n(C_*) \implies f_{n*} = g_{n*}.$$

This proves that

$$f_* = g_* : H_n(C_*) \to H_n(C'_*), \forall n \geq 0.$$

□

**Theorem 3.5.13** (a) *Chain complexes and chain maps form a category, called* ***chain complex category,*** *denoted by* Comp.
(b) *$H_n$ is a covariant functor from the category* Comp *of chain complexes and chain maps to the category* Ab *of abelian groups and homomorphisms for every $n \geq 0$, i.e., $H_n :$ Comp $\to$ Ab is a covariant functor for every integer $n \geq 0$.*
(c) *$H_n :$ Top $\to$ Ab is also a covariant functor for every integer $n$, where* Top *is the category of topological spaces and their continuous maps.*

**Proof** (a) Here , chain complexes are taken as objects and chain maps are taken as morphisms. Then under the composition of chain maps coordinatewise : $\{g_n\} \circ \{f_n\} = \{g_n \circ f_n\}$, they form the category Comp.
(b) Here, the object function assigns to every chain complex the sequence of its homology groups, and morphism function assigns to every chain map $f$ between chain complexes the induced map $f_*$ between their homology group. This asserts that , $H_n :$ Comp $\to$ Ab is a covariant functor for every integer $n$, by Proposition 3.5.9.
(c) Here, the object function assigns to every topological space $X \in$ Top, its homology groups $H_n(X; G)$. To define the morphism function, consider a continuous map $f : X \to Y \in$ Top. If $\sigma : \Delta_n \to X$ is an $n$-simplex in the space $X$, then

$$f \circ \sigma : \Delta_n \to Y$$

is an $n$-simplex in the space $Y$. This gives rise to a homomorphism

$$f_\# : C_n(X; \mathbf{Z}) \to C_n(Y; \mathbf{Z}), \ \Sigma k_\sigma \sigma \mapsto \Sigma k_\sigma (f \circ \sigma)$$

(extending by linearity), which is independent of $n$. Since an element of $H_n(X; \mathbf{Z}) = Z_n(X; \mathbf{Z})/B_n(X; \mathbf{Z})$, is a coset $\alpha_n + B_n(X; \mathbf{Z})$, where $\alpha_n \in Z_n(X; \mathbf{Z})$. Then $f$ induces the map

$$H_n(f) = f_* : H_n(X; \mathbf{Z}) \to H_n(Y; \mathbf{Z}), \quad \alpha_n + B_n(X; \mathbf{Z}) \to f_*(\alpha_n) + B_n(Y; \mathbf{Z}).$$

This is well defined and is a homomorphism. Moreover, it follows from Definition of $f_*$ that

(i) $H_n(1_X) : H_n(X; \mathbf{Z}) \to H_n(X; \mathbf{Z})$ is the identity homomorphism and
(ii) For any two maps $f : X \to Y$ and $g : Y \to Z$ in $\mathcal{T}op$

$$H_n(g \circ f) = H_n(g) \circ H_n(f) : H_n(X; \mathbf{Z}) \to H_n(Z; \mathbf{Z}).$$

Hence it is proved that

$$H_n : \mathcal{T}op \to \mathcal{A}b$$

is a covariant functor for each integer $n \geq 0$. $\qquad\square$

**Corollary 3.5.14** *Homeomorphic spaces induce isomorphic homology groups.*

**Proof** To prove the corollary, it is sufficient to show that if $X, Y \in \mathcal{T}op$ are homeomorphic, then the corresponding homology groups

$$H_n(X; \mathbf{Z}) \cong H_n(Y; \mathbf{Z}), \quad \forall n \geq 0.$$

It follows from the functorial property of

$$H_n : \mathcal{T}op \to \mathcal{A}b$$

for each $n \geq 0$ proved in Theorem 3.5.13.

$\qquad\square$

**Remark 3.5.15** For any $X \in \mathcal{T}op$,

(i) Every homology group $H_n(X; \mathbf{Z})$ is a **topological invariant of** $X$ in the sense that its homeomorphic spaces have isomorphic homology group for every integer $n \geq 0$.
(ii) rank $H_n(X; \mathbf{Z})$ is an invariant of $X$ for every integer $n \geq 0$ in the sense that if the topological spaces $X$ and $Y$ are homeomorphic, then

$$rank\ H_n(X; \mathbf{Z}) = rank\ H_n(Y; \mathbf{Z}), \quad \forall n \geq 0.$$

This rank is called the $n$th **Betti number** of $X$.

**Definition 3.5.16** The sequence $\{H_n : \mathcal{T}op \to \mathcal{A}b\}$ of covariant functors $H_n$ given in Theorem 3.5.13 is called the **homology sequence** from the category $\mathcal{T}op$ to the category $\mathcal{A}b$.

### 3.5.2  Cochain Complex and Its Cohomology Groups

Cohomology groups are dual to homology groups defined for any topological space. The classical method of their construction consists of construction of cohomology groups associated with topological spaces. The basic difference between these two theories is that

(i)  Homology theory

$$H_n : \mathcal{T}op \to \mathcal{A}b$$

is a covariant functor from the category $\mathcal{T}op$ of topological spaces and their continuous maps to the category of abelian groups $\mathcal{A}b$ and their homomorphisms, for every integer $n \geq 0$.

(ii)  On the other hand, cohomology theory

$$H^n : \mathcal{T}op \to \mathcal{A}b$$

is a contravariant functor from the category $\mathcal{T}op$ of topological spaces and their continuous maps to the category of abelian groups $\mathcal{A}b$ and their homomorphisms, for every integer $n \geq 0$.

### 3.5.3  Cochain Complex and Its Cohomology Groups

Let $\mathcal{A}b$ denote the category of abelian groups and their homomorphisms and $\mathcal{T}op$ denote the category of topological spaces and their continuous maps. The motivation of cohomology theory comes from homological algebra.

**Definition 3.5.17** For any $X \in \mathcal{T}op$ and any abelian group $G \in \mathcal{A}b$, the singular cochain group $C^n(X; G)$ with coefficient group $G$, is defined by

$$C^n(X; G) = Hom(C_n(X, G), G),$$

where the right-hand side is the group of all homomorphisms from the singular chain group $(C_n(X; G)$ to the abelian group $G$.

**Definition 3.5.18** Let $G$ be an abelian group and $\beta \in (C^n(X; G)$ be an arbitrary element. Then

$$\beta : C_n(X; G) \to G$$

is a group homomorphism. Define a map

$$\delta : C_{n+1}(X; G) \to G, \ \beta \mapsto \beta \circ \partial$$

Then $\delta$ is the composite homomorphism

$$C_{n+1}(X;G) \xrightarrow{\partial} C_n(X;G) \xrightarrow{\beta} G.$$

This asserts that

$$\delta\beta(\sigma) = \Sigma_i(-1)^i(\sigma|_{[v_0,v_1,...,\hat{v}_i,...,v_{n+1}]})$$

for all singular $(n+1)$ simplexes $\sigma$, where the notation hat over $v_i$, i.e., $\hat{v}_i$ indicates that the vertex $v_i$ is omitted from the vertex set

$$v_0, v_1, \ldots, v_i, \ldots, v_n.$$

Then it follows that $\delta \circ \delta = 0$.

For simplicity, the notation $C^* = \{C^n, \delta^n\}$ stands for notation $C^*(X;G) = \{C^n(X;G), \delta^n\}$

**Definition 3.5.19** A sequence $C^* = \{C^n, \delta^n\}, n \in \mathbf{Z}$ of abelian groups $C^n$ and their homomorphisms $\delta^n : C^{n-1} \to C^n$ is said to be a **cochain complex** if

$$\delta^{n+1} \circ \delta^n = 0 \ \forall n \geq 0.$$

This asserts that the sequence (3.4) of abelian groups $\{C^n\}$ and their homomorphisms $\delta^n : C^{n-1} \to C^n$ is a cochain complex if $\delta^{n+1} \circ \delta^n = 0 \ \forall n \geq 0$:

$$C^* : \cdots \to C^{n-1} \xrightarrow{\delta^n} C^n \xrightarrow{\delta^{n+1}} C^{n+1} \to \cdots \qquad (3.4)$$

**Definition 3.5.20** In Definition 3.5.19,

(i)  $\delta^n$ is called a **coboundary homomorphism**.
(ii)  The elements of $Z^n = \ker \delta^{n+1}$ are called $n$-**cocycles**.
(iii)  The elements of $B^n = \text{Im } \delta^n$ are called $n$-**coboundaries**

of the cochain complex $C^*$ given in (3.4).

**Proposition 3.5.21** $Z^n$ and $B^n$ form groups for all n for the cochain complex $C^*$ given in (3.4).

**Proof** It follows from respective definitions. ☐

**Proposition 3.5.22** $B^n$ is a subgroup of $Z^n$ for all n for the cochain complex $C^*$ given in (3.4).

**Proof** The cochain complex $C^*$ given in (3.4) implies that $\delta^{n+1} \circ \delta^n = 0$. This proves that $B^n$ is a subgroup of $Z^n$ for all $n$. ☐

**Fig. 3.2** Cochain map
$f = \{f^n : C^n \to C'^n\}$

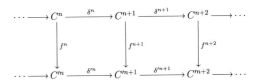

**Definition 3.5.23 (Cohomology group of $C^*$)** Given a cochain complex $C^*$ given in (3.4), the quotient group

$$Z^n(C^*)/B^n(C^*),$$

denoted by $H^n(C^*)$ (or simply $H^n$), is said to be the $n$-dimensional **cohomology group** of the cochain complex $C^*$.

### 3.5.4 Singular Cohomology

Definition 3.5.23 leads to singular cohomology group of an arbitrary topological space formulated in Definition 3.5.24.

**Definition 3.5.24 (singular cohomology group)** Given an abelian group $G$ and a topological space $X$, let $C^n(X; G)$ denote the group $Hom(C_n(X), G))$, $\forall n \geq 0$, where $C_n(X)$ denotes the group of singular $n$-chains in $X$ with coefficient group $G$ (see Definition 3.4.13).

(i)  An element of $C^n(X; G)$ is called an $n$-**dimensional singular cochain** with coefficients in G (in brief $n$-cochain).

(ii)  An element of $Z^n(X; G) = \ker \delta^{n+1}$ is called an $n$-**cocycle** with coefficient in $G$.

(iii)  An element of $B^n(X; G) = \mathrm{Im}\, \delta^n$ is called an $n$-**coboundary** with coefficient in G.

(iv)  The quotient group $Z^n(X; G)/B^n(X; G)$ denoted by the quotient group $H^n(X; G)$ is called the $n$-**dimensional singular cohomology group** of $X$ with coefficents in $G$.

(v)  An element of $H^n(X; G)$ is a coset $\alpha + B^n(X; G)$, where $\alpha$ is a cocycle. This is called the **cohomology class** of $\alpha$ written as $[\alpha] \in H^n(X; G)$.

**Definition 3.5.25** For $n \geq 0$, given two cochain complexes $C^* = \{C^n, \delta^n\}$ and $C'^* = \{C'^n, \delta'^n\}$, a sequence

$$f = \{f^n : C^n \to C'^n\}$$

is called a **cochain map** from $C^*$ to $C'^*$ if the diagram in Fig. 3.2 commutes,i.e., if

$$f^{n+1} \circ \delta^n = \delta'^n \circ f^n$$

holds for all $n \geq 0$.

### 3.5.5 Functorial Approach to Singular Cohomology Theory

This subsection presents a functorial approach to singular homology theory. Let $\mathcal{Ab}$ denote the category of abelian groups and their homomorphisms and $\mathcal{Top}$ denote the category of topological spaces and their continuous maps. The motivation of cohomology theory comes from homological algebra.

**Proposition 3.5.26** *Given an abelian group $G$, there is a* **contravariant functor**

$$Hom \, ( \, , \, G) : \mathcal{Ab} \to \mathcal{Ab} \, .$$

**Proof** The object function is defined by assigning to every group $A \in \mathcal{Ab}$, the group $Hom \, (A, G)$ of homomorphisms from $A$ to $G$ and morphism function is defined by assigning to a group homomorphism $f : A \to B \in \mathcal{Ab}$, the homomorphism

$$f^* : \, Hom \, (B, G) \to Hom \, (A, G), \, \psi \mapsto \psi \circ f.$$

This implies the proposition.

$\square$

**Theorem 3.5.27** *Given an integer $n \geq 0$, for every abelian group $G$, the cohomology*

$$H^n( \, ; G) : \mathcal{Top} \to \mathcal{Ab}$$

*is a contravariant functor.*

**Proof** Here, the object function is defined by assigning to every topological space $X \in \mathcal{Top}$, the cohomology group $H^n(X \, ; G) \in \mathcal{Ab}$, and morphism function is defined by assigning to a continuous map $f : X \to Y \in \mathcal{Top}$, the homomorphism

$$f^* : \, H^n(Y \, ; G) \to H^n(X \, ; G), \alpha + B^n(Y; G) \mapsto f^{\#}(\alpha) + B^n(X; G) = \alpha \circ f + B^n(X; G)$$

where $f^{\#}$ is defined by

$$f^{\#} : Hom(C_n(Y; G), G) \to Hom(C_n(X; G), G), \, k \mapsto k \circ f_{\#}$$

and $f_\#$ is defined in Theorem 3.5.13

$$f_\# : C_n(X; G) \to C_n(Y; G), \ \Sigma k_\sigma \sigma \mapsto \Sigma k_\sigma (f \circ \sigma).$$

Clearly, $f^*$ is well defined, and it is a homomorphism. Moreover, it follows from Definition of $f^*$ that

(i)  $H^n(1_X) : H^n(X; \mathbf{Z}) \to H^n(X; \mathbf{Z})$ is the identity homomorphism and
(ii) For any two maps $f : X \to Y$ and $g : Y \to Z$ in $\mathcal{T}op$

$$H^n(g \circ f) = H^n(f) \circ H^n(g) : H_n(Z; G) \to H^n(X; G).$$

Hence it is proved that $H^n : \mathcal{T}op \to \mathcal{A}b$ is a contravariant functor for each $n \geq 0$. $\square$

**Remark 3.5.28**  For any $X \in \mathcal{T}op$,

(i)  Every cohomology group $H^n(X; \mathbf{Z})$ is an invariant of $X$ in the sense that its homeomorphic spaces have isomorphic cohomology group for every integer $n \geq 0$;
(ii) Rank $H^n(X; \mathbf{Z})$ is an invariant of $X$ for every integer $n \geq 0$.

**Definition 3.5.29**  The sequence $\{H^n : \mathcal{T}op \to \mathcal{A}b\}$ of contravariant functors $H^n$ given in Theorem 3.5.27 is called the **cohomology sequence** from the category $\mathcal{T}op$ to the category $\mathcal{A}b$.

**Remark 3.5.30**  Dual concepts and results of homology mentioned in Section 3.5.1 hold in cohomology.

## 3.6  Homology Theory: Eilenberg and Steenrod Axioms

Eilenberg and Steenrod announced an outstanding new approach to **homology and cohomology theories** in 1945 by taking a small number of properties of the existing homology and cohomology theories (not focusing the tools used for construction of homology and cohomology groups) as axioms to characterize both homology and cohomology. This axiomatic approach was published in 1952 [Eilenberg and Steenrod 1952]. The most surprising result is the proof that on the category of all topological pairs having homotopy type of finite CW-complex pairs all homology (resp. cohomology) theories satisfying these axioms have isomorphic groups. This result proves that there is only one concept of homology on that category [Eilenberg and Steenrod 1952]. **The Eilenberg and Steenrod axioms** (in brief, E-S axioms) for homology or cohomology functors provide an elegant and quick access to the further study of algebraic topology, and Dimension Axiom locates the coefficient group at the right dimension.

This section presents **Eilenberg and Steenrod axioms** on the category $\mathcal{T}op^2$ of topological pairs and their continuous maps, which is the largest admissible category in the sense that it contains other admissible categories as subcategories. The earliest definition of homology group given by Poincaré in 1895 stands matched with the Eilenberg and Steenrod axioms on the admissible category $\mathcal{S}imp$ defined in Example 3.6.5. However, for modern study of homology and cohomology theories (with generalized homology and cohomology theories), the books [Adams 1972; Adhikari 2016] are referred.

### 3.6.1 Admissible Categories for Eilenberg and Steenrod Axioms

This subsection specifies admissible category $\mathcal{C}$ in Definition 3.6.4 to formulate Eilenberg and Steenrod axioms for homology and cohomology theories.

**Proposition 3.6.1** (i) *The pairs of topological spaces and their continuous maps form a category, called the category of topological pairs, denoted by $\mathcal{T}op^2$.*
(ii) *The pairs of topological spaces and the homotopy classes of their continuous maps relative to a subspace form a category, called the homotopy category of topological pairs, denoted by $\mathcal{H}tp^2$.*

**Proof** It follows from the corresponding definitions. □

**Definition 3.6.2** *(Topological triple)* A triple $(X : A, B)$ is said to be a topological triple, if $X$ is a topological space such that $A$ and $B$ are subspaces of $X$ with $B \subset A$.

**Definition 3.6.3** *(Topological triad)* A triple $(X : A, B)$ is said to be a topological triad, if $X$ is a topological space such that $(A, B)$ is an ordered pair of subspaces $A$ and $B$ of $X$.

The concept of topological triad is used in Sect. 3.11.4 on Mayer-Vietoris theorem.

**Definition 3.6.4** A category $\mathcal{C}$ is said to be an **admissible category** for the Eilenberg and Steenrod axioms if $\mathcal{C}$ satisfies the following conditions:

(i) $\mathcal{C}$ consists of topological pairs $(X, A)$, as objects, i.e., if $(X, A) \in Ob\ \mathcal{C}$, then the pairs $(X, X), (A, A), X \equiv (X, \emptyset), A \equiv (A, \emptyset), \emptyset \equiv (\emptyset, \emptyset) \in Ob\ \mathcal{C}$.
(ii) $\mathcal{C}$ consists of morphisms $f : (X, A) \to (Y, B)$, for every pair of objects $(X, A)$ and $(Y, B) \in Ob\ (C)$ with all their possible inclusions, i.e., if $f : (X, A) \to (Y, B)$ is a morphism in $\mathcal{C}$, then all the inclusion maps induced by $f$ on the corresponding subpairs are also morphisms in $\mathcal{C}$.
(iii) For any object $(X, A) \in Ob\ \mathcal{C}$, the object $(X \times \mathbf{I}, A \times \mathbf{I}) \in Ob\ \mathcal{C}$ and the maps

$$H_t : X \to X \times \mathbf{I},\ x \mapsto (x, t)$$

are morphisms in $\mathcal{C}$.

(iv) There is one-point spaces $X_0 \in Ob\ C$, with the property that the constant map

$$c : X \to X_0, \ x \mapsto X_0$$

is a morphism in $C$.

**Example 3.6.5** **Some examples of admissible category for Eilenberg and Steenrod axioms for homology and cohomology theories:**

(i) All topological pairs $(X, A)$ and their continuous maps form an admissible category for Eilenberg and Steenrod axioms for homology and cohomology theories, denoted by $\mathcal{T}op^2$.

(ii) All the pairs $(X, A)$ of spaces where the spaces $X$ and $A$ have triangulations together with their simplicial maps form an admissible category for Eilenberg and Steenrod axioms for homology and cohomology theories, denoted by $Simp$.

(iii) All compact pairs $(X, A)$ in the sense that $X$ is compact and $A$ is closed in $X$ and all continuous maps of such pairs form an admissible category for Eilenberg and Steenrod axioms for homology and cohomology theories, denoted by $\mathcal{C}_c$.

### 3.6.2 Eilenberg and Steenrod Axioms for Homology Theory

This subsection presents Eilenberg and Steenrod axioms for homology theory on the category $C = \mathcal{T}op^2$ of all topological pairs $(X, A)$ of spaces and their continuous maps.

A **homology theory** $\mathcal{H}$ on the category $\mathcal{T}op^2$ consists of three functions $\mathcal{H} = \{H, *, \partial\}$ which satisfy the following axioms for every integer $n \geq 0.$ :

(i) The first function $H$ assigns to every topological pair $(X, A)$ of spaces in $\mathcal{T}op^2$ and every integer $n \geq 0$, an abelian group $H_n(X, A)$, **called the $n$-dimensional homology group** of the topological pair $(X, A)$ in the homology theory $\mathcal{H}$. For $A = \emptyset$, it is called $n$-**dimensional (absolute) homology group** of the space $X$.

(ii) The second function $*$ assigns to every continuous map $f : (X, A) \to (Y, B)$ in $\mathcal{T}op^2$ and every integer $n \geq 0$, a homomorphism

$$f_* = f_{n*} : H_n(X, A) \to H_n(Y, B),$$

called the **homomorphism induced** by the map $f$ in the homology theory $\mathcal{H}$.

(iii) The third function $\partial$ assigns to each topological pair $(X, A)$ in $\mathcal{T}op^2$ and every integer $n \geq 1$, a homomorphism

$$\partial = \partial_n : H_n(X, A) \to H_{n-1}(A),$$

called the **boundary operator** on the group $H_n(X, A)$ in the homology theory $\mathcal{H}$.

**Fig. 3.3** Diagram
connecting boundary
operator $\partial$ and induced
homomorphisms in $\mathcal{H}$

$$
\begin{array}{ccc}
H_n(X, A) & \xrightarrow{\ f_*\ } & H_n(Y, B) \\
\Big\downarrow{\scriptstyle \partial} & & \Big\downarrow{\scriptstyle \partial} \\
H_{n-1}(A) & \xrightarrow{\ g_*\ } & H_{n-1}(B)
\end{array}
$$

Moreover, these functions satisfy the following seven axioms, called the **Eilenberg–Steenrod axioms for homology theory** $\mathcal{H}$ on $\mathcal{T}op^2$;

**Axiom H(1) (Identity Axiom).** If $1_X : (X, A) \to (X, A)$ is the identity map on a topological pair $(X, A)$ in $\mathcal{T}op^2$, then its induced homomorphism

$$
1_{X*} : H_n(X, A) \to H_n(X, A)
$$

is the identity automorphism of the homology group $H_n(X, A)$ for every integer $n \geq 0$.

**Axiom H(2) (Composition Axiom).** If $f : (X, A) \to (Y, B)$ and $g : (Y, B) \to (Z, C)$ are two continuous maps in $\mathcal{T}op^2$, then

$$
(g \circ f)_{*n} = g_{*n} \circ f_{*n} : H_n(X, A) \to H_n(Z, C)
$$

for every integer $n \geq 0$.

**Remark 3.6.6** **In the language of category theory**, the above axioms **H(1) and H(2)** assert that for every fixed integer $n \geq 0$, the functions $H_n$ form a covariant functor from the category $\mathcal{T}op^2$ to the category $\mathcal{A}b$. Each

$$
H_n : \mathcal{T}op^2 \to \mathcal{A}b
$$

is called the **homology functor in the homology theory** $\mathcal{H}$ on the category $\mathcal{T}op^2$.

We use the notation $H_n(f) = f_{*n}$ (Fig. 3.3).

**Axiom H(3) (Commutativity Axiom).** If $f : (X, A) \to (Y, B)$ is a continuous map in category $\mathcal{T}op$ and if $g : A \to B$ is a continuous map in $\mathcal{T}op$, defined by $g(x) = f(x)$ for all $x \in A$, then the diagram in Fig. 3.3 is commutative in the sense that $g_* \circ \partial = \partial \circ f_*$ for every integer $n \geq 0$.

**Remark 3.6.7** This axiom connects the homology functor $H_n$ with boundary operator $\partial$ and induced homomorphisms in the homology theory $\mathcal{H}$.

**Axiom H(4) (Exactness Axiom).** If $(X, A)$ is a topological pair in $\mathcal{T}op^2$ and $i : A \hookrightarrow X, j : X \to (X, A)$ are the inclusion maps, then the beginningless sequence

$$
\cdots \to H_n(A) \xrightarrow{\ i_*\ } H_n(X) \xrightarrow{\ j_*\ } H_n(X, A) \xrightarrow{\ \partial\ } H_{n-1}(A) \to \cdots
$$

of groups and homomorphisms is exact, called the **homology exact sequence** of $(X, A)$.

**Remark 3.6.8** *The axioms formulated in* **H(1)-H(4)** *are all algebraic axioms.*

**Axiom H(5)(Homotopy Axiom)** If two continuous maps $f, g : (X, A) \to (Y, B)$ in $\mathcal{T}op^2$ are homotopic in $\mathcal{T}op^2$, then

$$f_{*n} = g_{*n}, \ \forall n \geq 0.$$

**AxiomH(6)(Excision Axiom).** Let $U$ be an open set of a topological space $X$, where its closure $\overline{U}$ is contained in the interior Int A= $\mathring{A}$ of a subspace $A$ of $X$ (i.e., $\overline{U} \subset \mathring{A}$). If the inclusion map
$$i : (X - U, A - U) \hookrightarrow (X, A)$$

is in $\mathcal{T}op^2$, then the induced homomorphism

$$i_* : H_n(X - U, A - U) \to H_n(X, A)$$

is an isomorphism for every integer $n \geq 0$. The map $i$ is called the **excision of the open set** $U$, and $i_*$ is called its $n$-**dimensional excision isomorphism**.
**Axiom H(7)(Dimension Axiom).** The $n$-dimensional homology group $H_n(X)$ of a one-point space $X = \{point\}$ in the homology theory $\mathcal{H}$ consists of a single element for every integer $n \neq 0$, in symbol, $H_n(point) = \{0\}$, for $n \neq 0$.

**Definition 3.6.9** If the homology theory $\mathcal{H}$ satisfies only the first six axioms **H(1)**-**H(6)**, then $\mathcal{H}$ is called a **generalized homology theory** on the category $\mathcal{T}op^2$. Such homology theories form the basic topics of modern algebraic topology.

**Definition 3.6.10** For a homology theory $\mathcal{H}$ on an admissible category $\mathcal{C}$, the 0-dimensional homology group
$$G = H_0(point)$$

is called the **coefficient group** of the homology theory $\mathcal{H}$.

**Example 3.6.11** The simplicial homology theory applies to the category $\mathcal{S}imp$ of pairs $(X, A)$ of spaces, where $X$ and $A$ have triangulations $K$ and $L$, respectively, for which $L$ is a subcomplex of $K$. On the other hand, the singular homology theory applies to all pairs of spaces $(X, A)$, where $X$ is a topological space and $A$ is a subspace of $X$.

## 3.7  The Uniqueness Theorem for Homology Theory

This section proves in Theorem 3.7.3 the **uniqueness of homology groups** in axiomatic homology theories satisfying E-S axioms on the category $\mathcal{T}op^2$ with isomorphic coefficient groups. A surprising result is that on the category $\mathcal{T}op^2$ of all topological pairs having homotopy types of finite CW-complex pairs (see Sect. 3.3),

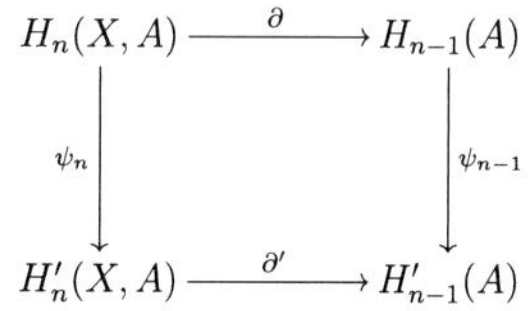

**Fig. 3.4** Natural
equivalence of homology
functors $H_n$ and $H'_n$

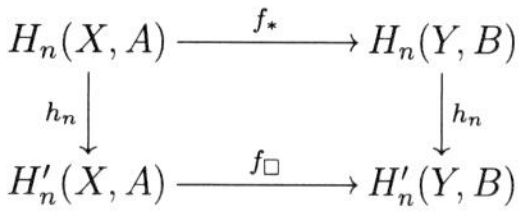

$$
\begin{array}{ccc}
H_n(X, A) & \xrightarrow{\ \partial\ } & H_{n-1}(A) \\
\downarrow{\psi_n} & & \downarrow{\psi_{n-1}} \\
H'_n(X, A) & \xrightarrow{\ \partial'\ } & H'_{n-1}(A)
\end{array}
$$

**Fig. 3.5** Commutative
diagram involving $h_n$ and
induced homomorphisms $f_*$
and $f_\square$

$$
\begin{array}{ccc}
H_n(X, A) & \xrightarrow{\ f_*\ } & H_n(Y, B) \\
\downarrow{h_n} & & \downarrow{h_n} \\
H'_n(X, A) & \xrightarrow{\ f_\square\ } & H'_n(Y, B)
\end{array}
$$

all homology theories satisfying the E-S axioms have isomorphic groups [Eilenberg and Steenrod 1952].

This uniqueness theorem is a key result in algebraic topology. For example, the simplicial and singular homology (cohomology) theories on the category of compact polyhedra coincide in the sense that for any compact polyhedra $|K|$, the simplicial homology groups $H_n^S(|K|; G)$ and the singular homology group $H_n(|K|; G)$ with coefficient group $G$ are isomorphic for all $n \geq 0$.

To prove Theorem 3.7.3, let $\mathcal{H} = \{H, *, \partial\}$ and $\mathcal{H}' = \{H', \square, \partial'\}$ be two arbitrary homology theories on the category $\mathcal{T}op^2$. Suppose $G = H_0(point)$, $G' = H'_0(point)$ are their coefficient groups.

**Definition 3.7.1** Let $\mathcal{H}$ and $\mathcal{H}'$ be two homology theories on $\mathcal{T}op^2$. **An isomorphism (natural)** $\psi : \mathcal{H} \to \mathcal{H}'$ is a sequence of natural equivalences

$$\psi_n : H_n \to H'_n, \quad \forall n \geq 0$$

such that the rectangular diagram in Fig. 3.4 is commutative in the sense that $\partial \circ \psi_{n-1} = \psi_n \circ \partial'$ for all pairs $(X, A)$ in $\mathcal{T}op^2$ and for all $n \geq 0$.

Theorem 3.7.2 proves that at some situation described in this theorem, there exists a unique homomorphism

$$h_n : H_n(X, A) \to H'_n(X, A).$$

**Theorem 3.7.2** *Let $G$ and $G'$ be abelian groups and $h : G \to G'$ be a homomorphism. Then for every pair $(X, A)$ in $\mathcal{T}op^2$ and every integer $n \geq 0$, there exists a unique homomorphism*

$$h_n : H_n(X, A) \to H'_n(X, A)$$

*such that*

*(i) $h_0 = h$ on $G = H_0(point)$.*
*(ii) For every map $f : (X, A) \to (Y, B)$ in $\mathcal{T}op^2$ and every integer $n \geq 0$, the diagram in Fig. 3.5 is commutative, i.e., $h_n \circ f_* = f_\square \circ h_n$.*

**Fig. 3.6** Diagram
connecting boundary
homomorphisms with
homomorphism $h_n$

$$\begin{array}{ccc} H_n(X,A) & \xrightarrow{\partial} & H_{n-1}(A) \\ h_n \downarrow & & \downarrow h_{n-1} \\ H'(X,A) & \xrightarrow{\partial'} & H'_{n-1}(A) \end{array}$$

*(iii) For every pair of spaces $(X, A)$ in $\mathcal{T}op^2$ and every integer $n \geq 0$, the diagram
in Fig. 3.6 is commutative, i.e., $h_{n-1} \circ \partial = \partial' \circ h_n$.*

**Proof** For proof, see [Eilenberg and Steenrod 1952].    □

Theorem 3.7.3 proves that there is only one concept of homology on that category.
This uniqueness theorem is very important in the development of algebraic topology.

**Theorem 3.7.3** (The uniqueness theorem for homology theory) *Let $G$ and $G'$ be
two abelian groups and $h : G \to G'$ be an isomorphism of groups. Then the unique
homomorphism*

$$h_n : H_n(X, A) \to H'_n(X, A)$$

*defined in Theorem 3.7.2 is also an isomorphism for every pair of spaces $(X, A)$ in
$\mathcal{T}op^2$ and every integer $n \geq 0$.*

**Proof** By hypothesis, $G$ and $G'$ are two abelian groups, and $h : G \to G'$ is an isomorphism of groups. Let $k : G' \to G$ be the isomorphism of groups defined by
$k = h^{-1}$. Then by Theorem 3.7.2, there exists a unique homomorphism

$$k_n : H'_n(X, A) \to H_n(X, A)$$

satisfying the conditions **(i)-(iii)** of Theorem 3.7.2 for every pair of topological spaces
$(X, A)$ in $\mathcal{T}op^2$ and every integer $n \geq 0$. This shows that

(i)  $k_n \circ h_n =$ identity automorphism of the groups $H_n(X, A)$ for every integer $n \geq 0$.
(ii) $h_n \circ k_n =$ identity automorphism of the groups $H'_n(X, A)$ for every integer $n \geq 0$.

.

This proves that
$$h_n : H_n(X, A) \to H'_n(X, A)$$

is an isomorphism of groups.    □

**Corollary 3.7.4** *Given a coefficient group $G$, **there exists only one homology theory**
in the category $\mathcal{T}op^2$.*

**Proof** It follows from Theorem 3.7.2.    □

**Remark 3.7.5** Corollary 3.7.4 justifies the name **the uniqueness theorem**.

## 3.8 Cohomology Theory: Eilenberg and Steenrod Axioms

This section presents an axiomatic approach to **cohomology theory** given by Eilenberg and Steenrod, which is dual (parallel) to the homology theory described in Section 3.6. In fact these two theories differ in only one point: Homology functors are covariant; on the other hand, cohomology functors are contravariant, and hence, most arrows in cohomology theory change their directions compared to the directions with homology theories. Therefore, one can expect a dual theorem in cohomology theory for every theorem established in homology theory. The Eilenberg–Steenrod axioms for cohomology functors make the subject algebraic topology quickly accessible.

The most surprising result is the proof that on the category of all topological pairs having homotopy type of finite CW-complex pairs, all cohomology theories satisfying these axioms have isomorphic groups. This basic theorem asserts that there is only one concept of cohomology on that category.

**A cohomology theory** $\mathcal{K}$ on the category $\mathcal{T}op^2$ consists of three functions $\mathcal{K} = \{H, *, \delta\}$ satisfying the following axioms for each integer $n \geq 0$:

(i): The first function $H$ assigns to each topological pair $(X, A)$ in the category $\mathcal{T}op^2$ and every integer $n \geq 0$, an abelian group $H^n(X, A)$, called the $n$-**dimensional cohomology group** of the topological pair $(X, A)$ in the cohomology theory $\mathcal{K}$. For $A = \emptyset$, it is called the $n$-dimensional (absolute) cohomology group of the space $X$.

(ii): The second function $*$ assigns to every $f : (X, A) \to (Y, B)$ in $\mathcal{T}op^2$ and every integer $n \geq 0$, a homomorphism

$$f^* = f_n^* : H^n(Y, B) \to H^n(X, A),$$

called the **homomorphism induced** by the map $f$ in the cohomology theory $\mathcal{K}$.

(iii): The third function $\delta$ assigns to each topological pair $(X, A)$ in $\mathcal{T}op^2$, a homomorphism

$$\delta = \delta(X, A, n) : H^{n-1}(A) \to H^n(X, A),$$

called the **coboundary operator** on the group $H^{n-1}(A)$ in the cohomology theory $\mathcal{K}$.

Moreover, these three functions satisfy the following axioms **C(1)–C(7)**, called the **Eilenberg–Steenrod axioms**, in brief, E-S axioms for cohomology theory on $\mathcal{T}op^2$ :

**Axiom C(1)(Identity Axiom).** If $1_X : (X, A) \to (X, A)$ is the identity map on a topological pair $(X, A)$ in $\mathcal{T}op^2$, then the induced homomorphisms

$$1_X^* : H^n(X, A) \to H^n(X, A)$$

are the identity automorphism of the cohomology group $H^n(X, A)$ for every integer $n \geq 0$. **Axiom C(2)(Composition Axiom).** If $f : (X, A) \to (Y, B)$ and $g : (Y, B) \to (Z, C)$ are maps in $\mathcal{T}op^2$, then

**Fig. 3.7** Diagram
connecting coboundary
operator $\delta$ with induced
homomorphisms in $\mathcal{K}$

$$
\begin{array}{ccc}
H^{n-1}(B) & \xrightarrow{\ g^*\ } & H^{n-1}(A) \\
\downarrow{\scriptstyle \delta} & & \downarrow{\scriptstyle \delta} \\
H^n(Y,B) & \xrightarrow{\ f^*\ } & H^n(X,A)
\end{array}
$$

$$(g \circ f)_n^* = f_n^* \circ g_n^* : H^n(Z,C) \to H^n(X,A)$$

for every integer $n \geq 0$.

**Remark 3.8.1** The above axioms **C(1) and C(2)** assert that for every fixed integer $n$, the function $H^n$ forms a contravariant functor from the category $\mathcal{T}op^2$ to the category $\mathcal{A}b$. The functor $H^n$ is called the $n$-**dimensional cohomology functor in the cohomology theory** $\mathcal{K}$.

We use the notation $H^n(f) = f_n^*$.

**Axiom C(3)(Commutativity Axiom).** If $f : (X,A) \to (Y,B)$ is a map in $\mathcal{T}op^2$ and if $g : A \to B$ is the map in $\mathcal{T}op$ defined by $g(x) = f(x)$ for all $x \in A$, then the diagram in Fig. 3.7 is commutative, i.e., $\delta \circ g^* = f^* \circ \delta$ for every integer $n \geq 0$.

**Remark 3.8.2** The Axiom C(3) connects the cohomology functors in the cohomology theory $\mathcal{K}$ with the coboundary operator $\delta$ and induced homomorphisms.

**Axiom C(4)(Exactness Axiom).** If $(X,A)$ is a topological pair of spaces in $\mathcal{T}op^2$ and

$$i : A \hookrightarrow X, \ and \ j : X \hookrightarrow (X,A)$$

are inclusion maps, then the cohomology sequence

$$\cdots \to H^{n-1}(A) \xrightarrow{\ \delta\ } H^n(X,A) \xrightarrow{\ j^*\ } H^n(X) \xrightarrow{\ i^*\ } H^n(A) \to \cdots$$

of the topological pair $(X,A)$ is exact.

**Remark 3.8.3** The above four axioms **C(1)-C(4)** are algebraic axioms.

**Axiom C(5)(Homotopy Axiom).** If two maps $f, g : (X,A) \to (Y,B)$ in $\mathcal{T}op^2$ are homotopic in $\mathcal{T}op^2$, then their induced transformations are such that

$$f_n^* = g_n^*$$

for every integer $n \geq 0$.

**Axiom C(6)(Excision Axiom).** Let $U$ be an open set of a topological space $X$ whose closure $\overline{U}$ is contained in the interior Int $A = \mathring{A}$ of a subspace $A$ of $X$ (i.e., $\overline{U} \subset \mathring{A}$.) If the inclusion map $i : (X - U, A - U) \hookrightarrow (X,A)$ is in $\mathcal{T}op^2$, then the induced homomorphism

$$i^* : H^n(X,A) \to H^n(X - U, A - U)$$

is an isomorphism for every integer $n$. The map $i$ is called the **excision of the open set** $U$, and $i^*$ is called its $n$-dimensional **excision isomorphism**.

**Axiom C(7)(Dimension Axiom)**. The $n$-dimensional cohomology group $H^n(X)$ of a one-point space $X = \{point\}$ consists of a single element for every integer $n \neq 0$, in symbol,

$$H^n(point) = \{0\}, \text{ for } n \neq 0.$$

**Definition 3.8.4** If $\mathcal{K}$ satisfies only the first six axioms **C(1)-C(6)**, then $\mathcal{K}$ is called a **generalized cohomology theory on the category** $\mathcal{T}op^2$.

**Definition 3.8.5** For a cohomology theory $\mathcal{K}$, the 0-dimensional cohomology group $G = H^0(point)$ is called the **coefficient group** of the cohomology theory $\mathcal{K}$.

**Theorem 3.8.6** *(the uniqueness theorem for cohomology theory) Let $G$ and $G'$ be two abelian groups and $h : G \to G'$ be an isomorphism of groups. Then $h$ induces an isomorphism*

$$h^n : H_n(X, A) \to H'_n(X, A)$$

*for every pair of spaces $(X, A)$ in $\mathcal{T}op^2$ and for every integer $n \geq 0$.*

**Proof** It follows from Theorem 3.7.2 by duality principle (see [Eilenberg and Steenrod 1952]). $\square$

**Remark 3.8.7** Theorem 3.8.6 justifies the name **the uniqueness theorem for cohomology theory.**

## 3.9 The Reduced Homology and Cohomology Groups

This section studies reduced homology and cohomology theories on the admissible category $\mathcal{T}op^2$. Let $(X, A)$ be a topological pair in $\mathcal{T}op^2$ and $i : A \hookrightarrow X$, $j : X \to (X, A)$ be the inclusion maps. If $f : (X, A) \to (Y, B)$ is a continuous map in $\mathcal{T}op^2$, then it defines two continuous maps

$$g : A \to B, \text{ and } h : X \to Y$$

such that their induced homomorphisms in homology theory $\mathcal{H} = \{H, *, \partial\}$ on the category $\mathcal{T}op^2$ satisfy the following relations:

(i)

$$i_* \circ g_* = h_* \circ i_*;$$

(ii)

$$j_* \circ h_* = f_* \circ j_*,$$

and

(iii)
$$\partial \circ f_* = g_* \circ \partial.$$

## 3.9.1 The Reduced Homology Groups

This subsection studies the reduced homology groups and their relations with homology groups in a homology theory $\mathcal{H}$ with coefficient group $G$.

**Definition 3.9.1** Let $(X, A)$ be a topological pair in $\mathcal{T}op^2$ and $i : A \hookrightarrow X$, $j : X \to (X, A)$ be the inclusion maps. If

$$f : (X, A) \to (Y, B)$$

is a continuous map in $\mathcal{T}op^2$, then it defines two continuous maps

$$g : A \to B, \ and \ h : X \to Y$$

such that

(i)  For the kernel $K_n(X, A)$ of

$$f_* : H_n(X, A) \to H_n(Y, B)$$

$$ker \, f_* = K_n(X, A) \subset H_n(X, A) \, for \, every \, integer \, n \geq 0.$$

(ii)  For the kernel $K_n(A)$ of

$$g_* : H_n(A) \to H_n(B)$$

$$ker \, g_* = K_n(A) \subset H_n(A) \, for \, every \, integer \, n \geq 0.$$

(iii)  For the kernel $K_n(X)$ of

$$h_* : H_n(X) \to H_n(Y)$$

$$ker \, h_* = K_n(X) \subset H_n(X) \, for \, every \, integer \, n \geq 0.$$

**Proposition 3.9.2** *The homomorphisms $i_*$, $j_*$ and $\partial$ have the following properties:*

(i) $i_*(K_n(A)) \subset H_n(X) \forall n \geq 0$;
(ii) $j_*(K_n(X)) \subset H_n(X, A) \forall n \geq 0$;
(iii) $\partial(K_n(X, A)) \subset H_{n-1}(A) \forall n \geq 0$.

**Proof** It follows from the respective definitions. $\quad\square$

**Definition 3.9.3** The beginningless infinite sequence (3.5) (obtained by using Proposition 3.9.2)

$$\cdots \to K_n(A) \xrightarrow{i_*} K_n(X) \xrightarrow{j_*} K_n(X, A) \xrightarrow{\partial} K_{n-1}(A) \to \cdots \quad (3.5)$$

of groups and homomorphisms is called the **kernel sequence** of the continuous map

$$f : (X, A) \to (Y, B)$$

in the homology theory $\mathcal{H}$ on the admissible category $\mathcal{T}op^2$.

Let $(X, A) \neq (\emptyset, \emptyset)$, $(Y, B) = (0, 0) \in \mathcal{T}op^2$ be two topological pairs in the homology theory $\mathcal{H}$ on category $\mathcal{T}op^2$, where 0 is the distinguished one-point space in this category. Then there is a unique map

$$\psi : (X, A) \to (0, 0).$$

**Definition 3.9.4** The kernel sequence of the map $\psi : (X, A) \to (0, 0)$ represented by

$$\cdots \to \tilde{H}_n(A) \xrightarrow{i_*} \tilde{H}_n(X) \xrightarrow{j_*} \tilde{H}_n(X, A) \xrightarrow{\partial} \tilde{H}_{n-1}(A) \to \cdots \quad (3.6)$$

is called the **reduced homology sequence** of $(X, A)$ in the homology theory $\mathcal{H}$.

Consider the homology sequence of the pair $(0, 0)$ in the homology theory $\mathcal{H}$ with coefficient group $G \neq \{0\}$

$$\cdots \to H_n(0) \xrightarrow{i_*} H_n(0) \xrightarrow{j_*} H_n(0, 0) \xrightarrow{\partial} H_{n-1}(0) \to \cdots \to H_0(0) \xrightarrow{i_*} H_0(0) \to \cdots .$$

All the groups of the above sequence are trivial except for $H_0(0) = G$. The groups $\tilde{H}_0(X)$ and $\tilde{H}_0(A)$ are called the **reduced homology 0-dimensional homology groups** the spaces $X$ and $A$, respectively, in the reduced homology sequence of $(X, A)$ in the homology theory $\mathcal{H}$.

The above discussion is summarized in the basic and important result.

**Theorem 3.9.5** *Let $\mathcal{H}$ be a homology theory with coefficient group $G \neq \{0\}$ on the category $\mathcal{T}op^2$ and $X \neq \emptyset$. Then*

(i)  $\widetilde{H}_n(X, A) = H_n(X, A)$, for every integer $n \geq 0$.
(ii)  $\widetilde{H}_n(A) = H_n(A)$ for every integer $n \neq 0$.
(iii)  $\widetilde{H}_n(X) = H_n(X)$ for every integer $n \neq 0$.
(iv)  $H_0(A) = \widetilde{H}_0(A) \oplus G$.
(v)  $H_0(X) = \widetilde{H}_0(X) \oplus G$.
(vi)  The identity map $1_X : X \to X$ induces the identity automorphism

$$1_{X^*} : \widetilde{H}_n(X) \to \widetilde{H}_n(X).$$

(vii)  For the continuous maps $f : X \to Y$ and $g : Y \to Z$ that are in the above category,

$$(g \circ f)_* = g_* \circ f_* : \widetilde{H}_n(X) \to \widetilde{H}_n(Z),$$

for every integer $n$.
(viii)  For the homotopic maps $f, g : X \to Y$ in the above category,

$$f_* = g_* : \widetilde{H}_n(X) \to \widetilde{H}_n(Y),$$

for every integer $n$.
(ix)  For the homotopically equivalent spaces $X$ and $Y$, in the above category, reduced homology groups

$$\widetilde{H}_n(X) \cong \widetilde{H}_n(Y),$$

for every integer $n$.
(x)  For the homeomorphic topological spaces $X$ and $Y$, in the above category, reduced homology groups

$$\widetilde{H}_n(X) \cong \widetilde{H}_n(Y),$$

for every integer $n$.
(xi)  The reduced homology group $\widetilde{H}_n(X)$ of a topological space $X$ is both a homotopy and a topological invariant of $X$.
(xii)  The reduced homology sequence of $(X, A)$

$$\cdots \to \widetilde{H}_n(A) \xrightarrow{i_*} \widetilde{H}_n(X) \xrightarrow{j_*} \widetilde{H}_n(X, A) \xrightarrow{\partial} \widetilde{H}_{n-1}(A) \to \cdots$$

is exact.

**Proof**  Left as an exercise.    □

**Proposition 3.9.6**  Let $X \neq \emptyset$ be a contractible space. Then $\widetilde{H}_n(X) = G$ for every integer $n$ in the homology theory $\mathcal{H}$ with coefficient group $G$.

**Proof** By hypothesis, $X$ is contractible. Then there is a unique map $f : X \to P$ for some one-point space $P$, which is a homotopy equivalence. Then it induces an isomorphism

$$f_* : H_n(X; G) \to H_n(P; G)$$

for every integer $n$. This asserts that

$$\tilde{H}_n(X) = ker\, f_* = G$$

for every integer $n$.                                                                    □

**Corollary 3.9.7** *Let $X \neq \emptyset$ be a contractible space. Then $H_n(X) = \{0\}$ for every integer $n \neq 0$ and $H_0(X) = G$.*

**Corollary 3.9.8** *Let $(X, A)$ be a topological pair.*

(i) *If $X \neq \emptyset$ is contractible, then*

$$\partial : H_n(X, A; G) = \tilde{H}_n(X, A; G) \to \tilde{H}_{n-1}(A; G)$$

   *is an isomorphism for every integer $n$.*
(ii) *If $A \neq \emptyset$ is contractible, then*

$$j_* : \tilde{H}_n(X; G) \to \tilde{H}_n(X, A : G) = H_n(X, A; G)$$

   *is an isomorphism for every integer $n$.*

**Proof** It follows from exactness of the reduced homology sequence for the pair $(X, A)$.                                                                    □

### 3.9.2  The Reduced Cohomology Groups

The reduced homology groups and their relations with homology groups in a homology theory $\mathcal{H}$ with coefficient group $G$. are studied in in Sect. 3.9.1.

**Remark 3.9.9** The assertions in Sect. 3.9.1 have duals in a cohomology theory $\mathcal{H}$ with coefficient group $G$, and they are left as exercises.

## 3.10  Invariance of Homology and Cohomology Groups

This section considers homology theory $\mathcal{H} = \{H, *, \partial\}$ with an abelian group $G$ as its coefficient group on the category $\mathcal{T}op^2$ of topological pairs and utilizes the Eilenberg and Steenrod axioms for homology and cohomology theories to prove **invariance of homology groups** in Sect. 3.10.1 and **invariance of cohomology groups** in Sect. 3.10.2.

### 3.10.1  Invariance of Homology Groups

This subsection proves invariance of homology groups in the sense that homology group is a topological invariant. Moreover, it is proved that it is also a homotopy invariant.

**Theorem 3.10.1** *Every homotopy equivalence* $f : (X,A) \to (Y,B)$ *in the category* $\mathcal{T}op^2$ *induces an isomorphism*

$$f_* : H_n(X,A) \to H_n(Y,B), \text{for every integer } n \geq 0.$$

**Proof** Let $f : (X,A) \to (Y,B)$ be a homotopy equivalence with its homotopy inverse $g : (Y,B) \to (X,A)$. Then

$$g \circ f \simeq 1_{(X,A)} \implies (g \circ f)_* = g_* \circ f_* = 1_d,$$

by using homotopy and identity axioms of Eilenberg and Steenrod. Similarly,

$$f \circ g \simeq 1_{(Y,B)} \implies f_* \circ g_* = 1_d.$$

This asserts that $f_*$ is an isomorphism of groups with its inverse $g_*$.     $\square$

**Corollary 3.10.2** *Every homeomorphism* $f : (X,A) \to (Y,B)$ *in the category* $\mathcal{T}op^2$ *induces an isomorphism*

$$f_* : H_n(X,A) \to H_n(Y,B), \text{for every integer } n \geq 0.$$

**Proof** Let $f : (X,A) \to (Y,B)$ be a homeomorphism with its inverse $g : (Y,B) \to (X,A)$. Then $g \circ f = 1_{(X,A)}$ and $f \circ g = 1_{(Y,B)}$. Hence the corollary follows by using Theorem 3.10.1.     $\square$

**Corollary 3.10.3** *(i) Every homology group is a topological invariant.*
*(ii) Every homology group is a homotopy invariant.*

**Proof** (i) $H_n(X,A)$ is a topological invariant by Corollary 3.10.2 in the sense that if $f : (X,A) \to (Y,B)$ is a homeomorphism in the category $\mathcal{T}op^2$, then it induces an isomorphism

$$f_* : H_n(X,A) \to H_n(Y,B) \text{ for every integer } n \geq 0.$$

(ii) $H_n(X, A)$ also a homotopy invariant by Theorem 3.10.1 in the sense that if $f : (X, A) \to (Y, B)$ is a homotopy equivalence in the category $\mathcal{T}op^2$, then it induces an isomorphism

$$f_* : H_n(X, A) \to H_n(Y, B) \text{ for every integer } n \geq 0.$$

$\square$

**Corollary 3.10.4** *If two topological spaces $X$ and $Y$ are homotopically equivalent, then the corresponding homology groups*

$$H_n(X) \cong H_n(Y) \ \forall n \geq 0.$$

**Proof** It follows from Theorem 3.10.1 by taking $A = \emptyset$ and $B = \emptyset$, in particular. $\square$

**Corollary 3.10.5** *If two topological spaces $X$ and $Y$ are homeomorphic, then the corresponding homology groups*

$$H_n(X) \cong H_n(Y) \ \forall n \geq 0.$$

**Proof** It follows from Corollary 3.10.2 by taking $A = \emptyset$ and $B = \emptyset$, in particular. $\square$

**Corollary 3.10.6** *If $X$ is a contractible space, then the homology group of $X$ with coefficient group $G$ is*

$$H_p(X; G) \cong \begin{cases} G, & \text{if } p = 0 \\ \{0\}, & \text{otherwise.} \end{cases}$$

**Proof** By hypothesis $X$ is contractible. Hence $X$ is homotopically equivalent to the singleton space $\{*\}$. This proves the corollary by using Corollary 3.10.4. $\square$

### 3.10.2  Invariance of Cohomology Groups

This subsection considers cohomology theory $\mathcal{H} = \{H, *, \delta\}$ with an abelian group $G$ as its coefficient group on the category $\mathcal{T}op^2$ of topological pairs proves invariance of cohomology groups in the sense that homeomorphic pairs of topological spaces have isomorphic cohomology groups.

**Theorem 3.10.7** *A homotopy equivalence $f : (X, A) \to (Y, B)$ in the category $\mathcal{T}op^2$ induces an isomorphism*

$$f_* : H_n(Y, B) \to H_n(X, A), \text{for every integer } n \geq 0.$$

**Proof** The proof is similar to that of Theorem 3.10.1 $\square$

**Corollary 3.10.8** *A homeomorphism $f : (X, A) \to (Y, B)$ in the category $\mathcal{T}op^2$ induces isomorphisms*

$$f^* : H^n(Y, B) \to H^n(X, A), \text{ for every integer } n \geq 0.$$

**Proof** The proof is similar to that of Corollary 3.10.2.    □

**Corollary 3.10.9** *(i) Cohomology group is a topological invariant.*
*(ii) Cohomology group is a homotopy invariant.*

**Proof** (i) $H^n(X, A)$ is a topological invariant by Corollary 3.10.8 .
(ii) $H^n(X, A)$ also a homotopy invariant by Theorem 3.10.7.    □

# 3.11  Consequences of the Exactness and Excision Axioms of Eilenberg and Steenrod

This section considers homology theory $\mathcal{H} = \{H, *, \partial\}$ on the category $\mathcal{T}op^2$ of topological pairs and establishes some immediate consequences of Eilenberg and Steenrod axioms by using Exactness Axiom **H(4)** and Excision Axiom **H(6)** on the homology theory $\mathcal{H}$.

## 3.11.1  Consequence of the Exactness Axiom

This section studies the effect of the Exactness Axiom: **H(4)** on homology groups in the homology theory $\mathcal{H}$ with coefficient group $G$.

**Proposition 3.11.1** *Let $X$ be a topological space, $A$ be a subspace of $X$ and the inclusion map $i : A \hookrightarrow X$ be a homotopy equivalence. Then $H_n(X, A) = \{0\}$ for every integer $n \geq 0$.*

**Proof** By hypothesis, $i : A \hookrightarrow X$ is a homotopy equivalence. Hence by Theorem 3.10.1 it follows that its induced homomorphism

$$i_* : H_n(A) \to H_n(X)$$

is an isomorphism for every integer $n \geq 0$. Consider the exact homology sequence (3.7) of the pair of topological spaces $(X, A)$ on the category $\mathcal{T}op^2$.

$$\cdots \to H_n(A) \xrightarrow{i_*} H_n(X) \xrightarrow{j_*} H_n(X, A) \xrightarrow{\partial} H_{n-1} \xrightarrow{i_*} H_{n-1}(X) \to \cdots \quad (3.7)$$

Two isomorphisms $i_*$ in the exact sequence (3.7) assert that $H_n(X, A)$ consists of exactly a singleton element for every integer $n \geq 0$ and hence $H_n(X, A) = \{0\}$ for every integer $n \geq 0$.                                                                    □

**Corollary 3.11.2**  *For any topological space $X$, the homology group*

$$H_n(X, X) = \{0\}$$

*for every integer $n \geq 0$.*

**Proof**  Taking in particular, $A = X$ in Proposition 3.11.1, the corollary follows.   □

**Remark 3.11.3**  By specifying the topological spaces $X$ and $A$ in the pair $(X, A)$ of topological spaces, some interesting relations among the homology groups of $(X, A)$, $X$ and $A$ are established.

**Proposition 3.11.4**  *Let $X$ be a topological space and $A$ be a retract of $X$. Then*

*(i)   The inclusion map $i : A \hookrightarrow X$ induces a monomorphism*

$$i_* : H_n(A) \to H_n(X) \text{ for each integer } n \geq 0.$$

*(ii)  The inclusion map $j : X \hookrightarrow (X, A)$ induces an epimorphism*

$$j_* : H_n(X) \to H_n(X, A) \text{ for each integer } n \geq 0.$$

*(iii)  The boundary operator*

$$\partial : H_n(X, A) \to H_{n-1}(A) \text{ is a  trivial homomorphism  for  each integer } n \geq 0.$$

*(iv)  $H_n(X) \cong H_n(A) \oplus H_n(X, A)$ for each integer $n \geq 0$.*

**Proof**  Since $A$ is a retract of $X$, there exists a retraction $r : X \to A$ such that $r \circ i : A \to A$ is the identity map $1_A$ on $A$. Hence the axioms **H(1)** and **H(2)** assert that the composite of the homomorphisms

$$H_n(A) \xrightarrow{\ i_*\ } H_n(X) \xrightarrow{\ r_*\ } H_n(A) \text{ is the identity automorphism of the group } H_n(A)$$
$$(3.8)$$

This asserts that $i_*$ is a monomorphism and $r_*$ is an epimorphism, and the abelian group $H_n(X)$ decomposes into the direct sum

$$H_n(X) = \operatorname{Im} i_* \oplus \ker r_* \text{ for each integer } n.$$

Consider the exact homology sequence (3.7) of the pair $(X, A)$:

$$\cdots \to H_n(A) \xrightarrow{\ i_*\ } H_n(X) \xrightarrow{\ j_*\ } H_n(X, A) \xrightarrow{\ \partial\ } H_{n-1}(A) \xrightarrow{\ i_*\ } H_{n-1}(X) \to \cdots$$

Since $i_* : H_{n-1}(A) \to H_{n-1}(X)$ is a monomorphism, the above exact sequence (3.7) asserts that $\partial$ is a trivial homomorphism and $j_*$ is an epimorphism.

Finally, since $i_* : H_n(A) \to H_n(X)$ is a monomorphism, $\operatorname{Im} i_* \cong H_n(A)$ for each integer $n \geq 0$. Moreover the exactness of the above sequence asserts that $\ker j_* = \operatorname{Im} i_*$. Since $H_n(X) = \operatorname{Im} i_* \oplus \ker r_*$ and $j_*$ is an epimorphism, it follows that

$$\ker r_* \cong H_n(X)/\operatorname{Im} i_* = H_n(X)/\ker j_*$$

by Isomorphism Theorem for groups. This proves that

$$H_n(X) = H_n(A) \oplus H_n(X, A) \text{ for each integer } n \geq 0.$$

$\square$

**Corollary 3.11.5** *Let $X$ be a topological space and $x_0$ is a point in $X$. Then in the homology theory $\mathcal{H}$ with coefficient group $G$,*

$$H_0(X) \cong G \oplus H_0(X, x_0),$$

$$H_n(X) \cong H_n(X, x_0) \text{ for } n \neq 0.$$

**Proof** As $\{x_0\}$ is a singleton subspace of the topological space $X$, it is a retract of $X$. Hence the corollary is proved by using Proposition 3.11.4 and Corollary 3.10.6. $\square$

### 3.11.2  Consequence of Excision Axiom

This subsection proves an Isomorphism Theorem 3.11.6 by using Excision Axiom **H(6)** for homology theory.

**Theorem 3.11.6** *Let $X$ be a topological space and $U$ be an open set of $X$ such that $U$ is contained in a subspace $A$ of $X$. Then in a homology theory $\mathcal{H}$ on $\mathbf{Top}^2$ with coefficient group $G$, the excision*

$$e : (X - U, A - U) \to (X, A)$$

*induces an isomorphism*

$$e_* : H_n(X - U, A - U; G) \to H_n(X, A; G) \text{ for each integer } n \geq 0,$$

*if there exists an open set $V$ in $X$ such that the closure $\overline{V}$ of $V$ is contained in $U$ and the inclusion map*

$$i : (X - U, A - U) \hookrightarrow (X - V, A - V) \text{ is a homotopy equivalence.}$$

***Proof*** Let $\mathring{A}$ denote the interior of $A$ in $X$. Since $\overline{V} \subset U \subset A$ by hypothesis, it follows that $\overline{V} \subset \mathring{A}$. Then the excision

$$\tilde{e} : (X - V, A - V) \to (X, A)$$

induces an isomorphism for each integer $n \geq 0$

$$\tilde{e}_* : H_n(X - V, A - V; G) \to H_n(X, A; G)$$

by Excision Axiom **H(6)**. Moreover, since

$$i : (X - U, A - U) \hookrightarrow (X - V, A - V)$$

is a homotopy equivalence, it induces an isomorphism

$$i_* : H_n(X - U, A - U; G) \to H_n(X - V, A - V; G) \text{ for each integer } n \geq 0.$$

Again, composite of maps

$$e = \tilde{e} \circ i : (X - U, A - U) \hookrightarrow (X - V, A - V) \xrightarrow{\ \tilde{e}\ } (X, A)$$

induces an isomorphism $e_*$ by Composition Axiom **H(2)**,

$$e_* = \tilde{e}_* \circ i_* : H_n(X - U, A - U; G) \to H_n(X, A; G) \text{ for each integer } n \geq 0.$$

$\square$

## 3.11.3   Additivity Property of Homology Theory

This subsection proves **additivity property of homology theory** in Proposition 3.11.7 by using Excision Axiom **H(6)**.

**Proposition 3.11.7** *Let $X$ and $Y$ be topological spaces with $X + Y$ be their topological sum. If*

$$i_X : X \hookrightarrow X + Y, \text{ and } i_Y : Y \hookrightarrow X + Y$$

*are the inclusion maps, then*

$$i_X^* \oplus i_Y^* : H_n(X; G) \oplus H_n(Y; G) \to H_n(X + Y; G)$$

*is an isomorphism for any homology theory $\mathcal{H}$ satisfying E-S axioms with coefficient group $G$.*

**Proof** Consider the exact homology sequence

$$\cdots \to H_p(X;G) \xrightarrow{i_{X*}} H_p(X+Y;G) \xrightarrow{j_*} H_p(X+Y,X;G) \xrightarrow{\partial} H_{p-1}(X;G) \to \cdots \tag{3.9}$$

of groups and homomorphisms for the pair $(X+Y,X;G)$. The inclusion map $i_Y$ :
$(Y,\emptyset) \hookrightarrow (X+Y,X)$ is an excision map by taking $U=X$ in Excision Axiom, and
hence by this axiom, there is an isomorphism

$$k_* : H_p(Y;G) \to H_p(X+Y;G) \forall p \geq 0,$$

where $k = j \circ 1_Y$. This asserts that $1_Y^* \circ k_*^{-1}$ is a splitting of the above long exact
sequence (3.9).                                                                        □

**Remark 3.11.8** The property of homology theory $\mathcal{H}$ proved in Proposition 3.11.7 is
taken as an axiom, called **the additivity axiom**. For example, the additivity axiom
added by J.W. Milnor (1931-) in 1962 with Eilenberg–Steenrod axioms to establish
the uniqueness of homology is not required for the category $\mathcal{T}op$, but it contributes
for the category of all polyhedra (compact or not) and hence for the category of all
CW-complexes (not discussed in this book) [Milnor 1962].

### 3.11.4  Mayer-Vietoris Theorem

This subsection proves Mayer-Vietoris theorem by using 'Excision Axiom' in homology theory $\mathcal{H} = \{H, *, \partial\}$, which provides a technique to compute homology groups.

**Definition 3.11.9** A topological triad $(X;A,B)$ is said to be a **proper triad with
respect to a homology theory** $\mathcal{H}$ if the inclusion maps

$$i : (A, A \cap B) \to (A \cup B, B)$$
$$j : (B, A \cap B) \to (A \cup B, A)$$

induce isomorphisms

$$i_* : H_n(A, A \cap B) \to H_n(A \cup B, B)$$
$$j_* : H_n(B, A \cap B) \to H_n(A \cup B, A)$$

in the homology theory $\mathcal{H}$ for every integer $n$.

**Theorem 3.11.10** (Mayer-Vietoris Theorem) *Let* $X, X_1, X_2$ *and* $A$ *be topological
spaces such that* $X = X_1 \cup X_2, A = X_1 \cap X_2$. *If the inclusion* $(X_1, A) \to (X, X_2)$ *is
an excision, then there is a long exact sequence in homology, called* **Mayer-Vietoris
sequence** *of the proper topological triad* $(X; X_1, X_2)$ :

$$\cdots \longrightarrow H_n(A) \xrightarrow{\alpha_1} H_n(X_1) \xrightarrow{\gamma_1} H_n(X_1, A) \xrightarrow{\delta_1} H_{n-1}(A) \longrightarrow \cdots$$

$$\cdots \longrightarrow H_n(X_2) \xrightarrow{\beta_2} H_n(X) \xrightarrow{\gamma_2} H_n(X_1, X_2) \xrightarrow{\delta_2} H_{n-1}(X_2) \longrightarrow \cdots$$

**Fig. 3.8** Diagram for Mayer-Vietoris theorem

$$\cdots \longrightarrow H_n(A) \xrightarrow{\alpha} H_n(X_1) \oplus H_n(X_2) \xrightarrow{\beta} H_n(X) \xrightarrow{\Delta} H_{n-1}(A) \longrightarrow \cdots$$

**Proof** Consider the commutative diagram with two long exact homology sequences as shown in Fig. 3.8 provided by axiom **H(4)**, where by assumption $\alpha : H_n(X_1, A) \to H_n(X, X_2)$ is an isomorphism by Exicison Axiom **H(6)**. Then use four lemma to the above diagram to complete the proof. □

## 3.12 Applications and Computations

This section presents some applications derived as further consequences of Eilenberg and Steenrod axioms. Moreover, this section computes the ordinary homology groups of $S^n$ with coefficients in an arbitrary abelian group $G$. Let $C_0$ be the full subcategory of $C$, whose objects are topological spaces with base points.

### 3.12.1 Computation of Homology Groups of $S^n$

Homology groups of $n$-sphere $S^n$ in a homology theory $\mathcal{H}$ with a coefficient $G$ play a key role in the development of topology. For example, degree of a spherical map is a useful concept in topology. It is defined in Sect. 3.13 by using homology theory $\mathcal{H}$ with the coefficient group $G$ as an infinite cyclic group with the help of $H_n(S^n; G)$ This subsection computes the homology groups of $S^n$ in a homology theory $\mathcal{H}$ with coefficient group $G$ in Theorem 3.12.9, which determines the degree of a spherical map $f : S^n \to S^n$ through homology theory.

### 3.12.2 Suspension Space and Suspension Functor

This subsection recalls the concepts of suspension space and suspension functor, which are used in some computations.

**Definition 3.12.1** Given a pointed topological space $X$ with base point $x_0$. its **suspension space,** denoted by $\Sigma X$, is defined to be the quotient space

$$\Sigma X = \frac{X \times \mathbf{I}}{(X \times 0) \cup (x_0 \times \mathbf{I}) \cup (X \times 1)}$$

endowed with quotient topology. For the point $(x, t) \in X \times \mathbf{I}$, the symbol $[x, t]$ denotes its corresponding point in $\Sigma X$ under the quotient map

$$p : X \times \mathbf{I} \to \Sigma X, (x, t) \mapsto [x, t].$$

This implies that

$$[x_0, 0] = [x_0, t] = [x, 1], \ \forall \ x \in X \ and \ \forall \ t \in \mathbf{I}.$$

The point $[x_0, 0] \in \Sigma X$ is also denoted by $x_0$. Hence the space $\Sigma X$ is a pointed space with base point $x_0$.

**Example 3.12.2** For any integer $n \geq 0$, the suspension space of the $n$-sphere is the $(n + 1)$-sphere $S^{n+1}$.

**Proposition 3.12.3** *The suspension* $\Sigma : \mathcal{T}op_* \to \mathcal{T}op_*$ *is a covariant functor.*

**Proof** The object function is defined by assigning to every object $X \in \mathcal{T}op_*$, its suspension space $\Sigma X \in \mathcal{T}op_*$, i.e., the object function is the assignment

$$X \mapsto \Sigma X \text{ in the category } \mathcal{T}op_*$$

in the category $\mathcal{T}op_*$. The morphism function is defined by assigning to every morphism $f : X \to Y \in \mathcal{T}op_*$,(which is a base point preserving continuous map $f : X \to Y$) the morphism is the assignment

$$\Sigma f : \Sigma X \to \Sigma Y, [x, t] \mapsto [f(x), t] \text{ in the category } \mathcal{T}op_*.$$

Then $\Sigma$ satisfies the following properties:

(i) If $f = 1_X : X \to X$ is the identity map in the category $\mathcal{T}op_*$, then $\Sigma(1_X)$ is also an identity map in the category $\mathcal{T}op_*$, and
(ii) If $f : X \to Y$ and $g : Y \to Z$ are morphisms in the category $\mathcal{T}op_*$, then

$$\Sigma(g \circ f) = \Sigma(g) \circ \Sigma(f).$$

This proves that $\Sigma : \mathcal{T}op_* \to \mathcal{T}op_*$ is a covariant functor. The functor $\Sigma$ is known as **suspension functor**.

□

Proposition 3.12.4 proves homotopy invariance property of the suspension functor $\Sigma$ in the sense that suspension functor sends homotopic maps to homotopic maps in the category $\mathcal{T}op_*$.

**Proposition 3.12.4** (*Homotopy invariance of $\Sigma$*) *In the category $\mathcal{T}op_*$, if*

$$f \simeq g : X \to Y,$$

*then*

$$\Sigma f \simeq \Sigma g : \Sigma X \to \Sigma Y.$$

**Proof** It follows from Definition of $\Sigma f$. □

Theorem 3.12.5 proves a basic theorem in reduced homology theory, known as suspension isomorphism theorem.

**Theorem 3.12.5** (*suspension isomorphism*) *Let $X$ be a nonempty topological space. Then there exists is an isomorphism*

$$\tilde{\sigma}_n : \tilde{H}_n(X; G) \to \tilde{H}_{n+1}(\Sigma X; G)$$

*in the reduced homology theory $\mathcal{H}$ with coefficient group $G$, where $\Sigma$ is the suspension functor.*

**Proof** Let $X$ be a nonempty topological space with a base point $x_0 \in X$. Then it can be embedded in its suspension space $\Sigma X$ by an embedding

$$i : X \hookrightarrow \Sigma X, \ x \mapsto p(x, 1/2),$$

where the standard projection map

$$p : X \times \mathbf{I} \to \Sigma X, (x, t) \mapsto [x, t] : [x, 0] = [x_0, t] = [x', 1], \ \forall x, x' \in X \text{ and } \forall t \in \mathbf{I}.$$

Let $U$ and $V$ be two subspaces of $\Sigma X$ defined by

$$U = \{p(x, t) : x \in X \text{ and } t \in [0, \frac{1}{2}], \ i.e., \ 0 \le t \le \frac{1}{2}\}$$

and

$$V = \{p(x, t) : x \in X \text{ and } t \in [\frac{1}{2}, 1], \ i.e., \ \frac{1}{2} \le t \le 1\}.$$

Then $U$ and $V$ are both contractible spaces such that

$$U \cup V = \Sigma X \text{ and } U \cap V = X.$$

Since $V$ is contractible, it follows by using the reduced homology exact sequence of the pair $(V, X)$ of topological spaces that

$$\partial : H_{n+1}(V, X; G) \to \tilde{H}_n(X; G)$$

is an isomorphism for every integer $n$. Again, since $U$ is contractible, it follows by using the reduced homology exact sequence of the pair $(\Sigma X, U)$ of topological spaces that

$$j_* : \tilde{H}_{n+1}(\Sigma X; G) \to H_{n+1}(\Sigma X, U; G)$$

is an isomorphism for every integer $n$. Consider the inclusion map

$$i : (V, X) \hookrightarrow (\Sigma X, U)$$

The inclusion map $i$ is the excision of the open set $Y = Int\, U = \Sigma X - V$ from the topological pair $(\Sigma X, U)$. Define the open set

$$W = \{p(x, t) : x \in X \text{ and } 0 \le t < \frac{1}{3}\}$$

of $\Sigma X$. Clearly, the closure $\overline{W}$ is contained in $Y$, and the inclusion map

$$i_1 : (V, X) \hookrightarrow (\Sigma X - W, U - X)$$

is a homotopy equivalence. This asserts by using the Excision Axiom that the induced homomorphism

$$i_* : H_{n+1}(V, X; G) \to H_{n+1}(\Sigma X, U; G)$$

is an isomorphism for every integer $n$. Consequently, the composite isomorphism

$$\tilde{\sigma}_n = j_*^{-1} \circ 1_* \circ \partial^{-1} : \tilde{H}_n(X; G) \to \tilde{H}_{n+1}(\Sigma X; G)$$

is an isomorphism for every integer $n$.

□

**Definition 3.12.6**  The isomorphism

$$\tilde{\sigma}_n : H_n(X; G) \to \tilde{H}_{n+1}(\Sigma X; G)$$

for every integer $n$ defined in Theorem 3.12.5 is called the **suspension isomorphism** on the reduced homology group $\tilde{H}_n(X; G)$.

**Proposition 3.12.7**  *Let $m, n$ be two integers with $m \ge 0$. Then in a homology theory $\mathcal{H}$ with coefficient group $G$,*

$$H_n(S^m, \{*\}; G) \cong H_{n+1}(S^{m+1}, \{*\}; G) \cong H_{n-m}(S^0, \{*\}; G) \cong H_{n-m}(\{*\}; G).$$

***Proof*** The first two isomorphisms follow from suspensions $S^{m+1} = \Sigma S^m = \Sigma^{m+1} S^0$. The other isomorphism follows from the triad $(S^0, \{-1\}, \{1\})$ and inclusion $(\{-1\}, \emptyset) \hookrightarrow (S^0, \{+1\})$. $\qquad\qquad\qquad\square$

Theorem 3.12.8 computes the reduced homology groups of some common spaces, from which the homology groups of spheres follow in Theorem 3.12.9.

**Theorem 3.12.8** *Let $\mathbf{D}_+^n$ be the closed upper half of the sphere $S^n$. Then for every integer $n \geq 0$, the homology groups with coefficient group $G$ are given by*

*(i)* $\tilde{H}_p(S^n; G) \cong \begin{cases} G, & \text{if } p = n \\ \{0\}, & \text{otherwise} \end{cases}$

*(ii)* $H_p(\mathbf{D}_+^n, S^{n-1}; G) \cong \begin{cases} G, & \text{if } p = n \\ \{0\}, & \text{otherwise} \end{cases}$

*(iii)* $H_p(S^n, \mathbf{D}_+^n; G) \cong \begin{cases} G, & \text{if } p = n \\ \{0\}, & \text{otherwise} \end{cases}$

***Proof*** Denote the statements (i), (ii) and (iii) by $\widetilde{(S_n)}$, $(D_n)$ and $(S_n)$, respectively. They are proved by recursive method. The statement $(S_0)$ follows by using the Excision and Dimension Axioms asserting that

$$H_p(S^0, \mathbf{D}_+^0; G) \cong H_p(point) \cong G$$

for $p = 0$ and it is $\{0\}$, otherwise.

The equivalence of the statements $\widetilde{(S_n)}$ and $(S_n)$ from the exact homology sequence of the inclusion map $\mathbf{D}_+^n \hookrightarrow S^n$

$$\{0\} = \tilde{H}_p(\mathbf{D}_+^n; G) \xrightarrow{i_*} \tilde{H}_p(S^n; G) \xrightarrow{j_*} H_p(S^n, \mathbf{D}_+^n; G) \xrightarrow{\partial} H_{p-1}(\mathbf{D}_+^n; G) = \{0\}. \tag{3.10}$$

The equivalence of the statements $(D_n)$ and $(S_n)$ follows similarly asserting that

$$H_p(S^n, \mathbf{D}_+^n; G) \cong H_p(S^n - V, \mathbf{D}_+^n - V; G) \cong H_p(\mathbf{D}_+^n, S^{n-1}; G),$$

where $V$ is some small nbd of the north pole of the sphere $S^n$. Use the Excision Axiom for the left-hand isomorphism and use the Homotopy Axiom for the right-hand isomorphism.

Consider the exact sequence of the pair $(\mathbf{D}^n, S^{n-1})$ of spaces in reduced homology theory

$$\{0\} = \tilde{H}_p(\mathbf{D}^n; G) \to H_p(\mathbf{D}^n, S^{n-1}; G) \to \tilde{H}_{p-1}(S^{n-1}; G) \to \tilde{H}_{p-1}(\mathbf{D}^n; G) = \{0\}.$$

since $\mathbf{D}^n$ is contractible. Hence it follows that

$$H_p(\mathbf{D}^n, S^{n-1}; G) \cong \tilde{H}_{p-1}(S^{n-1}; G)$$

The validity of the statement $(S_0)$ together with the equivalence of the statements $(D_0)$. $(S_0)$ and $(\tilde{S}_0)$ assert that

$$(D_1) \implies (S_1) \implies (\tilde{S}_1) \implies (D_2) \implies \cdots$$

□

The above discussion is summarized in the important Theorem 3.12.9.

**Theorem 3.12.9** (Homology Groups of Spheres) Let $\mathcal{H}$ be a homology theory with coefficient group $G$. Then

$$H_n(S^m; G) \cong \begin{cases} G \oplus G, & \text{if } n = m = 0 \\ G, & \text{if } n = m \neq 0 \text{ or } n = 0, m \neq 0 \\ \{0\}, & \text{otherwise.} \end{cases}$$

Theorem 3.12.10 is a basic result in homology theory, known as suspension isomorphism theorem.

**Theorem 3.12.10** *(suspension isomorphism) Let $X$ be a topological space and $x_0 \in X$. Then there is an isomorphism*

$$\sigma_n = \tilde{\Sigma}_n : H_n(X, \{x_0\}; G) \to H_{n+1}(\Sigma X, \{*\}; G)$$

*in the homology theory $\mathcal{H}$ with coefficient group $G$, where $\Sigma$ is the suspension functor.*

**Proof** Let $CX$ be the cone over $X$ with vertex $x_0$. Then $CX/X \approx \Sigma X$. This asserts that

$$H_n(CX/X; G) \cong H_n(\Sigma X; G)$$

and

$$H_{n+1}(CX, X; G) \cong H_n(X, \{x_0\}; G).$$

Now, use the projection map

$$(CX, X) \to (CX/X, \{*\})$$

to prove the theorem.    □

**Corollary 3.12.11** *There exists an isomorphism*

$$\sigma : H_n(S^n : G) \to H_{n+1}(\Sigma S^n; G) = H_{n+1}(S^{n+1}; G),$$

*called **suspension isomorphism** on $H_n(S^n; G)$ in the homology theory $\mathcal{H}$ with coefficient group $G$.*

**Proof** It follows from Theorem 3.12.10 by taking $X = S^n$, in particular.    □

## 3.13  Degrees of Spherical Maps from the Viewpoint of Homology Theory

This section is devoted to the study of **degrees of spherical maps by using homology theory** $\mathcal{H}$ with the coefficient group $G$ as an infinite cyclic group. A study of degrees of spherical maps by **using homotopy theory** has already been made in Chap. 2.

Historically, the concept of degrees of spherical maps was defined and studied by L. E. J. Brouwer (1881–1967) during the period 1910–1912. It provides an important tool in algebraic topology. This concept establishes a key link between homotopy and homology theories with various applications. Some of them are studied in this chapter, specially in its Sects. 3.13– 3.15. For example, Hopf classification theorem 3.13.7 characterizes homotopy of spherical maps with the help of their degrees, which are integers. Definition 3.13.1 formulates the **degree** of a spherical map through homology theory (which was defined in Chap. 2 by homotopy theory).

**Definition 3.13.1**  Let $f : S^n \to S^n$ be a continuous map for $n \geq 1$. Then it induces a homomorphism in the corresponding homology groups in a homology theory $\mathcal{H}$ with coefficient group $G$

$$f_* : H_n(S^n; G) \to H_n(S^n; G) \text{ with } H_n(S^n; G) \cong G.$$

By hypothesis, $G$ is an infinite cyclic group. If $\alpha$ is a generator of $H_n(S^n; G)$, then there is some integer $m$ such that

$$f_*(\alpha) = m\alpha.$$

Given an arbitrary element $x \in H_n(S^n; G)$, there is some integer $p$ such that $x = p\alpha$. Consequently,

$$f_*(x) = f_*(p\alpha) = pf_*(\alpha) = pm\alpha = mp\alpha = mx.$$

Moreover, $f_*(-\alpha) = m(-\alpha)$ and the integer $m$ is independent of the choice of the generator $\alpha$ of $G$. The integer $m$ is called the **degree** of $f$, denoted by

$$deg\, f = m.$$

It follows from Definition 3.13.1 that a continuous $f : S^n \to S^n$ is of degree $m$ iff the induced homomorphism

$$f_* : H_n(S^n; G) \to H_n(S^n; G)$$

has the property that

$$f_*(x) = mx, \quad \forall x \in H_n(S^n; G).$$

This asserts that if $deg\, f = m$, then the homomorphism $f_*$ is well defined by

$$f_* : H_n(S^n; G) \to H_n(S^n; G),\ x \mapsto mx.$$

**Example 3.13.2** Consider the map

$$f_n : S^1 \to S^1,\ z \mapsto z^n.$$

Then $f_{-1} : S^1 \to S^1$, $z \mapsto z^{-1}$ sends every point $z \in S^1$ to its reciprocal $z^{-1}$. This means that $f_{-1}$ is the reflection of $S^1$ with respect to real axis. This asserts that $deg\, f_{-1} = -1$. In general, $\deg f_n = n$.

### 3.13.1 Brouwer Degree Theorem

This subsection proves Brouwer degree theorem by using degree functions $d : f \mapsto deg\, f$ of spherical maps $f : S^n \to S^n$ through **homology theory** $\mathcal{H}$ instead of homotopy theory.

**Theorem 3.13.3 (Brouwer degree theorem)** *Let $f, g : S^n \to S^n$ be two continuous maps. If $f \simeq g$, then $deg\, f = deg\, g$.*

**Proof** Let $f \simeq g$. Then by Homotopy Axiom for homology theory

$$f_* = g_* : H_n(S^n; G) \to H_n(S^n; G)\, for\, all\, integers\, n.$$

This asserts that $deg\, f = deg\, g$.                                                  □

**Proposition 3.13.4** *Let $1_{S^n} : S^n \to S^n$, $x \mapsto x$ be the identity map on $S^n$. Then $deg\, 1_{S^n} = 1$.*

**Proof** Since $1_{S^n} : S^n \to S^n$, $x \mapsto x$ induces the identity homomorphis on the homology groups

$$1_{S^n*} : H_n(S^n; G) \to H_n(S^n; G)\, for\, all\, integers\, n.$$

□

This proves that $deg\, 1_{S^n} = 1$.

**Remark 3.13.5** Theorem 3.13.3 raises the problem: Is the converse of Brouwer degree theorem true? An affirmative answer is proved in Hopf classification Theorem 3.13.7.

**Fig. 3.9** Diagram
connecting the suspension
operator $\sigma$ and induced
homomorphisms in $\mathcal{H}$

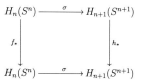

## 3.13.2   *Hopf Classification Theorem by Using Homology Theory*

This subsection solves the problem raised in Remark 3.13.5 by proving that the converse of Theorem 3.13.3 is also true by Hopf classification theorem 3.13.7. This theorem provides a complete homotopy classification of spherical maps by using their degrees defined by homology theory, which are integers.

Brouwer proved that if $f$ and $g$ are continuous maps on the 2-sphere which have the same degree, then $f$ and $g$ are homotopic. H. Hopf (1894–1971) generalized Brouwer theorem to an arbitrary dimension $n$ by proving in 1927 that the converse of Brouwer degree theorem is also true for arbitrary dimension $n$. These two results taken together are known as **Hopf classification theorem** obtained by degree function. It is defined **through homology theory** and is proved in Theorem 3.13.7. On the other hand, the same classification is also given in Chap. 2 by degree function defined **through homotopy theory**.

**Theorem 3.13.6**   *If a map $f : S^n \to S^n$ is of degree $m$, then its suspension map*

$$h = \Sigma f : S^{n+1} \to S^{n+1}, [x, t] \to [f(x), t]$$

*is also of degree $m$.*

**Proof** To prove the theorem, consider the commutative diagram of groups and homomorphisms, where $\sigma : H_n(S^n) \to H_{n+1}(S^{n+1})$ is the suspension isomorphism. Hence it follows that $h_* = \sigma \circ f_* \circ \sigma^{-1}$. We claim that $deg\ h = m$. By hypotheses, $deg\ f = m$. Then for any $x \in H_{n+1}(S^{n+1})$,

$$h_*(x) = (\sigma \circ f_* \circ \sigma^{-1})(x) = \sigma\ (f_* \circ (\sigma^{-1})(x)) = \sigma(m\sigma^{-1})(x)) = mx.$$

This implies that

$$h_* : H_{n+1}(S^{n+1}) \to H_{n+1}(S^{n+1}), x \mapsto mx.$$

This asserts that $deg\ h = m$ (Fig. 3.9).                                      □

Hopf classification theorem 3.13.7 characterizes homotopy of spherical maps with the help of their degrees, which are integers.

**Theorem 3.13.7 (Hopf classification theorem by using homology theory)**  *Let $f, g : S^n \to S^n$ be two continuous maps. Then $f \simeq g$, iff $deg\, f = deg\, g$.*

**Proof** First let $f \simeq g$. Then it follows by Brouwer degree theorem 3.13.3 that $deg f = deg\, g$. Conversely, let $deg\, f = deg\, g$. If $f, g : S^1 \to S^1$ are continuous maps such that $deg\, f = deg\, g$, then it follows that $f \simeq g$. This shows that the theorem is true for $n = 1$. To use induction on $n$, suppose that the theorem is true for $n - 1$, i.e., suppose for continuous maps $f, g : S^{n-1} \to S^{n-1}$, $deg f = deg\, g$. Then by using Theorem 3.13.6, it follows that $deg f = deg\, \Sigma f$ and $deg\, g = deg\, \Sigma g$. This asserts by induction hypothesis that $\Sigma f$ and $\Sigma g$ are homotopic.  □

### 3.13.3  More Properties of Degree Function

This subsection proves more properties of degree functions of spherical maps and relates to the fixed-point property of spherical maps. Their proofs are similar to the proofs given in Chap. 2.

**Proposition 3.13.8** *For every integer $k \in \mathbf{I}_n = \{0, 1, 2, 3, \ldots, n\}$, the reflection $r_k$ of $S^n$ is about the $x_k - axis$.*

$$r_k : S^n \to S^n, \ x \mapsto (x_0, x_1, x_2, \ldots, x_{k-1}, -x_k, x_{k+1}, \ldots, x_n)$$

*has degree* $-1$.

**Proof** For $n = 0$, $S^0 = \{x, y\}$ consists of two points $x = 1$ and $y = -1$ on the real axis. Then $f : S^0 \to S^0$ interchanges $x$ and $y$. Now use additive property 3.11.7 of homology theory. Clearly, $deg\, r_0 = -1$, because $r_0 = \Sigma^{n-1}(f_{-1})$, where $f_n : S^1 \to S^1$, $z \mapsto z^n$. Since for each $k = 1, 2, 3, \ldots, n$, every map

$$r_k : S^n \to S^n, \ x \mapsto (x_0, x_1, x_2, \ldots, x_{k-1}, -x_k, x_{k+1}, \ldots, x_n)$$

is homotopic to $r_0$ by rotation of $S^n$, it follows that

$$deg\, r_0 = -1 = deg\, r_1 = deg\, r_2 = \cdots = deg\, r_n.$$

This proves that every reflection $r_k$ of $S^n$ about the $x_k$- axis has degree $-1$.  □

**Proposition 3.13.9** *Let $g, h : S^n \to S^n$ be two continuous maps. Then*

$$deg\, (g \circ h) = deg\, g \, deg\, h.$$

**Proof** Let $\psi = g \circ h : S^n \to S^n$. Then its induced homomorphism

$$\psi_* : H_n(S^n) \to H_n(S^n)$$

is such that

$$\psi_*(x) = g_*(h_*(x)) = (deg\ g\ deg\ f)(x), \quad \forall x \in H_n(S^n).$$

It proves that $deg\ \psi = deg\ g\ deg\ h$. $\qquad\square$

**Definition 3.13.10** The continuous map

$$f : S^n \to S^n,\ x = (x_0, x_1, x_2, \ldots, x_n) \mapsto (-x_0, -x_1, -x_2, \ldots, -x_n) = -x$$

is said to be **antipodal**.

**Proposition 3.13.11** *The antipodal map*

$$f : S^n \to S^n,\ x = (x_0, x_1, x_2, \ldots, x_n) \mapsto (-x_0, -x_1, -x_2, \ldots, -x_n) = -x$$

*has the degree $deg\ f = (-1)^{n+1}$.*

**Proof** The antipodal map $f$ is the product of $(n + 1)$ reflections

$$f = r_0 \circ r_1 \circ r_2 \circ \cdots \circ r_n.$$

This proves that by Proposition 3.13.8 that

$$deg\ f = (-1)^{n+1}.$$

$\qquad\square$

**Remark 3.13.12** Proposition 3.13.13 is proved by using the concept of antipodal map If $x$ is a point of $S^n$, then its antipode is the point $-x \in S^n$. If $f : S^n \to S^n$ is an antipodal map, then $f(x) = -x$, $\forall x \in S^n$, and if $f$ is antipode preserving, then $f(-x) = -f(x)$, $\forall x \in S^n$.

**Proposition 3.13.13** *For any positive integer $n$, the sphere $S^n$ has no fixed-point property.*

**Proof** Let $f : S^n \to S^n$ be an arbitrary continuous map. Since the antipodal map $f : S^n \to S^n$, $x \mapsto -x$ has no fixed point, it follows that $S^n$ has no fixed-point property.
$\square$

**Proposition 3.13.14** *For any positive integer $n$, if two continuous maps $f, g : S^n \to S^n$ are such that $f(x) \neq g(x)$, $\forall x \in S^n$, then*

$$deg\ f + (-1)^n deg\ g = 0.$$

**Proof** By the given condition,

$$(1 - t)f(x) - tg(x) \neq 0, \quad \forall x \in S^n \text{ and } \forall t \in \mathbf{I}.$$

Because, if $(1 - t)f(x) - tg(x) = 0$, for some $x \in S^n$ and for some $t \in I$, then

$$1 - t = \|(1 - t)f(x)\| = t\|g(x)\|, \text{ since } f(x), \ g(x) \in S^n \implies 1 - t = t \implies f(x) = g(x) \text{ for some } x \in S^n,$$

which contradicts our hypothesis. This asserts that the map

$$H : S^n \times I \to S^n, \ (x, t) \mapsto \frac{(1 - t)f(x) - tg(x)}{\|(1 - t)f(x) - g(x)\|}$$

is well defined. $H$ is also a continuous map such that

$$H : f \simeq k, \text{ where } k = r \circ g \text{ and } r : S^n \to S^n \text{ is the antipodal map.}$$

Hence

$$deg f = deg(r \circ g) = deg \, r \, deg \, g = (-1)^{n+1} deg \, g \implies deg f + (-1)^n deg \, g = 0.$$

$\square$

**Proposition 3.13.15** *Let* $f : S^n \to S^n$ *be a continuous map such that* $f$ *has no fixed point. Then*

$$deg \, f = (-1)^{n+1}, \quad \forall n \geq 1.$$

**Proof** By the given condition, it follows that

$$(1 - t)f(x) - tx \neq 0. \forall t \in \mathbf{I}$$

otherwise, $1 - t = t$, $\forall t \in \mathbf{I}$ would imply $f(x) = x$, contradicting the hypothesis that $f$ has no fixed point. Then the map

$$H : S^n \times \mathbf{I} \to S^n, \ (x, t) \mapsto \frac{(1 - t)f(x) - tx}{\|(1 - t)f(x) - tx\|}$$

is well defined and continuous and is such that $f$ is homotopic to the antipodal map

$$A : S^n \to S^n, x \mapsto -x$$

under $H$ and hence $deg f = deg \, A = (-1)^{n+1}$.

The proposition also follows from Proposition 3.13.14 by taking $g$ as identity map on $S^n$. $\square$

**Corollary 3.13.16** *Let* $f : S^n \to S^n$ *be a continuous map such that*

$$deg \, f \neq (-1)^{n+1}.$$

*Then f has a fixed point.*

**Proof** It follows from Proposition 3.13.15 by way of contradiction.                    □

**Corollary 3.13.17** *Let f : $S^{2n} \to S^{2n}$ be continuous map such that f is homotopic to the identity map. Then f has a fixed point.*

**Proof** As $f$ is homotopic to the identity map, $\deg f = +1$. If possible, $f$ has no fixed point, then by Proposition 3.13.15 $\deg f = (-1)^{2n+1} = -1$. This contradiction shows that $f$ has a fixed point.                    □

**Corollary 3.13.18** *Let $I_d : S^n \to S^n$ be the identity map and n be an even integer. Then $I_d$ is not homotopic to a map free from fixed points.*

**Proof** It follows from Corollary 3.13.17.                    □

## 3.14   More Applications of Homology Theory

This section is devoted to prove some classical results : Brouwer fixed-point theorem for any degree, invariance of dimensions of spheres and Euclidean planes. It also solves some extension problems by using homology theory.

### 3.14.1   *Brouwer Fixed-Point Theorem for Dimension n*

This subsection proves Brouwer fixed-point theorem and its immediate consequences by using homology theory. Historically, L.E.J. Brouwer (1881–1967) took the first step toward connecting homotopy and homology by demonstrating in 1912 that two continuous mappings of a two-dimensional sphere into itself can be continuously deformed into each other if and only if they have the same degree (i.e., if they are equivalent from the point of view of homology theory).

The papers of H. Poincaré (1854–1912) during 1895–1904 can be considered as blue prints for theorems to come. The results of Brouwer during 1910–1912 may be considered the first one of the proofs in algebraic topology. He proved the celebrated theorem 'Brouwer fixed-point theorem 3.14.3' by using the concept of degree of a continuous spherical map defined by Brouwer himself. This subsection proves Brouwer fixed-point theorem for finite dimension $n \geq 0$ by using the tools of homology theory. On the other hand, the same theorem is proved in Chap. 2 using the tools of homotopy theory.

**Proposition 3.14.1** *Let $\mathbf{D}^n$ be the n-disk in $\mathbf{R}^n$ and $S^{n-1}$ be its boundary. The identity map*

$$1_{S^{n-1}} : S^{n-1} \to S^{n-1}$$

*has no continuous extension over $\mathbf{D}^n$ for every $n \geq 1$.*

**Proof** Clearly, $S^{n-1} \subset \mathbf{D}^n$. Suppose that there exists some continuous extension

$$f : \mathbf{D}^n \to S^{n-1}$$

of the identity map $1_{S^{n-1}}$ for every $n \geq 1$. Consider the two possible cases:

**Case I** For $n = 1$, the sphere $S^{n-1} = S^0 = \{\pm 1\}$ is a disconnected space. On the other hand, $\mathbf{D}^n$ is connected for every $n \geq 1$. In this particular case, $\mathbf{D}^n = \mathbf{D}^1$ is connected, and hence, the image $f(D^1)$ is also connected. Since by assumption, $f : \mathbf{D}^n \to S^{n-1}$ is a continuous extension of the identity map $1_{S^{n-1}}$, it follows that $f(\mathbf{D}^n) = S^{n-1}$. But in this case, since $S^0$ is not connected, we have a contradiction.

**Case II** For every $n > 1$, take $k = n - 1 > 0$, and consider the homology theory $\mathcal{H}$ with coefficient group $\mathbf{Z}$. Then it follows that

$$H_k(\mathbf{D}^n; \mathbf{Z}) = H_{n-1}(\mathbf{D}^n; \mathbf{Z}) = \{0\} \text{ and } H_k(S^{n-1}; \mathbf{Z}) = H_{n-1}(S^{n-1}; \mathbf{Z}) \cong \mathbf{Z}.$$

Since by hypothesis, $1_{S^{n-1}} : S^{n-1} \to S^{n-1}$ is the identity map, it follows that its induced homomorphism

$$1_{S^{n-1}*} : H_k(S^{n-1}) \to H_k(S^{n-1})$$

is the identity automorphism of the group $H_k(S^{n-1})$. Since $H_k(S^{n-1}) \cong \mathbf{Z}$, it follows that $1_{S^{n-1}*}$ is not the trivial homomorphism. Consider the inclusion map $i : S^{n-1} \hookrightarrow \mathbf{R}^n$. Then, the map

$$f \circ i : S^{n-1} \to S^{n-1}$$

is the composite of the continuous maps

$$S^{n-1} \xrightarrow{\ i\ } \mathbf{D}^n \xrightarrow{\ f\ } S^{n-1}$$

and is the identity map $1_{S^{n-1}}$. Hence it follows that

$$\mathbf{Z} \cong H_{n-1}(S^{n-1}; \mathbf{Z}) \xrightarrow{\ i_*\ } H_{n-1}(\mathbf{D}^n ; \mathbf{Z}) \xrightarrow{\ f_*\ } H_{n-1}(S^{n-1}\mathbf{Z}) \cong \mathbf{Z}$$

is the identity automorphism on the group $\mathbf{Z}$ in the homology theory $\mathcal{H}$ with coefficient group $\mathbf{Z}$. But this is not possible, since

$$H_{n-1}(S^{n-1};\ \mathbf{Z}) \cong \mathbf{Z} \neq \{0\} \text{ and } H_{n-1}(\mathbf{D}^n;\ \mathbf{Z}) = \{0\}.$$

This produces a contradiction. This contradiction proves the proposition     □

**Remark 3.14.2** Theorem 3.14.3 proves **Brouwer fixed-point theorem for any finite dimension** $n$ by using the tools of homology theory used in Proposition 3.14.1. An alternative proof of this theorem is also given by using the functorial property of homology functor in a homology theory $\mathcal{H}$. In Chap. 2, the same theorem is proved by homotopy theory.

**Theorem 3.14.3** *(Brouwer fixed-point theorem for dimension n) Every continuous map* $f : \mathbf{D}^{n+1} \to \mathbf{D}^{n+1}$ *has a fixed point for every* $n \geq 0$.

**Proof** If possible, $f$ has no fixed point for any $n \geq 0$. This implies that $f(x) \neq x$ for each $x \in \mathbf{D}^n$. For $n = 0$, as it an immediate contradiction, it is well assumed from now that $n \geq 1$. By assumption, for each $x \in \mathbf{D}^n$, the points $x$ and $f(x)$ are distinct. For any $x \in \mathbf{D}^n$, we now consider the half line in the direction from $f(x)$ to $x$. Let h(x) denote the point of intersection of this ray with $S^n$. Then the map $h : \mathbf{D}^{n+1} \to S^n$ is continuous. Moreover, $h(x) = x$ for every $x \in S^n$. This implies that $h : \mathbf{D}^{n+1} \to S^n$ is an extension of the identity $1_{S^n} : S^n \to S^n$. But this contradicts Proposition 3.14.1. This contradiction proves that $f$ has a fixed point for every $n \geq 0$.

**Alternative proof**: Using the functorial property of homology functor in a homology theory $\mathcal{H}$ with a nonzero coefficient group $G$, it follows from the above discussion that

$$h_* \circ i_* : H_n(S^n : G) \to H_n(S^n; G)$$

is the identity automorphism on the group $G$ in the homology theory $\mathcal{H}$. But this is not possible, since

$$H_n(S^n;\ G) \cong G \neq \{0\} \text{ and } H_n(\mathbf{D}^{n+1};\ G) = \{0\}.$$

$\square$

**Remark 3.14.4** For more consequences of Eilenberg and Steenrod axioms for homology and cohomology theories, see Exercises of Section 3.16 .

## 3.14.2 Hurewicz Homomorphism: Relation Between Homology and Homotopy Groups

Hurewicz homomorphism defined by W. Hurewicz (1904–1956) in 1935 [Hurewicz 1935] establishes a close relation between homology and homotopy groups of a specified class of topological spaces. His original definition is now simplified as follows.

**Definition 3.14.5** Let $(X, x_0)$ be a pointed topological space and $g_n$ be the standard generator of the homology group $H_n(S^n; \mathbf{Z})$ for $n = 1, 2, 3, \ldots$. If $f$ represents an element $\alpha \in \pi_n(X, x_0)$, then

(i) The induced homomorphism

$$f_* : H_n(S^n; \mathbf{Z}) \to H_n(X; \mathbf{Z})$$

defines an element $f_*(g_n) \in H_n(X; \mathbf{Z})$ and

(ii) $f_*$ defines a homomorphism

$$h : \pi_n(X, x_0) \to H_n(X; \mathbf{Z}), \quad \alpha \mapsto f_*(g_n), \quad \forall n = 1, 2, 3, \ldots.$$

The homomorphism $h$ is well defined, because, it is independent of the choice of the representative $f$. This is called the **Hurewicz homomorphism**.

For $n = 1$, the fundamental group $\pi_1(X, x_0)$ is not abelian in general, but its abelization is isomophic to $H_1(X; \mathbf{Z})$, i.e., $H_1(X; \mathbf{Z}) \cong \pi_1(X, x_0)/[\pi_1(X, x_0), \pi_1(X, x_0)]$, where $[\pi_1(X, x_0), \pi_1(X, x_0)]$ is the commutator subgroup of $\pi_1(X, x_0)$, $X$ is path connected.

Hurewicz isomorphism 3.14.6 provides a sufficient condition under which Hurewicz homomorphism $h : \pi_n(X, x_0) \to H_n(X; \mathbf{Z})$ between the homotopy and homology groups are isomorphic.

**Theorem 3.14.6** (Hurewicz isomorphism)

(a) *If $X$ is a simply connected pointed topological space, then the following statements are equivalent:*

   (i) $\pi_k(X) = 0, \forall k < n;$
   (ii) $\tilde{H}_k(X) = 0, \forall k < n.$

(b) *Moreover, (a) implies that the Hurewicz homomorphism $h$*

$$h : \pi_n(X, x_0) \to H_n(X; \mathbf{Z}), \quad \alpha \mapsto f_*(g_n)$$

*formulated in Definition 3.14.5 is an 1 $(n + 1)$ isomorphism .*

**Proof** See [Gray 1975; Spanier 1966]    □

**Corollary 3.14.7** *Let $X$ be pointed simply connected space. Then $X$ is n-connected iff*

$$H_m(X) = 0, \quad \forall m = 2, 3, \ldots n.$$

**Remark 3.14.8** For special properties of Hurewicz homomorphism, see Exercises 21–23 of Sect. 3.16.

## 3.15 The Lefschetz Number and Fixed-Point Theorems

This section introduces the concept of Lefschetz number defined by S.Lefschetz (1884–1972) in 1923 corresponding to each continuous map $f : |K| \to |K|$ from a polyhedron into itself. The number is denoted by $\Lambda_f$. It is an integral-valued topological invariant and generalizes the Euler characteristic. It is closely linked with the degree of a spherical map by Theorems 3.15.5. More properties of the Lefschetz number are given in Exercise 24 of Section 3.16.

**Definition 3.15.1** *(Lefschetz number)* Let $X$ be a compact triangulable space with a given triangulation $K$ of dimension $n$ and $f : X \to X$ be a continuous map. Then there exists a homeomorphism $\psi : |K| \to X$. The homology groups $H_q(K; \mathbf{Q})$ with rational coefficients are all vector spaces over $\mathbf{Q}$ and the homomorphisms

$$f_{p^*}^{\psi} = (\psi^{-1} \circ f \circ \psi)_* : H_p(K; \mathbf{Q}) \to H_p(K; \mathbf{Q})$$

are linear transformations. Then

$$\Lambda_f = \sum_{p=0}^{n} (-1)^p \operatorname{trace} f_{p^*}^{\psi}$$

is defined to be the Lefschetz number of $f$, which is well defined, since the trace of the corresponding matrices is independent of choice of the basis and the number does not dependent on the triangulation of $X$.

**Theorem 3.15.2** *Let $X$ be a compact triangulable space. If the identity map $1_X : X \to X$ is homotopic to a fixed-point free map, then its Euler characteristic $\kappa(X) = 0$.*

**Proof** By hypothesis, $X$ is a compact triangulable space. Then the Lefschetz number $\Lambda_{1_X}$ of the identity map $1_X : X \to X$ is the characteristic $\kappa(X)$ of $X$. Since the homotopic maps have the same Lefschetz number, it follows that $\kappa(X) = 0$. $\square$

**Theorem 3.15.3** *(Lefschetz fixed-point theorem) If $X$ is a compact triangulable space and $f : X \to X$ is a continuous map such that its Lefschetz number $\Lambda_f \neq 0$, then $f$ has a fixed point.*

**Proof** Consider $X$ as the polyhedron of a finite simplicial complex $K$ and $f : |K| \to |K|1$ as a simplicial map. If $f$ has no fixed point, then $f(\sigma) \neq \sigma$ for a simplex $\sigma$ of $K$. Orient each $p$-simplex of $K$ in some way to obtain a basis of the vector space $C_p(K; \mathbf{Q})$ over $\mathbf{Q}$. Then the linear map

$$f_p : C_p(K; \mathbf{Q}) \to C_p(K; \mathbf{Q})$$

with respect to the above basis has the matrix representation $M_{f_p}$ with its diagonal elements zeros, and hence, its trace is $trace f_p = 0$. Since the Lefschetz number of $f$ at the chain level and at homology level is the same, it follows that

$$\sum_{p=0}^{n} (-1)^p \operatorname{trace} f_{p^*} = \Lambda_f = \sum_{p=0}^{n} (-1)^p \operatorname{trace} f_p. \tag{3.11}$$

Then it follows from the relation (11.8) that $\Lambda_f = 0$, since $trace f_p = 0$ for each $p$. But it contradicts our hypothesis that $\Lambda_f \neq 0$. This contradiction asserts that $f$ has a fixed point.

$\square$

**Remark 3.15.4** Theorem 3.15.5 asserts that the degree of a spherical map $f : S^n \to S^n$ is closely related to its Lefschetz number $\Lambda_f$.

**Theorem 3.15.5** *For any spherical map $f : S^n \to S^n$, its Lefschetz number $\Lambda_f = 1 + (-1)^n \deg f$.*

**Proof** Given $f : S^n \to S^n$, the only nonzero groups $H_n(S^n; \mathbf{Q})$ are $\mathbf{Q}$ in dimension $n$ and 0, where in dimension $n$, the homomorphism $f_*$ induced by $f$

$$f_* : H_n(S^n; \mathbf{Q}) \to H_n(S^n; \mathbf{Q}) \cong \mathbf{Q}$$

is obtained just multiplication by $\deg f$. This proves the theorem.    □

Lefschetz number of any spherical map is characterized in Corollary 3.15.6 by its homotopic maps.

**Corollary 3.15.6** *For two spherical maps $f, g : S^n \to S^n$, their Lefschetz numbers $\Lambda_f$ and $\Lambda_g$ are the same iff $f$ is homotopic to $g$.*

**Proof** It follows from Theorem 3.15.5 by using Hopf classification theorem 3.13.7.
□

**Corollary 3.15.7** *If the degree of a spherical map $f : S^n \to S^n$ is neither 1 nor -1, then $f$ has a fixed point.*

**Proof** It follows from Theorems 3.15.5 and 3.15.3.    □

# 3.16   Exercises and Multiple Choice Exercises

As solving exercises plays an essential role of learning mathematics, various types of exercises and multiple choice exercises are given in this section. They form an integral part of the book series.

## 3.16.1   Exercises

This section considers exercises in a given homology theory $\mathcal{H}$ with coefficient group $G$ on the category $\mathcal{T}op^2$, unless stated otherwise.

1. Prove that every $CW$-complex is a normal space.
2. Show that

   (i) The torus $T$ is a $CW$-complex with one 0-cell, two 1-cells and one 2-cell.
   (ii) The real projective space $\mathbf{R}P^n$ is a $CW$-complex with 1-cell of each dimension $0, 1, \ldots, n$.

(iii) The complex projective space $\mathbf{C}P^n$ is a $CW$-complex with one cell of dimension $2k$ for each $k$ satisfying $0 \leq k \leq n$, i.e., $(n+1)$ cells

$$e^{2k} : 0 \leq k \leq n, \; symbolized, \; \mathbf{C}P^n = e^0 \cup e^2 \cup e^4 \cup \cdots \cup c^{2n}.$$

(iv) The quaternionic projective space $\mathbf{H}P^n$ is a $CW$-complex with one cell of dimension $4k$ for each $k$ satisfying $0 \leq k \leq n$, i.e., $(n+1)$ cells

$$e^{4k} : 0 \leq k \leq n, \; symbolized \; \mathbf{H}P^n = e^0 \cup e^4 \cup e^8 \cup \cdots \cup c^{4n}.$$

3. $\{X_n\}$ be a sequence of $CW$-complexes such that $X_n$ is a subcomplex of $X_{n+1}$. Show that the space $X = \bigcup X_n$ endowed with the weak topology is a $CW$-complex and every $X_n$ is a subcomplex of $X$. Hence prove that

(i) $\mathbf{R}P^\infty = \bigcup_{n=1}^\infty \mathbf{R}P^n$ endowed with the weak topology is a $CW$-complex.
(ii) $\mathbf{C}P^\infty = \bigcup_{n=1}^\infty \mathbf{C}P^n$ endowed with the weak topology is a $CW$-complex.
(iii) $\mathbf{H}P^\infty = \bigcup_{n=1}^\infty \mathbf{H}P^n$ endowed with the weak topology is a $CW$-complex.
[ Hint: Consider $\mathbf{R}P^n \subset \mathbf{R}P^{n+1}$ as a subcomplex. Then $\mathbf{R}P^\infty$ endowed with weak topology is a closure finite cell complex and hence $\mathbf{R}P^\infty$ is a $CW$-complex. ]

4. Prove that the space of an infinite one-point union of circles is a $CW$-complex. In particular, the figure-eight space $F_8$ is a $CW$-complex.
[ Hint: Let $X = \vee_i^\infty S_n^1$ endowed with the weak topology as a subset of $\Pi_i^\infty S_n^1$ having the product topology. $X$ is a closure finite cell complex. Consider $F_8 = S^1 \vee S^1$. ]

5. Using Dimension Axiom, show that the homology groups of any singleton space $X$ in the homology theory $\mathcal{H}$ with coefficient group $G$, are

$$H_p(X; G) \cong \begin{cases} G & \text{if } p = 0 \\ \{0\}, & \text{otherwise.} \end{cases}$$

6. Show that the homology groups of a discrete topological space $X = \{x_1, x_2, \ldots, x_n\}$ in the homology theory $\mathcal{H}$ with coefficient group $G$, are

$$H_p(X; G) \cong \begin{cases} \overbrace{G \oplus G \oplus \cdots \oplus G}^{n} & \text{if } p = 0 \\ \{0\}, & \text{otherwise.} \end{cases}$$

7. In the homology theory $\mathcal{H}$ with coefficient group $G$, show that for every integer $n \geq 0$,

(i) $\tilde{H}_n(S^n) \cong G$.
(ii) $\tilde{H}_m(S^n) \cong \{0\}$, if $m \neq n$.

8. In the homology theory $\mathcal{H}$ with coefficient group $G$, show that for every integer $n \geq 0$,

   (i) $H_n(S^n; G) \cong G$, if $n \neq 0$;
   (ii) $H_0(S^0; G) \cong G \oplus G$;
   (iii) $H_m(S^n; G) \cong \{0\}$, if $m \neq n \neq 0$.

9. For the pair $(\mathbf{R}^n, S^{n-1})$ of spaces, show that the homology groups of $(\mathbf{R}^n, S^{n-1})$ in the homology theory $\mathcal{H}$ with coefficient group $G$ are

$$H_p(\mathbf{R}^n, S^{n-1}; G) \cong \begin{cases} G, & \text{if } p = n \\ \{0\}, & \text{otherwise.} \end{cases}$$

10. Show that the homology groups of an $n$-dimensional manifold $M^n$ $(n \geq 1)$ in the homology theory $\mathcal{H}$ with coefficient group $G$ are

$$H_p(M^n, \{x\}; G) \cong \begin{cases} G, & \text{if } p = n \\ \{0\}, & \text{otherwise.} \end{cases}$$

for every point $x \in M^n$.

11. Given an integer $k$, show that there is a continuous map $f : S^n \to S^n$ such that $deg f = k$.

12. Let $f : S^n \to S^n$ be a continuous map for $n > 0$. Show that

   (i) If $deg f = m$ and $\Sigma f : S^{n+1} \to S^{n+1}$ is its suspension map, then

$$deg f = deg \Sigma(f) = m, \quad \forall n > 0.$$

   (ii) The suspension map $\Sigma f : S^{n+1} \to S^{n+1}$ of $f$ has degree $m$ iff

$$deg f = m.$$

13. Show that the antipodal map on $S^n$ is not homotopic to the identity map on $S^n$ for every even integer $n > 1$.

14. If $n$ is an even integer, by using homology theory $\mathcal{H}$, show that $S^n$ admits no nonzero continuous tangent vector field.
    [ Hint: To prove it by contradiction method, use Exercise 13. ]

15. Let $X$ be a topological space and $X_1$, $X_2$ be two subsets of $X$ such that

   (i) $X_1$ is closed and
   (ii) $X = IntX_1 \cup IntX_2$.

   Show that the inclusion map

$$i : (X_1, X_1 \cup X_2) \hookrightarrow (X, X_2)$$

induces an isomorphism

$$i_* : H_n(X_1, X_1 \cup X_2) \to H_n(X, X_2)$$

is an isomorphism for every integer $n$.

[ Hint: Apply Excision Axiom $\mathbf{H}(6)$.]

16. Construct a statement of Mayer-Vietoris theorem 3.11.10 in the cohomology theory and prove it.

17. For any topological pair $(X, A)$ with $X \neq \emptyset$ and $A \neq \emptyset$, show that the reduced homology sequence of $(X, A)$

$$\cdots \to \widetilde{H}_p(A) \xrightarrow{i_*} \widetilde{H}_p(X) \xrightarrow{j_*} \widetilde{H}_p(X, A) \xrightarrow{\partial} \widetilde{H}_{p-1}(A) \to \cdots$$

of groups and homomorphisms is exact in the homology theory $\mathcal{H}$.

18. Let $f : (X, A) \to (Y, B)$ be a map of pair of spaces. If both $f : X \to Y$ and $f|_A : A \to B$ are homotopy equivalences, show that $f_* : H_n(X, A) \to H_n(Y, B)$ is an isomorphism for all $n$.

[Hint. Consider the commutative diagram of two rows of exact sequences as shown in Fig. 3.10, and use five lemma result

19. Show that the kernel sequence ( 3.5) of a map $f : (X, A) \to (Y, B)$ is exact if there exists a map

$$g : (Y, B) \to (X, A) \text{ such that} f \circ g \text{homotopic to identity map on } (Y, B).$$

20. Show that for every pair of topological spaces $(X, X)$

$$H_n(X, X) = \{0\}, \quad \forall n \geq 0.$$

[ Hint: Apply 'Exactness Axiom' for the pair $(X, X)$.]

21. (**Hurewicz**) Given a pointed topological space $(X, x_0)$, if

$$\pi_0(X, x_0) = \{0\}, \ \pi_1(X, x_0) = \{0\}, \ \pi_2(X, x_0) = \{0\}, \ \ldots, \ \pi_{n-1}(X, x_0) = \{0\}, \text{ for all } n \geq 2,$$

show that

(i)

$$H_1(X; \mathbf{Z}) = \{0\}, \ H_2(X; \mathbf{Z}) = \{0\}, \ \ldots, \ H_{n-1}(X; \mathbf{Z}) = \{0\}, \text{ for all } n \geq 2;$$

(ii)  Hurewicz homomorphism

$$h : \pi_n(X, x_0) \to H_n(X; \mathbf{Z}), \ \alpha \mapsto f_*(g_n),$$

given in Definition 3.14.5, is an isomorphism for all $n \geq 2$.

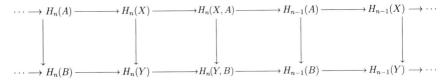

**Fig. 3.10** Commutative diagram of exact sequences

Hint : See [ Adhikari, 2016]

22. Given a path-connected topological $X$, show that under usual notation, there exists an epimorphism

$$h : \pi_1(X) \to H_1(X; \mathbf{Z}),$$

which determines an isomorphism

$$h_* : \pi_1(X)/ker\, h \to H_1(X; \mathbf{Z})$$

with $ker\, h$ the commutator subgroup of $\pi_1(X)$. Hence prove that $h$ is an isomorphism iff the fundamental group $\pi_1(X)$ is abelian.

23. A pointed topological space $(X, x_0)$ is said to be $n$-connected, if $\pi_k(X, x_0) = 0$, $\forall k \leq n$. Given an $n$-connected topological space $(X, x_0)$, show that

(i) $H_k(\tilde{X}; \mathbf{Z}) = \{0\}$, $\forall k \leq n$.
(ii) The Hurewicz homomorphism $h : \pi_{n+1}(X, x_0) \to \tilde{H}_{n+1}(X; \mathbf{Z})$ is an isomorphism for all $n \geq 1$.

24. Let $X$ be a compact triangulable space and $f : X \to X$ be a continuous map with its Lefschitz number $\Lambda_f$. Show that

(i) $\Lambda_f \neq 0$ implies $f$ has a fixed point.
(ii) Every contractible compact triangulable space has the fixed-point property (Brouwer fixed-point theorem).
(iii) If the identity map $1_X : X \to X$ is homotopic to a fixed-point free map $f : X \to X$, the Euler characteristic $\kappa(X)$ of $X$ is zero.

### 3.16.2   Multiple Choice Exercises

Identify the correct alternative (s) (there may be none or more than one) from the following list of exercises:

1. (i) The spheres $S^{10}$ and $S^{11}$ are homeomorphic.
   (ii) The spheres $S^{100}$ and $S^{111}$ are homotopically equivalent.
   (iii) The spheres $S^{11}$ and $S^{12}$ are neither homeomorphic nor homotopically equivalent.

2. (i) The Euclidean spaces $\mathbf{R}^{10}$ and $\mathbf{R}^{11}$ are homeomorphic.
   (ii) The Euclidean spaces $\mathbf{R}^{10}$ and $\mathbf{R}^{11}$ are homotopically equivalent.
   (iii) The Euclidean spaces $\mathbf{R}^{10}$ and $\mathbf{R}^{11}$ are neither homeomorphic nor homotopically equivalent.
3. (i) Let $\mathbf{R}^5 - \{0\}$ be punctured five-dimensional Euclidean space and $S^4$ be the four-dimensional sphere. Then the inclusion map

$$i : S^4 \hookrightarrow \mathbf{R}^5 - \{0\}$$

   is a homotopy equivalence.
   (ii) The homology group of $H_n (\mathbf{R}^5 - \{0\})$ of the punctured Euclidean space $\mathbf{R}^5 - \{0\}$ and the homology group of $H_n (S^4)$ of the sphere $S^4$ are isomorphic for all $n > 1$.
   (iii) The cohomology group of $H^n (\mathbf{R}^5 - \{0\})$ of the punctured Euclidean space $\mathbf{R}^5 - \{0\}$ and the homology group of $H^n (S^4)$ of the sphere $S^4$ are isomorphic for all $n > 1$.
4. Let $\mathcal{H}$ be a homology theory with coefficient group $G \neq \{0\}$. Then for the subspace $S^4$ of the punctured Euclidean space $\mathbf{R}^5 - \{0\}$ ,

   (i) $\tilde{H}_n(S^4) = H_n(S^4)$ for every integer $n \neq 0$.
   (ii) $H_0(S^4) = \tilde{H}_0(S^4) \oplus G$.
   (iii) For $X = \mathbf{R}^5 - \{0\}$, the identity map $1_X : X \to X$ induces the identity automorphism of reduced homology groups

$$1_{X*} : \tilde{H}_n(X; G) \to \tilde{H}_n(X; G), \quad \forall n > 1.$$

5. (i) Homology group of a topological space is a topological invariant.
   (ii) Cohomology group of a topological space is a topological invariant.
   (iii) If $X$ and $Y$ are two homotopically equivalent topological spaces, then their homology groups $H_n(X; G)$ and $H_n(Y; G)$ are isomorphic.
6. (i) The 101-dimensional sphere $S^{101}$ admits a nowhere nonvanishing tangent vector field.
   (ii) Every tangent vector field on the 100-dimensional sphere $S^{100}$ vanishes.
   (iii) If $v : S^{51} \to \mathbf{R}^{52}$ is a nowhere vanishing tangent vector field on the 51-dimensional sphere $S^{51}$, then the map

$$f : S^{51} \to S^{51}, \ x \mapsto \frac{v(x)}{||v(x)||}$$

   is homotopic to the identity map.

# References

Adams JF. Algebraic topology: a student's guide. Cambridge: Cambridge University Press; 1972.

Adhikari, M.R., Adhikari, A.: Groups. Rings and Modules with Applications. Universities Press, Hyderabad (2003)

Adhikari, M.R., Adhikari, A.: Textbook of Linear Algebra: An Introduction to Modern Algebra. Allied Publishers, New Delhi (2006)

Adhikari A, Adhikari MR. Basic topology, vol. 1: metric spaces and general topology. India: Springer; 2022a.

Adhikari A, Adhikari MR, Basic topology, vol. 2: topological groups, topology of manifolds and lie groups. India: Springer; 2022b.

Adhikari MR, Adhikari A. Basic modern algebra with applications. New Delhi, New York, Heidelberg: Springer; 2014.

Adhikari MR. Basic algebraic topolgy and its applications. India: Springer; 2016.

Bredon GE. Topology and geometry. New York: Springer; 1993.

Dieudonné J. A history of algebraic and differential topology, 1900–1960. Modern Birkhäuser; 1989.

Eilenberg S, Steenrod N. Foundations of algebraic topology. Princeton: Princeton University Press; 1952.

Gray B. Homotopy theory. An introduction to algebraic topology. New York: Academic Press; 1975.

Hatcher A. Algebraic topology. Cambridge University Press; 2002.

Hilton PJ, Wylie S. Homology theory. Cambridge: Cambridge University Press; 1960.

Hu ST. Homology theory. Oakland CA: Holden Day; 1966.

Hurewicz W. Beitrage de Topologie der Deformationen. Proc. K. Akad. Wet. Ser. A 1935;38:112–119, 521–528.

Milnor JW. On axiomatic homology theory . Pacific J. Math 1962;12:337-341.

Poincaré H. Analysis situs. J. l'Ecole Polyt. 1895;1:1–121

Poincaré H. Papers on topology: analysis situs and its five supplements. Amer. Math. Soc. 2010;37. (Translated by Stillwell J, History of Mathematics)

Rotman JJ. An introduction to algebraic topology. New York: Springer; 1988.

Spanier E. Algebraic topology. McGraw-Hill; 1966.

Switzer RM. Algebraic topology-homotopy and homology. Berlin, Heidelberg, New York: Springer; 1975.

# Chapter 4
# Topology of Fiber Bundles: General Theory of Bundles

A bundle in topology is a union of fibers parametrized by its base space and glued together by the topology of the total space. On the other hand, a **fiber bundle** is a bundle with an additional structure derived from the action of a topological group on the fibers. A fiber bundle is a locally trivial fibration having covering homotopy property. The theory of fiber bundles, in particular, vector bundles, establishes a very strong link between algebraic topology and differential topology. Topology of fiber bundles is studied in Chaps. 4 and 5. This chapter studies **General Theory of Bundles**. On the other hand, Chap. 5 develops the theory of bundles based on their **homotopy properties.** The topology of fiber bundles has created general interest and promises for more work, because it is involved of interesting applications of topology to other areas such as algebraic topology, geometry, physics and gauge groups.

**Historically**, the theory of fiber bundles was first recognized during the period 1935–1940 through the work of H. Whitney (1907–1989), H. Hopf (1894–1971), E. Stiefel (1909–1978), J. Feldbau (1914–1945) and some others.

For this chapter, the books (Adhikari, 2022a), (Bredon, 1993), (Husemoller, 1966), (Steenrod, 1951), (Adhikari, 2022b), (Adhikari, 2014), (Armstrong, 1983), (Hu, 1959), (Mayer, 1972), (Massey, 1991), (Maunder, 1970), (Munkers, 1984), (Rotman, 1988), (Spanier, 1966).

## 4.1 General Properties of Bundles in Topology

This section introduces the concept of bundles in topology and studies bundles. Bundles in topology play a key role in the theory of both fiber bundles and vector bundles. So we start with the study of preliminaries of the concept of bundles. Since a bundle in topology is a union of fibers parametrized by its base space and glued together by the topology of its total space, it provides a basic underlying structure

© The Author(s), under exclusive license to Springer Nature Singapore Pte Ltd. 2022
M. R. Adhikari, *Basic Topology 3*,
https://doi.org/10.1007/978-981-16-6550-9_4

for both the fiber bundles and vector bundles. They are special families of bundles with additional structures and are closely related to topology. Their deep study uses homotopy theory (see Chap. 5). The concept of fiber spaces is the most fruitful generalization of covering spaces, and its importance was first realized during 1935–1950 which facilitated to solve several problems involving homotopy and homology theories. On the other hand, a vector bundle is a bundle admitting an additional vector space structure on each of its fibers.

### 4.1.1 The Concept of Bundles in Topology

A bundle is a triple consisting of two topological spaces, one is called total space and the other is called base space connected by a continuous map from the total space to the base space, called the projection of the bundle space. This concept is formalized in Definition 4.1.1.

**Definition 4.1.1** A **bundle** $\xi = (X, p, B)$ is an ordered triple which consists of a topological space $X$, called the **total space** of $\xi$, a topological space $B$, called the **base space** of $\xi$, and a continuous onto map $p : X \to B$, called the **projection** of the bundle $\xi$. The space $F_b = p^{-1}(b) \subset X$ with subspace topology inherited from the topology of $X$ is called the **fiber** of $\xi$ over $b$ for each point $b \in B$. It is also sometimes abbreviated as $X_b$.

**Remark 4.1.2** The total space $X = \bigcup_{b \in B} X_b = \bigcup_{b \in B} p^{-1}(b)$ and any two fibers $X_b$ and $X_a$ are disjoint if $b \neq a$. This implies that every point of $X$ lies in exactly one fiber. The notation $X(\xi)$ is sometimes used to denote the total space and $B(\xi)$ the base space of the bundle $\xi$ to avoid any confusion.

**Example 4.1.3** The fibers of a bundle may be of different types. For example, consider the bundle $\xi = (X, p, B)$ displayed in Fig. 4.1. The total space $X$ of the bundle $\xi = (X, p, B)$ is decomposed into four different types of fibers such as a line segment, two line segments, a point together with a line segment and a point.

**Fig. 4.1** Bundle $\xi$ having four types of fibers

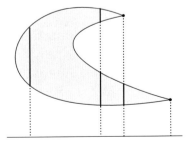

**Definition 4.1.4** A bundle $\eta = (Y, q, B)$ is called a **subbundle** of a bundle $\xi = (X, p, B)$ if

(i) $Y$ is a subspace of $X$ and
(ii) $q = p|_Y : Y \to B$ is its projection.

Moreover, if $A$ is a subspace of $B$, then the bundle $(Y, q, A)$ is a particular subbundle of $\xi$.

**Example 4.1.5** Given a bundle $\xi = (X, p, B)$, if $\eta = (Y, q, A)$ is a bundle such that there are inclusion maps

(i) $i : Y \hookrightarrow X$ and
(ii) $j : A \hookrightarrow B$ with $q = p|_Y$ is the restriction of $p$ over $Y$,

then $\eta$ is a subbundle of the bundle $\xi$.

**Definition 4.1.6** (*Induced bundle*) Given a bundle $\xi = (X, p, B)$ and a continuous map $f : A \to B$ from a space $A$, the induced bundle $f^*(\xi) = (Y, q, A)$ of $\xi$ over $A$ under $f$ is the bundle $(Y, q, A)$, where the total space $Y$ is defined by

$$Y = \{(a, x) \in A \times X : f(a) = p(x)\}$$

and the projection $q$ is defined by

$$q : Y \to A, \ (a, x) \mapsto a.$$

**Proposition 4.1.7** *Given a bundle $\xi = (X, p, B)$ and a continuous map $f : A \to B$ from a space $A$, let $f^*(\xi) = (Y, q, A)$ be the induced bundle of $\xi$ over $A$ under $f$. If*

$$p : X \to B$$

*is an open map, then*

$$q : Y \to A$$

*is also an open map.*

**Proof** It follows from the Definition 4.1.6 of the induced bundle. $\square$

**Definition 4.1.8** (*Product bundle*) The bundle $(B \times F, p, B)$, where $p : B \times F \to B$, $(b, x) \mapsto b$ is the projection from the product space $B \times F$ on the first factor, is called the product bundle over $B$ with fiber $F$.

**Definition 4.1.9** A **trivial bundle** $\xi = (X, p, B)$ with fiber $F$ is a bundle such that

(i) Its total space $X$ is homeomorphic to the product space $B \times F$.
(ii) $p : B \times F \to B$, $(b, x) \mapsto b$ is the projection from the product space $B \times F$ on the first factor.

**Definition 4.1.10** Let $\xi = (X, p, B)$ be a bundle with fiber $F$. An open covering $\{U_i : i \in \mathbf{K}\}$ of its base space $B$ with a family of homeomorphisms

$$\psi_i : p^{-1}(U_i) \to U_i \times F$$

is said to be a **local trivialization** if

$$p \circ \psi_i^{-1}(b, x) = b, \quad \forall b \in U_i, \ \forall x \in F \text{ and } \forall i \in \mathbf{K}.$$

***Example 4.1.11 (Möbius band)*** It is obtained from the product space $L \times F$ by identifying the two ends with a twist described in **Basic Topology, Volume 1, Chap. 3**, of the present series of books, where $L$ is a line segment. Consider projection

$$p : L \times F \to L, \ (x, y) \to x.$$

Then the Möbius band is a bundle whose base space is the circle $B$, obtained from the line segment $L$ by identifying its end points. Its fiber $F$ is a line segment. From the viewpoint of manifold, the Möbius band is the quotient manifold $\mathbf{R}^2 \ mod \ \mathbf{Z}$, as described in **Basic Topology, Volume 2** of the present series of books.

***Example 4.1.12 (Klein bottle)*** It is a bundle obtained from the product space $L \times F$ by identifying the two ends of the cylinder $L \times F$ described in **Basic Topology, Volume 1, Chap. 3**, of the present series of books, where $L$ is a line segment. Its base space is the circle $B$, obtained from a line segment $L$ by identifying its end points. Its fiber $F$ is a circle. Klein bottle cannot be topologically embedded in $\mathbf{R}^3$. From the viewpoint of manifold, Klein bottle is the quotient manifold $\mathbf{R}^2 \ mod \ (\mathbf{Z} \times \mathbf{Z})$ described in **Basic Topology, Volume 2** of the present series of books.

### 4.1.2  General Properties of Cross Sections of Bundles

This subsection studies the cross sections of a bundle, some of them are identified with familiar geometric objects.

**Definition 4.1.13** A **cross section** $s$ of a bundle $\xi = (X, p, B)$ is a continuous map

$$s : B \to X$$

such that the composite

$$p \circ s = 1_B,$$

where $1_B : B \to B$ is the identity map on $B$.

**Proposition 4.1.14** *For a bundle* $\xi = (X, p, B)$, *every cross section*

$$s : B \to X$$

*is injective and* $s(b) \in p^{-1}(b)$, $\forall b \in B$.

**Proof** Let $\xi = (X, p, B)$ be a bundle and $s : B \to X$ be a cross section. Then for every

$$b \in B, \ (p \circ s)(b) = 1_B(b) = b$$

implies that $s(b) \in p^{-1}(b)$. It asserts that $s(b) \in p^{-1}(b)$ for every $b$ of the base space $B$. Again the condition $p \circ s = 1_B$ proves that the map $s : B \to X$ is injective.  □

**Proposition 4.1.15** *Let* $\xi = (B \times F, p, B)$ *be a product bundle. Then its cross sections s are precisely of the form*

$$s : B \to B \times F, \ b \mapsto (b, f_s(b)),$$

*such that* $f_s : B \to F$ *is a continuous map which is uniquely determined by the cross section s.*

**Proof** Every cross section $s : B \to B \times F$ takes the form

$$s : B \to B \times F, \ b \mapsto (g_s(b), f_s(b)) : p \circ s = 1_B,$$

where the maps $f_s : B \to F$ and $g_s : B \to B$ are uniquely determined by $s$. Thus $p(s(b)) = 1_B(b)$, $\forall b \in B$ and hence

$$p(g_s(b), f_s(b)) = b \text{ asserts that } g_s(b) = b, \forall b \in B,$$

which implies that $s(b) = (b, f_s(b))$, $\forall b \in B$.

Conversely, given a continuous map $s : B \to B \times F$, suppose that

$$s(b) = (b, f_s(b)), \ \forall b \in B.$$

Then

$$(p \circ s)(b) = b, \ \forall \ b \in B \text{ asserts that } p \circ s = 1_B,$$

which proves that $s$ is a cross section of the product bundle $\xi = (B \times F, p, B)$.  □

**Remark 4.1.16** The map $f_s : B \to F$ defined in Proposition 4.1.15 has the property given in Corollary 4.1.17.

**Corollary 4.1.17** *The cardinality of the set* $\mathcal{S}(\xi)$ *of all cross sections of a product bundle* $\xi = (B \times F, p, B)$ *and the set* $\mathcal{C}(\xi)$ *of all continuous maps* $B \to F$ *are same.*

**Proof** Define a map $\phi$ using the map $f_s : B \to F$ formulated in Proposition 4.1.15

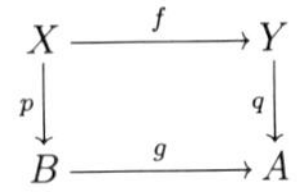

**Fig. 4.2** Rectangle
representing morphism of
bundles

$$X \xrightarrow{\quad f \quad} Y$$
$$p \downarrow \qquad \qquad \downarrow q$$
$$B \xrightarrow{\quad g \quad} A$$

$$\phi : \mathcal{S}(\xi) \to \mathcal{C}(\xi), \, s \mapsto f_s.$$

Since the map $\phi : \mathcal{S}(\xi) \to \mathcal{C}(\xi), \, s \mapsto f_s$ is a bijection, the corollary follows.    □

Proposition 4.1.18 characterizes the cross sections of a subbundle of a bundle in terms of cross sections of its mother bundle, which follows from the above discussion.

**Proposition 4.1.18** *Given a subbundle $\xi' = (X', p', B)$ of a bundle $\xi = (X, p, B)$, a continuous map $s : B \to X$ is a cross section of $\xi'$ iff $s(b) \in X'$ for every $b \in B$.*

## 4.1.3 General Properties of Bundle Morphisms

This subsection conveys the concept of bundle morphisms needed for the study of fiber and vector bundles specially, for comparing fiber bundles or vector bundles over the same or different base spaces likewise group homomorphisms (for example, see Proposition 4.1.21).

**Definition 4.1.19** Given two bundles $\xi = (X, p, B)$ and $\eta = (Y, q, A)$, a **bundle morphism or a fiber map** consists of a pair of continuous maps

$$(f, g) : \xi \to \eta : f : X \to Y \text{ and } g : B \to A$$

such that $q \circ f = g \circ p$, which means that in mapping diagram Fig. 4.2, the rectangle is commutative.

**Remark 4.1.20** Proposition 4.1.21 proves that the continuous map $f : X \to Y$ in Fig. 4.2 is fiber preserving in the sense that $f(p^{-1}(b)) \subset q^{-1}(g(b))$, $\forall b \in B$.

**Proposition 4.1.21** *The map $f : X \to Y$ in Fig. 4.2 is fiber preserving*

*Proof* For each $b \in B$, the map $f$ sends fibers of $\xi$ over $b$ into the fibers of $\eta$ over $f(b)$. Because, for each $x \in X$, the equality $(q \circ f)(x) = (g \circ p)(x)$ holds. This shows that the pair of maps $(f, g)$ sends the pair $(x, p(x))$ into the pair $(f(x), g(p(x)))$ by $(f, g)$. Hence, it follows that for every $b \in B$

$$f(p^{-1}(b)) \subset q^{-1}(g(b)).$$

□

**Fig. 4.3** Triangle representing $B$-morphism of bundle $f : \xi \to \eta$

**Fig. 4.4** Cross section of bundle

Definition 4.1.22 formalizes the interesting particular concept of $B$-morphism, when $\xi$ and $\eta$ are both bundles over the same base space $B$.

**Definition 4.1.22** Let $\xi = (X, p, B)$ and $\eta = (Y, q, B)$ be two bundles over the same base space $B$. Then a continuous map $f : X \to Y$ is said to be a **bundle $\mathbb{B}$-morphism** (in brief, $B$-morphism) written as $f : \xi \to \eta$ if the triangle in the Fig. 4.3 is commutative in the sense that $p = q \circ f$.

**Proposition 4.1.23** *The map $f : X \to Y$ given in Definition 4.1.22 is fiber preserving.*

**Proof** By hypothesis $q \circ f = p$ and hence $f(p^{-1}(b)) \subset q^{-1}(b)$ for every $b \in B$. □

Proposition 4.1.24 shows that a cross section of a bundle satisfies the general property of a bundle morphism.

**Proposition 4.1.24** *Let $\xi = (X, p, B)$ be a bundle over $B$. Then its cross sections are precisely the $B$-morphisms $s : (B, 1_B, B) \to (X, p, B)$.*

**Proof** For any cross section $s$ of $\xi$, since $p \circ s = 1_B$, it follows that the triangle in Fig. 4.4 is commutative. This shows that $s$ is a $B$-morphism. For its converse, let $s : (B, 1_B, B) \to (X, p, B)$ be a $B$-morphism. Then it follows that $s$ is a cross section of $\xi$. □

**Example 4.1.25** For a subbundle $(Y, q, A)$ of a bundle $(X, p, B)$ with inclusion maps $i : Y \hookrightarrow X$ and $j : A \hookrightarrow B$, the pair $(i, j) : (Y, q, A) \to (X, p, B)$ is a bundle morphism.

**Definition 4.1.26** Given two bundles $\xi = (X, p, B)$ and $\eta = (Y, q, A)$, a bundle map $(f, g) : \xi \to \eta$ is said to be a **bundle isomorphism** if both the maps

$$f : X \to Y \text{ and } g : B \to A$$

are homeomorphisms.

***Example 4.1.27*** Let $\xi = (X, p, B)$ and $\eta = (Y, q, A)$ be two bundles and $(f, g)$ : $\xi \to \eta$ be a bundle morphism. If $(f, g)$ is a bundle isomorphism, then the pair $(f^{-1}, g^{-1}) : \eta \to \xi$ is also a bundle isomorphism with both the pairs of maps

$$(f \circ f^{-1}, \ g \circ g^{-1}) \text{ and } (f^{-1} \circ f, g^{-1} \circ g)$$

are identity bundle morphisms.

**Definition 4.1.28** Given a bundle $\xi = (X, p, B)$ and a nonempty subset $A$ of $B$, the **restricted bundle** of $\xi$ to $A$, written as $\xi|A$, is the bundle $(Y, q, A)$, such that $Y = p^{-1}(A)$ and $q = p|Y$.

***Example 4.1.29*** Let $\xi = (X, p, B)$ be a bundle. If $X'$ is a nonempty subspace of $X$ and $p' = p|X' : X' \to B$, then $(X', p', B)$ is a restricted bundle of $\xi$.

**Definition 4.1.30** Given two bundles $\xi = (X, p, B)$ and $\eta = (Y, q, B)$ over the same base space $B$, if there exists a homeomorphism $f : X \to Y$, then $f$ is called a $B$-**isomorphism.** If for every point $b \in B$, there is an open nbd $U_b$ of $b$ and an $U_b$-isomorphism between the restricted bundles $\xi|U_b$ and $\eta|U_b$, then the given bundles are said to be **locally isomorphic.**

**Proposition 4.1.31** *A bundle which is locally isomorphic to a locally trivial bundle is also locally trivial.*

***Proof*** Let $\mathcal{B}$ be the set of all bundles over $B$ and $\sim$ be the equivalence relation on the set $\mathcal{B}$ defined by two bundles $\xi \sim \eta$ in $\mathcal{B}$ iff they are locally isomorphic. This asserts that if $\xi$ is locally isomorphic to a locally trivial bundle, then $\xi$ is locally trivial.  □

**Definition 4.1.32** *(Canonical morphism)* Given a bundle $\xi = (X, p, B)$ and a continuous map $f : A \to B$, if the bundle $f^*(\xi) = (Y, q, A)$ is the induced bundle under $f$ in the sense of Definition 4.1.6, then the pair of maps $(g, f)$, where $g : Y \to X, (a, x) \mapsto x$ forms a bundle morphism

$$(g, f) : f^*(\xi) \to \xi$$

called the canonical morphism of the induced bundle.

***Remark 4.1.33*** Proposition 4.1.34 establishes an interesting result asserting that if $f^*(\xi) = (Y, q, A)$ is the induced bundle of $\xi = (X, p, B)$ by a continuous map $f : A \to B$, then the fibers $q^{-1}(a)$ and $p^{-1}(f(a))$ are homeomorphic for every $a \in A$.

**Proposition 4.1.34** *Let $\xi = (X, p, B)$ be a bundle and $f : A \to B$ be a continuous map. If $f^*(\xi) = (Y, q, A)$ is the induced bundle under $f$ and $(g, f) : f^*(\xi) \to \xi$ is the canonical bundle morphism, then for every $a \in A$, the restricted map*

$$g|q^{-1}(a) : q^{-1}(a) \to p^{-1}(f(a))$$

*is a homeomorphism.*

**Fig. 4.5**  Construction of
cross section of the bundle
$(Y, q, A)$ by using $(f, g)$

**Fig. 4.6**  Construction of
induced cross section of
$(Y, q, A)$ by using $s$ and $f$

**Proof**  It follows from the property of the induced bundle.                    □

**Proposition 4.1.35**  *Given a bundle $\xi$ over $B$ and two continuous maps $g : A \to B$
and $h : C \to A$,*

(i)  *The induced bundle $1_B^*(\xi)$ and the bundle $\xi$ are B-isomorphic.*
(ii)  *The induced bundles $h^*(g^*(\xi))$ and $(g \circ h)^*(\xi)$ are C-isomorphic.*

**Proof**  It follows from definition of the induced bundle.                       □

**Proposition 4.1.36**  *Given a bundle morphism $(f, g) : (X, p, B) \to (Y, q, A)$ and a
cross section $s : B \to X$ of $(X, p, B)$, if the map $g : B \to A$ is a homeomorphism,
then*

$$s^* = f \circ s \circ g^{-1} : A \to Y$$

*is a cross section of the bundle $(Y, q, A)$ (Fig. 4.5).*

**Proof**  It follows from the construction of $s^*$.                              □

**Corollary 4.1.37**  *Given two bundles $(X, p, B)$ and $(Y, q, B)$ over the same base
space $B$ and a bundle B-morphism $f : (X, p, B) \to (Y, q, B)$, for every cross section
$s$ of $(X, p, B)$, there exists an induced cross section $s^*$ of $(Y, q, B)$ satisfying the
property $s^* = f \circ s$.*

**Proof**  It follows from Proposition 4.1.36 by taking $g = 1_B$ in diagram in Fig .4.6.□

**Definition 4.1.38**  *(Trivial bundle)*  A topological space $F$ is said to be the **fiber
space** of a bundle $\xi = (X, p, B)$ if for every $b \in B$, the fiber $p^{-1}(b)$ with subspace
topology inherited from the topology of $X$, is homeomorphic to $F$. The bundle $\xi$
is called a **trivial bundle** with fiber $F$ if $\xi$ is B-isomorphic to the product bundle
$(B \times F, q, B)$ with

$$q : B \times F \to B, (b, f) \mapsto b$$

the projection. It is said to be **locally trivial** if it is locally a product.

**Example 4.1.39**  The normal bundle $N(S^n)$ over $S^n$ is an 1-dimensional real trivial
bundle for every integer $n \geq 1$ (see Definition 4.4.9).

## 4.2  Fiber Bundles

This section discusses basic concepts of fiber bundles which form a special class of topological spaces widely used in different areas. A fiber bundle is a topological space which looks locally a direct product of two topological spaces. The concept of fiber bundles arose through the study of some problems in topology and geometry of manifolds around 1930. It is locally the product of its base space and a discrete space. The motivation of fiber bundles came through the study of the covering space

$$p : \mathbf{R} \to S^1, t \mapsto e^{2\pi it}$$

over $S^1$. This study leads to introduce the concept of fiber bundles in this section. For its general theory, see (Steenrod, 1951).

### 4.2.1  Introductory Concepts

This subsection introduces the concept of a fiber bundle and illustrates it by several examples.

**Definition 4.2.1**  An ordered quadruple $\xi = (X, p, B, F)$ is said to be a **fiber bundle** if

(i)   $X$ is a topological space, called the **total space** of $\xi$.
(ii)  $B$ is a topological space, called the **base space** of $\xi$.
(iii) $F$ is a topological space, called the **fiber** of $\xi$.
(iv)  $p : X \to B$ is a continuous surjective map, called the **fiber bundle projection** or simply projection of $\xi$.
(v)   the inverse image $p^{-1}(b) = F_b$ is homeomorphic to $F$ for each point $b \in B$. The subspace $F_b$ is called the **fiber over** $b$.
(vi)  $B$ has an open covering $\{U_a\}_{a \in \mathbf{A}}$ such that for each $a \in \mathbf{A}$, there is a homeomorphism

$$\psi_a : U_a \times F \to p^{-1}(U_a)$$

with the composite map $p \circ \psi_a$ is the projection

$$p_{U_a} : U_a \times F \to U_a$$

to the first factor in the sense that $p \circ \psi_a = p_{U_a}$, which means that

$$(p \circ \psi_a)(x, y) = x, \ \forall x \in U_a, \ \forall y \in F.$$

It is sometimes symbolized as

$$F \hookrightarrow X \xrightarrow{\;p\;} B$$

and it is called a covering of $B$ if $F$ is a discrete space. In that case, the space $X$ is called a **covering space** over $B$ with $p$ a covering projection. Then $p^{-1}(b_0)$ is discrete for $b_0 \in B$. The space $F$ is homeomorphic to $p^{-1}(b)$ for each $b \in B$.

**Definition 4.2.2** A **trivial fiber bundle** is a fiber bundle $\xi = (X, p, B, F)$ such that its total space $X$ is homeomorphic to the product space $B \times F$, and the projection is given by

$$p : B \times F \to B, \ (b, x) \to b.$$

**Remark 4.2.3** Since a fiber $F$ of a fiber bundle $\xi = (X, p, B, F)$ is a topological space, the set $\mathcal{H}omeo(F)$ of all homeomorphisms of $F$ forms a group $G$ under usual composition of functions. This is called **the structure group** $G$ for the fiber bundle $\xi = (X, p, B, F)$, and it leads to the concept of $G$-**bundles** (see Sect. 4.8). In particular, if $B$ is paracompact, then the projection

$$p : X \to B$$

is a **fibration** (see Chap. 5).

**Proposition 4.2.4** *For the exponential map*

$$p : \mathbf{R} \to S^1, t \mapsto e^{2\pi i t},$$

*the ordered quadruple* $\xi = (\mathbf{R}, p, S^1, \mathbf{Z})$ *is a fiber bundle.*

**Proof** Consider the unit circle $S^1 = \{z \in \mathbf{C} : |z| = 1\}$. To prove the proposition, consider the open sets $U_1 = S^1 - \{1\}$ and $U_2 = S^1 - \{-1\}$. Then $p^{-1}(U_1) = \mathbf{R} - \mathbf{Z}$. Consider the map

$$\psi_1 : U_1 \times \mathbf{Z} \to p^{-1}(U_1), \ (z, n) \mapsto n + (1/2\pi i) \log z$$

where $\log z$ is the principal value of the logarithm function on $\mathbf{C} - \{t \in \mathbf{R} : t \geq 0\}$. Then its inverse map exists and is given by

$$\psi_1^{-1} : p^{-1}(U_1) \to U_1 \times \mathbf{Z}, \ t \mapsto (e^{2\pi i t}, [t]),$$

where $[t]$ denotes the greatest integer contained in $t$ for $t \in p^{-1}(U_1) = \mathbf{R} - \mathbf{Z}$. Hence, it follows that $\psi_1$ is a homeomorphism such that the triangle in Fig. 4.7 is commutative. Consider the map

$$\psi_2 : U_2 \times \mathbf{Z} \to p^{-1}(U_2), \ (z, n) = n + (1/2\pi i) \log z,$$

where $\log z$ is the principal value of the logarithm function on $\mathbf{C} - \{t \in \mathbf{R} : t \leq 0\}$ with its inverse map

**Fig. 4.7** Homeomorphism
$\psi_1$ commuting the triangle

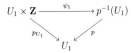

**Fig. 4.8** Homeomorphism
$\psi_2$ commuting the triangle

$$\psi_2^{-1} : p^{-1}(U_2) \to U_2 \times \mathbf{Z}, \ t \mapsto (e^{2\pi i t}, [t + \frac{1}{2}]).$$

Proceeding as before it follows that $\psi_2$ is a homeomorphism such that the triangle in Fig. 4.8 is commutative. This proves that the ordered quadruple $\xi = (\mathbf{R}, p, S^1, \mathbf{Z})$ is a fiber bundle.  $\square$

### Example 4.2.5 *(More examples of fiber bundles)*

(i) Given a connected space $B$ and a point $b_0 \in B$, if $p : X \to B$ is a covering projection, then

$$(X, p, B, p^{-1}(b_0))$$

is a fiber bundle.

(ii) The $n$-dimensional torus

$$T^n = \overbrace{S^1 \times S^1 \times \cdots \times S^1}^{n}$$

and the map

$$p : \mathbf{R}^n \to T^n, \ (t_1, t_2, \ldots, t_n) \mapsto (e^{2\pi i t_1}, e^{2\pi i t_2}, \ldots, e^{2\pi i t_n})$$

form a fiber bundle $(\mathbf{R}^n, p, T^n, F)$ with fiber $F$ which is the set of integer lattice points in $\mathbf{R}^n$.

(iii) $\xi = (SO(n, \mathbf{R}), p, S^{n-1})$ is a fiber bundle over $S^{n-1}$ with fiber $SO((n-1), \mathbf{R})$. To show it, consider $SO((n-1), \mathbf{R})$ as a subgroup of $SO(n, \mathbf{R})$, which consists of matrices $A \in SO(n, \mathbf{R})$ that keep the vector $e_n = (0, 0, \ldots, 1)$ fixed. The continuous map

$$p : SO(n, \mathbf{R}) \to S^{n-1}, \ A \mapsto Ae_n$$

is the projection map of $\xi$. Then $p(AB) = A, \ \forall B \in SO(n-1, \mathbf{R})$ shows that $p^{-1}(x)$ homeomorphic to the group $SO(n-1, \mathbf{R})$ for every $x \in S^{n-1}$.

## 4.2.2 Fiber Map and Mapping Fiber

This subsection studies fiber maps, induced maps, trivial fibering and mapping fiber by imposing certain conditions on bundles studied in Sect. 4.1. A fiber bundle $\xi = (X, p, B)$ is sometime written simply as $p : X \to B$. Given two fiberings $p : X \to B$ and $q : Y \to A$, a continuous map $f : X \to Y$ is said to be a fiber map if $f$ sends fibers into fibers. This concept is formulated in Definition 4.2.6.

**Definition 4.2.6** Let $p : X \to B$ and $q : Y \to A$ be two fiberings. A continuous map

$$f : X \to Y$$

is said to be a **fiber map** if for every point $b \in B$, there exists a point $a \in A$, such that

$$f(p^{-1}(b)) \subset q^{-1}(a).$$

Given two fiberings, Definition 4.2.7 formulates the concept of an induced map on their base spaces.

**Definition 4.2.7** (**Induced map**) Let $p : X \to B$ and $q : Y \to A$ be two fiberings, and $f : X \to Y$ be a fiber map. Then $f$ induces a map

$$f_* : B \to A, b \mapsto (q \circ f)(p^{-1}(b)).$$

For an arbitrary subset $V \subset A$,

$$f_*^{-1}(V) = (p \circ f^{-1})(q^{-1}(V))$$

asserts that $f_*$ is continuous if the map $p$ is either open or closed. The map $f_*$ is called the induced map of the given fiber map $f$.

**Remark 4.2.8** Let $p : X \to B$ be a bundle. Then $p$ is open, and hence, the induced map $f_* = g : B \to A$ is a continuous map by Definition 4.2.7 such that the diagram in Fig. 4.9 is commutative in the sense that $q \circ f = g \circ p$.

**Example 4.2.9** (i) (**Trivial fibering**) The fibering $q : Y \to A$ is said to be trivial over the base space $B$, with the projection $p : X \to B$ as a fiber map, where $A = B = Y$, $q = 1_B$ (identity map on B) and the induced map $g = 1_B$.

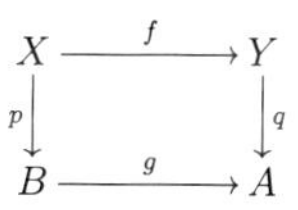

**Fig. 4.9** Rectangle involving fiber map

$$
\begin{array}{ccc}
X & \xrightarrow{f} & Y \\
{\scriptstyle p}\downarrow & & \downarrow{\scriptstyle q} \\
B & \xrightarrow{g} & A
\end{array}
$$

(ii) Let $p : X \to B$ be the trivial fibering over the base space $B$. Then every map $f : X \to Y$ is a fiber map, where $B = X$ and $p = 1_X$.

**Definition 4.2.10 (Mapping Fiber)** Let $f : (X, x_0) \to (Y, y_0)$ be a base point preserving continuous map. Its mapping fiber $F_f$ is the pointed topological space

$$F_f = \{(x, \alpha) \in X \times Y^{\mathbf{I}} : \alpha(0) = y_0 \text{ and } \alpha(1) = f(x)\} \subset X \times Y^{\mathbf{I}}$$

with product topology and base point $(x_0, \alpha_0)$, where $\alpha_0$ is the constant path at $y_0$.

**Remark 4.2.11** Definition 4.2.10 implies that the elements of the mapping fiber $F_f$ are precisely the ordered pairs $(x, \alpha)$ such that $\alpha$ is a path in $Y$ from the point $y_0$ to the point $f(x)$. In particular, the loop space $\Omega(Y, y_0)$ based at the point $y_0 \in Y$ consists of exactly of all ordered pairs of the form $(x_0, \alpha)$. The map

$$h : \Omega(Y, y_0) \to F_f : \alpha \mapsto (x_0, \alpha)$$

is clearly injective. The map

$$p : F_f \to X, \ (x, \alpha) \mapsto x$$

is the projection map.

## 4.3  Vector Bundles over Topological Spaces and Manifolds

A vector bundle is a special type of a fiber bundle for which every fiber admits a vector space structure compatible on its neighboring fibers and the structure group is a group of linear automorphism of the corresponding vector space. The concept of vector bundles was born through the study of tangent vector fields to smooth geometric objects such as manifolds, viz., spheres, projective spaces, etc. Roughly speaking, a vector bundle over a topological $B$ is a family of vector spaces continuously parameterized by the base space $B$. This section studies vector bundles over topological spaces in Subsect. 4.3.1 as well as over manifolds in Sect. 4.3.2 as base spaces. The theory of vector bundles over a manifold provides a convenient language to study many problems on the manifold.

For vector bundles, $\mathbf{F}$ denotes one of the fields $\mathbf{R}, \mathbf{C}$ or the division ring $\mathbf{H}$ of quaternionic numbers and $\mathbf{F}^n$ denotes the $n$-dimensional vector space over $\mathbf{F}$.

### 4.3.1  Vector Bundles over Topological Spaces

This subsection communicates the concept of vector bundles over topological spaces and studies it. In particular, vector bundles over manifolds are studied in Sect. 4.3.2.

**Definition 4.3.1   (*Vector bundle*)** A fiber bundle $\xi = (X, p, B, \mathbf{F}^n)$ together with the structure of an $n$-dimensional vector space $\mathbf{F}^n$ over $\mathbf{F}$ on every fiber $X_b = p^{-1}(b)$ for $b \in B$ is said to be an $n$-dimensional $\mathbf{F}$-vector bundle over a space $B$ if

(i)  There is an open covering $\{U_a : a \in \mathbf{A}\}$ of $B$.
(ii) For each $a \in \mathbf{A}$, there is a homeomorphism $\psi_a$ such that

$$\psi_a : U_i \times \mathbf{F}^n \to p^{-1}(U_a) : p \circ \psi_a = p_{U_a}$$

and

$$(\psi_a|\{b\} \times \mathbb{F}^n) : \{b\} \times \mathbf{F}^n \to p^{-1}(b)$$

is an isomorphism of vector spaces over $\mathbf{F}$ for every $b \in U_a$.

Each $\psi_a$ is called a **coordinate transformation.** An $\mathbf{F}$-vector bundle $\xi$ is said to be a real, complex or quaternionic vector bundle if $\mathbf{F} = \mathbf{R}, \mathbf{C}$ or $\mathbf{H}$. The vector bundle $\xi = (X, p, B)$ is also called an $n$-dimensional $F$-vector bundle over $B$.

**Example 4.3.2**  (i)  If the fiber of a vector bundle $\xi$ is $\mathbf{R}^n$, then $\xi$ is said to be **finite dimensional** with dim $\xi = n$.
(ii) If the fiber of $\xi$ is an infinite-dimensional Banach space and the structure group is the group of invertible bounded operators of the Banach space, the bundle $\xi$ is said to be **infinite dimensional.**

**Example 4.3.3   (*Vector bundle* $\gamma_r^n = (X, p, G_r(\mathbf{F}^n), \mathbf{F}^n)$ )** Let $G_r(\mathbf{F}^n)$ be the Grassmann manifold of $r$-dimensional subspaces of $\mathbf{F}^n$. Then the bundle $\gamma_r^n = (X, p, G_r(\mathbf{F}^n), \mathbf{F}^n)$, where

$$X = \{(V, y) \in G_r(\mathbf{F}^n) \times \mathbf{F}^n\}$$

and

$$p : X \to \mathbf{F}^n, \ (V, y) \mapsto y \text{ is the orthogonal projection of } y \text{ into } V,$$

is an $n$-dimensional $\mathbf{F}$ vector bundle.

**Remark 4.3.4**  Proposition 4.3.5 proves that the set of all cross sections of an $\mathbf{F}$-vector bundle $\xi = (X, p, B)$ form a module $\mathcal{M}_F$ over the ring $R$ of $\mathbf{F}$-vector-valued continuous maps on $B$.

**Proposition 4.3.5**  *Given an $n$-dimensional $\mathbf{F}$-vector bundle $\xi = (X, p, B)$ over $B$, the set of all cross sections of $\xi$ forms a module $\mathcal{M}_F$ over the ring $R$ of $\mathbf{F}$-vector-valued continuous maps on $B$.*

**Proof**  Given cross sections $s, s'$ of $\xi$ and a continuous map $f : B \to \mathbf{F}$, the map

$$s + s' : B \to X, b \mapsto s(b) + s'(b)$$

is a cross section of $\xi$ and the map

$$fs : B \to X, b \mapsto f(b)s(b)$$

is also a cross section of $\xi$ for all continuous maps $f : B \to \mathbf{F}$. Again, the map

$$\theta : B \to X, \ b \mapsto 0 \in p^{-1}(b)$$

is a cross section (zero cross section). Let $\psi : U \times \mathbf{F}^n \to p^{-1}(U)$ be a local coordinate of $\xi$ over $U$. If

(i) $\psi^{-1}(s(b)) = (b, g(b))$, $\forall b \in B$ and
(ii) $\psi^{-1}(s'(b)) = (b, g'(b))$, $\forall b \in B$

for some continuous maps $g : U \to \mathbf{F}^n$ and $g' : U \to \mathbf{F}^n$, then

(i)
$$\psi^{-1}((s + s')(b)) = (b, g(b) + g'(b)), \ \forall b \in B;$$

(ii)
$$\psi^{-1}((fs)(b)) = (b, f(b)g(b)), \ \forall b \in B$$

and
(iii)
$$\psi^{-1}(0)(b) = (b, 0) \ \forall b \in B.$$

Hence, it follows that $s + s'$, $fs$ and $0$ are continuous maps such that they are all cross sections. This proves that $\mathcal{M}_F$ is a module over the ring $R$.                  □

**Definition 4.3.6** Given two vector bundles $\xi = (X, p, B)$ and $\eta = (Y, q, A)$, a vector bundle morphism

$$(f, g) : \xi \to \eta$$

is a pair of continuous maps

$$f : X \to Y \text{ and } g : B \to A$$

such that the diagram in the Fig. 4.10 is commutative in the sense that $q \circ f = g \circ p$ and

$$f|_{p^{-1}(b)} : p^{-1}(b) \to q^{-1}(g(b))$$

is a linear map (transformation) for every $b \in B$. In particular for $B = A$, the given vector bundles become $\xi = (X, p, B)$ and $\eta = (Y, q, B)$. Then a morphism of the form $(f, 1_B) : \xi \to \eta$ as shown in diagram Fig. 4.11 is called a **B-morphism of vector bundles** $f : \xi \to \eta$.

**Fig. 4.10**  Morphism of
vector bundles

**Fig. 4.11**  B-morphism
$f : \xi \to \eta$ of vector bundles

**Remark 4.3.7**  Let $f : \xi \to \eta$ is a B-morphism of vector bundles. Then it follows from Definition 4.3.6 that

(i)  $q \circ f = p$ and
(ii)  $f|_{p^{-1}(b)} : p^{-1}(b) \to q^{-1}(b)$ is a linear map for every $b \in B$.

**Definition 4.3.8** (*Isomorphism of vector bundles*)  Given two $n$-dimensional **F**-vector bundles $\xi = (X, p, B, \mathbf{F}^n)$ and $\eta = (Y, q, A, \mathbf{F}^n)$, a vector bundle morphism

$$(f, g) : \xi \to \eta$$

is said to be an **isomorphism or an equivalence** if there exists a homeomorphism $f : X \to Y$ such that

$$f|_{p^{-1}(b)} : p^{-1}(b) \to q^{-1}(g(b))$$

is a linear isomorphism for every $b \in B$. In particular, if $B = A$ and $g = 1_B$, then the given vector bundles become $\xi = (X, p, B, \mathbf{F}^n)$ and $\eta = (Y, q, B, \mathbf{F}^n)$ and

$$f : \xi \to \eta$$

is called an equivalence over $B$, and in this case, $\xi$ and $\eta$ are said to be **equivalent** over the base space $B$, denoted by $\xi \cong \eta$.

**Remark 4.3.9**  Any two equivalent vector bundles have the same dimension.

**Proposition 4.3.10**  *Let $\mathcal{V}_\mathbf{F}$ be the set of all $\mathbf{F}$-vector bundles over $B$. Equivalent relation $\cong$ of $\mathbf{F}$-vector bundles over the same base space $B$ is an equivalence relation on $\mathcal{V}_\mathbf{F}$.*

**Proof**  The reflexivity and transitivity of the given relation follow from Definition 4.3.8. To prove the symmetry of the relation, take any two bundles $\xi = (X, p, B, \mathbf{F}^n)$ and $\eta = (Y, q, B, \mathbf{F}^n)$ with $f : \xi \to \eta$ an equivalence. Then $f : X \to Y$ is an injective continuous map. To show that $f$ is open, it suffices to prove that $f|_{p^{-1}(U_\alpha)}$ is open, for an open covering $\{U_\alpha\}$ of $B$. If it is expressed in terms of local coordinates as $(x, v) \to (x, A_x v)$, where $A_x$ is a nonsingular linear transformation depending continuously on $x$ then map has a continuous inverse, because matrix inversion is continuous. This shows that $f|_{p^{-1}(U_\alpha)}$ is a homeomorphism. This implies that the $f$ is open.  □

**Definition 4.3.11** *(Trivial vector bundle)* Let $\xi = (X, p, B, \mathbf{F}^n)$ be a vector bundle. If $\xi$ is isomorphic to the product bundle $B \times \mathbf{F}^n \to B$, then it is said to be a trivial vector bundle.

**Proposition 4.3.12** *Given an n-dimensional product bundle $\xi = (B \times \mathbf{F}^n, p, B, \mathbf{F}^n)$ and an m-dimensional product bundle $\eta = (B \times \mathbf{F}^m, p, B, \mathbf{F}^m)$, the B-morphisms $f : \xi \to \eta$ are expressed in the form*

$$f(b, x) = (b, g(b, x)), i.e., f : B \times \mathbf{F}^n \to B \times \mathbf{F}^m, (b, x) \mapsto (b, g(b, x)),$$

*where the map*

$$g : B \times \mathbf{F}^n \to \mathbf{F}^m, \ (b, x) \mapsto g(b, x) \text{ is linear in } x.$$

**Proof** Let $\mathcal{L}(\mathbf{F}^n, \mathbf{F}^m)$ be the vector space of all linear transformations $T : \mathbf{F}^n \to \mathbf{F}^m$. Then the two vector spaces $\mathcal{L}(\mathbf{F}^n, \mathbf{F}^m)$ and $\mathbf{F}^{mn}$ are isomorphic and the map

$$g : B \times \mathbf{F}^n \to \mathbf{F}^m$$

is continuous iff the map

$$\alpha : B \to \mathcal{L}(\mathbf{F}^n, \mathbf{F}^m), b \mapsto g(b, -)$$

is continuous. Hence the proposition follows.                                                                      □

**Theorem 4.3.13** *(i)  Given a vector bundle $\xi = (X, p, B)$ over $B$, the vector bundle $1_B^*$ induced by the identity map $1_B : B \to B$ and the vector bundle $\xi$ are B-isomorphic.*
*(ii)  Again, given a pair of continuous maps $f$ and $g$*

$$A \xrightarrow{\ f\ } B_1 \xrightarrow{\ g\ } B,$$

*their two induced vector bundles $f^*(g^*(\xi))$ and $(g \circ f)^*(\xi)$ are A-isomorphic.*

**Proof** By hypothesis, $f : A \to B_1$ and $g : B_1 \to B$ are two continuous maps, and $g \circ f$ is their composite map

$$A \xrightarrow{\ f\ } B_1 \xrightarrow{\ g\ } B,$$

which is also continuous. From the definition of the induced bundles, it follows that if $g^*(\xi) = (X_1, p_1, B_1)$, then

$$X_1 = \{(b_1, x) \in B_1 \times X : g(b_1) = p(x)\}$$

and

$$p_1 : X_1 \to B_1, (b_1, x) \mapsto b_1.$$

In particular, if $g = 1_B : B \to B$, then $1_B^*(\xi)$ and $\xi$ are $B$-isomorphic. Consider the $A$-vector bundle induced by the map $g \circ f : A \to B$ with its total space $X_2 = E((g \circ f)^*(\xi))$ defined by

$$E((g \circ f)^*(\xi)) = \{(a, x) \in A \times X : (g \circ f)(a) = p(x)\}$$

and the total space of $E(f^*(g^*(\xi)))$ is defined by
$$\begin{aligned} E(f^*(g^*(\xi))) &= \{(a, y) \in A \times X_1 : f(a) = p_1(y)\} \\ &= \{(a, (b_1, x)) : g(b_1) = p(x) \text{ and } f(a) = p_1(y) = b_1\} \\ &= \{(a, (f(a), x)) : (g \circ f)(a) = p(x)\}. \end{aligned}$$
Hence, it follows that the two induced vector bundles $f^*(g^*(\xi))$ and $(g \circ f)^*(\xi)$ are $A$-isomorphic, because the map

$$\alpha : (g \circ f)^*(\xi) \to f^*(g^*(\xi)), (a, x) \mapsto (a, (f(a), x))$$

is a vector bundle isomorphism from $(g \circ f)^*(\xi)$ to $f^*(g^*(\xi))$.                          □

The above discussion can be expressed in the language of category theory formulated in Theorem 4.3.14.

**Theorem 4.3.14**  *Let $\mathcal{T}op$ be the category of topological spaces and their continuous maps and $\mathcal{S}et$ be the category of sets and set functions. If $\mathrm{Vect}_n(B)$ denotes the set of isomorphism classes of n-dimensional vector bundles over B, then*

$$\mathrm{Vect}_n : \mathcal{T}op \to \mathcal{S}et$$

*is a contravariant functor.*

**Proof**  (i) The object function is defined as follows: $\mathrm{Vect}_n$ assigns to every object $B \in \mathcal{T}op$, the vector bundle $\mathrm{Vect}_n(B)$, which is the set of isomorphism classes of n-dimensional vector bundles over $B$.

(ii) The morphism function is defined as follows: Let $\{\xi\}$ denotes the isomorphism class of a vector bundle $\xi$ over $B$. Then for every morphism $f : B_1 \to B$ in the category $\mathcal{T}op$, the function

$$f^* : \mathrm{Vect}_n(B) \to \mathrm{Vect}_n(B_1), \ \{\xi\} \mapsto \{f^*(\xi)\}$$

defines the morphism function.

(iii) Finally, using Theorem 4.3.13, it follows that

$$\mathrm{Vect}_n : \mathcal{T}op \to \mathcal{S}et$$

is a contravariant functor.

                                                                                                □

### 4.3.2   Vector Bundles over Manifolds

The theory of vector bundles over a manifold $M$ provides a convenient language to describe many ideas in manifold. Given a manifold $M$ of dimension $n$, and an open covering $\{U_a : a \in \mathbf{A}\}$ of $M$, a vector bundle over $M$ is constructed by gluing together a family of product bundles $\{U_a \times \mathbf{R}^n\}$ by an action of the group of $GL(n, \mathbf{R})$ on $\mathbf{R}^n$.

**Definition 4.3.15**  Let $M$ be a smooth manifold of dimension $n$ and $N$ be a smooth manifold of dimension $m$. Then a smooth map $f : M \to N$

(i) is said to be an **immersion** at a point $x \in M$, if $n \le m$ and rank of $f$ at $x$ is $n$. It is said to be an immersion if it is an immersion at every point $x \in M$.
(ii) is said to be a **submersion** at a point $x \in M$, if $n \ge m$ and rank of $f$ at $x$ is $m$. It is said to be a submersion if it is an submersion at every point $x \in M$.
(iii) is an **embedding** if it is an immersion and a homeomorphism onto its image $f(M)$.
(iv) is a **diffeomorphism**, if $n = m$ and $f$ is a surjective embedding.

**Example 4.3.16**  The natural inclusion map $i : S^2 \hookrightarrow \mathbf{R}^3$ is an immersion, since at each point of $S^2$, the map $i$ has rank 2, the dimension of $S^2$.

**Definition 4.3.17**  An $n$-dimensional $\mathbf{R}$-vector bundle over a manifold $B$ is an ordered triple $\xi = (X, p, B)$ consisting of a pair of smooth manifolds $X$ and $B$ connected by a smooth surjective map $p : X \to B$ such that the following conditions are satisfied:

(a) **(VB$_1$)** For every $b \in B$, the inverse image $X_b = p^{-1}(b)$ is an $n$-dimensional vector space over $\mathbf{R}$.
(b) **(VB$_2$)** For every $b \in B$, there exists an open nbd $U$ of $b$ and a diffeomorphism $\psi : U \times \mathbf{R}^n \to p^{-1}(U)$ with the properties that

(i) The triangle in Figure 4.12 is commutative, where

$$p_1 : U \times \mathbf{R}^n \to U, \ (y, v) \mapsto y$$

is the projection map onto the first factor.
(ii) For every $y \in B$, there exists a linear isomorphism

$$\psi_y : \mathbf{R}^n \to p^{-1}(y), \ v \mapsto \psi(y, v)$$

The vector bundle $\xi = (X, p, B)$ is also denoted by $p : X \to B$.

(i) The manifold $X$ is called the **total space** and the manifold $B$ is called the **base space** and the map $p$ is called the projection of the vector bundle $\xi$.
(ii) The inverse image $X_b = p^{-1}(b)$ is called its **fiber** over $b$.
(iii) The defining condition **VB$_2$** is called the **local triviality**, the pair $(U, \psi)$ is called a **vector bundle chart**, $U$ is called a **trivializing open set** and the family $\{(U, \psi)\}$ is called a **vector bundle atlas** of the vector bundle $\xi$.

**Fig. 4.12** Commutative
diagram of vector bundles
involving $\psi$ and projection
maps

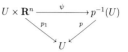

(iv) The **dimension of** $\xi$ is the dimension of its fiber.

**Remark 4.3.18** Definition 4.3.17 asserts that a vector bundle $\xi = (X, p, B)$ of dimension $n$ over a manifold $B$ is a product bundle $B \times \mathbf{R}^n$ obtained by gluing together a family of product bundles $\{U_a \times \mathbf{R}^n\}$, where $\{U_a : a \in \mathbf{A}\}$ is an open covering of $B$ and is obtained by an action of the linear group $GL(n, \mathbf{R})$ on $\mathbf{R}^n$. The dimension of a vector bundle $\xi = (X, p, B, \mathbf{R}^n)$ is actually the dimension of its fibers (instead of dimension of $X$ or $B$). It is well defined, because, for any fiber $X_b$ over $b \in B$, the map

$$\psi : B \to \mathbf{R}, b \mapsto dim \ X_b$$

is locally constant, and it is constant on each component of $B$. If $\psi$ is constant on the entire $B$, then $dim \ X_b$ is constant for all $b \in B$ and this value is the dimension of $\xi$.

**Example 4.3.19** Let $V$ be an $n$-dimensional vector space over $\mathbf{R}$. If $B$ is a manifold, then the projection $p : B \times V \to B$, $(b, v) \to b$ defines a vector bundle, called a **product bundle**.

**Proposition 4.3.20** *The projection map*

$$p : p^{-1}(U) \to U$$

*given in Definition 4.3.17 is an submersion.*

**Proof** It follows from commutativity of triangle in Figure 4.12 that $p = p_1 \circ \psi^{-1}$ holds locally. Since $\psi^{-1}$ is a diffeomorphism and $p_1$ is an submersion and hence $p$ is a submersion (see **Basic Topology, Volume 2**). ☐

**Example 4.3.21** Let $\xi = (X, p, B)$ be a vector bundle over a manifold $B$ and $S$ be a submanifold of $B$. Then the **restricted bundle** $\xi|S$ is vector bundle $p : p^{-1}(S) \to S$.

**Example 4.3.22 (Whitney sum of vector bundles)** Let $\xi_1 = (X_1, p_1, B)$ and $\xi_2 = (X_2, p_2, B)$ be two vector bundles over $B$. Then their Whitney sum, denoted by $\xi_1 \oplus \xi_2$, is the vector bundle $\xi = (X_1 \oplus X_2, p, B)$, where the total space is given by

$$X_1 \oplus X_2 = \{(x_1, x_2) \in X_1 \times X_2 : p_1(x_1) = p_2(x_2)\}$$

and the projection $p$ is defined by

$$p : X_1 \oplus X_2 \to B, (x_1, x_2) \mapsto p_1(x_1) = p_2(x_2)$$

and its fiber over $b \in B$, is the direct sum $p_1^{-1}(b) \oplus p_2^{-1}(b)$.

There is a natural question: does there exist any similarity between the concepts of a vector bundle and a manifold? Remark 4.3.23 gives its answer.

**Remark 4.3.23** **Similarity between a vector bundle and a manifold.**

(i) A vector bundle is analogous to a manifold in the sense that both of them are built up from elementary objects by gluing together by specified maps. For example, the elementary objects for vector bundle are trivial bundles $U \times \mathbf{R}^n$, and the gluing maps are morphisms of the form

$$\psi : U \times \mathbf{R}^n \to U \times \mathbf{R}^n, (x, y) \mapsto (x, g(x)) \text{ for } \text{ some } g : U \to GL(n, \mathbf{R}).$$

(ii) On the other hand, the elementary objects for manifolds are open subsets of $\mathbf{R}^n$, and the gluing maps are homeomorphisms.

(iii) More generally, a vector bundle over a topological space $B$ consists of a family $\{X_b\}_{b \in B}$ of disjoint vector spaces parametrized by the space $B$. The union $X = \bigcup_{b \in B} X_b$ of these vector spaces is a space $X$ and the map

$$p : X \to B, X_b \mapsto b$$

is continuous, and it is locally trivial in the sense that $X$ looks locally like the product $U \times \mathbf{R}^n$.

(iv) On the other hand, given a manifold $M$ of dimension $n$, and an open covering $\{U_i : i \in \mathbf{A}\}$, a vector bundle over $M$ is constructed by gluing together a family of product bundles $\{U_i \times \mathbf{R}^n\}_{i \in \mathbf{A}}$ by an action of the group of $GL(n, \mathbf{R})$ on $\mathbf{R}^n$.

## 4.4  Tangent and Normal Bundles over Manifolds

This section studies tangent and normal bundles over manifolds specially over $S^n$.

### 4.4.1  Tangent Bundle of a Smooth Manifold

This subsection introduces the concept of tangent bundle over a smooth manifold and studies it. If $M$ is an $n$-dimensional smooth manifold and $T_x(M)$ is the tangent space at $x \in M$, then $T(M) = \bigcup_{x \in M} \{T_x(M)\}$ (disjoint union) together with a projection $p :$ $T(M) \to M$ forms a tangent bundle of the manifold $M$. It is formalized in Definition 3.5.26. The tangent bundle $T(M)$ is a smooth manifold of dimension $2n$ by Theorem 4.4.3.

**Definition 4.4.1** *(The tangent bundle of a manifold)* The tangent bundle of a smooth $n$ manifold $M$ is the bundle $(T(M), p, M)$, where

(i) The total space $T(M)$ is the disjoint union of all tangent spaces $T_x(M)$ as $x$ runs over $M$. This is the set of all ordered pairs $(x, v)$ such that $x \in M$ and $v \in T_x(M)$.
(ii) The map

$$p : T(M) \to M, \ (x, v) \to x$$

is called the **projection map** of the tangent bundle. $T_x(M)$ is called the **tangent space** at $x$, and $v \in T_x(M)$ is called the **tangent vector** with initial point $x$.

More precisely, let $M$ be an $n$-dimensional smooth manifold, $T_x(M)$ be the vector space of all tangent vectors to $M$ at any point $x \in M$ and $T(M) = \bigcup_{x \in M} \{T_x(M)\}$
(disjoint union). Then the set $T(M)$ is the disjoint union set of all tangent spaces $T_x(M)$, and there is projection map

$$p : T(M) \to M, \ (x, v) \mapsto x.$$

The subspace $F_x = p^{-1}(x) = T_x(M)$ with topology inherited from $M$ is called the **fiber** over $x$. If $M$ is a smooth manifold of dimension $n$, then $T(M)$ is a $2n$-dimensional manifold (see Theorem 4.4.3). A **cross section of the tangent bundle** $(T(M), p, M)$ is a smooth map $s : M \to T(M)$ such that $p \circ s = 1_M$ and hence $(p \circ s)(x) = x, \ \forall x \in M$.

**Remark 4.4.2** If $M$ is a smooth manifold of dimension $n$, then $T(M)$ is the set of all tangent vectors $(x, v)$ at all points $x \in M$. If $p$ assigns to every vector $(x, v)$ its initial point $x$, then $T_x(M) = p^{-1}(x) = F_x$ is the tangent plane at the point $x$, which is a linear space. A cross section of the tangent bundle $T(M)$ of $M$ is a **vector field** over $M$.

Theorem 4.4.3 gives a unique differential structure on $T(M)$ and proves if $dim\ M = n$, then $dim\ T(M) = 2n$.

**Theorem 4.4.3** *Let $M$ be a smooth manifold of dimension $n$, then its tangent bundle $T(M)$ is a smooth manifold of dimension $2n$.*

**Proof** Let $p : T(M) \to M, (x, v) \to x$ be the projection map of the tangent bundle $T(M)$ and $(\psi, U)$ be a chart of $M$. Then it determines a map

$$T_\psi : p^{-1}(U) \to \psi(U) \times \mathbf{R}^n \subset \mathbf{R}^n \times \mathbf{R}^n, (x, v) \mapsto (\psi(x), d\psi_x(v))$$

Clearly, $T_\psi$ is a bijective map with its inverse

$$T_\psi^{-1} : \psi(U) \times \mathbf{R}^n \to p^{-1}(U), (a, w) \mapsto (b.d\psi_x^{-1}(w) : b = \psi^{-1}(a).$$

If $(\psi, U)$ and $(\phi, V)$ be two compatible charts of $M$, then the map

$$T_\phi \circ T_\psi^{-1} : \psi(U \cap V) \times \mathbf{R}^n \to \phi(U \cap V) \times \mathbf{R}^n, (a, w) \mapsto T_\phi(b, d_\psi^{-1}(W)) = (\phi(b), d\phi_b \circ d\psi_b^{-1}(w))$$

$$= (\phi \circ \psi^{-1}(a, d\phi_b \circ d\psi_b^{-1}(w)).$$

It asserts that $T_\phi \circ T_\psi^{-1}$ is a homeomorphism. Hence, $T(M)$ has a unique topology such that

(i)  Each $T_\psi$ is a homeomorphism.
(ii)  $T(M)$ is second countable and Hausdorff.
(iii)  The projection map $p : T(M) \to M$, $(x, v) \to x$ is continuous.

Again, since $T_\phi \circ T_\psi^{-1}$ is a diffeomorphism, the family of charts $\{(p^{-1}(U), T_\psi)\}$ forms a smooth atlas on $\mathcal{M}$. Hence, it follows that $T(M)$ is a smooth manifold of dimension $2n$.

□

### 4.4.2  Tangent Bundle over $S^n$

This section continues the study of tangent bundle over a very particular smooth manifold $S^n$.

**Definition 4.4.4  (*Tangent bundle over* $S^n$ )**  The tangent bundle over the $n$-sphere $S^n$ in $\mathbf{R}^{n+1}$ for $n \geq 1$, is a fiber bundle

$$\xi = T(S^n) = (X, p, S^n, \mathbf{R}^n),$$

where $X = \{(x, y) \in \mathbf{R}^{n+1} \times \mathbf{R}^{n+1} : ||x|| = 1$ and $\langle x, y \rangle = 0\}$ and the projection map

$$p : X \to S^n, (x, y) \mapsto x.$$

Then $\xi$ is an $n$-dimensional real vector bundle. To show the local triviality condition **VB₂**, take the open covering $\{U_i\}$ of $S^n$ defined by

$$U_i = \{x = (x_1, x_2, \ldots, x_i, \ldots, x_{n+1}) \in \mathbf{R}^{n+1} : ||x|| = 1, x_i \neq 0, 1 \leq i \leq n + 1\}.$$

Then $U_i \subset S^n$ is the open set $U_i = \{x \in \mathbf{R}^{n+1} : ||x|| = 1, x_i \neq 0, 1 \leq i \leq n + 1\}$ and it is not connected, because it has two components corresponding to $x_i > 0$ and $x_i < 0$. Define the map

$$\psi_i : U_i \times \mathbf{R}^n \to p^{-1}(U_i), (x, y) \mapsto (x, f_i(y) - \langle x, f_i(y) \rangle x),$$

where

$$f_i : \mathbf{R}^n \to \mathbf{R}^{n+1}, \ y = (y_1, y_2, \ldots, y_n) \mapsto (y_1, y_2, \ldots, y_{i-1}, 0, y_i, \ldots, y_n).$$

This asserts that $p \circ \psi = p_1$ and every $\psi_i$ is a linear map on each fiber. The point $x = (x_1, x_2, \ldots, x_i, \ldots, x_{n+1}) \in \mathbf{R}^{n+1}$ lies outside $Im \ f_i$, since $x_i \neq 0$ and hence if $y \neq 0$, then

$$f_i(y) - \langle x, f_i(y) \rangle x \neq 0.$$

This implies that $\psi_i$ is injective and it is an isomorphism on each fiber. It follows that $\psi_i$ is also a diffeomorphism such that

$$p \circ \psi_i = p_{U_i}.$$

**Remark 4.4.5** The tangent bundle over the $n$-sphere $S^n$ in the Euclidean $(n + 1)$-space $\mathbf{R}^{n+1}$ is the subbundle $\xi = (T(S^n), p, S^n)$ of the product bundle $(S^n \times \mathbf{R}^{n+1}, p, S^n)$, whose total space $T(S^n)$ is defined by

$$T(S^n) = \{(b, x) \in S^n \times \mathbf{R}^{n+1} : \langle b, x \rangle = 0\},$$

and the projection map $p$ is defined by

$$p : T(S^n) \to S^n, (b, x) \mapsto b.$$

(i)  An element of the total space $T(S^n)$ is said to be a tangent vector to $S^n$ at the point $b \in S^n$.
(ii)  The fiber $p^{-1}(b) \subset T(S^n)$ is a vector space of dimension $n$.
(iii)  A cross section of the tangent bundle $\xi$ over $S^n$ is said to be a tangent vector field (or simply vector field) over $S^n$.

**Remark 4.4.6** Consider the three-sphere $\mathbf{S}^3 = \{(z, w) \in \mathbf{C}^2 : |z|^2 + |w|^2 = 1\}$ as the particular case of $S^n$, when $n = 3$. The two-sphere $S^2$ may be considered as the extended complex plane. Then the map

$$p : S^3 \to S^2, \ (z, w) \mapsto z/w$$

is well defined. It has a fiber space structure (see Remark 4.4.5). This asserts that its fiber space structure is related to the properties of complex numbers.

## 4.4.3  Normal Bundle over a Manifold

This subsection studies normal bundles over manifolds

**Definition 4.4.7** Given an $n$-dimensional submanifold $M$ of $\mathbf{R}^m$, the **normal space** $\mathcal{N}_x(M)$ of $M$, at $x \in M$, defined by

$$\mathcal{N}_x(M) = \{(x, v) \in M \times \mathbf{R}^m : v \perp T_x(M)\}.$$

Its **normal bundle** $(\mathcal{N}(M), p, M)$ is the vector bundle, $(\mathcal{N}(M), p, M)$, where the total space

$$\mathcal{N}(M) = \bigcup_{x \in M} \{\mathcal{N}_x(M)\}$$

(disjoint union) and the projection

$$p : \mathcal{N}(M) \to M, \ (x, v) \to x.$$

**Remark 4.4.8** Given an $n$-dimensional submanifold $M$ of $\mathbf{R}^m$, its normal bundle $(\mathcal{N}(M), p, M)$ is a vector bundle with fiber dimension $m - n$. Moreover,

(i)  $\mathcal{N}(M)$ is a manifold of dimension $m$.
(ii) The projection

$$p : \mathcal{N}(M) \to M, \ (x, v) \to x$$

is a submersion.

### 4.4.4  Normal Bundle over $S^n$

This section continues the study of normal bundle over a smooth manifold $M$ by considering in particular, $M = S^n$.

**Definition 4.4.9** (*Normal bundle over $S^n$*) The normal bundle $N(S^n)$ over $S^n$ for every integer $n \geq 1$ is the fiber bundle $N\xi = (X, q, S^n, \mathbf{R}^1)$, where

$$X = \{(x, y) \in S^n \times \mathbf{R}^{n+1} : y = rx, \ for \ some \ r \in \mathbf{R}\}$$

and

$$q : X \to S^n, (x, y) \mapsto x.$$

The maps

$$\psi : S^n \times \mathbf{R}^1 \to X, (x, r) \mapsto (x, rx)$$

and

$$\phi : X \to S^n \times \mathbf{R}^1, (x, y) \mapsto (x, \langle x, y \rangle)$$

are homeomorphisms such that $\psi$ is a homeomorphism with its inverse $\phi$. This asserts that $\xi = N(S^n)$ is an 1-dimensional real trivial bundle.

**Remark 4.4.10** The normal over $S^n$ is the subbundle $(N(S^n), q, S^n)$ of the product bundle $(S^n \times \mathbf{R}^{n+1}, p, S^n)$ whose total space $N(S^n)$ is defined by

$$N(S^n) = \{(b, x) \in S^n \times \mathbf{R}^{n+1} : x = tb \text{ for some } t \in \mathbf{R}\}$$

and the projection $q$ is defined by

$$q : N(S^n) \to S^n, (b, x) \mapsto b.$$

(i) An element of $N(S^n)$ is called a **normal vector** to $S^n$ at the point $b \in S^n$.
(ii) The fiber $q^{-1}(b) \subset T(S^n)$ is a vector space of dimension 1. A cross section of the normal bundle $\xi_N$ over $S^n$ is called a **normal vector field** on $S^n$.

## 4.4.5 Orthonormal K-Frames over Spheres

This subsection describes orthonormal $k$-frames over spheres as a subbundle $(X, p, B)$ of the product bundle $(S^n \times (S^n)^k, p, S^n)$.

**Example 4.4.11** (**Orthonormal k-frames over spheres**) The bundle $\xi_{ok}$ of orthonormal $k$-frames over $S^n$ for $k \leq n$ is the subbundle $(X, p, B)$ of the product bundle $(S^n \times (S^n)^k, p, S^n)$, where total space $X$ is given by

$$X = \{(b, x_1, x_2, \dots, x_k) \in S^n \times (S^n)^k : \langle b, x_i \rangle = 0 \text{ and } \langle x_i, x_j \rangle = \delta_{ij}, 1 \leq i, j \leq k\},$$

where the **Kronecker** $\delta_{ij}$ is defined by

$$\delta_{ij} = \begin{cases} 1, & \text{if } i = j \\ 0, & \text{otherwise} \end{cases}$$

and the projection $p$ is given by

$$p : X \to S^n, (b, x_1, x_2, \dots, x_k) \mapsto b.$$

(i) An element $(b, x_1, x_2, \dots, x_k)$ of the total space $X$ of an orthonormal $k$-frames over $S^n$ for $k \leq n$ is an orthonormal system of $k$-tangent vectors to $S^n$ at $b \in S^n$.
(ii) A cross section of $\xi_{ok}$ over $S^n$ is said to be a **field of k-frames.**

### 4.4.6  Canonical Vector Bundle $\gamma^n$

This subsection describes real, complex and quaternionic canonical vector bundle $\gamma^n$ over real, complex and quaternionic Grassman manifolds $G_r(\mathbf{R}^n)$, $G_r(\mathbf{C}^n)$ and $G_r(\mathbf{H}^n)$.

**Example 4.4.12  (Canonical vector bundle $\gamma^n$)** The real Grassman manifold $G_r(\mathbf{R}^n)$ is the set of all $r$-dimensional vector subspaces (or $r$-planes through the origin) of $\mathbf{R}^n$. Then $G_1(\mathbf{R}^n) = \mathbf{R}P^{n-1}$. The canonical $r$-dimensional vector bundle $\gamma_r^n = (X, p, G_r(\mathbf{R}^n))$ over the Grassmann manifold $G_r(\mathbf{R}^n)$ of $r$-frames in $\mathbf{R}^n (r \leq n)$ is the subbundle of the product bundle $(G_r(\mathbf{R}^n) \times \mathbf{R}^n, p, G_r(\mathbf{R}^n))$ with the total space $X$ consisting of the subspace of pairs $(V, x) \in G_r(\mathbf{R}^n) \times \mathbf{R}^n$ with $x \in \mathbf{R}^n$ and the orthogonal complement vector bundle of $\gamma_r^n$, denoted by $\gamma_r^{*^n}$ is the subbundle of $(G_r(\mathbf{R}^n) \times \mathbf{R}^n, p, G_r(\mathbf{R}^n))$ defined by $\gamma_r^{*^n} = (Y, p, G_r(\mathbf{R}^n))$, where $Y = \{(V, x) \in G_r(\mathbf{R}^n) \times \mathbf{R}^n : \langle V, x \rangle = 0 \text{ ( i.e., } x \perp V)\}$. Then for $r = 1$, in particular, $\gamma_1^n$ on $\mathbf{R}P^{n-1} = G_1(\mathbf{R}^n)$, is one dimensional and is called the canonical line bundle.

By natural inclusion

$$G_r(\mathbf{R}^n) \subset G_r(\mathbf{R}^{n+1}) \subset G_r(\mathbf{R}^{n+2}) \subset \cdots .$$

The topological space $G_r(\mathbf{R}^\infty)$ is defined by $G_r(\mathbf{R}^\infty) = \bigcup_{r \leq n} G_r(\mathbf{R}^n)$ endowed with weak topology.

Similarly, $G_r(\mathbf{C}^n)$, $G_r(\mathbf{H}^n)$ and $G_r(\mathbf{C}^\infty)$ and $G_r(\mathbf{H}^\infty)$ are defined, where $\mathbf{H}$ is the division ring of quaternions. In general, for $F = \mathbf{R}, \mathbf{C}$ or $\mathbf{H}$, the canonical vector bundle over $G_r(\mathbf{F}^\infty)$ is written as $\gamma_r$. Since $G_r(\mathbf{R}^n) \subset G_r(\mathbf{R}^{n+t})$ for integers $t \geq 1$, the vector bundle $\gamma_r^n$ can be viewed as

$$\gamma_r^n = \gamma_r^{n+1}(G_r(\mathbf{R}^n)).$$

Analogously, the complex and quaternionic canonical vector bundles over $G_r(\mathbf{C}^n)$ (or $G_r(\mathbf{H}^n)$) are defined.

## 4.5  Covering Spaces and Covering Homomorphism

This section introduces the concepts of covering spaces and covering homomorphisms which are basic concepts in topology of fiber bundles. For example, they are used to compute fundamental groups of some important topological spaces (see Chap. 5), and the algebraic features of these groups are expressed in the geometric language of the corresponding covering spaces. All topological spaces in this section are assumed to be Hausdorff.

## 4.5.1 Covering Spaces and Covering Projections

This subsection presents covering spaces, which form another class of important bundles. The covering projections and covering spaces play a key role in homotopy theory.

**Definition 4.5.1** A topological space $X$ is called **locally path connected** if for every point $x \in X$ and every nbd $U$ of $x$, there is an open set $V$ such that

(i) $x \in V \subset U$.
(ii) Every pair of elements in $V$ can be joined by a path in $U$.

On the other hand, the space $X$ is called **semi locally path connected** if for every point $x \in X$, there is an open nbd $U$ of $x$ such that every closed path in $U$ based at the point at $x$ is null homotopic in the sense that it is homotopic to a constant map in $X$.

**Example 4.5.2** Every path-connected space is not locally path connected. For example, consider the topological space $X \subset \mathbf{R}^2$, which is the union of the graph $\{(x, \sin 1/x) : x \in (0, 1]\}$ and a path connecting the points $(1, 0)$ and $(0, 1)$ with subspace topology of $\mathbf{R}^2$. Then $X$ is path connected, but it is not locally path connected.

Proposition 4.5.3 characterizes locally path connectedness in terms of path components.

**Proposition 4.5.3** *Let $X$ be a topological space. Then it is locally path connected iff every path component of any open subset of $X$ is open.*

**Proof** It follows from Definition 4.5.1. $\qquad\square$

**Definition 4.5.4** Let $X$ and $B$ be two path connected, locally path-connected spaces and the map $p : X \to B$ be continuous. Then $p$ is said to be a **covering projection (map)** if

(i) $p$ is a surjection;
(ii) for each point $b \in B$, there is an open set $U$ in the space $B$ containing the point $b$ such that $p^{-1}(U)$ is a disjoint union of open sets in $X$, each of which is mapped homeomorphically by $p$ onto $U$ called sheets. Then the ordered pair $(X, p)$ is called a **covering space** of $B$. Every open set in the disjoint union of open sets in $X$ is sometimes called a **sheet or an admissible open set** for the covering space $(X, p)$ and $U$ is called **evenly covered** by $p$.

**Remark 4.5.5** Definition 4.5.4 says that given a continuous surjective map $p : X \to B$, an open set $U$ of $B$ is evenly covered by $p$ if $p^{-1}(U)$ is a union of disjoint open sets $V_i$, called sheets such that $p|_{V_i} : V_i \to U$ is a homeomorphism for each $i$ and the open set $U$ is an admissible open set in $B$.

**Definition 4.5.6** *(Trivial covering projection)* A covering projection $p : X \to B$ is called trivial, if it is isomorphic to the projection $p : B \times F \to B$ from the product space $B \times F$ onto $B$, where $F$ is any set endowed with discrete topology, i.e.,all of its points are closed.

**Example 4.5.7** Let $S^1$ be the unit circle in the complex plane. Then the map

$$p : \mathbf{R} \to S^1, x \mapsto e^{2\pi i x}$$

is a covering projection and $(\mathbf{R}, p)$ is a covering space. Consider the open sets $U_1 = S^1 - \{1\}$ and $U_2 = S^1 - \{-1\}$. They are evenly covered by $p$. Because, $p^{-1}(U_2) = \bigcup_{n \in \mathbf{Z}} (n - \frac{1}{2}, n + \frac{1}{2})$. This asserts that the sheets are open intervals. Similarly, $U_1$ is also evenly covered by $p$.

**Example 4.5.8** For every nonzero real number $t$, the map

$$p : \mathbf{R} \to S^1, \ x \mapsto e^{itx}$$

is a covering projection and $(\mathbf{R}, p)$ is a covering space of $S^1$.

**Example 4.5.9** For every $n \in \mathbf{N}$, the map

$$p_n : S^1 \to S^1, z \mapsto z^n$$

is a covering map with $(S^1, p_n)$ a covering space of $S^1$.

**Example 4.5.10** There are many noncovering spaces. For example, let $B = S^1$ and $X$ be a finite open spiral over $S^1$ as shown in Fig. 4.13. Let $p : X \to B$ be the map , which projects every point $x \in X$ to the point on the circle $S^1$ directly below it. Then $(X, p)$ is not a covering space, because the condition (ii) of Definition 4.5.4 is not satisfied in this case.

**Proposition 4.5.11** *Let $\tilde{X}$ be a locally path-connected space and $p : \tilde{X} \to X$ be a projection map. Then $p$ is an open mapping*

**Proof** Let $V$ be an open set in $\tilde{X}$ and $x \in p(V)$ and $\tilde{x} \in p^{-1}(x)$ and $U$ be an admissible nbd of $x$. This shows that $\tilde{x}$ is a point of $V$ such that $p(\tilde{x}) = x$. Let $W$ be the component of $p^{-1}(U)$ which contains the point $\tilde{x}$. Since by hypothesis, $\tilde{X}$ is locally path connected, it follows that $W$ is open in $\tilde{X}$ by Proposition 4.5.3 . Clearly, $p$ maps the open set $W \cap V$ to the open subset $p(W \cap V)$ in $X$, because $p$ maps $W$ homeomorphically onto $U$. Then $x \in p(W \cap V) \subset p(V)$. Since $x$ is an arbitrary point of $p(V)$, it follows that $p(V)$ is a union of open sets, which is an open set and hence $p$ is an open map.                                                                                      □

Theorem 4.5.12 is an interesting result without any assumption of path connectedness or local path connectedness property of $Y$.

**Fig. 4.13**  Finite spiral $X$
with projection $p : X \rightarrow S^1$

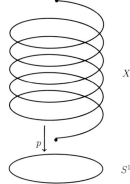

**Fig. 4.14**  Triangular
diagram involving $f$, $g$ and $p$

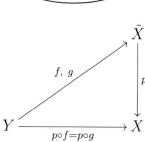

**Theorem 4.5.12**  *Given a covering space $(\tilde{X}, p)$ of $X$ and a topological space $Y$, if $f, g : Y \rightarrow \tilde{X}$ are two continuous maps such that $p \circ f = p \circ g$, as shown in Fig. 4.14, then the set $A = \{y \in Y : f(y) = g(y)\}$ is both open and closed in $Y$.*

**Proof**  We first claim that $A$ is open in $Y$. Given any point $y \in A$ and an evenly covered nbd $U$ of $(p \circ f)(y)$, the path component $V$ of $p^{-1}(U)$ in which $f(y)$ lies is an open set in $\tilde{X}$. This shows that $f^{-1}(V)$ and $g^{-1}(V)$ are both open in $Y$. Since by hypothesis, $f(y) \in V$ and $f(y) = g(y)$, then $y \in f^{-1}(V) \cap g^{-1}(V)$. Let $z \in f^{-1}(V) \cap g^{-1}(V)$. Then $f(z), g(z) \in V$ and $(p \circ f)(z) = (p \circ g)(z)$. Since $p$ maps $V$ homeomorphically onto $U$, it follows that $f(z) = g(z)$ and hence $z \in A$. Thus, it follows that $A$ is an open set.

We also claim that $A$ is closed in $Y$. If possible suppose $A$ is not closed in $Y$. Let $z$ be a limit point of $A$ which is not in $A$. Then there is an elementary nbd $U$ of $(p \circ f)(z) = (p \circ g)(z)$ such that the points $f(z)$ and $g(z)$ must lie in different path components $V_1$ and $V_2$, say, of $p^{-1}(U)$. The open set $f^{-1}(V_1) \cap g^{-1}(V_2)$ in $Y$ contains a point $y \in A$. But this is not possible, because $V_1 \cap V_2 = \emptyset$ and $f(y) = g(y) \in V_1 \cap V_2$, This asserts that all limit points of $A$ are in $A$ and hence $A$ is closed. $\square$

Corollary 4.5.13 and Corollary 4.5.14 prove the uniqueness of the lifting of a continuous map. Its independent proof is given in Proposition 4.5.25.

**Corollary 4.5.13** *Given a covering space* $(\tilde{X}, p)$ *of* $X$, *a connected space* $Y$ *and two continuous maps* $f, g : Y \to \tilde{X}$ *such that* $p \circ f = p \circ g$, *if* $f(y) = g(y)$ *at some point* $y \in Y$, *then* $f = g$.

**Proof** Since by hypothesis, $Y$ is a connected space, the only sets that are both open and closed in $Y$ are $Y$ and $\emptyset$. Hence, it follows by using Theorem 4.5.12 that either $A = Y$ or $A = \emptyset$. This shows that either $f(y) = g(y)$ at every $y \in Y$ or $f(y) \neq g(y)$ at every $y \in Y$. Since by hypothesis $f(y) = g(y)$ at some $y \in Y$, it follows that $A \neq \emptyset$ and hence $A = Y$. This implies that $f(y) = g(y)$, $\forall y \in Y$. It proves that $f = g$. □

Corollary 4.5.14 proves the uniqueness of lifting. Its alternative form is given in Proposition 4.5.25.

**Corollary 4.5.14** (**Uniqueness of lifting**) *Let* $p : \tilde{X} \to X$ *be a covering space,* $\tilde{Y}$ *be a path-connected space and* $f : \tilde{Y} \to X$ *be a continuous map having two liftings* $\tilde{f}_1, \tilde{f}_2 : \tilde{Y} \to \tilde{X}$. *If* $\tilde{f}_1$ *and* $\tilde{f}_2$ *agree at some point of* $\tilde{Y}$, *then* $\tilde{f}_1 = \tilde{f}_2$.

**Proof** By hypothesis, $p : \tilde{X} \to X$ is a covering space, $\tilde{Y}$ be a connected space and $f : \tilde{Y} \to X$ is a continuous map having two liftings $\tilde{f}_1, \tilde{f}_2 : \tilde{Y} \to \tilde{X}$. Then $p \circ \tilde{f}_1 = p \circ \tilde{f}_2$. Again, since, $\tilde{f}_1(y) = \tilde{f}_2(y)$ at some point $y \in \tilde{Y}$, by hypothesis, it follows from Corollary 4.5.13 that $\tilde{f}_1 = \tilde{f}_2$.. □

**Example 4.5.15** (i) Given a covering map $p : X \to B$, if $B'$ is any subspace of $B$, then its restriction $X' = p^{-1}(B') \to B'$ is a covering map.

(ii) If $p : X \to B$ is a covering map of an open subset $B$ in the Euclidean plane $\mathbf{R}^2$, then $X$ can be endowed with the structure of a differentiable surface with $p$ a local diffeomorphism. For a connected open set $U$ in $B$, the components $C$ of $p^{-1}U$ over which the covering is trivial along with the homeomorphisms of $U$ onto $C$ obtained by the inverse of $p$ determine charts covering $X$.

(iii) Every covering of a manifold has a natural manifold structure.

(iv) If $B$ is connected, all the fibers of the covering space $p : X \to B$ have the same cardinality.

**Definition 4.5.16** Let $p : X \to B$ be covering map such that every fiber $p^{-1}(b)$ has a finite cardinality $n$. Then the covering is said to be $n$-**sheeted.** One fails to distinguish its different $n$-sheets unless the covering is trivial.

**Proposition 4.5.17** (**Product of covering spaces**) *Given two covering spaces* $(\tilde{X}_1, p_1)$ *and* $(\tilde{X}_2, p_2)$, *the map*

$$p_1 \times p_2 : \tilde{X}_1 \times \tilde{X}_1, \to X_1 \times X_2, \ (\tilde{x}_1, \tilde{x}_2) \mapsto (p_1(\tilde{x}_1), p_2(\tilde{x}_2))$$

*is a covering map with* $(\tilde{X}_1 \times \tilde{X}_2, p_1 \times p_2)$ *a covering space of* $X_1 \times X_2$.

**Proof** Let $(x_1, x_2) \in X_1 \times X_2$ and $U_1$ be an open nbd of $x_1$ evenly covered by $p_1$ and $U_2$ be an open nbd of $x_2$ evenly covered by $p_2$. Then $U_1 \times U_2$ is a nbd of $(x_1, x_2)$ in $X_1 \times X_2$ evenly covered by $p_1 \times p_2$. Hence it follows that $p_1 \times p_2$ is a covering map with $(\tilde{X}_1 \times \tilde{X}_2, p_1 \times p_2)$ the corresponding covering space of $X_1 \times X_2$. □

**Example 4.5.18** (i) The map

$$p : \mathbf{R}^2 \to S^1 \times S^1, \ (x_1, x_2) \mapsto (e^{2\pi i x_1}, e^{2\pi i x_2})$$

from the Euclidean plane to the torus is a covering projection with $(\mathbf{R}^2, p)$ a covering space of $S^1 \times S^1$. Because, given a point $(z_1, z_2) \in S^1 \times S^1$, a small rectangle $U$ constructed by the product of two open arcs in $S^1$ containing $z_1$ and $z_2$, respectively, is an admissible nbd having its inverse image consisting of a countably infinite family of open rectangles in the Euclidean plane $\mathbf{R}^2$.

(ii) The exponential map

$$p : \mathbf{R} \to S^1, x \mapsto e^{2\pi i x}$$

defines the projection map

$$(p, p) : \mathbf{R} \times \mathbf{R} \to S^1 \times S^1$$

in the natural way. For every positive integer $n$, by induction on $n$, the product map

$$(p, p, \dots, p) = p^n : \mathbf{R}^n \to T^n$$

is a covering projection with the corresponding covering space $(\mathbf{R}^n, \ p^n)$ over the n-dimensional torus $T^n = \prod_1^n S^1$.

## 4.5.2 Automorphism Group of Covering Spaces and Lifting Problem

This subsection starts with the concepts of covering homomorphisms and deck transformations.

**Definition 4.5.19** (*Covering homomorphism*) Given two covering spaces $(\tilde{X}, p)$ and $(\tilde{Y}, q)$ over the same base space $X$, a covering homomorphism $h$ from $(\tilde{X}, p)$ to $(\tilde{Y}, q)$ is a continuous map $h : \tilde{X} \to \tilde{Y}$ such that the diagram in Fig. 4.15 is commutative in the sense that $q \circ h = p$. The homomorphism $h$ is said to be an **isomorphism** if $h$ is a homeomorphism. If there exists an isomorphism between the covering spaces $(\tilde{X}, p)$ to $(\tilde{Y}, q)$, then they are called **isomorphic or equivalent covering spaces**. If there exists no isomorphism between them, they are called distinct covering spaces. In particular, an isomorphism of a covering space onto itself is called an **automorphism or a deck transformation**.

**Example 4.5.20** Every homomorphism of path connected covering spaces is a covering projection. Because, if $h : \tilde{X} \to \tilde{Y}$ is a homomorphism of covering spaces, then $(\tilde{X}, h)$ is a covering space of $\tilde{Y}$.

**Fig. 4.15** Covering
homomorphism

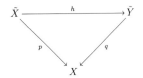

**Proposition 4.5.21** *The collection of all covering spaces of a base space $X$ and their homomorphisms form a category, denoted by* **Cov**.

**Proof** Take the collection of all covering spaces of a base space $X$ as the class of objects and their homomorphisms as the class of morphisms. Let $(\tilde{X}, p)$ be a covering space of $X$. Then $1_{\tilde{X}} : \tilde{X} \to \tilde{X}$ is a covering homomorphism. For the covering spaces $(\tilde{X}, p)$, $(\tilde{Y}, q)$ and $(\tilde{Z}, r)$ of $X$ and covering homomorphisms $h : \tilde{X} \to \tilde{Y}, g : \tilde{Y} \to \tilde{Z}$, their composite $g \circ h : \tilde{X} \to \tilde{Z}$ is also a covering homomorphism from $(\tilde{X}, p)$ to $(\tilde{Z}, r)$.                                                                                □

**Example 4.5.22** Isomorphisms in this category **Cov** are precisely the isomorphisms of covering spaces as defined above.

Proposition 4.5.23 discusses the algebraic structure of the set $Aut(\tilde{X}/X)$ of all automorphisms of the covering space $X$.

**Proposition 4.5.23** *Let $Aut(\tilde{X}/X)$ be the set of all automorphisms of the covering spaces of $X$. Then $(Aut(\tilde{X}/X), \circ)$ admits a group structure under usual composition of maps.*

**Proof** The identity map $1_{\tilde{X}} : \tilde{X} \to \tilde{X}$ is itself an automorphism and the inverse of an automorphism is again an automorphism. Hence under usual composition of maps, $Aut(\tilde{X}/X)$ is a group.                                                                                □

**Definition 4.5.24** The group $(Aut(\tilde{X}/X), \circ)$ defined in Proposition 4.5.23 is called the **automorphism group of covering spaces** of $X$, and their elements are called the **deck or covering transformations.**

Proposition 4.5.25 proves the uniqueness of liftings. Its alternative form is given in Corollary 4.5.14.

**Proposition 4.5.25 (Uniqueness of liftings)** *Let $(\tilde{X}, p)$ and $(\tilde{Y}, q)$ be two covering spaces of the same base space $X$, where $\tilde{X}$ is connected. If $f, g : \tilde{X} \to \tilde{Y}$ are two covering homomorphisms such that $f(\tilde{x}_0) = g(\tilde{x}_0)$ for some $\tilde{x}_0 \in \tilde{X}$, then $f = g$.*

**Proof** Let $(\tilde{X}, p)$ and $(\tilde{Y}, q)$ be two covering spaces of the same base space $X$. Suppose that $f, g : \tilde{X} \to \tilde{Y}$ are two covering homomorphisms such that $f(\tilde{x}_0) = g(\tilde{x}_0)$ for some $\tilde{x}_0 \in \tilde{X}$. Then each of $f$ and $g$ is considered as liftings of the map $p : \tilde{X} \to X$ with respect to the covering projection $q : \tilde{Y} \to X$. Since by hypothesis, $\tilde{X}$ is connected and $f, g$ both agree at the same point $\tilde{x}_0 \in \tilde{X}, g = h$. This proves the uniqueness of the lifting.                                                                                □

**Fig. 4.16** Commutativity of the triangle for the covering space $(X, p)$

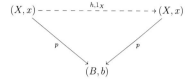

**Theorem 4.5.26** *Let $(X, p)$ be a covering space of $B$.*

(i) *If $h \in \mathrm{Cov}(X/B) = \mathcal{A}ut(X/B)$ and $h \neq 1_X$, then $h$ has no fixed point.*
(ii) *If $h, g \in \mathcal{A}ut(X/B)$ and $\exists\, x \in X$ with $h(x) = g(x)$, then $h = g$.*

**Proof** (i) Let $x \in X$ be a point such that $h(x) = x$. Then $x$ is a fixed point of $h$. For $b = p(x)$, consider the commutative triangle shown in Fig. 4.16 Since both $h$ and $1_X$ complete the diagram in Fig. 4.16, it follows that $h = 1_X$. This implies a contradiction. This contradiction proves that $h$ has no fixed point.
(ii) The map $h^{-1}g \in \mathcal{A}ut(X/B)$ has a fixed point, namely $x$ and hence by (i) it follows that

$$h^{-1}g = 1_X,$$

which proves that $h = g$. $\qquad\square$

## 4.6 Construction of Fiber Bundles and Their Local Cross Sections

This section describes construction of fiber bundle by continuous action by a discrete topological group.

### 4.6.1 Construction of Fiber Bundle by Continuous Action

This subsection constructs a fiber bundle by using a continuous action

$$X \times G \to X, \, (x, g) \mapsto xg$$

of a discrete topological group $G$ on a topological space $X$.

Let $G$ be a topological group and $X$ be a topological space. Then $G$ is said to act on $X$ (from the right) properly discontinuously if for any point $x \in X$ and any sequence $\{g_n\}$ of distinct points in $G$, the sequence $\{xg_n\}$ of points in $X$ does not converge to any point of $X$ in the sense that each orbit $xG$ is a closed discrete subset of $X$. This is formulated in Definition 4.6.1.

**Definition 4.6.1** Let $G$ be a discrete topological group with the identity element $e$ and $X$ be a topological space. Then $G$ is said to act (from the right) **properly discontinuously** on $X$ if

PD(i)   For every point $x \in X$, there is a nbd $U_x$ of $x$ in $X$ with the property:
        $U_x g \cap U_x \neq \emptyset$ implies $g = e$.
PD(ii)  For any two elements $x, y \in X$, $y \notin xG$, there are nbds $V_x$ and $V_y$ of $x$ and
        $y$, respectively, such that

$$V_x g \cap V_y = \emptyset, \ \forall g \in G.$$

The topological space $X/G = X$ mod $G$, topologized by the identification map

$$p : X \to X \text{ mod } G, \ x \mapsto xG$$

is the orbit space obtained by the action of $G$ on $X$.

**Remark 4.6.2** If an action of topological group $G$ on a topological space $X$ is properly discontinuous, then it is evenly covered by $p$ in the sense that any point in $X$ has a nbd $U$ such that $Ug \cap Uk = \emptyset$ for every pair of distinct points $g, k \in G$.

**Example 4.6.3** Every free action of a finite group on a Hausdorff space is properly discontinuous.

**Proposition 4.6.4** *The projection* $p : X \to X$ mod $G$, $x \mapsto xG$ *is a covering map, where $X$ is a Hausdorff space.*

**Proof** By hypothesis, $X$ is assumed to be Hausdorff. Hence the Hausdorff property of $X$ mod $G$ follows from the condition **PD(ii)** that distinct points $x, y \in X$, which are not equivalent, have nbds $U$ and $V$ of $x$ and $y$ respectively such that

$$Ug \cap Vk = \emptyset, \ \forall g, k \in G,$$

since, $Ug \cap Vk = (k^{-1}g)U \cap V$. Hence it follows that the projection

$$p : X \to X \text{ mod } G, x \mapsto xG$$

is a covering map.                                                                              □

Proposition 4.6.5 prescribes an important method in which fiber bundles arise in a natural way.

**Proposition 4.6.5** *Let $G$ be a discrete topological group acting properly discontinuously on a topological space $X$. Then $(X, p, X$ mod $G, G)$ forms a fiber bundle over $X$ mod $G$.*

**Proof** For an element $y \in X$ mod $G$,

(i) There exists an element $x \in X$ such that $p(x) = y$.

(ii) There exists a nbd $U_x$ of $x$ in $X$ by the condition **PD(i)**.

If $V_y = p(U_x)$, then

$$p^{-1}(V_y) = \bigcup_{g \in G} U_x g$$

asserts that $V_y$ is open and $y \in V_y$. Again, $p_{U_x} : U_x \to V_y$ is a homeomorphism. Then the map

$$\psi_y : V_y \times G \to p^{-1}(V_y), \ (z, g) \mapsto ((p|U_x)^{-1}(z))g,$$

is a homeomorphism with its inverse

$$\psi_y^{-1} : p^{-1}(V_y) \to V_y \times G, \ (z'g) \mapsto (p(z'), g).$$

This implies that $\psi_y$ is a homeomorphism such that $p \circ \psi_y = p_{V_y}$. This asserts that the open sets $\{V_y\}$ form a covering of the space $X$ mod $G$. It proves that $(X, p, X$ mod $G, G)$ forms a fiber bundle over $X$ mod $G$. □

For convenience of future study, Definition 4.6.1 is reformulated in Definition 4.6.6.

**Definition 4.6.6** Let $G$ be the group of homeomorphisms of a topological space $X$. Then $G$ is said

(i) To act (from the right) **properly discontinuously** on $X$ if for any $x \in X$, there is a nbd, $U_x$ of $x$ in $X$ such that for $g, g' \in G$ if $U_x g$ intersects $U_x g'$, then $g = g'$.
(ii) To **act without fixed points** if the only element of $G$ having fixed point is its identity element $1_X : X \to X$.

### 4.6.2 Local Cross Sections of Fiber Bundles

This subsection studies the local cross sections of fiber bundles $\xi$.

**Definition 4.6.7** Given a bundle $\xi = (X, p, B)$, the map $p : X \to B$ is said to have a **local cross section** $s$ at a point $b \in B$ if there is a nbd $U_b$ in $B$ and a continuous map

$$s : U_b \to X$$

is such that $p \circ s = 1_{U_b}$.

*Example 4.6.8* **There are bundles $\xi$ having no cross section.** For example, consider the bundle $\xi = (\mathbf{D}^2, p, S^1)$, where $p : \mathbf{D}^2 \to S^1$ is the map letting $p(x)$ to be the point of $S^1$, where the ray in $\mathbf{R}^2$ starting from origin and passing through trough the point $x$. If $\xi$ has a cross section $s : S^1 \to \mathbf{D}^2$, then $p \circ s = 1_{S^1}$ would imply that the map $s : S^1 \to \mathbf{D}^2$, is injective. Again since $s : S^1 \to \mathbf{D}^2$ is continuous and injective, it contradicts the **Borsuk-Ulam Theorem** (see Chap. 2) which asserts that every continuous map $f : S^1 \to \mathbf{D}^2$ maps at least one pair of antipodal points of $S^1$ to the same point.

**Example 4.6.9** The bundle defined in Example 4.6.8 is not a fiber bundle, because every fiber bundle admits a cross section.

Proposition 4.6.10 gives a sufficient condition for the existence of local cross section of a bundle at a point.

**Proposition 4.6.10** *Given a topological group $G$ and a closed subgroup $H$ of $G$, the projection map*

$$p : G \to G/H, g \mapsto gH$$

*admits a local cross section at every point of $G/H$.*

**Proof** Consider the quotient space $G/H$ topolozied by quotient topology. Define an action $\sigma$ of $G$ on the space $G/H$, formulated by

$$\sigma : G \times G/H \to G/H, (g, g'H) \mapsto (gg')H.$$

For the proof of the given proposition, it is sufficient to show that the continuous map

$$p : G \to G/H, g \mapsto gH$$

admits a local cross section at the coset $H$. Because, if $(U, \sigma)$ is a local cross section for $p$ at the point $H$, then for any other point $gH$ of $G/H$, the set $gU$ is a nbd of $gH$ and the map

$$\sigma_g : gU \to G, \ g'H \mapsto g(\sigma(g^{-1}g'H))$$

is continuous such that

$$p \circ \sigma_g = 1_{gU}.$$

This proves that $\sigma_g$ is a local cross section.                                    □

**Theorem 4.6.11** *Given a topological group $G$ and a closed subgroup $K$ of $G$, if the map*

$$p : G \to G/K, \ g \mapsto gK$$

*admits a local cross section at $K$, then for every closed subgroup $A \subset K$, the natural projection*

$$q : G/A \to G/K, gA \mapsto gK,$$

*is a fiber bundle having its fiber $K/A$.*

**Proof** By hypothesis, $G$ is a topological group and $K$ is a closed subgroup of $G$. Then by using Proposition 4.6.10, it follows that the projection

$$p : G \to G/K, \ g \mapsto gK$$

has a local cross section at every point of $G/K$. Take any point $x \in G/K$ and a local cross section $(U, \alpha)$ of $p$ at $x$, define a map

$$\phi : U \times K/A \to G/A, \ (y, hA) \mapsto \alpha(y)hA.$$

Then $\phi$ is a continuous map having the property that for all $y \in U$ and for all $k \in K$,

$$(q \circ \phi)(y, hA) = q(\alpha(y)hA) = \alpha(y)hA = \alpha(y)K = p(\alpha(y)) = y = 1_U(y).$$

Again, the map

$$\psi : q^{-1}(U) \to U \times KA, \ gA \mapsto (gK, \alpha(gK)^{-1}gA), \ \forall \, gA \in q^{-1}(U)$$

is continuous. This implies that

$$\psi \circ \phi = 1_d \ \text{and} \ \phi \circ \psi = 1d,$$

where $1_d$ is the identity maps on $U \times K/A$ and $q^{-1}(U)$ in the respective cases. Hence the theorem follows. □

**Example 4.6.12** Fiber bundles obtained from the decomposition of compact Lie groups modulo their closed subgroups provide a rich supply of fiber bundles, and it serves as a valuable source of examples of fiber bundles, see (Samelson, 1952) and (Borel, 1955).

## 4.7 Hopf Fibering of Spheres

This section studies the three fiberings of spheres:

$$p : S^{2n-1} \to S^n, \ for \ n = 2, 4, 8,$$

the early examples of bundles spaces, of which, the map $p : S^3 \to S^2$ of the 3-sphere on the 2-sphere defined by Hopf in 1935 is known as a Hopf map, is the simplest.

### 4.7.1 Hopf Map $p : S^3 \to S^2$

This subsection is devoted to study the Hopf map $p : S^3 \to S^2$.

**Theorem 4.7.1** *The Hopf fibering $(S^3, p, S^2)$ has fibers a family of great circles.*

**Proof** To construct the Hopf fibering $(S^3, p, S^2)$, consider $S^3$ and $S^2$ defined as follows:

$$S^3 = \{(z_1, z_2) \in \mathbf{C}^2 : z_1\bar{z}_1 + z_2\bar{z}_2 = |z_1|^2 + |z_2|^2 = 1\}$$

and the 2-sphere $S^2$ represented as the complex projective line consisting of pairs $[z_1, z_2]$ of complex numbers, not both zero, given by the equivalence relation

$$[z_1, z_2] \sim [\lambda z_1, \lambda z_2] \text{ for } \lambda \neq 0.$$

Consider the map $p : S^3 \to S^2$ defined by

$$p : S^3 \to S^2, (z_1, z_2) \mapsto [z_1, z_2]/(|z_1|^2 + |z_2|^2)^{1/2}.$$

Since every pair $[z_1, z_2]$ is normalized on dividing by $z_1\bar{z}_1 + z_2\bar{z}_2$, it follows that $p$ is well defined and continuous. Again, consider the circle $S^1 = \{z \in \mathbf{C} : |z| = 1\}$ and two points $\alpha = [1, 0]$ and $\beta = [0, 1]$ of $S^2$. Then $U_1 = S^2 - \{\alpha\}$ and $U_2 = S^2 - \{\beta\}$ are two open sets which form an open covering of $S^2$. Since every point of $U_1$ can be expressed by a pair $[u, 1]$, the map

$$\psi_{U_1} : U_1 \times S^1 \to S^2, ([u, 1], z) \mapsto \left(\frac{zu}{(z_1\bar{z}_1 + z_2\bar{z}_2)^{1/2}}, \frac{z}{(z_1\bar{z}_1 + z_2\bar{z}_2)^{1/2}}\right)$$

is well defined and it maps $U_1 \times S^1$ homeomorphically onto $p^{-1}(U_1)$ with the property

$$p\psi_{U_1}(u, z) = u, \forall u \in U_1, z \in S^1.$$

This constructs $\psi_{U_1}$. Similarly, construct $\psi_{U_2}$. This completes the construction of the Hopf fibering $(S^3, p, S^2)$. For $(z_1, z_2) \in S^3$, the fiber $p^{-1}[z_1, z_2] = \{(\lambda z_1, \lambda z_2) : \lambda \in S^1\}$. This implies that the fibers of the Hopf fibering $(S^3, p, S^2)$ are precisely the great circles of $S^3$. Hence, it follows that the Hopf fibering $(S^3, p, S^2)$ has fibers a family of great circles, because the inverse image of $p$ of a point of $S^2$ is just a great circle of $S^3$ and this fibering has fibers a family of great circles. $\square$

**Corollary 4.7.2** *The 3-sphere $S^3$ is decomposed into a family of great circles with the 2-sphere $S^2$ as a quotient space of the Hopf fibering $(S^3, p, S^2)$.*

**Proof** Consider the Hopf fibering $(S^3, p, S^2)$. It follows from the proof of Theorem 4.7.1 that the 3-sphere is decomposed into a family of great circles, which are called fibers, having the 2-sphere as a quotient space (or **decomposition space**). $\square$

**Remark 4.7.3** The decompositions of compact Lie groups modulo their closed subgroups produce fiber bundles, which serve as a valuable source of examples. We recommend the survey articles by [H. Samelson, Topology of Lie groups, Bull. Amer. Math. Soc.52 (1952), 2–37] and [A. Borel, Topology of Lie groups and characteristic classes, Bull. Amer. Math. Soc.61 (1955), 397–432].

## 4.7.2 A Generalization of the Hopf Map $p : S^3 \to S^2$

This subsection generalizes the Hopf map $p : S^3 \to S^2$ through the study of some spaces that arise in projective geometry.

For $\mathbf{F} = \mathbf{R}, \mathbf{C}$ or $\mathbf{H}$, the right vector space $\mathbf{F}^n$ consists of elements, which are ordered sets of $n$ elements of $\mathbf{F}$. If $x = (x_1, x_2, \ldots, x_n) \in \mathbf{F}^n$ and $\beta \in \mathbf{F}$, then $x\beta = (x_1\beta, \ldots, x_n\beta)$. Using the usual inner product $x$ and $y$ in $\mathbf{F}^n$ by $\langle x, y \rangle = \sum_1^n \overline{x_i} y_i$, where $\overline{x_i}$ is the conjugate of $x_i$, it follows that

$$\langle y, x \rangle = \overline{\langle x, y \rangle}, \quad \langle x\beta, y \rangle = \overline{\beta}\langle x, y \rangle \quad \text{and} \quad \langle x, (y\beta) \rangle = \langle x, y \rangle\beta.$$

Two nonzero elements $x, y \in \mathbf{F}^n$ are said to be orthogonal denoted by $x \perp y$ iff $\langle x, y \rangle = 0$. The orthogonality relation is symmetric, because, $\langle x, y \rangle = 0$ iff $\langle y, x \rangle = 0$. Let $S$ be the unit sphere in $\mathbf{F}^n$ defined by the locus $\langle x, x \rangle = 1$ and $G_n$ be the orthogonal, unitary or symplectic group according as $\mathbf{F} = \mathbf{R}, \mathbf{C}$ or $\mathbf{H}$, then each $G_n$ is a compact Lie group. Let $\mathbf{F}P^n$ be the projective space associated with $\mathbf{F}$ and it is topolozied by considering it as a quotient space of $\mathbf{F}^{n+1} - \{0\}$. Geometrically, it may be considered as the set of all lines through the origin in $\mathbf{F}^{n+1} = \overbrace{\mathbf{F} \oplus \mathbf{F} \oplus \cdots \oplus \mathbf{F}}^{n+1}$, since, every point of $\mathbf{F}^{n+1} - \{0\}$ determines a line through the origin 0 and if $x$ and $y$ are nonzero elements of $\mathbf{F}^{n+1}$, then $x \sim y$ iff there is an element $\beta(\neq 0) \in \mathbf{F}$ such that $y = x\beta$. As it is an equivalence relation, it defines $\mathbf{F}P^n$ as the quotient set of equivalence classes endowed with the quotient topology. In particular,

(i) $\mathbf{R}P^n$ is called the $n$-dimensional real projective space.
(ii) $\mathbf{C}P^n$ is called the $n$-dimensional complex projective space.
(iii) $\mathbf{H}P^n$ is called the $n$-dimensional quaternionic projective space.

The natural projection map $\mathbf{F}^{n+1} - \{0\} \to \mathbf{F}P^n$, $w \mapsto [w]$ is continuous and defines maps on restriction to the unit sphere in $\mathbf{F}^n$ which are $S^n \subset \mathbf{R}^n$, $S^{2n+1} \subset \mathbf{C}^n$ and $S^{4n+3} \subset \mathbf{H}^n$. These maps are

$$p_n : S^n \to \mathbf{R}P^n,$$

$$q_n : S^{2n+1} \to \mathbf{C}P^n,$$

and

$$r_n : S^{4n+3} \to \mathbf{H}P^n.$$

**Remark 4.7.4**  Usually, the common notation $p$ is used instead of $p_n$, $q_n$ or $r_n$, unless there is any confusion. Example 4.7.5 plays a key role in computing the homotopy groups of sphere (results are only partly known) (see Chap. 5) and hence it reflects the importance of bundle theory.

**Example 4.7.5  (Real, complex and Quaternionic Hopf bundles)**

   (i) (Real Hopf bundle) $\xi = (S^n, p, \mathbf{R}P^n, \mathbf{Z}_2)$ is a locally trivial fiber bundle with fiber $\mathbf{Z}_2$.
  (ii) (Complex Hopf bundle) $\eta = (S^{2n+1}, p, \mathbf{C}P^n, S^1)$ is a trivial fiber bundle with fiber $S^1$.
 (iii) (Quaternionic Hopf bundle) $\gamma = (S^{4n+3}, p, \mathbf{H}P^n, S^3)$ is a locally trivial fiber bundle with fiber $S^3$.

## 4.8   *G*-bundles and Principal *G*-bundles

This section introduces the concepts of $G$-bundles and principal $G$-bundles obtained by an action of a topological group $G$ on a topological space and studies these bundles. A $G$-bundle is a bundle with an additional structure obtained from an action of the topological group $G$ on a topological space. Transformation groups are also derived from an action of topological groups on topological spaces.

The principal $G$-bundles for a Lie group $G$ named after Sophus Lie (1842–1899) are studied in Sect. 4.10. The main result of this section is the Corollary 4.8.21 which asserts that if $X$ is a simply connected space and $G$ is a properly discontinuous group of homeomorphisms of $X$, then the fundamental group $\pi_1(X \bmod G)$ of the orbit space $X \bmod G$ is isomorphic to $G$. The homotopy property of numerable principal $G$-bundles and Milnor construction are studied in Chap. 5. In this section, $G$ denotes a topological group and $X$ denotes a topological space.

### 4.8.1   *G-spaces*

This subsection introduces the concept of $G$-spaces obtained by an action of a topological group on a topological space.

**Definition 4.8.1**  A right action $\sigma$ of a topological group $G$ with identity element $e$ on a topological space $X$ is a continuous map

$$\sigma : X \times G \to X,$$

the image $\sigma(x, g)$ denoted by $x \cdot g$ or simply $xg$ such that

(i) $x \cdot (gh) = (x \cdot g) \cdot h$, $\forall\, g, h \in G$ and $x \in X$ and
(ii) $x \cdot e = x$, $\forall x \in X$.

Then the pair $(X, G)$ endowed with an action $\sigma$ is called a right *G*-**space or a topological transformation group**. Similarly, a left action of $G$ on $X$ is defined, and it gives a left *G*-space. One can convert a right *G*-space into a left *G*-space by defining

$$g \cdot x = x \cdot g^{-1}$$

So, it is sufficient to consider either right or a left *G*-space. Any such space is called simply a *G*-**space or a topological transformation group**.

**Definition 4.8.2** Let $G$ be a topological group with identity $e$ and $X$ be topological group. An action

$$\sigma : X \times G \to X, \ (x, g) \mapsto x \cdot g$$

of $G$ on $X$ is said to be

(i) **free** if for all $g(\neq e) \in G$, $x \cdot g \neq x$ for every $x \in X$ and
(ii) **effective or trivial** if for all $g(\neq e) \in G$, there exists an element $x \in X$ such that $x \cdot g \neq x$.
(iii) **transitive**, if given any two $x, y \in X$, there exists an element $g \in G$, such that $x \cdot g = y$.

**Remark 4.8.3** Let $\sigma : X \times G \to X$ be an action of a topological group $G$ on a topological space $X$. Then the action

(i) $\sigma$ is free if the **isotropy group** $G_x = \{g \in G : g(x) = x \cdot g = x\} = \{e\}$ for every $x \in X$, which means that $g(x) = x$ for some $x \in X$ asserts that $g = e$.
(ii) $\sigma$ is effective or trivial, if $g(x) = x$ for every $x \in X$, then $g = e$, i.e., if the homomorphism

$$f : G \to \mathbf{homeo}(X), \ g \mapsto \psi_g,$$

is a monomorphism, where $homeo(X)$ is the group of homeomorphisms of $X$.
(iii) is transitive, if only one orbit is generated by this action $\sigma$.

**Example 4.8.4** The condition PD(i) of Definition 4.6.1 implies that the topological group $G$ acts on $X$ freely. Because, if $x \in X$ and $x \cdot g = x$ for some $g \in G$, then, for any nbd $U$ of $x$, the point $x \in Ug \cap U$ and so $Ug \cap U \neq \emptyset$, which is possible only when $g = e$.

**Definition 4.8.5** Given a right G-space $X$, the set $X \bmod G = \{xG : x \in X\}$ of all orbits of $X$ under an action $\sigma$ of $G$ on $X$, with the quotient topology, which is the largest topology such that the projection map

$$p : X \to X \bmod G, x \mapsto xG$$

is continuous. The quotient space $X$ mod $G$ is called the **orbit space** of $X$ modulo $G$.

**Remark 4.8.6** Let $X$ be $G$-space. Then $x, y \in X$ are said to be $G$**-equivalent**, denoted by $x \sim y$ iff $y = x \cdot g$ for some $g \in G$. This is an equivalence relation on $X$ and the equivalent class determined by $x$ is an orbit of $x$, denoted by $xG$.

**Proposition 4.8.7** *Given a right $G$-space $X$, for every point $g \in G$, the map*

$$\psi_g : X \to X, x \mapsto x \cdot g$$

*is a homeomorphism and the projection*

$$p : X \to X \text{ mod } G, x \mapsto xG$$

*is an open map.*

**Proof** The map $\psi_g$ is a homeomorphism with its inverse $\psi_{g^{-1}}$ for every $g \in G$. To prove the second part, let $U$ be an open subset of $X$. Then $p^{-1}(p(U)) = \bigcup_{g \in G} Ug$ is a union of open sets in $X$ and hence $\bigcup_{g \in G} Ug$ is an open set in $X$ mod $G$. This asserts that $p(U)$ is an open set of $X$ mod $G$ for every open set $U$ of $X$.                    $\square$

**Remark 4.8.8** The quotient maps $X \to X \text{mod } G, x \mapsto G(x)$ are identified with the covering maps. We now use the symbol $gx$ or $g(x)$ for the symbol $g \cdot x$.

Theorem 4.8.9 is an important theorem in topology. Its geometrical applications are available in Section 4.12.

**Theorem 4.8.9** *Let $G$ be a compact topological group and $X$ be a Hausdorff space. If $G$ acts (from left) on $X$, and $G_x$ is the isotropy group at $x \in X$ with orbit $orb(x) = G(x)$, the orbit of $x$, then the map*

$$\psi : G/G_x \to G(x), \ gG_x \mapsto g(x) = gx$$

*is a homeomorphism.*

**Proof** Let $\{gG_x\}$ be the set of all left cosets of $G_x$ in $G$. Clearly, the map $\psi$ is continuous by quotient topology on $G/G_x$. It is onto, because for any $y \in G(x)$ it can be expressed as $y = g_y(x)$ for some $g_y \in G$ and hence $\psi(g_y G_x) = g_y(x) = y$. Again, for $g, h \in G$ if $g(x) = h(y)$, then $g^{-1}h \in G_x$, and hence $gG_x = hG_x$ implies that $\psi$ is injective. Consequently, $\psi$ is continuous one-one and onto map from a compact space to a Hausdorff space. This proves that $\psi$ is a homeomorphism.     $\square$

**Definition 4.8.10** (*G-morphism*) Given two right $G$-spaces $X$ and $Y$, a continuous map $f : X \to Y$ is said to be a $G$-morphism if $f(xg) = f(x)g$ holds for all $x \in X$ and for all $g \in G$.

***Remark 4.8.11*** Given two two right *G*-spaces $X$ and $Y$, a *G*-morphism map $f :$ $X \to Y$ sends $xG$ into $f(x)G$ such that

$$f(xG) \subset f(x)G, \quad \forall x \in X.$$

### 4.8.2  *G-Coverings*

This subsection studies *G*-coverings obtained by the action of a topological group as orbit spaces.

**Definition 4.8.12**  A covering $p : X \to B$ is said to be a *G*-**coverings,** if it is obtained as a properly discontinuous action of a topological group $G$ on $X$. It is said to be a **trivial covering** of $B$, if this covering is the product

$$G \times X \to X.$$

**Definition 4.8.13**  Let $p : X \to B$ and $q : Y \to B$ be two *G*-coverings. An isomorphism between them is a homeomorphism

$$\psi : X \to Y$$

such that

(i)  $q \circ \psi = p$ and hence the diagram in Fig. 4.17 is commutative and
(ii)  $\psi(gx) = g\psi(x)$, $\forall g \in G$ *and* $\forall x \in X$.

**Proposition 4.8.14**  *Let $\xi$ be an arbitrary G-covering . Then it is locally trivial as a G-covering.*

**Proof**  $\xi : p : X \to B$ be a *G*-covering. Then every point in *B* has a nbd *U* such that the *G*-covering $q_U : p^{-1}(U) \to U$ is isomorphic to the trivial *G*-covering $U \times G \to U$. Let $U = p(V)$. Then a local trivialization is defined by

$$p^{-1}(U) \to U \times G, \quad gv \mapsto (p(v), g).$$

This proves that $\xi$ is locally trivial as a *G*-covering.                              □

**Fig. 4.17**  Isomorphism $f$ of
*G*-coverings

### 4.8.3  G-bundles

This subsection is devoted to the study of $G$-bundles which admits an additional structure obtained from the action of a topological group $G$ on a topological space.

**Definition 4.8.15** Let $(X, p, B)$ be a bundle. Then it is said to be a **G-bundle** if there exists some G-space structure on $X$ such that the two the bundles $(X, p, B)$ and $(X, p_X, X \bmod G)$ are isomorphic in the sense that there exists a homeomorphism $f : X \bmod G \to B$ with the property that the pair of maps

$$(1_X, f) : (X, p_X, X \bmod G) \to (X, p, B)$$

makes the diagram in Fig. 4.18 commutative (see Definition 4.1.26).

**Proposition 4.8.16** *(Existence of a bundle morphism) Let $X$ and $Y$ be two G-spaces and $f : X \to Y$ be a G-morphism. Then there exists a bundle morphism*

$$(f, \tilde{f}) : (X, p_X, X \bmod G) \to (Y, p_Y, Y \bmod G).$$

**Proof** Corresponding to $G$-spaces $X$ and $Y$, let $\xi(X) = (X, p_X, X \bmod G)$ and $\xi(Y) = (Y, p_Y, Y \bmod G)$ be the $G$-bundles. Then the $G$ morphism $f : X \to Y$ produces a quotient map

$$\tilde{f} : X \bmod G \to Y \bmod G, \ xG \mapsto f(x)G.$$

**Fig. 4.18** Rectangular commutative diagram for $G$-bundle $(X, p, B)$

**Fig. 4.19** $G$-bundle morphism

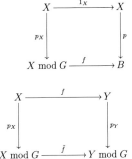

**Fig. 4.20** Covering
transformation $\psi$

The commutativity of the rectangle in the Fig. 4.19 proves that the pair of maps $(f, \tilde{f})$
is a bundle morphism.                                                                   □

**Definition 4.8.17** Let $p : X \to B$ be a covering space. Then a **covering transformation** is a homeomorphism

$$\psi : X \to X : p \circ h = p,$$

i.e., $\psi$ is a homeomorphism such that the triangular diagram in Fig. 4.20 is commutative.
If $\mathbf{Cov}(X/B)$ denotes the set of all covering transformations of $\psi : X \to X$, then it
is group under usual composition of maps, called the **group of covering transformations** of the covering space $p : X \to B$.

**Definition 4.8.18** A covering space $p : X \to B$ is said to be **regular** if the image
$p_* \pi_1(X, x_0)$ of induced homomorphism

$$p_* : \pi_1(X, x_0) \to \pi_1(B, b_0)$$

is a normal subgroup of $\pi_1(B, b_0)$.

**Remark 4.8.19** Let $G = \mathbf{Cov}(X/B)$ be the group of covering transformations of a
covering space $p : X \to B$. This group resembles $\pi_1(B, b_0)$. Then the action of $G$ on
$X$ is properly discontinuous. It is a natural question: is its converse true? Its answer
is available in Theorem 4.8.20, which considers action of properly discontinuous
group $G$ of homeomorphisms of a space $X$ in the sense of Definition 4.6.6.

Theorem 4.8.20 answers the problem raised in Remark 4.8.19.

**Theorem 4.8.20** *Let $G$ be a properly discontinuous group of homeomorphisms of
a topological space $X$ and $X$ mod $G$ be its orbit space.*

*(i)   Then the projection $p : X \to X$ mod $G, x \mapsto xG$ is a covering projection.*
*(ii)  If $X$ is connected, then this covering projection is regular and $G$ is its group of
covering transformations.*

*Proof* i) follows from Proposition 4.6.4. The proof the theorem is also given as
follows: by hypothesis, $p : X \to X$ mod $G$, $x \mapsto xG$ is the usual projection, then $p$
is continuous and open by Proposition 4.8.7. If $U$ is an open subset of $X$ satisfying
the condition that whenever $Ug$ meets $Ug'$, then $g = g'$, then $p(U)$ is evenly covered
by $p$. By hypothesis on $U$, it follows that $\{Ug\}_{g \in G}$ is a disjoint family of open sets
whose union is $p^{-1}(p(U))$. To prove the theorem, it suffices to show that

$$p|Ug : Ug \to p(U)$$

is a bijection. For $x \in U$,

$$p(xg) = p(x) \implies p(Ug) = p(U).$$

Again, if $p(xg) = p(x'g)$ for $x, x' \in U$, then there exists some $s$ in $G$ such that $xg = xsg$. This implies that $Ug$ intersects $x'sg$ and $g = sg$. Hence $s = e$ and $xg = x'g$ and $p : Ug \to p(U)$. Since by hypothesis, $G$ is properly discontinuous, the sets $p(U)$ are evenly covered by $p$ and form an open covering of $X$ mod $G$. Since $p(xg) = p(x)$, it follows that $G$ is contained in the group of covering transformations of $p$. Hence, the group $G$ and the group of covering transformations are same. Since the group of covering transformations is transitive on each fiber, it is proved that the covering projection $p$ is regular. It proves (ii).                                                    □

Corollary 4.8.21 identifies properly discontinuous group $G$ of homeomorphisms of a simply connected space $X$ with the fundamental group of the orbit space $X$ mod $G$.

**Corollary 4.8.21** *Given a simply connected space $X$ and a properly discontinuous group $G$ of homeomorphisms of $X$, the fundamental group $\pi_1(X$ mod $G)$ and the group $G$ are isomorphic.*

**Proof** Using Theorem 4.8.20, it follows that $G$ is the group of covering transformations of the regular covering projection $p \to X$ mod $G$, $x \mapsto xG$. Hence, the Corollary follows from Theorem 4.8.20.                                                    □

### 4.8.4  Principal G-bundles

This subsection continues the study of $G$-bundles by introducing the concept of principal $G$-bundles. In particular, principal $G$-bundles for Lie groups are studied in Sect. 4.10.

**Definition 4.8.22  (Principal G-bundle)** Given a topological group $G$ and a topological space $B$, a principal $G$-bundle over $B$ consists of

(a)  a fiber bundle $p : X \to B$ and
(b)  an action $X \times G \to X$, $(x, g) \mapsto xg$

such that

PG(i)     the **shearing map**

$$T : X \times G \to X \times X, (x, g) \mapsto (x, xg)$$

maps the product space $X \times G$ homeomorphically to its image space $T(X \times G)$;

PG(ii)   for the space $B = X$ mod $G$, the projection $p : X \to X$ mod $G$ is the quotient map;

PG(iii)  for every point $b \in B$, there exists an open nbd $U$ of $b$ such that the bundle

$$p : p^{-1}(U) \to U$$

is $G$-bundle isomorphic to the trivial bundle

$$q : U \times G \to U$$

in the sense that there exists a homeomorphism

$$\psi : p^{-1}(U) \to U \times G$$

with $q \circ \psi = p$, where the action is given by $(x, g')g = (x, gg')$.

**Remark 4.8.23** (i) The shearing map $T$ formulated in PG(i) is injective iff the action of $G$ on $X$ is free. This implies by PG(i) that the action of $G$ on the total space $X$ of a principal bundle is always free. If $G$ and $X$ are compact, then a free action satisfies PG(i).

(ii) Every free action produces a translation function $\alpha : Y \to G$, where $Y = \{(x, x \cdot g) \in X \times X\}$ is the image of the shearing map $T$. Condition PG(i) is equivalent to a free action with a continuous translation function.

**Definition 4.8.24** (*Locally trivial principal G-bundle*) A fiber bundle $\xi = (X, p, B, G)$ with a continuous action $\sigma : X \times G \to X$ is said to be a (locally trivial) principal $G$-bundle, if given an open covering $\{U_i\}$ of $B$ and for every $i$, there is a homeomorphism

$$\psi_i : U_i \times G \to p^{-1}(U_i)$$

such that for all $b \in U_i$ and $g \in G$.

(i)  $(p \circ \psi_i)(b, y) = b$ and
(ii) $\psi_i(b, g) = \psi_i(b, e) \cdot g$.

**Example 4.8.25** (*Product principal G -bundle*) The product $G$-space $B \times G$ is principal under the action of $G$ given by $(b, t)s = (b, ts)$.

**Remark 4.8.26** For more study of principal $G$-bundles, see Chap. 5.

## 4.9  Charts and Transition Functions

This section continues the study of principal $G$-bundles corresponding to a topological group $G$ by introducing the concepts of charts and transition functions of principal

$G$-bundles and proves a bijective correspondence in Theorem 4.9.8 between the sets of the equivalence classes of principal $G$-bundles $\xi$ over a fixed base space and the equivalence classes of sets of transition functions determined by an atlas of the $G$-bundles $\xi$. This result together with Theorem 4.9.11 establishes a close connection between $n$-dimensional $\mathbf{F}$-vector bundles over $B$ and principal $G$-bundles over $B$, where $\mathbf{F} = \mathbf{R}, \mathbf{C}, \mathbf{H}$ and $G = GL(n, \mathbf{F})$.

**Definition 4.9.1** Given a topological group $G$ with identity $e$ and a principal $G$-bundle $\xi = (X, p, B, G)$, a **chart** $(\psi, U)$ of $\xi$ is a pair, which consists of

(a)  an open set $U \subset B$ and
(b)  a homeomorphism $\psi : U \times G \to p^{-1}(U)$

such that

(i)
$$p \circ \psi = p_U$$

and
(ii)
$$\psi(b, g) = \psi(b, e)g, \ \forall\, b \in U \ \text{ and } \ \forall\, g \in G$$

Given an open covering $\{U_\alpha : \alpha \in \mathbf{K}\}$ of $B$, the family of charts $\{(\psi_\alpha, U_\alpha) : \alpha \in \mathbf{A}\}$ of $B$ is said to be an **atlas of** $\xi$ if each homeomorphism

$$\psi_\alpha : U_\alpha \times G \to p^{-1}(U_\alpha)$$

satisfies the properties

$$(p \circ \psi_\alpha) = p_{U_\alpha} \ \text{ and } \ \psi_\alpha(b, g) = \psi_\alpha(b, e)g, \ \forall\, b \in U_\alpha, \ \forall\, g \in G.$$

If an atlas of $\xi$ includes all its charts, then it is said to be a **complete atlas**.

*Example 4.9.2* Let $\xi$ be an arbitrary principal $G$-bundle. Then it has at least one atlas.

**Definition 4.9.3** Given a topological group $G$ with identity $e$ and a principal $G$-bundle $\xi = (X, p, B, G)$, a collection of transition functions $\mathcal{T}$ for $\xi$ consists of

(a)  an open covering $\{U_\alpha : \alpha \in \mathbf{K}\}$ of $B$ and
(b)  a family of continuous functions $\{g_{\alpha\beta} : U_\alpha \cap U_\beta \to G, \ \forall\, \alpha, \beta \in \mathbf{K}\}$ for $U_\alpha \cap U_\beta \neq \emptyset$ such that

$$g_{\alpha\beta}(b)g_{\beta\gamma}(b) = g_{\alpha\gamma}(b), \ \forall\, b \in U_\alpha \cap U_\beta \cap U_\gamma \ (\neq \emptyset).$$

Each function $g_{\alpha\beta} \in \mathcal{T}$ is called a **transition function** on $U_\alpha \cap U_\beta$ and it is sometimes written by the pair $(U_\alpha \cap U_\beta, g_{\alpha\beta})$.

**Proposition 4.9.4** *The family of transition functions $\{g_{\alpha\beta} : \alpha, \beta \in \mathbf{K}\}$ formulated in Definition 4.9.3 have the following properties.*

(i)  $g_{\alpha\alpha}(b) = g_{\alpha\alpha}(b)g_{\alpha\alpha}(b) \in G$, $\forall b \in B$, $\alpha \in \mathbf{K}$.

(ii)  $g_{\alpha\alpha}(b) = e$, $\forall b \in B$, $\alpha \in \mathbf{K}$.

(iii)  $g_{\alpha\beta}(b) = [g_{\beta\alpha}(b)]^{-1} \forall b \in U_\alpha \cap U_\beta$, $\alpha, \beta \in \mathbf{K}$.

**Proof**  Consider the relation $g_{\alpha\beta}(b)g_{\beta\gamma}(b) = g_{\alpha\gamma}(b)$, $\forall b \in U_\alpha \cap U_\beta \cap U_\gamma$ ($\neq \emptyset$).

(i)  In particular, take $\alpha = \beta = \gamma$ in the relation $g_{\alpha\beta}(b)g_{\beta\gamma}(b) = g_{\alpha\gamma}(b)$. Then it follows that $g_{\alpha\alpha}(b) = g_{\alpha\alpha}(b)g_{\alpha\alpha}(b) \in G$, $\forall b \in B$, $\alpha \in \mathbf{K}$.

(ii)  (*i*) asserts that $g_{\alpha\alpha}(b) = e$, $\forall b \in B$, $\alpha \in \mathbf{K}$, since $g_{\alpha\alpha}(b) \in G$ and $G$ is a group.

(iii)  In particular, take $\gamma = \alpha$ in the relation $g_{\alpha\beta}(b)g_{\beta\gamma}(b) = g_{\alpha\gamma}(b)$. Then using (i), it follows that

$$g_{\alpha\beta}(b) = [g_{\beta\alpha}(b)]^{-1} \forall b \in U_\alpha \cap U_\beta, \ \alpha, \beta \in \mathbf{K}.$$

$\square$

**Definition 4.9.5**  Let $\{(U_\alpha \cap U_\beta, \ g_{\alpha\beta}) : \alpha, \beta \in \mathbf{K}\}$ and $\{(U_{\alpha'} \cap U_{\beta'}, g'_{\alpha'\beta'}) : \alpha', \beta' \in A'\}$ be two sets of transition functions of two principal $G$-bundles $\xi = (X, p, B, G)$ and $\xi' = (X', p', B, G)$ over the same base space $B$. Then they are said to be **equivalent** if there exist a family of continuous maps

$$f_{\alpha'\alpha} : U_\alpha \cap U'_{\alpha'} \to G \ \text{ with } \ U_\alpha \cap U'_{\alpha'} \neq \emptyset$$

such that

$$g'_{\alpha'\beta'}(b) = f_{\alpha'\alpha}(b)g_{\alpha\beta}(b)[f_{\beta'\beta}(b)]^{-1}, \ \forall b \in U_\alpha \cap U_\beta \cap U'_{\alpha'} \cap U'_{\beta'} : \alpha, \beta \in \mathbf{K} \text{ and } \alpha', \beta' \in \mathbf{K}$$

**Definition 4.9.6**  Given two $n$-dimensional $\mathbf{F}$-vector bundles $\xi = (X, p, B, \mathbf{F}^n)$ and $\xi' = (X', p', B', \mathbf{F}^n)$, a morphism

$$\psi : \xi \to \xi'$$

consists of a pair of continuous maps $\psi : X \to X'$ and $\tilde{\psi} : B \to B'$ such that

(i)  $p' \circ \psi = \tilde{\psi} \circ p$ and

(ii)  $\psi|p^{-1}(b) : p^{-1}(b) \to p'^{-1}(\tilde{\psi}(b))$ is a linear map for every $b \in B$.

In particular. the identity morphism $1_d : \xi \to \xi$ consists of the pair of the identity map $1_X : X \to X$ and the identity map $1_B : B \to B$.

Theorem 4.9.7 establishes a close relation between vector bundles and principal $G$-bundles.

**Theorem 4.9.7**  (*i*)  *For every principal $G$-bundle $\xi = (X, p, B, G)$ over the same base space $B$, there exists a unique set of transition functions $\mathcal{T} = \{(U_\alpha \cap U_\beta, \ g_{\alpha\beta}) : \alpha, \beta \in \mathbf{K}\}$ with the property*

$$\psi_\beta : U_\beta \times G \to p^{-1}(U_\beta) : (b, g) \mapsto \psi_\alpha(b, g_{\alpha\beta}(b)g),\ \forall \alpha, \beta \in \mathbf{K},\ \forall b \in U_\alpha \cap U_\beta, g \in G.$$

(ii) *For two G-bundles* $\xi = (X, p, B, G)$ *and the* $\xi' = (X', p', B', G)$ *and the corresponding sets* $\tilde{\xi}$ *and* $\tilde{\xi}'$ *of transition functions if* $\psi : \xi \to \xi'$ *is a bundle morphism, then it induces a unique morphism of sets of transition functions*

$$f : \tilde{\xi} \to \tilde{\xi}'$$

*such that*

$$\tilde{f} = \tilde{\psi} : B \to B'$$

*and*

$$\psi \circ \psi_\alpha(b, g) = \psi'_{\alpha'}(\tilde{\psi}(b), f_{\alpha'\alpha}(b)g),\ \forall b \in U_\alpha \cap \tilde{\psi}^{-1}(U'_{\alpha'}),\ \forall g \in G.$$

**Proof** (i) For $\alpha, \beta \in \mathbf{K}$, the map

$$\psi_{\alpha\beta} = \psi_\alpha{}^{-1} \circ (\phi_\beta|(U_\alpha \cap U_\beta) \times G) : (U_\alpha \times U_\beta) \times G \to (U_\alpha \times U_\beta) \times G : p_{U_\alpha \cap U_\beta} \circ \psi_{\alpha\beta} = p_{U_\alpha \cap U_\beta}$$

Writing $\psi_{\alpha\beta}$ in the form $\psi_{\alpha\beta}(b, f_{\alpha\beta}(b, g))$ for some $f_{\alpha\beta} : (U_\alpha \cap U_\beta) \times G \to G$.
Hence

$$\psi_\beta(b, g) = \psi_\alpha(b, f_{\alpha\beta}(b, g)),\ \forall b \in U_\alpha \cap U_\beta, g \in G.$$

This asserts that

$$\psi_\alpha(b, f_{\alpha\beta}(b, g)) = \psi_\beta(b, g) = \psi_\beta(b, e)g = \psi_\alpha(b, f_{\alpha\beta}(b, e))g = \psi_\alpha(b, f_{\alpha\beta}(b, e)g).\ \forall b \in U_\alpha \cap U_\beta, g \in G.$$

This implies that

$$f_{\alpha\beta}(b, g) = f_{\alpha\beta}(b, e)g,\ \forall b \in U_\alpha \cap U_\beta, g \in G.$$

Now, taking, $g_{\alpha\beta}(b) = f_{\alpha\beta}(b, e),\ \forall b \in U_\alpha \cap U_\beta$ and $\forall \alpha, \beta \in \mathbf{K}. g \in G$, it follows that $\phi_\beta$ has the requisite property. Moreover, for any $b \in U_\alpha \cap U_\beta \cap U'_{\alpha'}$.

$$\psi_\alpha(b, g_\alpha(b)g) = \psi_{\alpha'}(b, g) = \psi_\beta(b, g_{\beta\alpha'}(b)g) = \psi_\alpha(b, g_{\alpha\beta}(b)g_{\beta\alpha'}(b)g).$$

This asserts that $g_{\alpha\alpha'}(b) = g_{\alpha\beta}(b)g_{\beta\alpha'}(b)$. This proves that $\tilde{\xi} = \{(U_\alpha, g_{\alpha\beta}) : \alpha, \beta \in \mathbf{A}\}$ is a set of transition functions.

(ii) For any $\alpha, \beta, \alpha', \beta' \in \mathbf{K}$, consider the map

$$\psi_{\alpha'\alpha} = \psi'^{-1}_{\alpha'} \circ \psi(\psi_\alpha|(U_\alpha \cap \tilde{\psi}^{-1}(U'_{\alpha'} \times G)) : U_\alpha \cap \tilde{\psi}^{-1}(U_\alpha \cap \tilde{\psi}^{-1}(U_{\alpha'})) \times G \to U_{\alpha'} \times G$$

satisfies the relation

$$p_{U_{\alpha'}} \circ \theta_{\alpha'\beta} = \tilde{\psi} \circ p_{U_\alpha} \cap \tilde{\psi}^{-1}(U_{\alpha'}),$$

where $\theta_{\alpha'\beta}$ is given by

$$\theta_{\alpha'\beta}(b, g) = (\tilde{\psi}(b), h_{\alpha'\beta}(b, g))$$

for some $h_{\alpha'\beta} : (U_\beta \cap \tilde{\phi}^{-1}(U_{\alpha'})) \times G \to G$. This shows that

$$\psi \circ \psi_\alpha(b, g) = \psi'_{\alpha'}(\tilde{\psi}(b), h_{\alpha'\alpha}(b, g)), \ \forall b \in U_\alpha \cap \tilde{\psi}^{-1}(U'_{\alpha'}), \ \forall g \in G.$$

Hence it follows that

$$\psi'_{\alpha'}(\tilde{\psi}(b), h_{\alpha'\alpha}(b, g)) = (\psi \circ \psi_j(b, e))g = \psi'_{\alpha'}(\tilde{\psi}(b), h_{\alpha j}(b, e)g) = \psi'_{\alpha'}(\tilde{\psi}(b), h_{\alpha'\alpha}(b, e)g).$$

This shows that

$$h_{\alpha'\alpha}(b, g) = h_{\alpha'\alpha}(b, e)g.$$

Proceed as in (i) to prove (ii) except its uniqueness. To show its uniqueness, suppose $\tilde{\xi} = \{(U_\alpha \cap U_\beta, g_{\alpha\beta}) : \alpha.\beta \in \mathbf{K}\}$ and $\tilde{\xi}' = \{(U_{\alpha'} \cap U_{\beta'}, g_{\alpha'\beta'}) : \alpha', \beta' \in A'\}$. Then

$$\psi'_{\alpha'}(\tilde{\psi}(b), f_{\alpha'\alpha}(b)g_{\alpha\beta}(b)g) = \psi \circ \psi_j(b, g_{\beta\gamma}(b)g) = \psi \circ \psi_\gamma(b, g) = \psi_{\beta'}(\tilde{\psi}(b), f_{\beta'\beta}(b))$$

$$= \psi_\alpha(\tilde{\psi}(b), g_{\alpha'\beta'}(\tilde{\psi}(b))f_{\beta'\beta}(b)g).$$

Hence it follows that

$$f_{\alpha'\alpha}(b)g_{\alpha\beta}(b) = g_{\alpha'\beta'}(\tilde{\psi}(b))f_{\beta'\beta}(b).$$

This proves that $\{f_{\beta'\beta}\}$ is a morphism of sets of transition functions.

☐

Theorem 4.9.8 relates the set of the equivalence classes of principal $G$-bundles $\xi$ over a fixed base space $B$ and the equivalence classes of sets of transition functions associated with an atlas of $\xi$.

**Theorem 4.9.8** *Let $G$ be a topological group and $\xi$ be any principal $G$-bundles over a fixed base space $B$. Then there exists a $(1\text{–}1)$-correspondence between the set $SP$ of the equivalence classes of principal $G$-bundles $\xi$ and the set $ST$ of equivalence classes of sets of transition functions associated with an atlas of $\xi$.*

***Proof*** Let $\xi$ and $\xi'$ be any two equivalent principal $G$-bundles over the same base space $B$ with

$$\psi : \xi \to \xi'$$

an equivalence of $G$-bundles. Then there exists a morphism $f(\psi) : \tilde{\xi} \to \tilde{\xi}'$ by Proposition 4.9.7(b). Since $\xi$ and $\xi'$ are $G$-bundles over the same base space, the map $\tilde{\psi} : B \to B$ coincides with $1_B$. This defines a correspondence

$$\theta : \mathcal{SP} \to \mathcal{ST}, \ \psi \mapsto f(\psi).$$

To show that $\theta$ is surjective, take any set of transition functions $\tilde{\xi} \in \mathcal{ST}$. Then by Ex.4.13.1 of Sect. 4.13 there exist a principal $G$-bundle $\xi$ and an atlas $\{(U_j : \psi_j) : j \in \mathbf{J}\}$ of $\xi$ such that $\tilde{\xi}$ is the corresponding set of transition functions. This proves that $\theta$ is surjective. To show that $\theta$ is injective, take two principal $G$-bundles $\xi$ and $\xi'$ such that the corresponding sets $\tilde{\xi}$ and $\tilde{\xi}'$ of transition functions are equivalent, by an equivalence $f : \tilde{\xi} \to \tilde{\xi}'$. Then there exists a morphism $\phi : \xi' \to \xi$ by Exercise 4.13.1 of Sect. 4.1.3 inducing $f$. As before, $\tilde{\phi} : B \to B$ is $1_B$. Define the morphism

$$f^{-1} : \tilde{\xi}' \to \tilde{\xi}$$

by the rule

$$f^{-1} = \{f_{aj}^{-1} : \alpha \in \mathbf{A}, j \in \mathbf{K}\}.$$

Then morphism $\phi : \xi' \to \xi$ of $G$-bundles is the inverse of $\psi$. Because, $\psi \circ \phi$ and $\phi \circ \psi$ are both identity maps. This implies that the principal $G$-bundles $\xi$ and $\xi'$ are equivalent. This proves the map $\theta$ is injective. Hence, it follows that the map $\theta$ is bijection.                                                                                $\square$

Consider the topological group $GL(n, \mathbf{F}) = G$ of the group of all nonsingular $n \times n$ matrices over $\mathbf{F}$. There is a natural problem:

(i)  does there exist a (1–1)-correspondence between the set of the equivalence classes of $n$-dimensional $\mathbf{F}$-vector bundles over a topological space $B$ and the set of the equivalence classes of sets of transition functions for $B$ and $G = GL(n, \mathbf{F})$?
(ii) does there exist a (1–1)-correspondence between the equivalence classes of sets of transition functions for $B$ with principal $GL(n, F)$-bundles ?

We now define chart and atlas of a vector bundle in a way analogous to Definition 4.9.1.

**Definition 4.9.9** A chart $(U, \psi)$ of an $n$-dimensional vector bundle $\xi = (X, p, B, \mathbf{F}^n)$ is a pair consisting of

(a) an open set $U \subset B$ and
(b) a homeomorphism $\psi : U \times \mathbf{F}^n \to p^{-1}(U)$

such that

(i) $p \circ \phi = p_U$ and
(ii) $\phi$ is linear on all fibers $p^{-1}(b)$ for $b \in B$.

On the other hand, an atlas is a family $\{(U_k, \psi_k) : k \in \mathbf{K}\}$ of charts such that $\{U_k : k \in \mathbf{K}\}$ is an open covering of $B$.

**Construction 4.9.10** Let $\xi = (X, p, B, \mathbf{F}^n)$ be a given $n$-dimensional $\mathbf{F}$-vector bundle with an atlas $\{(U_k, \psi_k) : k \in \mathbf{K}\}$. **Construction of a set of transition functions** $\{(U_i, g_{ik}) : i, k \in \mathbf{K}\}$ for $B$ and the group GL $(n, F)$ is now described :
  The maps

$$\psi_{ik} = \phi_i^{-1} \circ (\phi_k|(U_i \cap U_k) \times \mathbf{F}^n) : (U_i \cap U_k) \times \mathbf{F}^n \to (U_i \cap U_k) \times \mathbf{F}^n, \ \forall i, k \in \mathbf{K}$$

are well defined and take the form

$$\psi_{ik}(b, u) = (b, f_{ik}(b, u)) \text{ for some } f_{ik} : (U_i \cap U_k) \times \mathbf{F}^n \to \mathbf{F}^n.$$

Hence it follows that for a fixed $b \in U_i \cap U_k$, the map $f_{ik}(b, -) : \mathbf{F}^n \to \mathbf{F}^n$ is a linear isomorphism and hence $f_{ik}(b, -)$ is in GL $(n, \mathbf{F})$. If $g_{ik}(b) = f_{ik}(b, -)$, then

$$f_{ik}(b, b) = g_{ik}(b)u.$$

It proves that

$$\phi_k(b, u) = \phi_i(b, g_{ik}(b)u), \ \forall b \in U_i \cap U_k, u \in \mathbf{F}^n.$$

  Summarizing the above construction and discussion a basic result interlinking $n$-dimensional $F$-vector bundles over $B$ with principal G-bundles over $B$, where $G = GL(n, \mathbf{F})$ is formulated in Theorem 4.9.11.

**Theorem 4.9.11** *There is a (1–1) correspondence between the set of equivalence classes of n-dimensional $\mathbf{F}$-vector bundles over a fixed base space $B$ and the set of equivalence classes of the set of transition functions for $B$ and the general linear group* GL $(n, \mathbf{F})$.

## 4.10  Principal *G*-bundles for Lie Groups *G*

This section continues the study of principal $G$-bundles over differentiable manifolds when $G$ is a Lie group. Throughout this section $G$ denotes an arbitrary Lie group.

**Definition 4.10.1** A principal (differentiable) $G$-bundle is a triple $(E, p, M)$ such that $p : E \to M$ is a differentiable mapping of differentiable manifolds. Furthermore, $E$ is given a differentiable right $G$-action $E \times G \to E$ such that the following conditions hold:

**Fig. 4.21** Local
trivialization

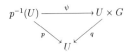

(i) $E_x = p^{-1}(x), x \in M$ are the orbits for the $G$-action.
(ii) **(Local trivialization)** Every point in $M$ has an open neighbourhood $U$ and a diffeomorphism $\psi : p^{-1}(U) \to U \times G$ such that the diagram in Fig. 4.21 commutes, i.e., $\psi_x = \psi|E_x$ maps $E_x$ to $\{x\} \times G$; and $\psi$ is equivariant, i.e.,

$$\psi(xg) = \psi(x)g, \ \forall x \in p^{-1}(U), g \in G,$$

where $G$ acts on $U \times G$ by

$$\sigma : (U \times G) \times G \to U \times G, \ ((x, g'), g) \mapsto (x, g'g).$$

$E$ is called the total space, $M$ the base space and $E_x = p^{-1}(x)$ the fiber at $x \in M$. Sometimes we use the notation $E$ to denote the $G$-bundle $(E, p, M)$.

**Remark 4.10.2**  (i)  Let $(E, p, M)$ be a principal $G$-bundle. Then $p$ is surjective and open.
(ii)  The orbit space $E$ mod $G$ is homeomorphic to $M$.
(iii)  The $G$-action is free, i.e., $x \cdot g = x \Rightarrow g = e, \ \forall x \in E, g \in G$.
(iv)  For each $x \in E$, the mapping $G \to E_x$ given by $g \mapsto x \cdot g$, is a diffeomorphism.
(v)  If $N \subset M$ is a submanifold (e.g., if $N$ is an open subset), then the restriction to $N$ $E|N = (p^{-1}(N), p, N)$ is again a principal $G$-bundle with base space $N$.

**Example 4.10.3**  (i)  For an $n$-dimensional real vector bundle $(V, p, M)$ the associated frame bundle $(F(V), \tilde{p}, M)$ is a principal $G = \mathrm{GL}\,(n, \mathbf{R})$-bundle.
(ii)  For an $n$-dimensional real vector bundle $V$ equipped with Riemannian metric, $(F_0(V), \tilde{p}, M)$ is a principal $O(n, \mathbf{R})$-bundle.
(iii)  Let $G$ be any Lie group and $M$ be a differential manifold. Then $(M \times G, p, M)$ with $p$ the projection onto the first factor, is a principal $G$-bundle called the product bundle.

**Definition 4.10.4**  Let $G$ be a Lie group and $(E, p, B)$ and $(F, q, B)$ be two principal $G$-bundles over the same base space $B$. An isomorphism $\psi : E \to F$ is a diffeomorphism of the total spaces such that

(i)  The diagram in Fig. 4.22 commutes, i.e., $\psi_b = \psi|E_b$ maps $E_b = p^{-1}(b)$ to $F_b = q^{-1}(b)$ and
(ii)  $\psi$ is equivariant, i.e., $\psi(xg) = \psi(x)g, \ \forall x \in E, \forall g \in G$.

**Remark 4.10.5**  The map $\psi_b : E_b \to F_b$ is also a diffeomorphism for each $b \in B$.

**Fig. 4.22** Isomorphism of
principal G-bundles

## 4.11 Applications

This section studies covering spaces of real projective spaces $\mathbf{R}P^n$ and figure-eight $F_8$. Chap. 5 computes their fundamental groups.

### 4.11.1 Covering Spaces of $\mathbf{R}P^n$

This subsection constructs covering spaces of real projective spaces $\mathbf{R}P^n$ We start with the real projective plane $\mathbf{R}P^2$ obtained as a quotient space of the 2-sphere $S^2$ by identifying every pair of antipodal points with $p : S^2 \to \mathbf{R}P^2$ the natural projection. It is topologized by defining $U$ to be open in $\mathbf{R}P^2$ iff $p^{-1}(U)$ is open in $S^2$. By using this covering space the fundamental group $\pi_1(\mathbf{R}P^n)$ is computed in Chap. 5.

**Theorem 4.11.1** $(S^2, p)$ *is a covering space of* $\mathbf{R}P^2$ *and* $\mathbf{R}P^2$ *is a surface.*

**Proof** Consider the natural projection map $p : S^2 \to \mathbf{R}P^2$. Let $y \in \mathbf{R}P^2$, $x \in p^{-1}(y)$ and

$$A : S^2 \to S^2, \ z \mapsto -z$$

be the antipodal map. Given an $\epsilon < 1$, choose an $\epsilon$ nbd $U$ of $x$ in $S^2$ such that $U$ contains no pair $\{z, A(z)\}$ of antipodal points of $S^2$, this choice is possible, since $d(z, A(z)) = 2$, where $d$ is the Euclidean metric of $\mathbf{R}^3$. This implies that $p : U \to p(U)$ is a bijective map and

$$A : S^2 \to S^2, \ z \mapsto -z$$

is a homeomorphism. Then $A(U)$ is an open set in $S^2$. Moreover, $p(U)$ is open in $\mathbf{R}P^2$, *because*, $p^{-1}(p(U)) = U \cup A(U)$ is open in $S^2$. This shows that $p$ is an open map. Consequently, it follows that

$$p : U \to p(U)$$

is a homeomorphism, since it is bijective, continuous and open. It follows similarly that

$$p : A(U) \to p(A(U)) = p(U)$$

is a homeomorphism.

$p(U)$ is a nbd of $p(x) = y$, which is evenly covered by $p$, since $p^{-1}(p(U))$ is the union of two open sets $U$ and $A(U)$ each of them is mapped homeomorphically by $p$ onto $p(U)$. This proves that $(S^2, p)$ is a covering space of $\mathbf{R}P^2$.

To prove the second part, consider a countable basis $\{U_n\}$ for $S^2$. This implies that $\{p(U)\}$ is a countable basis of $\mathbf{R}P^2$, Consider the Hausdorff space $\mathbf{R}P^2$. Given two distinct points $x_1$, $x_2 \in \mathbf{R}P^2$, the set $p^{-1}(x_1) \cup p^{-1}(x_2)$ consists of four points. If $2\epsilon$ is the minimum distance between them and $U_1$ is the $\epsilon$-nbd of one of the points $p^{-1}(x_1)$ and $U_2$ is the $\epsilon$-nbd of one of the points $p^{-1}(x_2)$, then the sets $U_1 \cup A(U_1)$ and $U_2 \cup A(U_2)$ are disjoint. This shows that $p(U_1)$ and $p(U_2)$ are disjoint ends of $x_1$ and $x_2$ respectively in $\mathbf{R}P^2$. The space $\mathbf{R}P^2$ is also a surface, because, $S^2$ is a surface and every point of $\mathbf{R}P^2$ has a nbd homeomorphic to an open subset of $S^2$.    □

Theorem 4.11.2 gives a generalization of Theorem 4.11.1 for $n > 1$.

**Theorem 4.11.2** $(S^n, p)$ *is a covering space of* $\mathbf{R}P^n$, *where* $p$ *is the map identifying antipodal points of* $S^n$ *for* $n > 1$.

**Proof** To show that $(S^n, p)$ is a covering space of $\mathbf{R}P^n$, consider open sets

$$U_i^+ = \{(x_1, x_2, \ldots, x_{n+1}) \in S^n : x_i > 0\}$$

and

$$U_i^- = \{(x_1, x_2, \ldots, x_{n+1}) \in S^n : x_i < 0\},$$

which cover $S^n$. For $U_i = p(U_i^+) = p(U_i^-)$, clearly $p^{-1}(U_i) = U_i^+ \cup U_i^-$ and the open sets $U_i^+$ and $U_i^-$ are disjoint and homeomorphic to $U_i$, since the map $p|_{U_i^+}$ is 1–1, continuous and open. This proves that $(S^n, p)$ is a covering space of $\mathbf{R}P^n$, for every integer $n > 1$.    □

**Definition 4.11.3** Given a covering space $(X, p)$ of $B$, the multiplicity of the covering space is defined to be the cardinal number of a fiber. If its **multiplicity** is $n$, covering space $(X, p)$ is said to be an $n$-**sheeted covering space** of $B$ or that $(X, p)$ is an $n$ **-fold cover** of $B$.

**Example 4.11.4** (i) $(S^2, p)$ is a double covering of $\mathbf{R}P^2$, Because, $p$ identifies pairs of antipodal points of $S^2$, and hence, the number of sheets of this covering is 2.
(ii) Consider the covering space $(\mathbf{R}, p)$ of $S^1$ defined in Example 4.5.7. Its number of sheets is countably infinite. Because, the covering projection $p : \mathbf{R} \to S^1$ sends every integer to the point $1 \in S^1$. This asserts that $p^{-1}(1) = \mathbf{Z}$. This proves that the number of sheets of the covering space $(\mathbf{R}, p)$ of $S^1$ is countably infinite.

**Fig. 4.23** Covering space of figure-eight

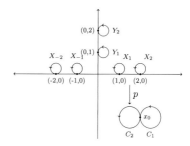

## 4.11.2 Covering Space of Figure-Eight $\mathbf{F}_8$

This section constructs a covering space of figure-eight $\mathbf{F}_8$ and computes its fundamental group in Chap. 5 by using this covering space.

***Example 4.11.5 (Covering of Figure-eight $\mathbf{F}_8$ )*** the figure eight $\mathbf{F}_8$ is the union of two circles $C_1$ and $C_2$ with a point $x_0$ in common. We construct a covering space $(X, p)$ of $\mathbf{F}_8$. Let $X$ be the subspace of the Euclidean plane $\mathbf{R}^2$, which consists of the $x$-axis and the y-axis, along with the small circles tangent to these axes, one circle tangent to the $x$-axis at each nonzero integer point and one circle tangent to the y-axis at each nonzero integer point as shown in Fig. 4.23.

   **Geometrical construction of the projection map** $p : X \to \mathbf{F}_8$. The map $p$ wraps the $x$-axis around the circle $C_1$, and it wraps the y-axis around the other circle $C_2$ such that the integer points are mapped by $p$ in each case into the base point $x_0$ of the figure-eight $\mathbf{F}_8$. Hence, every circle tangent to an integer point on the $x$-axis is mapped by $p$ onto $X$ homeomorphically, and every circle tangent to an integer point on the y-axis is mapped by $p$ homeomorphically onto $C_1$, In either of these cases the point of tangency is mapped onto the point $x_0$. Then $p$ is a covering map.

   This shows that $(X, p)$ is a covering space of the figure-eight $\mathbf{F}_8$.

## 4.12 More Geometrical Applications

This section communicates geometrical applications of Theorem 4.8.9.

**Theorem 4.12.1** *For every integer* $n \geq 2$, *the sphere* $S^{n-1}$ *and the factor space* $O(n, \mathbf{R})/O(n - 1, \mathbf{R})$ *are homeomorphic.*

***Proof*** **Proof I**: Consider the orthogonal (real) topological space $G = O(n, \mathbf{R})$ for $n \geq 2$. It is a compact topological group and $X = S^{n-1} \subset \mathbf{R}^n$ is Hausdorff space. A matrix $M \in O(n, \mathbf{R})$ is a transformation of the Euclidean space $\mathbf{R}^n$ and it preserves lengths of the vectors and hence it is a map of $S^{n-1}$ to itself. Again, $O(n - 1, \mathbf{R})$ is regarded as the subgroup of $G = O(n, \mathbf{R})$ obtained by keeping the last coordinate fixed. Let $s_0 = (0, 0. \ldots, 1)$. This point is kept fixed by $O(n - 1, \mathbf{R})$. This gives a mapping from $O(n, \mathbf{R})$ into $X = S^{n-1}$. More precisely, define the map

$$\psi : M \mapsto M(0, 0. \ldots, 1)^t.$$

If $A \in O(n - 1, \mathbf{R})$, then $\psi(MA) = \psi(M)$ asserts that the map $\psi$ factors via the left coset space $O(n, \mathbf{R})/O(n - 1, \mathbf{R})$. The topological group $O(n, \mathbf{R})$ acts on $S^{(n-1)}$ transitively and the induced map

$$\psi^* : O(n, \mathbf{R})/O(n - 1, \mathbf{R}) \to S^{n-1}$$

a continuous bijective map from a compact space to a Hausdorff space. Hence $\psi^*$ is a homeomorphism. This proves the theorem.

**Proof II**: Every orthogonal matrix $M \in O(n, \mathbf{R})$ represents a transformation of $\mathbf{R}^n$ that preserves length of vectors. This implies that it is a map from $S^{n-1}$ to itself. Consider $O((n - 1), \mathbf{R})$ as a subgroup of the orthogonal group $O(n, \mathbf{R})$, where the last coordinate is fixed. The isotropy group at the point $(0, 0, \ldots .0, 1)$ is $O((n - 1), \mathbf{R})$. Hence, the corollary follows from Theorem 4.8.9                        □

**Theorem 4.12.2** *For every integer $n \geq 2$, the sphere $S^{2n-1}$ and the factor space $U(n, \mathbf{C})/U(n - 1, \mathbf{C})$ are homeomorphic.*

***Proof*** Since $S^{2n-1}$ is the set of all unit vectors in $\mathbf{C}^n$, the topological group $U(n, \mathbf{C})$ acts on $S^{(2n-1)}$ transitively. Because, there is a unitary matrix sending any vector of length 1 sending any other vector of length 1. Now proceed as in Theorem 4.8.9 to prove the theorem..                        □

**Theorem 4.12.3** *For every integer $n \geq 2$, the sphere $S^{4n-1}$ and the factor space $Sp(n, \mathbf{H}) / Sp(n - 1, \mathbf{H})$ are homeomorphic.*

***Proof*** Since $S^{4n-1}$ is the set of all unit vectors in $\mathbf{H}$, the topological group $Sp(n, \mathbf{H})$ acts on $S^{(4n-1)}$ transitively. Because, there is a unitary matrix sending any vector of length 1 sending any other vector of length 1. Now proceed as in Theorem 4.8.9 to prove the theorem.                        □

## 4.13  Exercises and Multiple Choice Exercises

As solving exercises plays an essential role of learning mathematics, various types of exercises and multiple choice exercises are given in this section. They form an integral part of the book series.

## 4.13.1  Exercises

In this section, **F** denotes $\mathbf{F} = \mathbf{R}, \mathbf{C}$ or **H**.

1. Let $p : X \to B$ be covering space. Show that the cardinality of any fiber $p^{-1}(b)$ does not depend on the choice of the point $b \in B$.
2. Let $p : X \to B$ be a covering map, with $B$ a locally connected space (i.e., every nbd of a point of $B$, contains a connected nbd of the point ). Show that

   (i)  $B$ is a union of connected open sets $U$ such that every component of $p^{-1}(U)$ is mapped homeomorphically onto $U$ by $p$.
   (ii)  Given a component $X'$ of $X$, the image $B' = p(X')$ is also a component of $B$.
   (iii)  The restriction $p|X' : X' \to p(X')$ is also a covering map.

3. Let $p : X \to B$ and $p' : X' \to B$ be two covering spaces and $\psi : X \to X'$ be a continuous map such that $p' \circ \psi = p$. If $X$, $X'$ and $B$ are connected and $B$ is locally connected, show that $\psi$ is a covering map.
4. Let $S$ be a compact surface and $p : X \to S$ be an $n$-sheeted covering space. Show that

   (i)  $X$ is also a compact surface.
   (ii)  The Euler characteristic $\kappa(X) = n \times$ Euler characteristic $\kappa(S)$.
   (iii)  If $S$ is a sphere with $g$ handles and $X$ is a sphere with $k$ handles, then

$$k = ng - n + 1.$$

5. Consider the action $\sigma : \mathbf{Z} \times \mathbf{R}$, $(n, t) \mapsto n + t$ (translation). Show that the quotient map $p : \mathbf{R} \to \mathbf{R} \bmod \mathbf{Z}$, $t \mapsto \mathrm{orb}(t)$ is identified with the covering map $p : \mathbf{R} \to S^1$, $t \mapsto e^{it}$.
6. Let $G = \{-1, +1\} \cong \mathbf{Z}_2$ be a group of two elements. Consider the action

$$\sigma : G \times S^n, (-1, x) \mapsto -x, (1, x) \mapsto x.$$

   Show that

   (i)  The quotient group $S^n \bmod G$ is the real projective space $\mathbf{R}P^n$.
   (ii)  The quotient map $p : S^n \to \mathbf{R}P^n$ is two-sheeted.

7. Let $G_n$ be the group of the $n$th roots of unity. Show that the action

   (i)  $\sigma$ of $G_n$ on **C**

$$\sigma : G_n \times \mathbf{C} \to \mathbf{C}, (g, z) \mapsto gz \text{ (usual multiplication of complex numbers)}$$

   is not properly discontinuous;

(ii) On the other hand, the action $\psi$ of $G_n$ on the open subset

$$\psi : G_n \times \mathbf{C} - \{0\} \to \mathbf{C} - \{0\}, \ (g, z) \mapsto gz \ \text{(usual multiplication of complex numbers)}$$

is properly discontinuous.

8. Show the action

$$\sigma : \mathbf{Z}^n \times \mathbf{R}^n \to \mathbf{R}^n, \ ((n_1, n_2, \ldots, n_n), (x_1, x_2, \ldots, x_n)) \mapsto (n_1 + x_1, n_2 + x_2, \ldots, n_n + x_n)$$

is properly is continuous. Identify its orbit space $\mathbf{R}^n mod \ \mathbf{Z}^n$ with the $n$-torus $T^n$.

9. Let $X$ be the quotient space obtained from the disk $\mathbf{D}^2$ by identifying its antipodal points. Show that the space $X$ is homeomorphic the real projective space $\mathbf{R}P^2$.

10. For the group $G = \mathbf{Z}_2$, show that any 2-sheeted covering has the unique $G$-covering structure.

11. Let $p : X \to B = X mod \ G$ be a $G$-covering and $f, g : X \to X$ be two $G$-isomorphisms. If $B$ is connected and $f(x_0) = g(x_0)$, for some $x_0 \in X$, that $f = g$.

12. **(Klein bottle as an orbit space)** Let $G$ be the group of homeomorphisms of $\mathbf{R}^2$ and $H$ be the subgroup of $G$ generated by the translation

$$T : \mathbf{R}^2 \to \mathbf{R}^2, (x, y) \mapsto (x + 1, y + 1)$$

and by the mapping

$$f : \mathbf{R}^2 \to \mathbf{R}^2, (x, y) \mapsto (-x, y + 1).$$

Show that

(i) this action of $H$ on $\mathbf{R}^2$ is properly discontinuous;
(i) the orbit space $\mathbf{R}^2 mod \ H$ is homeomorphic to the Klein bottle.

13. **(Möbius band as an orbit space)** Let $G$ be the group of homeomorphisms of $\mathbf{R}^2$ and $H$ be the subgroup of $G$ generated by the translation

$$T : \mathbf{R}^2 \to \mathbf{R}^2, (x, y) \mapsto (x + 1, -y).$$

(i) this action $H$ on $\mathbf{R}^2$ is properly discontinuous;
(ii) the orbit space $\mathbf{R}^2 mod \ H$ is homeomorphic to the Möbius band.

14. Let $G$ be a connected Lie group and $p : X \to G$ be a covering space. Show that $X$ is also a Lie group.

15. Let $t \neq 0, 1 \ or - 1$ be a real number and $X = \mathbf{R}^n - \{0\}$. Consider the action

$$\sigma : \mathbf{Z} \times X \to X, \ (m, x) \mapsto t^m x.$$

Show that

    (i)  The action $\sigma$ is properly discontinuous.

    (ii)  The orbit space $X \ mod \ \mathbf{Z}$ is homeomorphic to the product space $S^1 \times S^{n-1}$.

16.  Let $G$ be the group of diffeomorphism of a differentiable manifold $X$ and $G$ acts
    on $X$ properly discontinuously. Endow the orbit space $X \ mod \ G$ a differentiable
    manifold structure such that the projection

$$p : X \to X \ mod \ G$$

    is a local diffeomorphism.

17.  (**Spin group**) Let $S^3$ be the topological group of unit quaternions in $\mathbf{H}$

$$S^3 = \{\mathbf{q} = a + bi + cj + dk : a, b, c, d \in \mathbf{R} \text{ and } a^2 + b^2 + c^2 + d^2 = 1\}.$$

    This group $S^3$ in $\mathbf{H}$ is called the spin group, denoted by **spin(3)**. Identifying the set
    $\{bi + cj + dk\}$ with $\mathbf{R}^3$ and taking $\mathbf{q} \in S^3$, consider an orthogonal transformation
    of $\mathbf{R}^3$ having determinant 1, which is an element of $SO(3, \mathbf{R})$ defined by the
    mapping $\mathbf{v} \mapsto \mathbf{v} \cdot \mathbf{q} \cdot \mathbf{q}^{-1}$. Show that

    (i)  The quotient map

$$p : S^3 \to SO(3, \mathbf{R})$$

      is an epimprphism of groups having $ker \ p = \{-1, +1\}$.

    (ii)  $p : S^3 \to SO(3, \mathbf{R})$ is a 2-sheeted covering.

18.  Let $G$ be a Lie group with identity element $e$ and $G^e$ denote the connected
    component of $G$, which contains $e$. Show that

    (i)  $G^e$ is a normal subgroup of $G$ and it is itself a Lie group.

    (ii)  The quotient group $G/G^e$ is discrete.

    [Hint: $G^e$ is closed under both the operations of multiplication and inversion
    inherited from the Lie group $G$. Moreover, $G^e$ is sent to $G^e$ by conjugation,
    which is continuous is used to prove the normality of $G^e$.]

19.  Let $\xi = (X, p, B, \mathbf{F}^n)$ be an $n$-dimensional vector bundle over the base space $B$.
    Prove the following statements:

    (i)  If $\theta = (Y, q, B, \mathbf{F}^n)$ is also an $n$-dimensional vector bundle over the same
      base space $B$ of $\xi$, then a B-morphism $f : \xi \to \theta$ is a vector bundle isomor-
      phism iff the map

$$f : p^{-1}(b) \to q^{-1}(b)$$

      is a linear isomorphism for every $b \in B$.

    (ii)  If $f : D \to B$ is a continuous map, then its induced bundle $f^*(\xi) = (Y, q, B, \mathbf{F}^n)$ is a vector bundle over $D$ such that

$$(f_\xi, f) : f^*(\xi) \to \xi$$

is a morphism of vector bundles, with the map $f_\xi$ defined by

$$f_\xi : Y \to X, (d, x) \mapsto x.$$

(iii) If the base space $B$ of $\xi$ is such that $B = B_1 \cup B_2$ with $B_1 = A \times [a, c]$
and $B_2 = A \times [c, b]$, where $a, b, c \in \mathbf{R} : a < c < b$ and the restricted bun-
dles $\xi|B_1 = (X_1, p_1, B_1)$ and $\xi|B_2 = (X_2, p_2, B_2)$ are trivial bundles, then the
bundle $\xi$ is also trivial.

20. Prove the following statements

   (i) If $\tilde{\eta} = \{(U_k, g_{ik}) : i, k \in K\}$ is a set of transition functions for a given topological space $B$ and a topological group $G$, there exists a principal $G$-bundle
$\eta = (X, p, B, G)$ and an atlas $\{(U_k, \psi_k) : k \in K\}$ for $\eta$ such that $\tilde{\eta}$ is the set
of transition functions for the atlas.
   (ii) Let $\eta = (X, p, B, G)$ and $\eta' = (X', p', B', G)$ be two principal $G$-bundles
with atlases $\{(U_k, \psi_k) : k \in K\}$ and $\{(U'_a, \psi'_a) : a \in \mathbf{A}\}$ and the corresponding sets of transition functions of $\tilde{\eta}$ and $\tilde{\eta}'$. If $f : \tilde{\eta} \to \tilde{\eta}'$ is a morphism of
sets of transition function, then there is a morphism $\psi : \eta \to \eta'$ of principal
$G$-bundles inducing $f$.

21. Let $G$ be a topological group (with identity e) acting on a topological space $X$
(from left). An open set $U$ in $X$, is said to be proper if

$$gU \cap U = \emptyset, \ \forall g \in G - \{e\}.$$

If every point $x \in X$ has a proper open nbd, then it is said that $G$ **acts properly**
on $X$. Let $\xi : (X, p, B)$ be a covering space. Then $\xi$ is said to be a **regular
covering** if the subgroup $p_*\pi_1(B, b_0)$ of $\pi_1(B, b_0)$ is normal. Let $X$ be a locally
path-connected space and $G$ act properly on $X$. For the natural projection

$$p : X \to mod \ G, x \mapsto xG,$$

prove the following statements:

   (i) $(X, p, Xmod \ G)$ is a regular covering space.
   (ii) If $X$ is semi locallyy and 1-connected, then the two groups $Cov(X/Xmod \ G)$
and $G$ are isomorphic, i.e., $Cov(X/Xmod \ G) \cong G$.
   (iii) If $X$ is a simply connected space, then the fundamental $\pi_1(Xmod \ G, *)$ is
isomorphic to the group $G$.

22. Let $X$ and $B$ be path-connected spaces. Prove the following statements:

(i) If $(X, p)$ is regular covering space of $B$, then $p$ induces the same group $p_* \pi_1(X, x_0) = p_* \pi_1(X, x_1)$ for every pair of points $x_0, x_1$ lying in the same fiber.

(ii) For every simply connected space $X$, any covering space $(X, p)$ of $B$ is regular.

(iii) For any covering space $(X, p)$ of $B$, if the fundamental group $\pi_1(B, b_0)$ is abelian, then the covering space $(X, p)$ of $B$ is regular.

(iv) A covering space $(X, p)$ of a connected and locally path-connected space $B$ with a base point $b_0 \in B$ is regular if and only if the group $Aut(X/B)$ acts transitively on the fiber $p^{-1}(b_0)$ over the point $b_0$.

23. Let $(X, p)$ be a covering space of $B$ and $b_0 \in B$. Prove the following statements

(i) If $X$ is locally path connected, then given any two points $x_0, x_1 \in Y = p^{-1}(b_0)$, then there exists an $f \in Cov(X/B)$ with $f(x_0) = x_1$ iff there exists an $h \in Aut(Y)$ with $h(x_0) = x_1$.

(ii) If $B$ is locally path connected and the fiber $p^{-1}(b_0) = Y$ is considered as a $G = \pi_1(B, b_0)$-set, then the map

$$\phi : Cov(X/B) \to Aut(Y), h \mapsto h|Y$$

is an isomorphism.

(iii) Let a group $G$ act transitively on the fiber $p^{-1}(b_0) = Y$ and let $y_0 \in Y$. For the normalizer $N_G(G_0)$ of the isotropy group $G_0$ of $y_0$,

$$Aut(Y) \cong N_G(G_0)/G_0.$$

(iv) If $B$ is locally path connected and $x_0 \in p^{-1}(b_0)$, then

$$Aut(X/B) \cong N_G(p_* \pi(X, x_0))/p_* \pi_1(X, x_0)$$

and in particular, the fundamental group

$$\pi_1(S^1, 1) \cong \mathbf{Z}.$$

(v) If the base space $B$ is an H-space, then every covering space of $B$ is regular.
[ Hint : By hypothesis, $B$ is an H-space. It implies that $\pi_1(B, b_0)$ is abelian for $b_0 \in B$ and hence every covering space of $B$ is regular. ]

24. If $(X, p)$ and $(Y, q)$ are covering spaces of the same path-connected and locally path-connected base space $B$ and if $b_0 \in B, x_0 \in X$ and $y_0 \in Y$ be points such that

$$p(x_0) = b_0 = q(y_0) \text{ and } q_* \pi_1(Y, y_0) \subset p_* \pi_1(X, x_0),$$

then show that

(i) There is a unique continuous map $f : (Y, y_0) \to (X, x_0)$ with $p \circ f = q$.
(ii) $X$ is a quotient space of $Y$ in the sense that $(Y, f)$ forms a covering space of $X$.

25. Given three path connected and locally path-connected spaces $X$, $Y$ and $B$ and a covering space $(X, p)$ of $B$ with $x_0 \in X$, $y_0 \in Y$ and $b_0 \in B$ and $p(x_0) = b_0$, if $f : (Y, y_0) \to (B, b_0)$ is a continuous map such that $f_* \pi_1(Y, y_0) \subset p_* \pi_1(B, b_0)$, show that there is a continuous map

$$\tilde{f} : (Y, y_0) \to (X, x_0) : p \circ \tilde{f} = f.$$

26. If $\xi = (X, p, B, \mathbf{F}^n)$ and $\eta = (Y, q, B, \mathbf{F}^n)$ are two vector bundles and $f : X \to Y$ is a continuous map such that

$$f | p^{-1}(b) : p^{-1}(b) \to q^{-1}(b)$$

is a linear isomorphism for each $b \in B$, show that $f$ is an isomorphism of vector bundles.

27. Prove the following statements

(i) If $A$ is a closed subgroup of a Lie group $G$, show that every normal subgroup $N \subset A$ determines a fiber bundle with bundle map

$$p : G/N \to G/A, \ gN \mapsto gA.$$

(ii) Let $G = \{+1, -1\}$ be the two-element group and the $n$-sphere $S^n$ be the $G$-space with action given by the relation $x(+1) = x, x(-1) = -x$. Then the principal $\mathbf{Z}_2$-space produces a principal $\mathbf{Z}_2$-bundle with the real $n$-dimensional projective space $\mathbf{R}P^n$ as its base space.

(iii) Every covering projection $p : X \to B$ is a principal $G$-bundle, with $G$ the group of covering transformations with the discrete topology.

(iv) Every fiber bundle $p : X \to B$ is an open map.

(v) If $X$ is a $G$-space, then automorphisms of the trivial $G$-bundle $p_2 : X \times B \to B$ are in (1–1)-correspondence with continuous functions $f : B \to G$

## 4.13.2   Multiple Choice Exercises

Identify the correct alternative (s) (there may be more than one) from the following list of exercises:

1. Consider the quadruple $\xi$ formulated below:

(i) The quadruple $\xi = (\mathbf{R}, p, S^1, \mathbf{Z})$ with the map $p : \mathbf{R} \to S^1, t \mapsto e^{2\pi i t}$ is a fiber bundle with fiber $\mathbf{Z}$.

(ii) The quadruple $\xi = (\mathbf{R}^n, p, T^n, F)$, where $T^n$ is the $n$-dimensional torus

$$T^n = S^1 \times S^1 \times \cdots \times S^1,$$

with the map

$$p : \mathbf{R}^n \to T^n, \; (t_1, t_2, \ldots, t_n) \mapsto (e^{2\pi i t_1}, e^{2\pi i t_2}, \ldots, e^{2\pi i t_n})$$

form a fiber bundle with fiber $F$ which is the set of integer lattice points in $\mathbf{R}^n$.

(iii) The quadruple $\xi = (SO(n, \mathbf{R}), p, S^{n-1})$ is a fiber bundle over $S^{n-1}$ with fiber $SO((n-1), \mathbf{R})$, where $SO((n-1), \mathbf{R})$ is a subgroup of $SO(n, \mathbf{R})$, which consists of matrices $A \in SO(n, \mathbf{R})$ that keep the vector $e_n = (0, 0, \ldots, 1)$ fixed and the map

$$p : SO(n, \mathbf{R}) \to S^{n-1}, \; A \mapsto Ae_n$$

is the projection map of $\xi$.

2. Let $M$ be a smooth manifold of dimension $n$ with its tangent bundle $T(M)$ and the projection map $p : T(M) \to M$, $(x, v) \to x$. Given a chart $(\psi, U)$ of $M$ consider the map

$$T_\psi : p^{-1}(U) \to \psi(U) \times \mathbf{R}^n \subset \mathbf{R}^n \times \mathbf{R}^n, \; (x, v) \mapsto (\psi(x), d\psi_x(v))$$

(i) $T(M)$ is a smooth manifold of dimension $2n$.
(ii) each $T_\psi$ is a homeomorphism.
(iii) $T(M)$ is second countable and Hausdorff.

3. The tangent bundl over the $n$-sphere $S^n$ in the Euclidean $(n+1) - space$ $\mathbf{R}^{n+1}$ is the subbundle $\xi = (T(S^n), p, S^n)$ of the product bundle $(S^n \times \mathbf{R}^{n+1}, p, S^n)$, whose total space $T(S^n)$ is defined by

$$T(S^n) = \{(b, x) \in S^n \times \mathbf{R}^{n+1} : \langle b, x \rangle = 0\},$$

and the projection map $p$ is defined by

$$p : T(S^n) \to S^n, (b, x) \mapsto b.$$

(i) An element of the total space $T(S^n)$ is a tangent vector to $S^n$ at the point $b \in S^n$.
(ii) The fiber $p^{-1}(b) \subset T(S^n)$ is a vector space of dimension $n$.
(iii) A cross section of the tangent bundle $\xi$ over $S^n$ is a vector field over $S^n$.

4. The normal bundle over $S^n$ in the Euclidean $(n+1) - space$ $\mathbf{R}^{n+1}$ is the subbundle $\xi = (N(S^n), q, S^n)$ of the product bundle $(S^n \times \mathbf{R}^{n+1}, p, S^n)$ whose total

space $N(S^n)$ is defined by

$$N(S^n) = \{(b, x) \in S^n \times \mathbf{R}^{n+1} : x = tb \text{ for some } t \in \mathbf{R}\}$$

and the projection $q$ is defined by

$$q : N(S^n) \to S^n, q(b, x) = b.$$

(i) An element of $N(S^n)$ is a normal vector to $S^n$ at the point $b \in S^n$.
(ii) The fiber $q^{-1}(b) \subset T(S^n)$ over the point $b \in S^n$ is a vector space of dimension 1.
(iii) A cross section of the normal bundle $\xi$ over $S^n$ is a normal vector field on $S^n$.

5. Consider the Hopf fibering $\mathcal{H} = (S^3, p, S^2)$.

(i) The fibering $\mathcal{H}$ has fibers a family of great circles.
(ii) The 3-sphere $S^3$ is decomposed into a family of great circles with the 2-sphere $S^2$ as a quotient space of the Hopf fibering $\mathcal{H}$.
(iii) The Hopf map $p : S^3 \to S^2$ for the Hopf fibering $\mathcal{H} = (S^3, p, S^2)$ has its generalization through some spaces that arise in projective geometry.

# References

Adhikari A, Adhikari MR. Basic topology, vol. 1: metric spaces and general topology. India: Springer; 2022a.

Adhikari A, Adhikari MR, Basic topology, vol. 2: topological groups, topology of manifolds and lie groups. India: Springer; 2022b.

Adhikari MR, Adhikari A. Groups. Rings and Modules with Applications. Universities Press, Hyderabad 2003

Adhikari MR, Adhikari A. Textbook of Linear Algebra: An Introduction to Modern Algebra. Allied Publishers, New Delhi 2006

Adhikari MR, Adhikari A. Basic modern algebra with applications. New Delhi, New York, Heidelberg: Springer; 2014.

Armstrong A. Basic Topology. New York: Springer-Verlag; 1983.

Bredon GE. Topology and Geometry. New York: Springer; 1993.

Hu ST. Homotopy Theory. New York: Academic Press; 1959.

Mayer J. Algebraic Topology. New Jersy: Prentice-Hall; 1972.

Massey WS. A Basic course in algebraic topology. New York, Berlin, Heidelberg: Springer; 1991.

Maunder CRF. Algebraic topology. London: Van Nostrand Reinhhold; 1970.

Munkers JR. Elements of algebraic topology. Addition-Wesley-Publishing Company; 1984.

Rotman JJ. An introduction to algebraic topology. New York: Springer; 1988.

Spanier E. Algebraic topology. New York: McGraw-Hill Book Company; 1966.

Steenrod N. The topology of fibre bundles. Prentice: Prentice University Press; 1951.

# Chapter 5
# Topology of Fiber Bundles: Homotopy Theory of Bundles

This chapter continues the study topology of fiber bundles, which has created general interest as it involves interesting applications of topology to other areas such as algebraic topology, geometry, physics, gauge groups and addresses the homotopy theory of bundles. Covering spaces provide tools to study the fundamental groups. On the other hand, fiber bundles provide similar tools to discuss higher homotopy groups, which are natural generalizations of fundamental groups. The concept of fiber spaces is the most important generalization of the concept of covering spaces. The importance of fiber spaces was greatly realized during 1935–1950 to solve several problems involving homotopy and homology.

The motivation of the study of fiber bundles and vector bundles was born through the distribution of signs of the derivatives of the plane curves at every point. **Historically,** the recognition of the theory of fiber bundles as a discipline of mathematics came through the work of H. Whitney (1907–1989), H. Hopf (1894–1971), E. Stiefel (1909–1978), J. Feldbau (1914–1945) and some others.

For this chapter the books [Adams, 1974], [Adhikari et al. 2022a, 2022b, 2003, 2006, 2014, 2016], [Arkowitz, 2011], [Bredon, 1993], [Dugundji, 1966], [Hopf, 1935], [Hu, 1959], [Mukherjee, 2015], [Steenrod, 1951] and some others are referred in the Bibliography.

## 5.1 Homotopy Properties of Vector Bundles Over Manifolds

This section extends the concept of homotopy for continuous maps to smooth maps and studies the homotopy properties of vector bundles. Intuitively, two smooth maps are said to be smoothly homotopic if one can be deformed to the other through smooth maps. The main result of this section is Theorem 5.1.7 asserting the homotopy invariance of smooth homotopic maps, in the sense that smooth homotopic maps

© The Author(s), under exclusive license to Springer Nature Singapore Pte Ltd. 2022
M. R. Adhikari, *Basic Topology 3*,
https://doi.org/10.1007/978-981-16-6550-9_5

induce the same vector bundle over smooth manifolds upto equivalence. Some results on smooth homotopy are also available in Exercises 5.24.1 of Sect. 5.24.

**Proposition 5.1.1** *Let $M$ and $N$ be two smooth manifolds and $\xi = (X, p, M)$ be a vector bundle over $M$. If $f : N \to M$ is a smooth map, then there exists*

*(i) a unique vector bundle $\eta = (Y, q, N)$ (up to equivalence) and*
*(ii) a bundle morphism*

$$(g, f) : \eta \to \xi$$

*such that for every $b \in N$, the fiber preserving map*

$$g_b = g|_{p^{-1}(b)} : p^{-1}(b) \to p^{-1}(f(b))$$

*is a linear isomorphism.*

***Proof*** Let $M$ and $N$ be two smooth manifolds and $\xi = (X, p, M)$ be a vector bundle over $M$ and $f : N \to M$ be a smooth map. Construct the bundle $\eta = (Y, q, N)$ with the total space $Y$ defined by

$$Y = \{(b, x) \in N \times X : f(b) = p(x).$$

and the projection map $q$ is defined by

$$q : Y \to N, (b, x) \mapsto b.$$

Define

$$g : Y \to X, (b, x) \mapsto x.$$

Consider the commutative rectangular diagram as shown in Fig. 5.1
    For local triviality of $\eta$, take a vector bundle chart for $\xi$

$$\psi : U \times \mathbf{R}^n \to p^{-1}(U).$$

Define

$$\psi_1 : f^{-1}(U) \times \mathbf{R}^n \to q^{-1}(f(U)), (b, v) \mapsto (b, \psi(f(b), v)).$$

Then $\psi_1$ is a vector bundle chart for $\eta$, with its inverse formulated by

$$\psi_1^{-1}(b, x) = (b, p_2 \circ \psi^{-1}(x)),$$

where $b \in f^{-1}(U)$, $x \in p^{-1}(U)$ so that $f(x) = p(x)$ and $p_2$ is the projection map onto the second factor. This proves that $g$ is a diffeomorphism and also it an isomorphism on every fiber.                                                                                $\square$

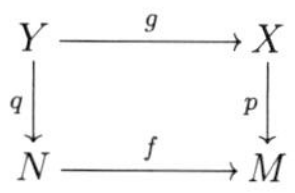

**Fig. 5.1**  Morphism of vector
bundles over manifolds

$$
\begin{array}{ccc}
Y & \xrightarrow{\;g\;} & X \\
{\scriptstyle q}\downarrow & & \downarrow{\scriptstyle p} \\
N & \xrightarrow{\;f\;} & M
\end{array}
$$

**Definition 5.1.2**  The bundle $\eta$ formulated in Proposition 5.1.1 is called the **induced bundle**. It is induced from the bundle $\xi$ by $f$, and it is denoted by $f^*\xi$ or by $f^*(\xi)$. The induced bundle $f^*\xi$ is also called the **pull-back** of $\xi$ by $f$. The pair of maps $(g, f)$ formulated in the same definition is called the **canonical bundle map** of the induced map $f^*\xi$.

**Definition 5.1.3**  Let $\xi = (X, p, M)$ be a vector bundle over a manifold $M$. Then $\eta = (Y, q, M)$ is said to be a **subbundle** of $\xi$ if $Y$ is a submanifold of $X$ and $q = p|_Y$. On the other hand, if $N$ a submanifold of $M$, the **restricted bundle** of $\xi$ to $N$, denoted by $\xi|_N$, is bundle $(Y, q, N)$, where $Y = p^{-1}(N)$ and $q = p|_Y$.

**Proposition 5.1.4**  *Let $\xi = (X, p, M)$ be a vector bundle over a manifold $M$.*

(i) *If $M_1$ be a submanifold of $M$ and $f : M_1 \hookrightarrow M$ is the inclusion map, then the restricted vector bundle $\xi|_{M_1}$ and the induced vector bundle $f^*\xi$ are isomorphic.*
(ii) *If $f_1 : M_1 \to M$ and $f_2 : M_2 \to M_1$ are smooth maps, then the induced bundles $(f_1 \circ f_2)^*(\xi)$ and $f_2^*(f_1^*(\xi))$ are isomorphic.*
(iii) *If $1_M : M \to M$ is the identity map on $M$, then its induced bundle $(1_M)^*(\xi)$ is isomorphic to $\xi$.*

**Proof**  It follows from the definition of induced vector bundle.  □

**Proposition 5.1.5**  *Let $\xi_1 = (X_1, p_1, M_1)$ and $\xi_2 = (X_2, p_2, M_2)$ be two vector bundles. Then every vector bundle morphism*

$$
(g, f) : \xi_1 \to \xi_2
$$

*can be expressed as a product*

$$
(g, f) = (h, f) \circ (k, 1_{M_1}) : g = h \circ k,
$$

*where $(k, 1_{M_1})$ is a homomorphism, and $(h, f)$ is the canonical bundle map of the induced bundle $f^*(\xi_2)$.*

**Proof**  Let $f^*(\xi_2) = (X, p, M_1)$. Consider the maps

$$
h : X \to X_2, (b, x) \mapsto x \quad \text{and} \quad k : X_1 \to X : x \mapsto (p_1(x), g(x)).
$$

Then $Im\ K \subset X$, because, $f \circ p_1 = p_2 \circ g$. This implies that $k$ is linear on every fiber and $g = h \circ k$.  □

**Remark 5.1.6**  Theorem 5.1.7 proves that smooth homotopic maps induce the same vector bundle up to equivalence. Its immediate consequence proves Corollary 5.1.8 saying that every vector bundle over a contractible manifold is trivial.

**Theorem 5.1.7**  (Homotopy invariance) *Let* $f, g : M \to N$ *be two smooth homotopic maps and* $\xi$ *be a vector bundle over* $M$. *Then their induced vector bundles over the manifold* $M$ *are isomorphic, in notation,*

$$f^*(\xi) \cong g^*(\xi).$$

**Proof**  By hypothesis, $f, g : M \to N$ be two smooth homotopic maps. Let $H : M \times \mathbf{R} \to N$ be a homotopy between $f$ and $g$ and $p : M \times \mathbf{R} \to M$ be the projection map. Then there is an isomorphism between vector bundles $H^*(\xi)$ and $p^* H_t^*(\xi)$ over the closed subinterval $M \times \{t\}$, because, $F = H_t \circ p$ on $M \times \{t\}$, which is closed in the product topology. Hence, by Exercise 5.24.1 of Sect. 5.24, there exists an isomorphism

$$\psi : H^*(\xi) \to p^* H_t^*(\xi)$$

over some vertical strip $N \times (t - \delta, t + \delta)$. This implies that the isomorphism class of $H_t^*(\xi)$ is a locally constant function of $t$ and it is constant, because the real line space $\mathbf{R}$ is connected. This proves that $f^*(\xi) \cong g^*(\xi)$.  □

Corollary 5.1.8 is an immediate consequence of Theorem 5.1.7.

**Corollary 5.1.8**  *Every vector bundle over a contractible manifold is trivial.*

**Proof**  Let $\xi = (X, p, M)$ be a vector bundle over a contractible manifold $M$ and $b_0 \in M$. Then the identity map $1_M : M \to M$ is homotopic to the constant map $c : M \to b_0$. Hence it follows by Theorem 5.1.7 that $f^*(\xi) \cong g^*(\xi)$. But the vector bundle $g^*(\xi)$ is trivial, because, it is a vector bundle over a point. This proves that the vector bundle $\xi = (X, p, M)$ over a contractible manifold $M$ is trivial.  □

## 5.2   Fibrations and Cofibrations

From topological viewpoint, the fibers over different points of the base space of a fiber bundle are homeomorphic. So, it is natural to study these fibers from viewpoint of homotopy theory. For this study, we investigate a structure in which they are precisely homotopy equivalent. This investigation generalizes the concept of fiber bundle to the concept of a fibration.

This section is devoted to the study of the concepts of fibrations and cofibrations and establishes a close connection between a fibration and a covering projection. They are dual concepts and well connected with the concepts of the homotopy lifting property (HLP) and the homotopy extension property (HEP). Duality principle in

**Fig. 5.2** Lifting $\tilde{f}$ of $f$

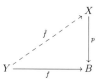

homotopy theory was born through these concepts. They play a key role in the study of
homotopy theory by providing strong mathematical tools to invade many problems.
It is interesting that any continuous map can be expressed as a fibration and also as a
cofibration up to homotopy. **Historically,** the concept of fibration born in geometry
and topology was implicitly found in 1937 in the work of K. Borsuk (1905–1982)
but explicitly first appeared in the work of J. H. C Whiteney (1904–1960) during the
period 1935–1940, while investigating sphere bundles.

## 5.2.1  Homotopy Lifting Problems

This subsection starts with recalling the homotopy lifting problem of a continuous
map. What is this problem? Given a continuous map $p : X \to B$, a topological space
$Y$ and a continuous map $f : Y \to B$, the lifting problem for $f$ is to examine whether
there is a continuous map $\tilde{f} : Y \to X$ (represented by an dotted arrow in the Fig. 5.2)
such that this diagram is commutative in the sense that $f = p \circ \tilde{f}$. The concept of
homotopy lifting property (HLP) and that of the homotopy extension property (HEP)
provide key tools in the study of homotopy theory. The first one leads to the concept
of fibration, the second one leads to the concept of cofibration, and they are dual
concepts.

The concept of homotopy lifting property (HLP) is very important in homotopy
theory, and it leads to the concept of fibration. On the other hand, the homotopy
extension property (HEP), the dual concept of the homotopy lifting property (HLP),
leads to the concept of cofibration, a dual concept of fibration.

**Definition 5.2.1** Let $p : X \to B$ and $f : Y \to X$ be two continuous maps and $H :
Y \times \mathbf{I} \to B$ be also continuous map such that

$$H(y, 0) = pf(y), \ \forall \, y \in Y.$$

Then the map $p : X \to B$ is said to have the **homotopy lifting property (HLP)** with
respect to $Y$, if there exists continuous map

$$\tilde{H} : Y \times \mathbf{I} \to X : \tilde{H}(y, 0) = f(y), \ \forall \, y \in Y \text{ and } H = p \circ \tilde{H}.$$

**Remark 5.2.2** Definition 5.2.1 asserts that the existence of $\tilde{H}$ is equivalent to
the existence of a continuous map $\tilde{H} : Y \times \mathbf{I} \to X$ indicated by the dotted arrow
in the diagram as shown in Fig. 5.3 such that this diagram is commutative. If

**Fig. 5.3**  Homotopy lifting
$\tilde{H}$ of $p$

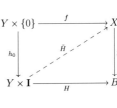

**Fig. 5.4**  Homotopy lifting
problem for $p$

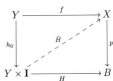

**Fig. 5.5**  Homotopy lifting
problem

$h_0(y) = (y, 0) \, \forall \, y \in Y$, then a homotopy lifting problem can be represented by the
commutative diagram in Fig. 5.4.

The two continuous maps $f : Y \to X$ and $H : Y \times \mathbf{I} \to B$ in the above diagram
are said to form the data for the given lifting problem. $\tilde{H}$ lifts the homotopy of $H$
of $p \circ f$ to a homotopy of $f$, because, the map $H$ is a homotopy of the composite
$p \circ f$ and a solution to the problem is to determine a homotopy

$$\tilde{H} : Y \times \mathbf{I} \to X \text{ of } f : p \circ \tilde{H} = H.$$

**Remark 5.2.3**  Proposition 5.2.4 proves that the lifting problem for continuous maps
$f : Y \to B$ to $X$ is a problem of homotopy category. Because, given two continuous
maps $p : X \to B$ and $f : Y \to B$ the map $f$ can or cannot be lifted to $X$ depends
on the homotopy class of $f$. This asserts that the lifting problem for continuous maps
$f : Y \to B$ to $X$ is a **problem of homotopy category**.

**Proposition 5.2.4**  *If a continuous map* $p : X \to B$ *has the HLP with respect to a
space $Y$, then one of the homotopic maps $f \simeq g : Y \to B$ can be lifted to $X$ if the
other one can also be lifted to $X$.*

**Proof**  By hypothesis the continuous map $p : X \to B$ has the HLP with respect to a
space $Y$. If $f$ is defined by

$$f : Y \to B, \ y \mapsto H(y, 0), \text{ then } \tilde{f} : Y \to X$$

is a lifting of $f$; if $g$ is defined by

**Fig. 5.6** Homotopy lifting
problem

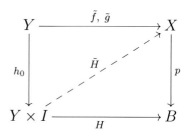

$$g : Y \to B, \; y \mapsto H(y, 1),$$

then $H : f \simeq g : Y \to B$. and $\tilde{H} : \tilde{f} \simeq \tilde{g} : Y \to X$, where $\tilde{g} : Y \to X : y \mapsto \tilde{H}(t, 1)$
is a lifting of $g$ to $X$. Consequently, it follows that if $H : f \simeq g : Y \to B$ and if $f$
has a lifting $\tilde{f}$, then the homotopy $H$ can be lifted to $\tilde{H} : Y \times \mathbf{I} \to X : \tilde{f} \simeq \tilde{g}$. This
implies if $f \simeq g : Y \to B$ and if the map $f$ has a lifting $\tilde{f} : Y \to X$, the homotopy
$H$ can be lifted to $\tilde{H}$ as shown in diagram in Fig. 5.6 and hence $g$ has a lifting
$\tilde{g} : Y \to X$ with $\tilde{H} : \tilde{f} \simeq \tilde{g}$. By symmetry, it follows that if $f \simeq g : Y \to B$ and if
the map $g$ has a lifting $\tilde{g}$, then $f$ has also a lifting $\tilde{f}$ such that $\tilde{f} \simeq \tilde{g}$.               $\square$

### 5.2.2 Fibration: Introductory Concepts

This subsection conveys the concept of a fibration which was explicitly first found
in the work of Whitney during 1935–1940 on sphere bundles. This concept is a
natural generalization of fiber bundles in the sense that in a fiber bundle, the fibers
over different points of its base space are homeomorphic; on the other hand in a
fibration, the fibers over different points of its base space are homotopy equivalent.
Fibrations (or Hurewicz fiber spaces) [Hurewicz, 1955] provide powerful tools to
study homotopy theory. For example, a covering map is a fibration, and a continuous
map $p : X \to B$ has the HLP with respect to a space $Y$ if every problem represented
by the commutative diagram in Fig. 5.5 has a solution (Fig. 5.6).

**Definition 5.2.5** Let $p : X \to B$ be a base point preserving continuous onto map.
Then it is said to be a fibration (or fiber map or Hurewicz fiber space) if $p$ has the
HLP with respect to each space $Y$ in the sense of Definition 5.2.1.

(i)   The space $X$ is called the **total space**, and the space $B$ is called the **base space
      of the fibration.**
(ii)  For every point $b \in B$, the space $p^{-1}(b) = F \subset X$ is called the fiber over $b$.
(iii) A Serre fibration is a continuous surjective map $p : X \to B$ such that it has
      the HLP with respect to disk $\mathbf{D}^n \subset \mathbf{R}^n$ for every $n \geq 1$. It is also called a weak
      fibration.

**Fig. 5.7** Diagram involving classifying space $K$ and classifying map $g$ for $p$

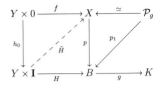

**Notation**: The fibration $p : X \to B$ with the fiber space $F$ over some point of $B$ and the inclusion $i : F \hookrightarrow X$ is denoted by

$$F \hookrightarrow X \xrightarrow{p} B.$$

**Example 5.2.6** The trivial projection

$$p : B \times F \to B, (b, f) \mapsto b$$

is an important example of fibration.

**Example 5.2.7** Every covering projection is a fibration by Theorem 5.2.15.

**Definition 5.2.8** Let $g : B \to K$ be a continuous map and $K^{\mathbf{I}}$ be the space of all continuous maps $\alpha : \mathbf{I} \to K$ topolozied by the compact open topology. The **path space** $\mathcal{P}_g$ of $g$ is defined by

$$\mathcal{P}_g = \{(b, \alpha) : \alpha(0) = k_0 \in K \text{ and } \alpha(1) = g(b)\} \subset B \times K^{\mathbf{I}}.$$

Definition 5.2.9 uses the notation of the diagram in Fig. 5.3.

**Definition 5.2.9** (*Principal fibration*) Let $p : X \to B$ be a fibration. It is said to be a principal fibration if there exist a topological space $K$ and a continuous map $g : B \to K$ such that the diagram in Fig. 5.7 is commutative with $p_1 : \mathcal{P}_g \to B, (b, \alpha) \mapsto b$ a projection. The space $K$ is called the **classifying space**, and $g$ is called the **classifying map** for the principal fibration $p : X \to B$ with $X \simeq \mathcal{P}_g$.

Theorem 5.2.10 solves a lifting problem associated with a principal fibration with the help of a classifying map.

**Theorem 5.2.10** *Let $p : X \to B$ be a principal fibration and $f : Y \to B$ be a continuous map. Then $f$ has a lifting $\tilde{f} : Y \to X$ iff $g \circ f$ is homotopic to a constant map, where $g : B \to K$ is the classifying map.*

**Proof** By hypothesis $p : X \to B$ is a principal fibration with a classifying map $g : B \to K$ and $f : Y \to B$ is a continuous map. Then $X \simeq \mathcal{P}_g$ with projection $p_1 : \mathcal{P}_g \to B, (b, \alpha) \mapsto b$. This implies that there exists two continuous maps $u : X \to \mathcal{P}_g$ and $v : \mathcal{P}_g \to X$ such that $v \circ u \simeq 1_X$ and $u \circ v \simeq 1_{\mathcal{P}_g}$. Again, $p_1 \circ u = p$ and $p \circ v = p_1$. First suppose that $f : Y \to B$ has a lifting $\tilde{f} : Y \to X$. Then by the given

**Fig. 5.8** Existence of lifting of $f$ for the principal fibration $p$

**Fig. 5.9** Diagram extending the classifying map $g$

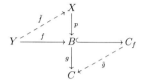

conditions and chasing the diagram as shown in Fig. 5.8 , there exists a homotopy $H : g \circ f \simeq c$, where $c : Y \to K$, $y \mapsto k_0$ is the constant map. Conversely, let $H : g \circ f \simeq c$. Then

$$H : Y \times \mathbf{I} \to K : H(y, 0) = (g \circ f)(y) = g(f(y)) \quad \text{and} \quad H(y, 1) = c(y) = k_0, \ \forall\, y \in Y.$$

Define two maps

$$H_y : \mathbf{I} \to K, t \mapsto H(y, t)$$

and

$$\tilde{f} : Y \to \mathcal{P}_g \to X, y \mapsto (f(y), H_y) \mapsto v(f(y), H_y).$$

This implies that $f$ has a lifting $\tilde{f} : Y \to X$, because,

$$(p \circ \tilde{f})(y) = (p \circ v)(f(y), H_y) = p_1(f(y), H_y) = f(y), \forall\, y \in Y$$

and hence $p \circ \tilde{f} = f$. ☐

Theorem 5.2.11 solves a lifting problem of a continuous map $f$ associated with a principal fibration with the help of classifying map from the mapping cone of $f$.

**Theorem 5.2.11** *Let $p : X \to B$ be a a principal fibration and $f : Y \to B$ be a continuous map. Then for this fibration, $f$ has a lifting $\tilde{f}$ iff there exists a continuous map $\tilde{g} : C_f \to K$ extending the classifying map $g$ in the diagram in Fig. 5.9, where $C_f$ is the mapping cone of $f$.*

**Proof** Consider the mapping cone $C_f$ of $f : Y \to B$ constructed from the mapping cylinder $M_f$ by identifying $Y \times \{0\} \cup \{*\} \times I$ with $*$ in $B$ in the category $\mathcal{T}op_*$ of pointed topological spaces. First suppose that there is a homotopy

$$H : c \simeq g \circ f : Y \to K,$$

where $c : Y \to k_0 \in K$ is a constant map. Construct the map

$$\tilde{g} : C_f \to K, \begin{cases} (y, t) \mapsto H(y, t), \\ b \mapsto g(b) \text{ for } b \notin f(Y). \end{cases}$$

Then

$$\tilde{g}(y, 0) = y_0 \text{ and } \tilde{g}(y, 1) = gf(y) = g \circ f, \; \forall \, y \in Y$$

proves that $\tilde{g}$ is the required extension of $g$. Next suppose that $\tilde{g}$ is an extension of $g$. Then there exists a homotopy

$$H : Y \times \mathbf{I} \to K, (y, t) \mapsto \tilde{g}(y, t) : H(y, 0) = \tilde{g}(y, 0) = k_0 \text{ and } H(y, 1) = \tilde{g}(y, 1) = (g \circ f)(y) \, \forall \, y \in Y.$$

This proves that $g \circ f \simeq c$.                                    □

**Proposition 5.2.12** *If $p : X \to B$ is a fibration and $\beta : \mathbf{I} \to B$ is any path in $B$ such that $\beta(0) \in p(X)$, then the path $\beta$ can be lifted to a path $\tilde{\beta} : \mathbf{I} \to X$ in $X$.*

**Proof** Under the given condition, the path $\beta$ can be regarded as a homotopy $\beta : \{y_0\} \times \mathbf{I} \to B$, where $\{y_0\}$ is a one-point space. If $x_0$ is a point in $X$ such that $p(x_0) = \beta(0)$, then there is a map $f : \{y_0\} \to X$ such that $pf(y_0) = \beta(y_0, 0)$. This implies by the HLP of $p$ that there exists a path

$$\tilde{\beta} : \mathbf{I} \to X : \tilde{\beta}(0) = x_0 \text{ and } p \circ \tilde{\beta} = \beta.$$

                                    □

**Example 5.2.13** Given a topological space $X$, let $\mathcal{P}(X)$ be the space of all paths $\beta : \mathbf{I} \to X$ in $X$ topolozied by the compact open topology. Consider the map

$$p : \mathcal{P}(X) \to X \times X, \beta \mapsto (\beta(0), \beta(1))$$

and two other maps $p_0, \; p_1$

$$p_0, \; p_1 : \mathcal{P}(X) \to X, \beta \mapsto \beta(0), \; \beta(1).$$

Then all these three maps $p$, $p_0$ and $p_1$ are fibrations.

**Example 5.2.14** Given two fibrations $p : X \to Y$ and $q : Y \to B$, then their composite $q \circ p : X \to B$ is also a fibration.

**Theorem 5.2.15** *Let $p : X \to B$ be an arbitrary covering projection. Then it is a fibration.*

**Fig. 5.10** Homotopy lifting
problem

**Fig. 5.11** Commutative
triangle representing a
cofibration $f$

**Proof** By hypothesis, $p : X \to B$ is a covering projection. Suppose the diagram
in Fig. 5.10 represents the corresponding homotopy lifting problem. Then for every
point $y \in Y$, there exists a unique path $\beta_y : \mathbf{I} \to X$ in $X$ such that $\beta_y(0) = f(y)$ and
$p\beta_y(t) = H(y, t)$. This defines a continuous map

$$\tilde{H} : Y \times \mathbf{I} \to X, (y, t) \mapsto \beta_y(t).$$

Hence it follows that $p$ is a fibration. □

### 5.2.3 Cofibration: Introductory Concepts

This subsection communicates the concept of cofibration, a dual concept of a fibration. This subsection works in the category $\mathcal{T}op_*$ of pointed topological spaces and
base point preserving continuous maps. The duality principle between fibration and
cofibration is based on the idea that while defining a fibration as a map satisfying
homotopy lifting property (HLP), if the directions of all arrows are reversed, a dual
concept is obtained, which is called a cofibration. This implies that a continuous map
$f : X \to Y$ is a cofibration if it satisfies the condition: given a map $\tilde{g} : Y \to Z$ in
$\mathcal{T}op_*$ and a homotopy $\tilde{F}_t : Y \to Z$ such that there is a continuous map $F_t : X \to Z$
making the triangle in Fig. 5.11 commutative, in the sense that $\tilde{F}_t \circ f = F_t$.

**Definition 5.2.16** Let $f : X \to Y$ be a continuous map in the category $\mathcal{T}op_*$. Then
it is said to be a **cofibration** if it satisfies the following conditions: given

(i)  any topological space $Z \in \mathcal{T}op_*$,
(ii)  a continuous map $g : Y \to Z$ and
(iii)  a homotopy $H : X \times \mathbf{I} \to Z$ starting from $g \circ f$,

there exists a homotopy $K : Y \times \mathbf{I} \to Z$, starting from $g$ such that $H = K \circ (f \times
1_d)$ making the three triangles shown in Fig. 5.12 commutative where $j_0 : Y \to Y \times
\mathbf{I}, y \mapsto (y, 0)$ and $j'_0 : X \to X \times \mathbf{I}, x \mapsto (x, 0)$.

**Remark 5.2.17** For any subspace $A \subset X \in \mathcal{T}op_*$

**Fig. 5.12** Diagram
representing cofibration of $f$

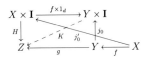

(i)  the inclusion $i : A \hookrightarrow X$ is a cofibration if the pair $(X, A)$ of spaces has the absolute homotopy extension property with respect to any space $Y$ in the sense that if given continuous maps $h : X \to Y$ and $H : A \times \mathbf{I} \to Y$ such that $h(x) = H(x, 0)$, $\forall\ x \in A$, there is a continuous map $G : X \times \mathbf{I} \to Y$ with the property

$$G(x, 0) = h(x), \ \forall\ x \in X \ and \ G|_{A \times \mathbf{I}} = H.$$

(ii)  On the other hand, its converse is not necessarily true. Because Definition 5.2.16 of a cofibration works in the category $\mathcal{T}op_*$ but the absolute homotopy extension property deals with maps and homotopies that are not necessarily in the category $\mathcal{T}op_*$.

**Theorem 5.2.18**  *Let $f : X \to Y$ be an arbitrary continuous map in the category $\mathcal{T}op_*$. Then it is the composite of a cofibration and a homotopy equivalence.*

**Proof**  By hypothesis, $f : X \to Y$ is an arbitrary continuous map in the category $\mathcal{T}op_*$. Consider the mapping cylinder $M_f$ in the category $\mathcal{T}op_*$ constructed from $Y$ and $(X \times \mathbf{I})/(x_0 \times \mathbf{I})$ by identifying the points $(x, 1)$ and $f(x)$ for all $x \in X$. This defines an inclusion map

$$g : X \hookrightarrow M_f : x \mapsto [(x, 0)].$$

Define another map $h : M_f \to Y$ by sending every element $y \in Y$ to $y$ itself and every other elements $[(x, t)]$ to $f(x)$. Then $f = h \circ g$. To prove the theorem it is sufficient to show that $g$ is a cofibration and $h$ is a homotopy equivalence. To show that $g$ is a cofibration, take any continuous map $u : M_f \to Z$ in $\mathcal{T}op_*$ and a homotopy

$$H : X \times \mathbf{I} \to Z : H(x, 0) = (u \circ g)(x), \ \forall x \in X,$$

which means that the homotopy starts from $u \circ g$.

Define two continuous maps $F_X$ and $F_Y$

$$F_X : (X \times \mathbf{I}) \times \mathbf{I} \to Z, \ (x, t, s) \mapsto \begin{cases} u(x, (2t - s)/(2 - s)), & 0 \le s \le 2t \\ H(x, s - 2t), & 2t \le s \le 1 \end{cases}$$

and

$$F_Y : Y \times \mathbf{I} \to Z, \ (y, s) \mapsto u(y), \ \forall s \in \mathbf{I}.$$

Then $F_X(x, 1, s) = u(x, 1) = (u \circ f)(x) = F_Y(f(x), s)$, and $F_X$, $F_Y$ give rise together a homotopy

$$F : M_f \times \mathbf{I} \to Z$$

such that it starts from $u$ and satisfies the relation

$$F \circ (g \times 1_d)(x, s) = F(x, 0, s) = H(x, s).$$

This implies that $G \circ (g \times 1_d) = H$ and hence it proves that $g$ is a cofibration. To show that $h$ is a homotopy equivalence, consider the continuous map $j : Y \to M_f$ defined to be (the restriction) of the identification map onto $M_f$. Then

$$h \circ j = 1_Y \quad \text{and} \quad j \circ h : M_f \to M_f, y \mapsto y \text{ and } (x, t) \mapsto f(x).$$

Define a homotopy

$$G : M_f \times \mathbf{I} \to M_f, \ (y, t, s) \mapsto y \text{ and } (x, t, s) \mapsto (x, t + s(1 - t)).$$

This asserts that $G : 1_{M_f} \simeq j \circ h$. This proves that $h$ is a homotopy equivalence. $\square$

**Remark 5.2.19** Since each continuous map $f : X \to Y$ in $\mathcal{T}op_*$ is also the composite of a homotopy equivalence and a fiber map, it implies that the dual of the Theorem 5.2.18 is also valid in $\mathcal{T}op_*$.

Theorem 5.2.20 and Corollary 5.2.21 both characterize cofibration in terms of retraction.

**Theorem 5.2.20** *Let $X$ be topological space and $K$ be a closed subset of $X$. Then the inclusion*

$$i : K \hookrightarrow X$$

*is a cofibration if the subspace*

$$X \times \{0\} \cup K \times \mathbf{I} \subset X \times \mathbf{I}$$

*is a retract of $X \times \mathbf{I}$.*

**Proof** By hypothesis, $K$ is a closed subset of a topological space $X$ and $i : K \hookrightarrow X$ is the inclusion map.
First suppose that

$$i : K \hookrightarrow X$$

is a cofibration. Then corresponding to a given pair of continuous maps

$$f : X \to X \times \{0\} \cup K \times \mathbf{I}, x \mapsto (x, 0)$$

and

$$H : K \times \mathbf{I} \to X \times \{0\} \cup K \times \mathbf{I}, (k, t) \mapsto (k, t),$$

there exists a retraction

$$r : X \times \mathbf{I} \to X \times \{0\} \cup K \times \mathbf{I}.$$

Since $r$ is a retraction, the first part is proved. Next suppose that there exists a retraction

$$r : X \times \mathbf{I} \to X \times \{0\} \cup K \times \mathbf{I}.$$

Corresponding to a given topological space $Y$, a continuous map $f : X \to Y$, and a homotopy $H : K \times \mathbf{I} \to Y$ such that $H(k, 0) = f(i(k))$, $\forall k \in K$, define a continuous map

$$F : X \times \mathbf{I} \to Y, (x, t) \mapsto \begin{cases} (f \circ p_X \circ r)(x, t), & \text{if } (x, t) \in r^{-1}(X \times \{0\}) \\ (H \circ r)(x, t), & \text{if } (x, t) \in r^{-1}(K \times \mathbf{I}). \end{cases}$$

Then $F$ is continuous, because the projection $p_X$ is continuous, $X \times \{0\}$ and $K \times \mathbf{I}$ are closed sets in the product space $X \times \mathbf{I}$. This proves that $i$ is a cofibration.    □

**Corollary 5.2.21** *Let $X$ be topological space and $K$ be a closed subset of $X$. Then the inclusion*

$$i : K \hookrightarrow X$$

*is a cofibration iff the inclusion*

$$j : X \times \{0\} \cup K \times \mathbf{I} \hookrightarrow X \times \mathbf{I}$$

*is a retraction.*

**Proof** It follows from Theorem 5.2.20.                                          □


## 5.3   Hurewicz Fiberings and Characterization of Fibrations

This section continues the study of fibrations, characterizes path liftings of fibrations with the help of their fibers and proves Hurewicz theorem. This theorem is due to W. Hurewicz (1904–1956). It gives a sufficient condition for a map $p : X \to B$ to be a fibration [Hurewicz, 1955]. The concept of fibering plays a key role in the homotopy theory of bundles specially in the application of homotopy theory to geometric problems and provide rich supply of tools to compute the homotopy groups of different spaces.

Definition 5.3.1 communicates the concept of Hurewicz fibering, named after W. Hurewicz and studies it.

**Definition 5.3.1** A continuous map $p : X \to B$ is said to be a **Hurewicz fibering** if it has the homotopy lifting property (HLP) with respect to any space $Y$.

**Proposition 5.3.2** *Given topological spaces $Y$ and $Z$, and a continuous map $f : Y \to Z$, consider the space $X_f = \{(y, w) \in Y \times Z^{\mathbf{I}} : w(0) = f(x)\} \subset Y \times Z^{\mathbf{I}}$ with subspace topology induced from the product topology on $Y \times Z^{\mathbf{I}}$, where $\mathbf{I} = [0, 1]$ has the subspace topology inherited from the real line space $\mathbf{R}$. It has the following properties.*

*(i) The spaces $X_f$ and $Y$ are homotopy equivalent.*
*(ii) The map*

$$\beta : X_f \to Z, (y, w) \mapsto w(1)$$

*is a Hurewicz fibering having fiber*

$$X_f = \{(y, w) \in Y \times Z^{\mathbf{I}} : w(0) = f(x), w(1) = z_* \in Z\}.$$

***Proof*** Define the three maps

$$\alpha : X_f \to Z, (y, w) \mapsto w(1),$$

$$\beta : X_f \to Y, (y, w) \mapsto y$$

and

$$\gamma : Y \to X_f, y \mapsto (y, c_{f(y)}),$$

where $c_{f(y)}$ denotes the constant path at $f(y)$ in $Z$. Then $\alpha \circ \gamma = f$ and $\beta \circ \gamma = 1_Y$, and the maps make the diagram in Fig. 5.13 commutative.

(i) The continuous map

$$F : X_f \times \mathbf{I} \to X_f, (y, w, t) \mapsto (y, w_t) : w_t(s) = w(st)$$

is a homotopy such that $\gamma \circ \beta \simeq 1_{X_f}$. Moreover, $\beta \circ \gamma = 1_Y$ implies that $\beta \circ \gamma \simeq 1_Y$. Hence it is proved that $Y \simeq X_f$.

(ii) Let $K \subset Y \times Z^{\mathbf{I}}$ and $h : K \times \{0\} \to X_f : h(k, 0) = (h_1(k), h_2(k)) \in X_f, \forall k \in K \subset Y \times Z^{\mathbf{I}}$. Using the commutativity of rectangular diagram in Fig. 5.14, define the map

$$H : K \times \mathbf{I} \to X_f \subset Y \times Z^{\mathbf{I}}, (k, t) \mapsto (H_1(k, t), H_2(k, t)),$$

where $H_1$ and $H_2$ are defined by

$$H_1(k, t) = h_1(k)$$

and

$$H_2(k, t)(s) = \begin{cases} h_2(k)(s(1 + t)), & 0 \le s \le 1/(1 + t) \\ F(k, (1 + t)s - 1), & 1/(1 + t) \le s \le 1. \end{cases}$$

**Fig. 5.13** Triangle for
Hurewicz fibering involving
$\alpha, \beta, \gamma$

**Fig. 5.14** Commutative
diagram for Hurewicz
fibering

Since the equality $(f \circ H_1)(k, t) = H_2(k, t)(0)$, $\forall k \in K$, $\forall t \in \mathbf{I}$ holds, the
map $H$ is continuous, because the maps $H_1$ and $H_2$ are both continuous.
Moreover, $H(k, 0)(s) = h(s)$ and $(\alpha \circ H)(k, t) = F(k, t)$. It asserts that $X_f$ is
its fiber.                                                                                      $\square$

**Theorem 5.3.3** (Hurewicz ) *Given a paracompact space B and a covering map* $p$ :
$X \to B$, *there is an open covering* $\{U_a : a \in \mathbf{A}\}$ *of B such that for every* $U_i \in \{U_a\}$

$$p|_{p^{-1}(U_i)} : p^{-1}(U_i) \to U_i$$

*is a fibration. Then* $p$ *is also a fibration.*

**Proof** The proof is long. See [Dugundji, 1966, p. 400].                                    $\square$

**Corollary 5.3.4** *Given a paracompact space B and a fiber bundle* $(X, p, B, F)$, *the
projection* $p \to B$ *is a fibration.*

**Proof** The corollary follows directly from Hurewicz Theorem 5.3.3.                          $\square$

## 5.4   Homotopy and Path Lifting Properties of Covering Spaces

This section continues the study of covering spaces initiated in Chap. 4 and gives
homotopy classification of covering spaces and proves Monodromy Theorem 5.4.7,
which characterizes equivalence of two lifted paths in a covering space in terms of its
projection map. It also studies **General Lifting Problems** from homotopy viewpoint.
Covering spaces of a pointed topological space $(X, x_0)$ carry a close relation with the
higher homotopy groups $\pi_n(X, x_0)$ for $n \geq 2$. For example, Theorem 5.4.26 asserts
that every covering space induces an isomorphism between the higher homotopy
group of its total space and that of its base space at every dimension . Finally, it is
proved in Theorem 5.4.27 that $\pi_n(S^1) = 0$ for every $n \geq 2$.

## 5.4.1 Introductory Concepts

**Definition 5.4.1** Let $X$ be a topological space.

(i) It is called **locally path connected** if for every point $x$ of $X$ and every nbd $U$ of $x$, there exists an open set $V$ with the property $x \in V \subset U$ such that every pair of points of $V$ can be joined by a path in $U$.

(ii) It is called **semilocally simply connected** if every point $x \in X$ has a nbd $U$ with the property that the homomorphism

$$i_* : \pi_1(U, x) \to \pi_1(X, x)$$

induced by the inclusion

$$i : U \hookrightarrow X$$

is trivial in the sense that every closed path in $U$ at $x$ is homotopic to a constant map in $X$.

Clearly, $X$ is locally path connected iff every path component of each open subset of $X$ is open.

**Definition 5.4.2** (i) Given a covering space $(\tilde{X}, p)$ of $X$ and a path $f : I \to X$, a path $\tilde{f} : I \to \tilde{X}$ is said to be a **lifting or a covering path** of $f$ if $p \circ \tilde{f} = f$, **diagrammatically,** if they make the triangle given in Fig. 5.15 commutative.

(ii) On the other hand, if $H : I \times I \to X$ is a homotopy, then a homotopy $\tilde{H} : I \times I \to \tilde{X}$ such that $p \circ \tilde{H} = H$, is called a **lifting or covering homotopy** of $H$.

**Theorem 5.4.3** (The Path Lifting Property) *Given a covering space $(\tilde{X}, p)$ of $X$ and a path $f : I \to X$ starting at a point $x_0 \in X$ with $\tilde{x}_0 \in p^{-1}(x_0)$, there exists a unique covering path $\tilde{f} : I \to \tilde{X}$ given in Fig. 5.16 of $f$ starting at the point $\tilde{x}_0$ with the property $p \circ \tilde{f} = f$.*

**Fig. 5.15** Lifting of $f$ in $\tilde{X}$

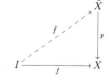

**Fig. 5.16** Path lifting property (PLP)

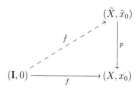

***Proof*** **Existence of** $\tilde{f}$: Let $[a, b] \subset \mathbf{I}$ be a subinterval of $\mathbf{I}$ such that $f([a, b]) \subset U$ for an admissible nbd $U$ of $x = f(a)$ in $X$. Then for any $\tilde{x} \in p^{-1}(x)$, the point $\tilde{x}$ will be an element of a unique sheet $V$, say. This defines a map

$$\tilde{g} : ([a, b], a) \rightarrow (\tilde{X}, \tilde{x}) : \tilde{g} = (p|_V)^{-1} \circ (f|_{[a,b]})$$

such that
$$p \circ \tilde{g} = f|_{[a,b]}.$$

For any $t \in \mathbf{I}$, let $U_t$ be an admissible nbd of $f(t)$. Then the family $\mathcal{C} = \{f^{-1}(U_t), t \in \mathbf{I}\}$ forms an open cover of the compact metric space $\mathbf{I}$, and hence $\mathcal{C}$ has a Lebesgue number $\lambda$. This asserts that for $0 < \delta < \lambda$ and a subset $\mathbf{I}_1$ with diameter less than $\delta$, the subinterval $\mathbf{I}_1 \subset f^{-1}(U_t)$ for some $t \in \mathbf{I}$. Hence $f(\mathbf{I}_1) \subset U_t$ partitions the interval $\mathbf{I}$ by the points $t_1 = 0, t_2, \ldots, t_k = 1$, where $t_{i+1} - t_i < \delta$ for $1 \leq i \leq k - 1$. This shows that there exists a continuous map $\tilde{g}_1 : [0, t_2] \rightarrow \tilde{X}$ such that $p \circ \tilde{g}_1 = f|_{[0,t_2]}$ and $\tilde{g}_1(o) = x_0$. Analogously, there is a continuous map $\tilde{g}_2 : [t_2, t_3] \rightarrow \tilde{X}$ such that $p \circ \tilde{g}_2 = f|_{[t_2,t_3]}$ and $\tilde{g}_2(t_2) = \tilde{g}_1(t_2)$ . Proceeding in this way, for $1 \leq i \leq k - 2$, there is a continuous map

$$\tilde{g}_{i+1} : [t_{i+1}, t_{i+2}] \rightarrow \tilde{X}$$

such that
$$p \circ \tilde{g}_{i+1} = f|_{[t_{i+1},t_{i+2}]} \text{ and } \tilde{g}_{i+1}(t_{i+1}) = \tilde{g}_i(t_{i+1}).$$

Using gluing lemma for continuity of functions, and assembling the functions $g_i$, there exists a continuous function

$$\tilde{f} : \mathbf{I} \rightarrow \tilde{X} : \tilde{f}_i(t) = \tilde{g}_i(t), \ \forall t \in [t_i, t_{i+1}].$$

**The uniqueness of** $\tilde{f}$: Since $\mathbf{I}$ is connected, and by assumption any two lifts of $f$ agree at the point $0 \in \mathbf{I}$, the uniqueness of $\tilde{f}$ follows from Exercise 5.24.1 of Sect. 5.24.                                                                                                                  □

**Corollary 5.4.4** *(Homotopy Lifting Property (HLP)) Given a covering space $(\tilde{X}, p)$ of $X$ and a homotopy $H : \mathbf{I} \times \mathbf{I} \rightarrow X$ such that $H(0, 0) = x_0$ and a point $\tilde{x}_0 \in p^{-1}(x_0)$, there exists a unique homotopy $\tilde{H} : \mathbf{I} \times \mathbf{I} \rightarrow \tilde{X}$ with the property that $\tilde{H}(0, 0) = \tilde{x}_0$.*

***Proof*** Subdividing $\mathbf{I} \times \mathbf{I}$ into rectangles and proceed as in Theorem 5.4.3 (here subdivide $\mathbf{I} \times \mathbf{I}$ in place of $\mathbf{I}$).                                                                                □

Theorem 5.4.5 gives a generalization of Corollary 5.4.4.

**Theorem 5.4.5** (The Generalized Homotopy Lifting Property) *Given a covering space $(\tilde{X}, p)$ of $X$ and a compact space $A$, if $f : A \rightarrow \tilde{X}$ is continuous and $H : A \times \mathbf{I} \rightarrow X$ is a homotopy starting from $p \circ f$, then there exists a homotopy $\tilde{H} : A \times \mathbf{I} \rightarrow \tilde{X}$ starting from $f$ lifts $H$. Moreover, if $H$ is a homotopy relative to a subset $A'$ of $A$, then $\tilde{H}$ has also the same property.*

**Proof** Proceed as above.                                                                                             □

**Theorem 5.4.6** *Given a covering space* $(\tilde{X}, p)$ *of* $X$, *if* $\tilde{x}_0 \in p^{-1}(x_0)$, *then for any path*

$$f : \mathbf{I} \to X : f(0) = x_0,$$

*there exists a unique path*

$$\tilde{f} : \mathbf{I} \to \tilde{X} : \tilde{f}(0) = \tilde{x}_0 \ \text{and} \ \ p \circ \tilde{f} = f.$$

**Proof** Let $B = \{b\}$ be a singleton space and $f : B \to X, b \mapsto x_0$ be a given map. Consider the homotopy

$$H : B \times \mathbf{I} \to X, \ (b, t) \mapsto f(b).$$

Then by the Homotopy Lifting Property, there is a continuous map

$$\tilde{H} : B \times \mathbf{I} \to \tilde{X} : \tilde{H}(b, 0) = \tilde{x}_0 \ \ \text{and} \ \ p \circ \tilde{H} = H.$$

Then,

$$\tilde{f} : \mathbf{I} \to \tilde{X}, t \mapsto \tilde{H}(b, t)$$

is a path in $\tilde{X}$ starting from $\tilde{x}_0$ such that

$$(p \circ \tilde{f})(t) = (p \circ \tilde{H}(b, t)) = H(b, t) = f(t), \forall t \in \mathbf{I} \Rightarrow p \circ \tilde{f} = f.$$

This shows that the path $\tilde{f} : \mathbf{I} \to \tilde{X}$ is unique.                                       □

   Theorem 5.4.7, known as 'Monodromy theorem', characterizes equivalence of two lifted paths in a covering space with the help of its projection map.

**Theorem 5.4.7** (The Monodromy Theorem) *Given a covering space* $(\tilde{X}, p)$ *of* $X$ *and two paths* $\tilde{f}$ *and* $\tilde{g}$ *with their common initial point* $\tilde{x}_0$, *they are equivalent in the sense* $\tilde{f} \simeq \tilde{g}$ *rel* $\mathbf{I}$ *iff* $p \circ \tilde{f}$ *and* $p \circ \tilde{g}$ *are equivalent paths in* $X$.

**Proof** First suppose that $\tilde{f}, \tilde{g} : \mathbf{I} \to \tilde{X}$ be two equivalent paths in $\tilde{X}$. Then there is a homotopy $H : \tilde{f} \simeq \tilde{g}$ rel $\dot{\mathbf{I}}$, and hence $p \circ H : \mathbf{I} \times \mathbf{I} \to X$ is a continuous map such that

$$p \circ H : p \circ \tilde{f} \simeq p \circ \tilde{g} \ \text{rel} \ \dot{\mathbf{I}}.$$

This proves that $p \circ \tilde{f}$ and $p \circ \tilde{g}$ are equivalent paths in $X$. Conversely, suppose that $p \circ \tilde{f}$ and $p \circ \tilde{g}$ are equivalent paths in $X$. Then there is a continuous map $F : \mathbf{I} \times \mathbf{I} \to X$ such that

$$F : p \circ \tilde{f} \simeq p \circ \tilde{g} \ \text{rel} \ \dot{\mathbf{I}}.$$

Hence by using Homotopy Lifting Property, it follows that there is a unique homotopy

$$\tilde{F} : \mathbf{I} \times \mathbf{I} \to X : \tilde{F}(0, 0) = \tilde{x}_0 \text{ and } p \circ \tilde{F} = F.$$

Then the restricted map $F|(\mathbf{I} \times \{0\})$ defines a path

$$\tilde{f} : \mathbf{I} \to \tilde{X} : t \mapsto \tilde{F}(t, 0) : \tilde{f}(0) = \tilde{x}_0,$$

Using the uniqueness property of the covering paths, it follows $\tilde{F}(t, 0) = \tilde{f}(t)$, $\forall t \in \mathbf{I}$. The restricted map $F|(\mathbf{I} \times \{0\})$ defines a path

$$\tilde{f} : \mathbf{I} \to \tilde{X} : t \mapsto \tilde{F}(t, 0) : \tilde{f}(0) = \tilde{x}_0.$$

Similarly, it follows that $\tilde{F}(t, 1) = \tilde{g}(t)$. Again, the restricted map $F|(\{0\} \times \mathbf{I})$ defines a path

$$\tilde{g} : \mathbf{I} \to \tilde{X} : s \mapsto \tilde{F}(0, s) : \tilde{g}(0) = \tilde{x}_0.$$

It projects by $p$ to the constant path at $x_0 = p(x_0)$. A constant path $s \mapsto x_0$ in $X$ also projects under $p$ to the constant path $s \mapsto x_0$ in $X$. Hence by uniqueness theorem $s \mapsto \tilde{F}(0, s)$ is a constant path based at $x_0$. Similarly, the path $s \mapsto \tilde{F}(s, t)$ is a constant path based at some point $\tilde{x}_1 \in p^{-1}(x_0)$. This shows that $\tilde{H} : \tilde{f} \simeq \tilde{g}$ rel $\dot{\mathbf{I}}$ and hence $\tilde{f}$ and $\tilde{g}$ are equivalent paths in $\tilde{X}$. $\qquad\square$

**Corollary 5.4.8** *Given a covering space* $(\tilde{X}, p)$ *of* $X$ *and* $x_0 \in X$, $\tilde{x}_0 \in p^{-1}(x_0)$, *the induced homomorphism*

$$p_* : \pi_1(\tilde{X}, \tilde{x}_0) \to \pi_1(X, x_0)$$

*is a monomorphism.*

**Proof** Suppose that $[\tilde{f}], [\tilde{g}] \in \pi_1(X, x_0)$ and $[\tilde{f}] \neq [\tilde{g}]$. This implies that $p_*([\tilde{f}]) = [p \circ \tilde{f}]$ and $p_*([\tilde{g}]) = [p \circ \tilde{g}]$. Hence it follows that

$$p \circ \tilde{f} \simeq p \circ \tilde{g} \text{ rel } \dot{\mathbf{I}} \Leftrightarrow \tilde{f} \simeq \tilde{g} \text{ rel } \dot{\mathbf{I}}$$

On the other hand

$$\tilde{f} \not\simeq \tilde{g} \text{ rel } \dot{\mathbf{I}} \Leftrightarrow p \circ \tilde{f} \not\simeq p \circ \tilde{g} \text{ rel } \dot{\mathbf{I}},$$

because, otherwise, a contradiction would be arrived by Theorem 5.4.7. This proves that $p_*$ is well defined and it is a monomorphism. $\qquad\square$

## 5.4.2  Characterization of Path Liftings of Fibrations

This subsection studies liftings of fibrations and their fibers. For example, Theorem 5.4.9 characterizes path liftings of fibrations with the help of their fibers.

**Theorem 5.4.9** *A fibration $p : X \to B$ has the unique path lifting property (PLP) iff its every fiber has no nonconstant paths.*

**Proof** First suppose that $p : X \to B$ is a fibration with unique path lifting. Let $b \in B$ and $f$ be a path in the fiber $p^{-1}(b)$ and $g$ be the constant path in $p^{-1}(b)$ such that $g(0) = f(0)$. Then

$$p \circ f = p \circ g \Rightarrow f = g.$$

It shows that $f$ is a constant path. Conversely, let $p : X \to B$ be a fibration such that every fiber has no nontrivial path. If $f$ and $g$ are two paths in $X$ such that $p \circ f = p \circ g$ and $f(0) = g(0)$, then for every $t \in \mathbf{I}$ define a path $h_t$ in $X$ by

$$h_t : \mathbf{I} \to X, t' \mapsto \begin{cases} f((1 - 2t')t), & 0 \le t' \le 1/2 \\ g((2t' - 1)t), & 1/2 \le t' \le 1. \end{cases}$$

It implies that

(i) $h_t : \mathbf{I} \to X$ is a path in $X$ from $f(t)$ to $g(t)$ for every $t \in \mathbf{I}$ and
(ii) $p \circ h_t$ is a closed path in $B$, which is homotopic to the constant path at $(p \circ f)(t)$ rel $\dot{\mathbf{I}}$.

Since $p$ has the HLP, there is a map

$$H : \mathbf{I} \times \mathbf{I} \to X : H(t', 0) = h_t(t')$$

and $H$ sends the subspace

$$\{0\} \times \mathbf{I} \cup \mathbf{I} \times \{1\} \cup \{1\} \times \mathbf{I} \subset \mathbf{I} \times \mathbf{I}$$

to the fiber $p^{-1}(pf(t))$. Since $p^{-1}(pf(t))$ has no nonconstant paths, $F$ maps $0 \times \mathbf{I}, \mathbf{I} \times 1$ and $1 \times \mathbf{I}$ to a single point. Hence it follows that $H(0, 0) = H(1, 0)$. This proves that $h_t(0) = h_t(1)$ and $f(t) = g(t)$. □

**Proposition 5.4.10** *Given a pointed topological space $X$ with base point $x_0$, let $\mathcal{P}(X)$ be the space of all paths in $X$ starting at $x_0$ and endowed with compact open topology. Then the map*

$$p : \mathcal{P}(X) \to X, \ \beta \mapsto \beta(1)$$

*is fibration with fiber $\Omega(X)$.*

**Proof** Given an arbitrary topological space $Y$ and a pair of continuous maps

$$f : Y \to \mathcal{P}(X) \text{ and } H : Y \times \mathbf{I} \to X \text{ with } H_0 = H(\ , 0) = p \circ f : Y \to X,$$

define

$$G : Y \times \mathbf{I} \times \mathbf{I} \to X, (y, t, s) \mapsto \begin{cases} (f(y))(s(t + 1)), & 0 \le s \le \frac{1}{t+1} \\ H(y, s(t + 1) - 1), & \frac{1}{t+1} \le s \le 1. \end{cases}$$

Then $G$ is a continuous map such that it defines a map

$$F : Y \times \mathbf{I} \to X^{\mathbf{I}} : F(y, t)(0) = f(y)(0) = x_0, \ \forall \, y \in Y, t \in \mathbf{I}.$$

Hence it follows that

$$F \in \mathcal{P}(X) \ and \ F(y, 0)(s) = f(y)(s), \ \forall \, y \in Y, \ \forall \, s \in \mathbf{I}.$$

This asserts that $F_0 = F(\ , 0) = f$ and $p \circ F = H$. This proves that

(i)  $F$ is the required lifting of $H$ and
(ii)

$$p^{-1}(x_0) = \{\beta \in \mathcal{P}(x) : \beta(1) = x_0\} = \Omega(X). \qquad \square$$


### 5.4.3   General Lifting Problems from Homotopy Viewpoint

It is a natural problem: given a covering space $(\tilde{X}, p)$ of $X$ and a continuous map $f : A \to \tilde{X}$, does there exist a continuous map $\tilde{f} : A \to \tilde{X}$ such that $p \circ \tilde{f} = f$ ? The answer is negative. In support consider the exponential map

$$p : \mathbf{R} \to S^1, \ t \mapsto e^{2\pi i t}$$

studied in Example 5.4.11. This subsection studies lifting problems of arbitrary continuous maps with the help of fundamental groups and solves a lifting problem in Theorem 5.4.12.

**Example 5.4.11**  The exponential map

$$p : \mathbf{R} \to S^1, \ t \mapsto e^{2\pi i t}$$

is a covering projection. The identity map $1_{S^1} : S^1 \to S^1$ cannot be lifted to a continuous map $\psi = \tilde{1}_{S^1} : S^1 \to \mathbf{R}$ such that $p \circ \psi = 1_{S^1}$, which makes the triangle in Fig. 5.17 commutative. Because, $p \circ \psi = 1_{S^1}$ would imply that $\psi$ is an embedding of $S^1$ into $\mathbf{R}$. Since $S^1$ is compact, $\psi(S^1)$ must be a compact connected subset $X$ of $\mathbf{R}$ such that it would be homeomorphic to $S^1$ under present situation by the embedding $\psi$. But it is not possible, because the fundamental groups $\pi_1(X) = 0$, $\pi_1(S^1) = \mathbf{Z}$ and $\pi_1$ is a topological invariant. This example asserts that the identity map $1_{S^1} : S^1 \to S^1$ cannot be lifted to a continuous map $\psi = \tilde{1}_{S^1} : S^1 \to \mathbf{R}$.

Theorem 5.4.12 provides a necessary and sufficient condition under which an arbitrary continuous map $f : A \to X$ can be lifted, which is proved by using the tools of homotopy theory.

**Fig. 5.17** Covering
projection for exponential
map $p$

**Fig. 5.18** Lifting of $f$ to $\tilde{f}$

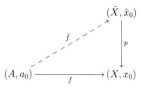

**Theorem 5.4.12** (Lifting Theorem) *Given a covering space* $(\tilde{X}, p)$ *of* $X$ *and a connected and locally path-connected space* $A$, *if* $f : A \to X$ *is any continuous map, then for arbitrary three points* $a_0 \in A$, $x_0 \in X$ *and* $\tilde{x}_0 \in \tilde{X}$ *such that* $f(a_0) = x_0$ *and* $p(\tilde{x}_0) = x_0$, *there exists a unique continuous map*

$$\tilde{f} : A \to \tilde{X}$$

*such that*

(i) $\tilde{f}(a_0) = \tilde{x}_0$ *and*
(ii) $p \circ \tilde{f} = f$ *if and only if* $f_*(\pi_1(A, a_0)) \subset p_*(\pi_1(\tilde{X}, \tilde{x}_0))$.

**Proof** Let there exist a continuous map $\tilde{f} : A \to \tilde{X}$ such that the given conditions hold. Then the triangle in Fig. 5.18 is commutative, and hence by using the functorial property of the fundamental group functor $\pi_1$, it follows that the triangle in Fig. 5.19 is also commutative. This asserts that

$$f_*(\pi_1(A, a_0)) = p_*(\tilde{f}_*(\pi_1(A, a_0))) \subset p_*(\pi_1(\tilde{X}, \tilde{x}_0)).$$

Conversely, let the algebraic property $f_*(\pi_1(A, a_0)) \subset p_*(\pi_1(\tilde{X}, \tilde{x}_0))$ hold. Since by hypothesis, $A$ is connected, it has only one component. Again, since $A$ is locally path connected, this component is a path component, and hence it follows that $A$ is path connected. For any point $a \in A$, construct a path

$$\alpha : \mathbf{I} \to A : \alpha(0) = a_0 \ and \ \alpha(1) = a.$$

Hence it follows that $f \circ \alpha : \mathbf{I} \to X$ is a path such that

$$(f \circ \alpha)(0) = f(\alpha(0)) = f(a_0) = x_0.$$

By path Lifting Property 5.4.3, there is a unique path $\tilde{\alpha} : \mathbf{I} \to \tilde{X}$ which lifts $f \circ \alpha$ in $\tilde{X}$ with $\tilde{\alpha}(0) = \tilde{x}_0$ ( see Fig. 5.16). Define a map

**Fig. 5.19** Lifting problem
for $f$ and $\tilde{f}$ converted in
homotopy theory

$$\pi_1(\tilde{X}, \tilde{x}_0)$$

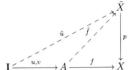

$$\pi_1(A, a_0) \xrightarrow{\quad f_* \quad} \pi_1(X, x_0)$$

**Fig. 5.20** Diagram
corresponding to lifting
Theorem 5.4.12

$$\tilde{f} : A \to X, a \mapsto \tilde{\alpha}(1).$$

Clearly, $\tilde{f}$ is well-defined. To prove the continuity, take any point $a \in A$ and an open nbd $U$ of $\tilde{f}(a)$. We now find an open nbd $W_a$ of $a$ with $\tilde{f}(W_a) \subset U$. We take an open admissible nbd $W$ of $p\tilde{f}(a) = f(a)$ with the property that $W \subset p(U)$. Let $C$ be the path component of $p^{-1}(W)$ which contains the point $\tilde{f}(a)$, and let $W'$ be an open admissible nbd of $f(a)$ such that

$$W' \subset p(U \cap C).$$

Then the path component of $p^{-1}(W')$ containing the point $\tilde{f}(a)$ is contained in $U$. Since $f$ is continuous and the path-connected set $A$ is locally connected, there is a path connected nbd $W_a$ of $a$ such that $f(W_a) \subset V$. This implies that $\tilde{f}(W_a) \subset U$ (Fig. 5.20).                                                                     □

**Corollary 5.4.13** *Suppose that $A$ is simply connected and locally path connected and $f : (A, a_0) \to (X, x_0)$ is continuous. If $(X, p)$ is a covering space of $X$ and if $\tilde{x}_0 \in p^{-1}(x_0)$, then $f$ has the unique lifting*

$$\tilde{f} : (A, a_0) \to (\tilde{X}, \tilde{x}_0).$$

**Proof** By hypothesis $A$ is simply connected. Hence $\pi_1(A, a_0)=0$ and $p_* \pi_1(A, a_0) = \{0\} \subset p_* \pi_1(\tilde{X}, \tilde{x}_0)$. This implies that there is a unique lifting $\tilde{f} : (A, a_0) \to (\tilde{X}, \tilde{x}_0)$ of $f$.                                                                     □

Corollary 5.4.14 follows immediately from above discussion.

**Corollary 5.4.14** *Let $X$ be a connected and locally path-connected space, and $(\tilde{X}, p)$ and $(Y, q)$ be two covering spaces of the same base space $X$. Let $x_0 \in X$ and $\tilde{x}_0 \in \tilde{X}$, $y_0 \in Y$ be base points such that*

$$p(\tilde{x}_0) = x_0 = q(y_0).$$

If $p_*\pi_1(\tilde{X}, \tilde{x}_0) = q_*\pi_1(Y, y_0)$, then there exists a unique continuous map

$$f : (Y, y_0) \to (\tilde{X}, \tilde{x}_0) : p \circ f = q,$$

i.e., $f$ satisfies the property: $p \circ f = q$.

**Example 5.4.15** $(S^n, p)$ is a covering space of the real projective space $\mathbf{R}P^n$ of multiplicity 2. Because, the $n$-sphere $S^n$ is simply connected for $n \geq 2$, and hence it follows that if $\tilde{x}_0 \in p^{-1}(x_0)$, then $x_0 \in \mathbf{R}P^n$. Hence it follows that for any continuous map

$$f : (S^n, s_0) \to (\mathbf{R}P^n, x_0),$$

there exists the unique lifting

$$\tilde{f} : (S^n, s_0) \to (S^n, \tilde{x}_0).$$

### 5.4.4  Homotopy Classification of Covering Spaces

This subsection completely classifies covering spaces of a fixed base space with the help of group theory and characterizes such covering spaces in terms of conjugate subgroups of the fundamental groups of base spaces.

We use in this subsection the following results of group theory .

(i) Two subgroups $A$ and $B$ of a group $G$ are conjugate subgroups iff $A = g^{-1}Bg$ for some $g \in G$.
(ii) For the subgroups $A$ and $B$ of a group $G$, the $G$-sets $G/A$ and $G/B$ are $G$-isomorphic iff $A$ and $B$ are conjugate subgroups of $G$.

**Theorem 5.4.16**  *Given a path-connected space* $X$, *let* $(\tilde{X}, p)$ *be a covering space of the base space* $X$, *where* $\tilde{X}$ *is also path connected. If* $x_0 \in X$, *then the family of groups* $p_*\pi_1(\tilde{X}, z)$, *as* $z$ *runs over* $F = p^{-1}(x_0)$, *forms a conjugacy class of subgroups of* $\pi_1(X, x_0)$.

**Proof** It is proved in two steps:

(i) **Step I** For any two points $z_0, z_1 \in F$, the subgroups $p_*\pi_1(\tilde{X}, z_0)$ and $p_*\pi_1(\tilde{X}, z_1)$ of $\pi_1(X, x_0)$ are conjugate.
(ii) **Step II**  For any subgroup $H$ of $\pi_1(X, x_0)$ conjugate to the subgroup $p_*\pi_1$ $(\tilde{X}, z_0)$, there exists some $z \in F$ such that

$$H = p_*\pi_1(\tilde{X}, z).$$

(i) Given a path $w : \mathbf{I} \to \tilde{X}$, from $z_0$ to $z_1$, define a map

$$\beta_w : \pi_1(\tilde{X}, z_0) \to \pi_1(\tilde{X}, z_1), \ [f] \mapsto [\bar{w} * f * w].$$

It is an isomorphism of groups (see Chap. 2). Hence it follows that

$$(p_* \circ \beta_w)\pi_1(\tilde{X}, z_0) = [p \circ w]^{-1} p_* \pi_1(\tilde{X}, z_0) [p \circ w].$$

which asserts that $p_* \pi_1(\tilde{X}, z_1)$ and $p_* \pi_1(\tilde{X}, z_0)$ are conjugate subgroups of $\pi_1(X, x_0)$,

(ii) By hypothesis, $H$ is a subgroup of $\pi_1(X, x_0)$ conjugate to the subgroup $p_* \pi_1(\tilde{X}, z_0)$. Then there exists some $[k] \in \pi_1(X, x_0)$ such that

$$H = [k]^{-1} p_* \pi_1(\tilde{X}, z_0)[k].$$

Suppose that the unique lifting $\tilde{k}$ of $k$ in $\tilde{X}$ starting at $z_0$ ends at the point $\tilde{k}(1) = z \in \tilde{X}$. Now proceeding as in $(i)$, it follows that

$$p_* \pi_1(\tilde{X}, z) = [p \circ \tilde{k}]^{-1} p_* \pi_1(\tilde{X}, z_0)[p \circ \tilde{k}] = [k]^{-1} p_* \pi_1(\tilde{X}, z_0)[k] = H.$$

This proves that the family $\{p_* \pi_1(\tilde{X}, z) : z \in F\}$ constitutes a complete conjugate class of subgroups of the group $\pi_1(X, x_0)$.

$\square$

**Definition 5.4.17** The family $\{p_* \pi_1(\tilde{X}, z) : z \in F\}$ constituting **conjugate class of subgroups** of the group $\pi_1(X, x_0)$ given in Theorem 5.4.16 is known as the conjugate class determined by the covering space $(\tilde{X}, p)$ of $X$.

Theorem 5.4.18 characterizes covering spaces of a base space $X$ in terms of conjugacy classes of subgroups of the fundamental group $\pi_1(X)$, and hence every conjugacy class of a subgroup of $\pi_1(X, x)$ determines uniquely the covering spaces of $X$ up to isomorphism (Fig. 5.22).

**Theorem 5.4.18** *Given a path connected and locally path-connected space $X$, let $(\tilde{X}, p)$ and $(\tilde{Y}, q)$ be two path connected covering spaces of $X$ such that $\tilde{x}_0 \in p^{-1}(x_0)$ and $\tilde{y}_0 \in q^{-1}(x_0)$, where $x_0 \in X$. Then the covering spaces $(\tilde{X}, p)$ and $(\tilde{Y}, q)$ are isomorphic iff they determine the same conjugacy class of subgroups of $\pi_1(X, x_0)$), i.e., iff*

$$p_* \pi_1(\tilde{X}, x_0) \text{ and } q_* \pi_1(\tilde{Y}, y_0)$$

*are conjugate subgroups of $\pi_1(X, x_0)$.*

**Proof** First suppose that the covering spaces $(\tilde{X}, p)$ and $(\tilde{Y}, q)$ are isomorphic with a homeomorphism $h : \tilde{Y} \to \tilde{X}$ such that the triangle in Fig. 5.21 is commutative. Hence $p \circ h = q$.

If $h(\tilde{y}_0) = \tilde{x}_1$, then $h$ induces an isomorphism

$$h_* : \pi_1(\tilde{Y}, \tilde{y}_0) \to \pi_1(\tilde{X}, \tilde{x}_1).$$

Consequently,

**Fig. 5.21** Isomorphisms
between covering spaces

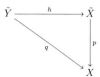

$$h_*(\pi_1(\tilde{Y}, \tilde{y}_0)) = \pi_1(\tilde{X}, \tilde{x}_1).$$

This implies that

$$(p_* \circ h_*)(\pi_1(\tilde{Y}, \tilde{y}_0)) = p_*(\pi_1(\tilde{X}, \tilde{x}_1)).$$

This asserts that

$$q_*(\pi_1(\tilde{Y}, \tilde{y}_0)) = p_*\pi_1(\tilde{X}, \tilde{x}_1).$$

Hence $p_*\pi_1(\tilde{X}, \tilde{x}_1)$ is a subgroup of $\pi_1(X, x_0)$, and it is conjugate to the subgroup $p_*\pi_1(\tilde{X}, \tilde{x}_0)$ by Theorem 5.4.16. This shows that $p_*\pi_1(\tilde{X}, \tilde{x}_0)$ and $q_*\pi_1(\tilde{Y}, \tilde{y}_0)$ are conjugate subgroups of $\pi_1(X, x_0)$. Conversely, let the two subgroups of $\pi_1(\tilde{X}, \tilde{x}_0)$ be conjugate. By Theorem 5.4.16 we can choose a different base point $\tilde{y}_0$ in $\tilde{Y}$ such that the two groups are equal. We now consider the lifting of $p$ in the diagram in Fig. 5.22, where $q$ is a covering map. Since by hypothesis, the space $X$ is path connected; it is also locally path connected. Now, $p_*\pi_1(\tilde{X}, \tilde{x}_0) \subset q_*\pi_1(\tilde{Y}, \tilde{y}_0)$ and these two groups subgroups of $\pi_1(X, x_0)$ are equal. Using Theorem 5.4.16 the map $p$ can be lifted to $\tilde{p} : \tilde{X} \to \tilde{Y}$ such that $\tilde{p}(\tilde{x}_0) = \tilde{y}_0$. This implies yhat $q \circ \tilde{p} = p$.
If we now reverse the role of $\tilde{X}$ and $\tilde{Y}$, then the map $q : \tilde{Y} \to X$ can also be lifted to $\tilde{q} : \tilde{Y} \to \tilde{X}$ such that $\tilde{q}(\tilde{y}_0) = \tilde{x}_0$ as shown in Fig. 5.23.
    To show that $\tilde{p}$ and $\tilde{q}$ in Fig. 5.24 are inverses of each other, consider the diagram in Fig. 5.25.
    Consider the two liftings of $p$ described below:

(i) $\tilde{q} \circ \tilde{p} : \tilde{X} \to \tilde{X}$ is a lifting of the map $p : \tilde{X} \to X$ such that $(\tilde{q} \circ \tilde{p})(\tilde{x}_0) = \tilde{x}_0$
    and
(ii) The identity map $1_{\tilde{X}} : \tilde{X} \to \tilde{X}$ is another such lifting of $p$.

    This asserts by uniqueness property of lifting that $\tilde{q} \circ \tilde{p} = 1_{\tilde{X}}$. Proceeding in a similar way, it follows that $\tilde{p} \circ \tilde{q} = 1_{\tilde{Y}}$. Hence it is proved that

(i) $\tilde{p} : \tilde{X} \to \tilde{Y}$ is a homeomorphism with its inverse $\tilde{q} : \tilde{Y} \to \tilde{X}$ and
(ii) the covering spaces $(\tilde{X}, p)$ and $(\tilde{Y}, q)$ are isomorphic.          □


*Example 5.4.19* (**Complete classification of covering spaces of** $S^1$) Consider the covering spaces of $S^1$. Since the fundamental group $\pi_1(S^1, 1) \cong \mathbf{Z}$ is abelian, its two subgroups are conjugate iff they are equal. Hence, two covering spaces of $S^1$ are isomorphic iff their corresponding subgroups of $\pi_1(S^1)$ are same. The subgroups in $\mathbf{Z}$ are the subgroups $< n >$, consisting of precisely all the multiples of $n$, for

**Fig. 5.22** Lifting of $p$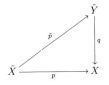

**Fig. 5.23** Lifting of $q$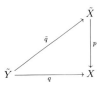

**Fig. 5.24** Liftings of $p$
and $q$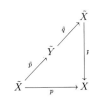

**Fig. 5.25** Lifting of $p$
involving $\tilde{p}$ and $\tilde{q}$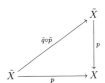

$n = 0, 1, 2, \ldots.$ Since **R** is simply connected, the subgroup corresponding to the usual covering space $(\mathbf{R}, p)$ of $S^1$ is the trivial subgroup of **Z**. On the other hand, if $p; S^1 \to S^1, z \mapsto z^n$, then the subgroup corresponding to the covering space $(S^1, p)$ of $S^1$ is the subgroup $< n >$ of **Z**. This asserts that every path-connected covering space of $S^1$ is isomorphic to one of these coverings. This asserts that any covering space of $S^1$ is isomorphic either

(i) to $(\mathbf{R}, p)$ or
(ii) to one of the coverings $(S^1, q_n)$, where $q_n : S^1 \to S^1 : z \mapsto z^n$, which geometrically wraps $S^1$ around itself $n$ times.

**Example 5.4.20** The double covering $(S^2, p)$ of $\mathbf{R}P^2$ determines the conjugacy class contains only the trivial subgroup. Because, $S^2$ is simply connected, and hence $\pi_1(S^2, s) = 0.$

**Example 5.4.21** The conjugacy class determined by the covering space $(\mathbf{R}^2, r)$ of the torus contains only the trivial subgroup. Because, the plane $\mathbf{R}^2$ is simply connected.

**Example 5.4.22** For the infinite spiral $X$, the projection map

**Fig. 5.26** Homotopy
diagram corresponding to $\tilde{f}_0$

$$q : X \rightarrow S^1$$

obtained by projecting each point on $X$ to the point on the circle directly below it is covering space of $S^1$. Since $X$ is contractible, $\pi_1(X) = 0$ and hence $(X, q)$ determines the conjugacy class of $\pi_1(S^1)$ consisting of only the trivial subgroup. Again, for the exponential map $p : \mathbf{R} \rightarrow S^1$, the covering space $(\mathbf{R}, p)$ also determines the conjugacy class of $\pi_1(S^1)$ consisting of only the trivial subgroup. This asserts that the covering spaces $(X, q)$ and $(\mathbf{R}, p)$ of $S^1$ are isomorphic by Theorem 5.4.18.

**Proposition 5.4.23** *Given a covering map $p : (\tilde{X}, \tilde{x}_0) \rightarrow (X, x_0)$,*

*(i)  its induced homomorphism*

$$p_* : \pi_1(\tilde{X}, \tilde{x}_0) \rightarrow \pi_1(X, x_0)$$

*is a monomorphism and*
*(ii)  the subgroup $p_*(\pi_1(\tilde{X}, \tilde{x}_0)) \subset \pi_1(X, x_0)$ has elements consisting of homotopy classes of loops in $X$ based at $x_0$ which lift to $\tilde{X}$ starting at the point $\tilde{x}_0 \in \tilde{X}$ are loops.*

**Proof** Given an element $\alpha \in \ker p_*$ represented by a loop $\tilde{f}_0 : \mathbf{I} \rightarrow \tilde{X}$ with a homotopy $H_t : \mathbf{I} \rightarrow X$ of $f_0 = p \circ \tilde{f}_0$ to the trivial loop $f_1$, there exists a lifted homotopy of loop $\tilde{H}_t : \mathbf{I} \rightarrow \tilde{X}$ ( see Fig. 5.26 ) started at $\tilde{f}_0$ and ending at a constant loop, since the lifted homotopy $\tilde{H}_t$ is a homotopy of paths fixing the end points and $t$ varies each point of $\tilde{H}_t$ gives a path lifting a constant path, which is clearly a constant path. This implies that $[\tilde{f}_0] = 0 \in \pi_1(\tilde{X}, \tilde{x}_0)$ and hence $p_*$ is a monomorphism. $\quad\square$

**Remark 5.4.24** If $p : \tilde{X} \rightarrow X$ is a covering map, then $p$ is also onto. But its induced homomorphism
$$p_* : \pi_1(\tilde{X}, \tilde{x}_0) \rightarrow \pi_1(X, x_0)$$

needs not be an epimorphism. However, $p_*$ is a monomorphism.

**Proposition 5.4.25** *Given a path-connected space $\tilde{X}$ and a covering projection*

$$p : (\tilde{X}, \tilde{x}_0) \rightarrow (X, x_0)$$

*(i)  there is a surjective map $\psi : \pi_1(X, x_0) \rightarrow p^{-1}(x_0)$.*
*(ii)  Moreover, if $\tilde{X}$ is simply connected, then $\psi$ is bijective.*

**Proof** It can be proved by using the technique for computation of $\pi_1(S^1, 1)$ described in Chap. 3. □

### 5.4.5  Isomorphism Theorem and Computation of $\pi_n(S^1) = 0$ for $n \geq 2$

This subsection proves that $\pi_n(S^1) = 0$ for $n \geq 2$ by applying Theorem 5.4.26 saying that every covering space induces an isomorphism between the higher homotopy group of its total space and that of its base space at every dimension $n \geq 2$.

**Theorem 5.4.26** *If $(\tilde{X}, p)$ is a covering space of $X$ such that*

*(i)  $x_0 \in X$ and*
*(ii)  $\tilde{x}_0 \in p^{-1}(x_0)$*

*then $p$ induces an isomorphism*

$$p_* : \pi_n(\tilde{X}, \tilde{x}_0) \to \pi_n(X, x_0), \ \forall n \geq 2.$$

**Proof** By hypothesis, $p : \tilde{X} \to X$ is a covering map with $p(\tilde{x}_0) = x_0$. Let $\tilde{\sigma} \in \pi_n(\tilde{X}, \tilde{x}_0)$ be an arbitrary element and let it is represented by a continuous map $\tilde{f} : (\mathbf{I}^n, \partial \mathbf{I}^n) \to (\tilde{X}, \tilde{x}_0)$. Then

$$\tilde{\sigma} = [\tilde{f}] \in \pi_n(\tilde{X}, \tilde{x}_0) = [S^n, \tilde{X}].$$

Consequently, $p$ induces a homomorphism

$$p_* : \pi_n(\tilde{X}, \tilde{x}_0) \to \pi_n(X, x_0), \ [\tilde{f}] \mapsto [p \circ \tilde{f}].$$

$p_*$ **is a monomorphism :** Let $\tilde{\sigma}$ and $\tilde{\tau}$ be two elements of $\pi_n(\tilde{X}, \tilde{x}_0)$ such that $p_*(\tilde{\sigma}) = p_*(\tilde{\tau})$. If they are represented by $\tilde{f}$ and $\tilde{g}$, respectively, where $\tilde{f}, \tilde{g} : S^n \to \tilde{X}$ are pointed maps, then $[p \circ \tilde{f}] = [p \circ \tilde{g}]$. This implies that $p \circ \tilde{f} \simeq p \circ \tilde{g}$. Hence it follows by covering homotopy theorem that $\tilde{f} \simeq \tilde{g}$. This implies $[\tilde{f}] = [\tilde{g}]$ and hence it is proved that $p_*$ is a monomorphism.

$p_*$ **is an epimorphism:** Let $\sigma = [f] \in \pi_n(X, x_0)$ be an arbitrary element. Consider the triangle as shown in diagram in Fig. 5.27.

Then there exists a unique lifting $\tilde{f} : S^n \to \tilde{X}$ of $f : S^n \to X$ such that $p \circ \tilde{f} = f$, because $S^n$ is simply connected for every integer $n \geq 2$. This proves that there exists an element $\tilde{\sigma} \in \pi_n(\tilde{X}, \tilde{x}_0)$ such $\tilde{\sigma} = [\tilde{f}]$. This implies that $p_*[\tilde{f}] = [f] = \sigma$ and hence it is proved that $p_*$ is an epimorphism. □

As an immediate application of Theorem 5.4.26, it is proved in Theorem 5.4.27 that $\pi_n(S^1) = 0$ for every integer $n \geq 2$.

**Fig. 5.27** Lifting of $f$ to $\tilde{f}$

**Theorem 5.4.27** $\pi_n(S^1) = \{0\}$ *for every integer $n \geq 2$.*

**Proof** Consider the well-known covering map

$$p : \mathbf{R} \to S^1, \; t \mapsto e^{2\pi it}$$

and apply Theorem 5.4.26. Then $p$ induces an isomorphism

$$p_* : \pi_n(\mathbf{R}) \to \pi_n(S^1), \; \forall n \geq 2.$$

Since $\mathbf{R}$ is a contractible space, it follows that

$$\pi_n(\mathbf{R}) = \{0\}, \; \forall n \geq 2.$$

This proves that $\pi_n(S^1) = \{0\}$ for every $n \geq 2$. $\qquad\square$

## 5.5 Galois Correspondence on Covering Spaces

This section continues to study the classification problem of covering spaces over a fixed base space $X$. The main tool of this classification is provided by the Galois correspondence between a special family of covering spaces of $X$ and subgroups of the fundamental group of the base space $\pi_1(X)$. The Galois correspondence $\psi$ is derived from the map that assigns the subgroup $p_*(\pi_1(\tilde{X}, \tilde{x}_0))$ of $\pi_1(X, x_0)$ corresponding to every covering space $p : (\tilde{X}, \tilde{x}_0) \to (X, x_0)$ under certain conditions prescribed in Theorem 5.5.6. The Galois correspondence $\psi$ is both injective and surjective.

**Theorem 5.5.1** *Given a path connected, locally path connected and semilocally path space $X$ and a subgroup $H$ of $\pi_1(X, x_0)$ there is a covering space $p : X_H \to X$ with*

$$p_*(\pi_1(X_H, \tilde{x}_0)) = H$$

*for some choice of the base point $\tilde{x}_0 \in X_H$.*

**Proof** By hypothesis $X$ is a semilocally path-connected space. Hence for every point $x \in X$,

(i) there is a nbd $W_x$ of $x$ such that each closed path in $W_x$ based at $x$ is nullhomotopic in $X$ by its semilocality property and

(ii) there is an open connected nbd $U_x$ of $x$ with

$$x \in U_x \subset W_x$$

by locally path connectedness property of $X$.

Then every closed path in $U_x$ at $x$ is null- homotopic in $X$, and $U_x$ is evenly covered by $p$.

**Construction of** $X_H$ : Let $\mathcal{P}(X, x_0)$ denote the family of all paths $f$ in $X$ with $f(0) = x_0$ and topologized by the compact open topology. Define an equivalence relation '$\sim$'on $\mathcal{P}(X, x_0)$ by the rule

$$f \sim g \mod H \text{ iff } f(1) = g(1) \text{ and } [f * g^{-1}] \in H.$$

Let $X_H$ be the set of all such equivalence classes [f], topologized by the quotient topology. For the constant loop $c \in \mathcal{P}(X, x_0)$, take $\tilde{x}_0 = [c] \in X_H$ and define

$$p : X_H \to X, [f] \mapsto f(1).$$

Then $p(\tilde{x}_0) = x_0$. Since any two paths in the basic nbds $U_{[f]}$ and $U_{[g]}$ are identified in $X_H$, the whole nbds are identified. This asserts that the natural projection

$$p : X_H \to X$$

is a covering space with $p(\tilde{x}_0) = x_0$. Since any loop $\beta$ in $X$ based at $x_0$ has its lifting to $X_H$ starting at $\tilde{x}_0$ ends at $[\beta]$ and the image of this lifted path in $X_H$ is a loop iff $[\beta] \sim [c_0]$ (equivalently, $[\beta] \in H$) the image of

$$p_* : \pi_1(X_H, \tilde{x}_0) \to \pi_1(X, x_0)$$

coincides with $H$. This concludes that $p_*(\pi_1(X_H, \tilde{x}_0)) = H$ for some choice of the base point $\tilde{x}_0 \in X_H$.                                    □

**Corollary 5.5.2** *Under the condition of Theorem 5.5.1, every subgroup $G$ of the fundamental group $\pi_1(X, x_0, )$ can be realized as the fundamental group of the topological space $X_G$.*

**Proof** Let $G$ be a subgroup of the fundamental group $\pi_1(X, x_0)$. Consider the topological $X_G$ constructed in the proof of Theorem 5.5.1. Then the corollary follows from Theorem 5.5.1.                                    □

**Corollary 5.5.3** *For any connected, locally path connected, semilocally simply connected space $X$, every covering space $q : Y \to X$ is isomorphic (equivalent) to a covering spaces of the form $p : X_G \to X$ for some group $G$.*

**Proof** Let $x_0 \in X$ be a base point of $X$ and $y_0 \in Y$ lie in the fiber $q^{-1}(x_0)$ over $x_0$. If $G = q_*\pi_1(Y, y_0)$, then $p_*\pi_1(X_G, x_0) = G$. Hence it follows Theorem 5.5.1 shows that the covering spaces $p : X_G \to X$ and $q : Y \to X$ are isomorphic.                                    □

**Fig. 5.28** Triangular
diagram for two isomorphic
coverings of $X$

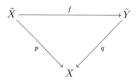

**Corollary 5.5.4**  *Given a connected, locally path connected, semilically simply connected space $X$, if $p : \tilde{X} \to X$ is a covering space of $X$, then every open contractible set $U$ in $X$ is evenly covered by $p$.*

**Corollary 5.5.5**  *Given a connected, locally path connected, semilically simply connected space $X$, it has a universal covering space $\tilde{X}$ (i.e., $\tilde{X}$ is simply connected ) iff $\tilde{X}$ is semilocally simply connected.*

Theorem 5.5.6 characterizes specified isomorphic coverings in terms of fundamental groups.

**Theorem 5.5.6**  (Classification theorem)  *Given a path connected and locally path-connected space $X$, the two coverings $p : \tilde{X} \to X$ and $q : \tilde{Y} \to B$ are isomorphic through a homeomorphism $f : \tilde{X} \to \tilde{Y}$ sending a base point $\tilde{x}_0 \in p^{-1}(x_0)$ to a base point $\tilde{y}_0 \in q^{-1}(x_0)$ iff*

$$p_*(\pi_1(\tilde{X}, \tilde{x}_0)) = q_*(\pi_1(\tilde{Y}, \tilde{y}_0)).$$

**Proof**  First suppose that there exists a homeomorphism $f : (\tilde{X}, \tilde{x}_0) \to (\tilde{Y}, \tilde{y}_0)$ making the triangular diagram as shown in Fig. 5.28 commutative. Then $p = q \circ f$ and $q = p \circ f^{-1}$ assert that $p_*(\pi_1(\tilde{X}, \tilde{x}_0)) = q_*(\pi_1(\tilde{Y}, \tilde{y}_0))$.

Conversely, let $p_*(\pi_1(\tilde{X}, \tilde{x}_0)) = q_*(\pi_1(\tilde{Y}, \tilde{y}_0))$. Then by the lifting property,

(i)  the map $p$ is lifted to $\tilde{p} : (\tilde{X}, \tilde{x}_0) \to (\tilde{Y}, \tilde{y}_0)$ such that

$$q \circ \tilde{p} = p;$$

(ii)  the map $q$ is also lifted to $\tilde{q} : (\tilde{Y}, \tilde{y}_0) \to (\tilde{X}, \tilde{x}_0)$ such that

$$p \circ \tilde{q} = q.$$

Then by unique lifting property, it follows that $\tilde{p} \circ \tilde{q} = 1_d$ and $\tilde{q} \circ \tilde{p} = 1_d$, because each of these composite lifts fix the base points. This asserts that the induced homomorphism $p_*$ is an isomorphism with the induced homomorphism $q_*$ its inverse isomorphism.                                                                    $\square$

Theorem 5.5.7 gives a generalization of the classification Theorem 5.5.6.

**Theorem 5.5.7**  (Generalized classification theorem)  *Given a path connected, locally path connected and semilocally simply connected space $X$, there is a bijection between the set of base point preserving isomorphism classes of path-connected covering spaces $p : (\tilde{X}, \tilde{x}_0) \to (X, x_0)$ and the set of subgroups of $\pi_1(\tilde{X}, \tilde{x}_0)$, obtained* *by assigning the subgroups $p_*(\pi_1(\tilde{X}, \tilde{x}_0))$ to the corresponding covering spaces of $(X, x_0)$. If the base points are ignored, this correspondence gives a bijection between isomorphism classes of path connected covering and locally path-connected space $X$, the covering spaces $p : \tilde{X} \to X$ and conjugacy classes of subgroups of $\pi_1(X, x_0)$.*

**Proof** By using Theorem 5.5.6 the first part follows. To prove the second part, let $\tilde{x}_0, \tilde{x}_1 \in p^{-1}(x_0)$ be two base points of $\tilde{X}$ and

$$\tilde{f} : \mathbf{I} \to \tilde{X} : \tilde{f}(0) = \tilde{x}_0, \ \tilde{f}(1) = \tilde{x}_1.$$

Then $p \circ \tilde{\alpha}$ determines a loop $f$ in $X$, and hence it represents an element, say $g \in \pi_1(X, x_0)$. For the pair of subgroups $H_1 = p_*(\pi_1(\tilde{X}, x_1))$ and $H_0 = p_*(\pi_1(\tilde{X}, x_0))$ of the group $G = \pi_1(X, x_0)$, there is an inclusion

$$g^{-1} H_0 \, g \subset H_1,$$

because, given a loop $\tilde{f}$ at $\tilde{x}_0$, the path $\tilde{f}^{-1} * \tilde{f} * \tilde{f}$ is a loop at $\tilde{x}_1$. It follows similarly that

$$g \, H_1 g^{-1} \subset H_0.$$

Using conjugation by $g^{-1}$ it follows that

$$H_1 \subset g^{-1} H_0 \, g.$$

Hence it follows that

$$g^{-1} H_0 \, g = H_1.$$

The above discussion asserts that a change of the base point from $\tilde{x}_0$ to $\tilde{x}_1$ produces a change from $H_0$ to the conjugate subgroup $H_1 = g^{-1} H_0 \, g$. Conversely, corresponding to a change $H_0$ to a conjugate subgroup $H_1 = g^{-1} H_0 \, g$ in $G$, take a loop $\beta$ representing $g$, which lifts to a path $\tilde{\beta}$ starting at $\tilde{x}_0$ and ending at $\tilde{x}_1 = \tilde{\beta}(1)$. Hence it follows as above that

$$H_1 = g^{-1} H_0 \, g. \qquad \square$$

**Theorem 5.5.8** (Galois correspondence) *Given be path connected and locally path-connected space $X$, the Galois correspondence $\psi$ arising from the function that assigns to each covering space $p : (\tilde{X}, \tilde{x}_0) \to (X, x_0)$ the subgroup $p_*(\pi_1(\tilde{X}, \tilde{x}_0))$ of $\pi_1(X, x_0)$ is a bijection.*

**Proof** Let $X$ be be a path connected and locally path-connected topological space and $\mathcal{C}(\tilde{X}, \tilde{x}_0)$ be the set of all covering spaces $\xi_p = p : (\tilde{X}, \tilde{x}_0) \to (X, x_0)$ and $\mathcal{S}G$ be the set of all subgroups of $\pi_1(X, x_0)$ of the form $p_*(\pi_1(\tilde{X}, \tilde{x}_0))$. Define a correspondence

$$\psi : \mathcal{C}(\tilde{X}, \tilde{x}_0) \to \mathcal{S}G, \ \xi_p \mapsto p_*(\pi_1(\tilde{X}, \tilde{x}_0)).$$

Then $\psi$ is a well-defined set function. By using Proposition 5.4.23 it follows that $\psi$ is injective. Again, since to each subgroup $G$ of $\pi_1(X, x_0)$, there is a covering space $p : (\tilde{X}, \tilde{x}_0) \to (X, x_0)$ with $p_* \pi_1(\tilde{X}, \tilde{x}_0) = G$, it follows by using classification Theorem 5.5.6 that $\psi$ is surjective. This shows that the correspondence $\psi$ is a bijection.                                                                                      $\square$

**Definition 5.5.9** $\psi$ defined in Theorem 5.5.8 is called a **Galois correspondence.**

## 5.6 Homotopy Property of Universal Covering Spaces

This section is devoted to the study universal covering spaces $(\tilde{X}, p)$ of the base space $(X, x_0)$. This study is based on homotopy theory and proves that all universal covering spaces of the same base space are isomorphic and establishes their relations with base spaces $X$ by considering the conjugacy class of the trivial subgroup $\{0\}$ of $\pi_1(X, x_0)$.

**Definition 5.6.1** Given a topological space $X$, a covering space $(\tilde{X}, p)$ of $X$, where $X$ is simply connected, ( if it exists in the sense that the space $\tilde{X}$ is path connected and $\pi_1(\tilde{X}, \tilde{x}_0) = \{0\}$ for every $\tilde{x}_0 \in \tilde{X}$ ) ) is called a **universal covering space** of $X$.

***Example 5.6.2*** (i) For $p : \mathbf{R} \to S^1$, $t \mapsto e^{2\pi t}$, the covering space $(\mathbf{R}, p)$ is a universal covering space of $S^1$, because the space $\mathbf{R}$ is simply connected.
(ii) If $p_n : S^n \to \mathbf{R}P^n$ is the map which identifies the antipodal points of $S^n$, then the covering space

$$(S^n, p_n)$$

is universal covering space of $\mathbf{R}P^n$.

Theorem 5.6.3 asserts that all universal covering spaces of the same base space are isomorphic proving its uniqueness up to isomorphism of the same base space, and hence it is called the 'universal covering space'. Consequently, henceforth a universal covering space of a base space is called the universal covering space.

**Theorem 5.6.3** *(i) All universal covering spaces of the same base space are isomorphic.*
*(ii) Given the universal covering space $(\tilde{X}, p)$ of $X$ and a covering space $(\tilde{Y}, q)$ of $X$, there exists a continuous map*

$$\tilde{p} : \tilde{X} \to \tilde{Y}$$

*with the property that $(\tilde{X}, \tilde{p})$ is a covering space of $\tilde{Y}$.*

***Proof*** (i) All universal covering spaces of the same base space $X$ are isomorphic, because any universal covering space of $X$, determines the conjugacy class of the trivial subgroup of $\pi_1(X, x_0)$ by Theorem 5.4.18.

**Fig. 5.29** Lifting $\tilde{p}$ of $p$

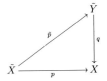

(ii) To prove it, consider the commutative triangle in Fig. 5.29, and take arbitrary base points $\tilde{x}_0 \in \tilde{X}$, $\tilde{y}_0 \in \tilde{Y}$ and $x_0 \in X$ with the property

$$p(\tilde{x}_0) = q(\tilde{y}_0) = x_0.$$

Clearly,

$$p_* \pi_1(\tilde{X}, \tilde{x}_0) \subset q_* \pi_1(\tilde{Y}, \tilde{y}_0),$$

because $\pi_1((\tilde{X}, \tilde{x}_0)) = \{0\}$. The existence of a continuous map $\tilde{p} : (\tilde{X}, \tilde{x}_0) \to (\tilde{Y}, \tilde{y}_0)$ with the property

$$q \circ \tilde{p} = p$$

is proved by using the Lifting Theorem 5.4.12. This shows that $\tilde{p}$ is a covering projection and hence $(\tilde{X}, \tilde{p})$ is a covering space of $\tilde{Y}$.  □

## 5.7 Homotopy Properties of Vector Bundles and Their Homotopy Classification

This section continues the study of vector bundles initiated in Chap. 4. The present study is devoted in homotopy properties of vector bundles and proves two theorems on the homotopy classification of vector bundles formulated in Theorem 5.7.8 and Corollary 5.7.10.

The strong structure of vector bundles facilitates to prove that there exists a natural bijective correspondence between the set of isomorphism classes of $n$-dimensional vector bundles over a paracompact space $B$ and the set of homotopy classes of maps from $B$ into a Grassmann manifold of $n$-dimensional subspaces in an infinite-dimensional space. This result leads to define topological K-theory by Grothendick in 1961. This theory closely relates algebraic topology with several other branches such as algebraic geometry, analysis, ring theory and number theory. Interested readers are referred to the books [Adhikari, 2016] and [Husemoller, 1966].

**Fig. 5.30** Commutative
diagram for the induced
bundle $(Y, q, A)$

## 5.7.1 Homotopy Properties of Vector Bundles

**Definition 5.7.1** (*Induced bundle*) Let $\xi = (X, p, B)$ be a vector bundle and $f : A \to B$ be a a continuous map from a topological space $A$. Then the induced vector bundle $f^*(\xi) = (Y, q, A)$ of $\xi$ over $A$ under $f$ is the vector bundle $(Y, q, A)$, where the total space $Y$ is defined by

$$Y = \{(a, x) \in A \times X : f(a) = p(x)\}$$

and the projection $q$ is defined by

$$q : Y \to A, (a, x) \mapsto a.$$

If $g : Y \to X, (a, x) \mapsto x$, then the pair of maps $(g, f) : (Y, q, A) \to (X, p, B)$ make the diagram in Fig. 5.30 commutative. The pair of maps

$$g : Y \to X, (a, x) \mapsto x \text{ and } f : A \to B$$

form a bundle morphism $(g, f) : f^*(\xi) \to \xi$ called the **canonical morphism** of the induced bundle. The map $q$ is a fiber preserving map.

Theorem 5.7.2 proves that for any vector bundle $\xi$ over a paracompact base space $A$, any pair of homotopic maps $f, g : B \to A$ induce $B$-isomorphic vector bundles, which is an important homotopy property of vector bundles.

**Theorem 5.7.2** *Given a paracompact space $B$, a vector bundle $\xi$ over $A$, and a pair of homotopic maps $f, g : B \to A$, there exists a $B$-isomorphism between the induced bundles $f^*(\xi)$ and $g^*(\xi)$ over the base $B$.*

**Proof** By hypothesis, $B$ is paracompact, $\xi$ is a vector bundle over $A$ and $f, g : B \to A$ is a pair of homotopic maps. Then there exists a homotopy

$$G : B \times \mathbf{I} \to A : G(x, 0) = f(x), G(x, 1) = g(x), \forall x \in B.$$

Hence it follows from Exercise 5.24.1 of Sect. 5.24 that

(i) the induced vector bundles $f^*(\xi)$ and $G^*(\xi)|(B \times \{0\})$ are $B$-isomorphic and
(ii) the induced bundles $g^*(\xi)$ and $G^*(\xi)|(B \times \{1\})$ are $B$-isomorphic.

Finally, since there exists a $B$-isomorphism

$$(\alpha, \beta) : G^*(\xi) : (B \times \{0\}) \to G^*(\xi)|(B \times \{1\})$$

by using the same exercise, it is proved that the induced vector bundles $f^*(\xi)$ and $g^*(\xi)$ are $B$-isomorphic.                                                        □

**Corollary 5.7.3** *Let $B$ be a contractible paracompact space. Then every vector bundle over $B$ is trivial.*

**Proof** By hypothesis, $B$ is a contractible paracompact space. By using the contractibility property of $B$, it follows that the identity map $1_B : B \to B$ is homotopic to a constant map $c : B \to B$. If $\xi$ is an $n$-dimensional vector bundle over $B$, then it follows that

(i)  $1_B^*(\xi)$ and $\xi$ are $B$-isomorphic and
(ii)  $f^*(\xi)$ is $B$-isomorphic to the product bundle $(B \times F^n, p, B)$.

Finally, since the maps $1_B \simeq c$, by using Theorem 5.7.2, it follows that the vector bundle $\xi$ is $B$-isomorphic to the product bundle $(B \times F^n, p, B)$ which is a trivial bundle.                                                                       □

## 5.7.2   Homotopy Classification of Vector Bundles

This subsection solves the homotopy classification problems of vector bundles by proving two basic results formulated in Theorem 5.7.8 and in Corollary 5.7.10. One result says that there is a bijective correspondence between isomorphism classes of $n$-dimensional vector bundles over a paracompact space $B$ and the homotopy classes of maps from $B$ to Grassmann manifold $G_n(F^\infty)$ and the other one says that every $n$-dimensional $\mathbf{F}$-vector bundle over a paracompact space $B$ is isomorphic to the vector bundle induced by a continuous map from the base space $B$ to the Grassmann manifold $G_n(\mathbf{F}^\infty)$.

This section uses the following notations

(i)  $\mathbf{F}$ stands for $\mathbf{R}$, $\mathbf{C}$ or $\mathbf{H}$.
(ii)  $G_n(\mathbf{F}^m)$ stands for Grassman manifold of $n$-dimensional subspaces of the vector space $\mathbf{F}^n$.
(iii)  $\mathcal{H}$ stands for the category of paracompact spaces and their homotopy classes.
(iv)  $\mathcal{S}et$ stands for the category of sets and their functions.
(v)  $\mathrm{Vect}_n(B)$ stands for the set of $B$-isomorphic classes of $n$-dimensional vector bundles over $B$.
(vi)  For an $n$-dimensional vector bundle $\xi$ over $B$, the family $\{\xi\}$ stands for the B-isomorphism classes in $\mathrm{Vect}_n(B)$ of $\xi$, and $[f]$ stands for the homotopy class of $f : A \to B$ between two paracompact spaces $A$ and $B$.

Theorem 5.7.4 proves the functorial property of $\mathrm{Vect}_n(-)$.

**Theorem 5.7.4** $\text{Vect}_n : \mathcal{H} \to \mathcal{Set}$ *is a contravariant functor from the category of paracompact spaces and their homotopy classes to the category of sets and set functions .*

**Proof** The assignment $B \mapsto \text{Vect}_n(B)$ for every object $B$ in $\mathcal{H}$ defines the object function

$$\text{Vect}_n : \mathcal{H} \to \mathcal{Set}, \ B \mapsto \text{Vect}_n(B)$$

and the assignment $\{\xi\} \mapsto \{f^*(\xi)\}$ for every morphism $[f]$ in $\mathcal{H}$, where $A, B$ are paracompact spaces and $f : A \to B$ is a continuous map, defines the morphism function

$$\text{Vect}_n([f]) : \text{Vect}_n(B) \to \text{Vect}_n(A) : \{\xi\} \mapsto \{f^*(\xi)\}.$$

The above functions are well-defined by Theorem 5.7.2. Moreover, they satisfy the properties

(i) For the identity map $1_B : B \to B$, the vector bundles $1_B^*(\xi)$ and $\xi$ are $B$-isomorphic.
(ii) If $[g]$ denotes the homotopy class of $g : C \to A$ between the paracompact spaces $C$ and $A$, then the induced vector bundles $g^*(f^*(\xi))$ and $(f \circ g)^*(\xi)$ are such that they are $C$-isomorphic.

These functorial properties of $\text{Vect}_n$ prove that

$$\text{Vect}_n : \mathcal{H} \to \mathcal{Set}$$

is a contravariant functor. □

**Definition 5.7.5** (*Grassmann manifold* $G_n(\mathbf{F}^\infty)$) The natural inclusion

$$G_n(\mathbf{F}^m) \subset G_n(\mathbf{F}^{m+1}) \subset G_n(\mathbf{F}^{m+2}) \subset \cdots$$

defines a topological space $G_n(\mathbf{F}^\infty) = \bigcup_{n \leq m} G_n(\mathbf{F}^m)$ with weak topology, called Grassmann manifold $G_n(\mathbf{F}^\infty)$.

**Remark 5.7.6** If $\gamma_n^\infty$ denotes the $n$-dimensional vector bundle over Grassmann manifold $G_n(\mathbf{F}^\infty)$, then $\phi_n = [-, G_n(\mathbf{F}^\infty)] : \mathcal{H} \to \mathcal{Set}$ is a contravariant functor.

Theorem 5.7.7 proves the equivalence of two contravariant functors $[-, G_n(\mathbf{F}^\infty)]$ and $\text{Vect}_n$ from the category $\mathcal{H}$ of paracompact spaces and their homotopy classes to the the category $\mathcal{Set}$ of sets and set functions. This equivalence plays a key role to solve a homotopy classification problem of vector bundles in Theorem 5.7.8.

**Theorem 5.7.7** *There exists a natural equivalence*

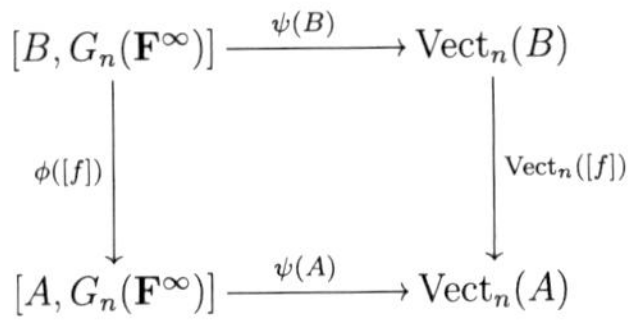

**Fig. 5.31** Rectangular
diagram involving natural
equivalence

$$[B, G_n(\mathbf{F}^\infty)] \xrightarrow{\psi(B)} \mathrm{Vect}_n(B)$$

$$\phi([f]) \downarrow \qquad\qquad\qquad \downarrow \mathrm{Vect}_n([f])$$

$$[A, G_n(\mathbf{F}^\infty)] \xrightarrow{\psi(A)} \mathrm{Vect}_n(A)$$

$$\psi : [-, G_n(F^\infty)] \to \mathrm{Vect}_n$$

*between the two contravariant functors.*

***Proof*** For every object $B \in \mathcal{H}$, define

$$\psi(B) : [B, G_n(\mathbf{F}^\infty)] \to \mathrm{Vect}_n(B), \ [f] \mapsto \{f^*(\gamma_n^\infty)\}.$$

Then $\psi(B)$ is a well-defined morphism in the category $\mathcal{S}et$.

To show that $\psi$ is a natural transformation, consider the homotopy class $[f]$ of the continuous map $f : A \to B$ between paracompact spaces $A$ and $B$ and also consider the rectangular diagram in Fig. 5.31

The diagram in Fig. 5.31 is commutative, because, for any $[h] \in [B, G_n(\mathbf{F}^\infty)]$,

$$(\mathrm{Vect}_n([f]) \circ \psi(B))([h]) = \mathrm{Vect}_n([f])\{h^*(\gamma_n^\infty)\} = \{(f^*h^*(\gamma_n))\}.$$

Consequently, $\psi$ is a natural transformation. To prove that $\psi$ is a natural equivalence, it is sufficient to prove that for every object $B \in \mathcal{H}$, the map $\psi(B)$ is a bijection.

Consider

$$\psi(A)\phi([f][h]) = \psi(A)([h \circ f]) = \{(h \circ f)^*(\gamma_n^\infty)\}.$$

Clearly, for every $B \in \mathcal{H}$, the map $\psi(B)$ is injective by Exercise 5.24.1 of Sect. 5.24 and is also surjective by Exercise Ex:5.13(4) 5.24.1 of Sect. 5.24, Hence it implies that for each $B \in \mathcal{H}$, the map $\psi(B)$ is a bijection. It proves that $\psi$ is an equivalence. $\qquad\square$

Theorem 5.7.8. solves a homotopy classification problem of vector bundles over a paracompact space $B$ in Theorem 5.7.8 by using the homotopy classes of maps from $B$ to Grassmann manifold $G_n(F^\infty)$.

**Theorem 5.7.8** (Homotopy classification of vector bundles) *There exists a bijective correspondence between the set of isomorphism classes of n-dimensional F-vector bundles on a paracompact space B and the set of homotopy classes of free continuous maps from B to Grassmann manifold $G_n(\mathbf{F}^\infty)$.*

***Proof*** For every object $B \in \mathcal{H}$, define the map

$$\psi(B) : [B, G_n(\mathbf{F}^\infty)] \to \mathrm{Vect}_n(B), \ [f]) \mapsto \{f^*(\gamma_n^\infty)\}.$$

Since $\psi(B)$ is a bijection by Theorem 5.7.7, the theorem follows. $\qquad\square$

**Definition 5.7.9** (*Representation of the functor* $\mathrm{Vect}_n$) The natural equivalence $\psi :$ $[-, G_n(\mathbf{F}^\infty)] \to Vect_n$, proved in Theorem 5.7.7, is called a **representation of the contravariant functor** $\mathrm{Vect}_n$.

**Corollary 5.7.10** (*Classification of vector bundles*) *Let* $\xi = (X, p, B, \mathbf{F}^n)$ *be an arbitrary $n$-dimensional $\mathbf{F}$-vector bundle over a paracompact space $B$. Then it is isomorphic to the vector bundle induced by a continuous map from the base space $B$ to the Grassmann manifold $G_n(\mathbf{F}^\infty)$.*

**Proof** By hypothesis, $\xi = (X, p, B, \mathbf{F}^n)$ is an $n$-dimensional $\mathbf{F}$-vector bundle over a paracompact space $B$. Since $\psi(B)$ is a bijection as proved in Theorem 5.7.7, it follows that exists a continuous map $f : B \to G_n(\mathbf{F}^\infty)$ such that $f^*(\{\gamma_n^\infty\}) = \{\xi\}$. This proves that vector bundles $f^*(\gamma_n^\infty)$ and $\xi$ are $B$-isomorphic. $\qquad\square$

## 5.8 Homotopy Properties of Numerable Principal $G$-Bundles

This section is devoted to the study of numerable principal $G$-bundles over $B$ associated with a given topological group $G$. This study is based on the viewpoint of homotopy theory and provides a contravariant functor

$$K_G : \mathcal{H}tp \to \mathcal{S}et,$$

which plays an key role in the study of homotopy theory.

**Theorem 5.8.1** *Let $\eta$ be a numerable principal $G$-bundle over $B \times \mathbf{I}$.*

(i) *Then the three bundles $\eta$, $(\eta|(B \times \{1\})) \times \mathbf{I}$ and $(\eta|(B \times \{0\})) \times \mathbf{I}$ are $G$-isomorphic.*

(ii) *If the map $h_t : B \to B \times \mathbf{I}, b \mapsto (b, t)$, then the principal $G$-bundles $h_0^*(\eta)$ and $h_1^*(\eta)$ are $B$-isomorphic.*

**Proof** Consider the map $f : B \times \mathbf{I} \to B \times \mathbf{I}$, $(b, t) = (b, 1)$ Then by Exercise 5.24.1 of Sect. 5.24, the bundles $\eta$ and $f^*(\eta)$ are isomorphic principal $G$-bundles over $B \times \mathbf{I}$. Since the bundles $f^*(\eta)$ and $(\eta|(B \times \{1\}) \times \mathbf{I})$ are isomorphic principal $G$-bundles over $B \times \mathbf{I}$, it follows that $\eta$ and $(\eta|(B \times \{1\}) \times \mathbf{I})$ are also isomorphic principal $G$-bundles. Analogously, the bundles $\eta$ and $(\eta|(B \times \{0\})) \times \mathbf{I})$ are also isomorphic principal $G$-bundles. This proves that the bundles $\eta$, $(\eta|(B \times \{1\})) \times \mathbf{I}$ and $(\eta|(B \times \{0\})) \times \mathbf{I}$ are $G$-isomorphic. This gives part (i). For part (ii), since $f \circ h_0 = h_1$ and the bundles $f^*(\eta)|(B \times \{0\})$ and $\eta|(B \times \{0\})$ are $G$-isomorphic, it is proved that the induced bundles $h_1^*(\eta) = h_0^* \circ f^*(\eta)$ and $h_0^*(\eta)$ are isomorphic principal $G$-bundles. $\qquad\square$

**Corollary 5.8.2** *Let $\eta = (X, p, B)$ be a numerable principal G-bundle and $f, g$ : $A \to B$ be two homotopic maps. Then their induced bundles $f^*(\eta)$ and $g^*(\eta)$ are isomorphic principal G-bundles over A.*

**Proof** By hypothesis, $f \simeq g : A \to B$, consider the homotopy $H_t : f \simeq g : A \to B$ and the map $h_t : A \to A \times \mathbf{I}$, $a \mapsto (a, t) : t = 0, 1$. Then $H_0 \circ h_0 = f$ and $H_1 \circ h_1 = g$ assert by Theorem 5.8.1 that the bundles $f^*(\eta)$ and $g^*(\eta)$ are isomorphic principal G-bundles over $A$. $\qquad\square$

**Theorem 5.8.3** *Given any topological space B, let $K_G(B)$ denote the set of isomorphism classes of numerable principal G-bundles over B. Then there exists a contravariant functor*

$$K_G : \mathcal{H}tp \to \mathcal{S}et$$

*from the homotopy category $\mathcal{H}tp$ of topological spaces and their homotopy classes of maps to the category $\mathcal{S}et$ of sets and their functions.*

**Proof** To prove the theorem the object function and morphism functions are defined as follows:

(i) For every object $B \in Htp$, the object $K_G(B) \in \mathcal{S}et$ is defined to be the set of isomorphism classes of numerable principal G-bundle over $B$. Hence the assignment

$$K_G : \mathcal{H}tp \to \mathcal{S}et, B \mapsto K_G(B)$$

defines the object function and

(ii) for the homotopy class $[f]$ of every continuous map $f : A \to B$, the function

$$K_G([f]) : K_G(B) \to K_G(A), [f] \mapsto \{f^*(\xi)\}$$

is well-defined by Corollary 5.8.2, and it defines the morphism function.

Finally, given any two continuous maps $f : A \to B$ and $g : B \to C$ and a numerable principal G-bundle $\eta$ over $C$, the induced bundle $(g \circ f)^*(\eta)$ and $f^*(g^*(\eta))$ are isomorphic over $C$. This implies that

$$K_G([g] \circ [f]) = K_G([f]) \circ K_G([g]).$$

Similarly, the function $K_G([1_C])$ is the same as the identity function on $K_G([C])$, since $\eta$ and $1_{C^*}(\eta)$ are isomorphic. This proves that $K_G$ is a contravariant functor. $\square$

**Corollary 5.8.4** *For any numerable principal G-bundle $\xi = (X, p, B, G)$, every homotopy equivalence $f : A \to B$ induces a bijection*

$$K_G([f]) : K_G(B) \to K_G(A).$$

**Proof** If $f : A \to B$ is a homotopy equivalence, then there exists a continuous map

$$g : B \to A : g \circ f \simeq 1_A \text{ and } f \circ g \simeq 1_B.$$

Since $K_G$ is a contravariant functor by Theorem 5.8.3, it is proved that

$$K_G([g \circ f]) = K_G([f]) \circ K_G([g]) : K_G(A) \to K_G(A)$$

is the identity map. This implies that $K_G([f])$ is a surjective map. Similarly, it follows that

$$K_G([f \circ g]) = K_G([g]) \circ K_G([f]) : K_G(B) \to K_G(B)$$

is the identity map. This implies the map $K_G([f])$ is injective. Consequently the map $K_G([f])$ is a bijection. $\qquad\square$

**Corollary 5.8.5**  *Every numerable principal $G$-bundle $\xi = (X, p, B, G)$ over a contractible space $B$ is trivial.*

**Proof**  By hypothesis, the base space $B$ of $\xi$ is contractible. Hence the space $B$ is homotopy equivalent to a point $\{*\}$. Since the set $K_G(\{*\})$ consists of only one point, which is precisely, the isomorphism class of the trivial bundle, corollary is proved by using Corollary 5.8.4. $\qquad\square$

**Theorem 5.8.6**  *Let $\xi_0 = (X, p_0, B_0, G)$ be a fixed numerable principal $G$-bundle. Then*

$$\psi_{\xi_0} : \mathcal{H}tp \to \mathcal{S}ets, \ [-, B_0] \to K_G$$

*is a covariant functor.*

**Proof**  By hypothesis, $\xi_0 = (X, p_0, B_0, G)$ is a fixed numerable principal $G$-bundle. Let $\xi = (X, p, B, G)$ be an arbitrary numerable principal $G$-bundle. Keeping $B_0$ fixed and varying $B$, define the object function

$$\mathcal{H}tp \to \mathcal{S}ets, \ B \mapsto K_G(B) = [B, B_0]$$

and the morphism function

$$\psi_{\xi_0}(B)([f]) : \mathcal{H}tp \to \mathcal{S}ets, \ [f] \mapsto [f^*(\xi_0)], \ \forall [f] \in [B, B_0].$$

Both the functions are well-defined by Theorem 5.8.3 and its Corollary 5.6.3. This proves that

$$\psi_{\xi_0} : [-, B_0] \to K_G$$

is a covariant functors from the homotopy category $\mathcal{H}tp$ to the category $\mathcal{S}et$. $\qquad\square$

**Definition 5.8.7**  *(Universal $G$-bundle)* A principal $G$-bundle $\xi_0 = (X, p_0, B_0, G)$ is said to be universal if

(i)  $\xi_0$ is numerable and

(ii) $\psi_{\xi_0} : [-, B_0] \to K_G$ defined in Theorem is a natural equivalence.

Theorem 5.8.8 characterizes universal G-bundles in terms of homotopic maps from one base space to the other base space.

**Theorem 5.8.8** *Let* $\xi_0 = (X_0, p_0, B_0, G)$ *be a numerable principal G-bundle. It is universal iff*

(i) *for every numerable principal G-bundle* $\xi = (X, p, B, G)$, *there exists a continuous map* $h : B \to B_0$ *with the property that* $\xi$ *and* $h^*(\xi_0)$ *are B-isomorphic; and*

(ii) *if* $f, g : B \to B_0$ *are two continuous maps with the property that* $f^*(\xi_0)$ *and* $g^*(\xi_0)$ *are isomorphic, then* $f \simeq g$.

**Proof** Under the condition (i) it follows that the map

$$\psi_{\xi_0}(B) : [B, B_0] \to K_G(B)$$

is surjective. Again, under condition (ii), it follows that

$$\psi_{\xi_0}(B) : [B, B_0] \to K_G(B)$$

is injective. Hence the theorem follows. □

## 5.9  Classifying Spaces: The Milnor Construction

This section describes Milnor method of construction of classifying spaces and also universal principal fiber spaces. He started with topological $G$ and take an join $X_G$ of $G$. He defines an action $\sigma : X_G \times G \to X_G$ and a topology on $X_G$ to make it a $G$-space. The orbit space $X_G mod G$ thus obtained is denoted by $B_G$, and then the projection map $p : X_G \to B_G$ is continuous and onto. The resulting bundle $\omega_G = (X_G, p, B_G)$ is a numerable (universal) principal $G-$ bundle. The space is called the classifying space of $K_G$. This method of construction is known as Milnor construction.

The construction of the bundle $\omega_G = (X_G, p, B_G)$ described in Theorem 5.9.1 is known as Milnor construction. The bundle $\omega_G = (X_G, p, B_G)$ provides an important family of numerable principal G-bundle.

**Theorem 5.9.1** (Milnor construction) *Given a topological group G, the universal fiber space* $X_G$ *is defined as an infinite join*

$$X_G = G * G * \cdots * G$$

*and an element* $\langle x, t \rangle \in X_G$ *is denoted by*

$$\langle x, t \rangle = (t_0 x_0, t_1 x_1, \ldots, t_r x_r, \ldots),$$

*where each $x_i \in G$ and $t_i \in [0, 1] = \mathbf{I}$ such that only a finite number $t_i \neq 0$ and*
$$\sum_{t_i \geq 0} t_i = 1.$$

*Two elements $\langle x, t \rangle = \langle x', t' \rangle \in X_G$ are said to be equal iff*

*(i)  $t_i = t_i'$ for every $i$ and*
*(ii)  $x_i = x_i'$ for every $i$ with $t_i = t_i > 0$.*

*Define an action $\sigma$ of $G$ on $X_G$ from the right*

$$\sigma : X_G \times G \to X_G, \langle x, t \rangle g = \langle xg, tg \rangle \text{ or } (t_0 x_0, t_1 x_1, \ldots)g = (t_0 x_0 g, t_1 x_1 g_1, \ldots)$$

*and topologize the set $X_G$ in such a way that $X_G$ admits a G-space structure. Consider two families of maps*
$$f_i : X_G \to \mathbf{I}, \forall i \geq 0,$$

*which assigns to the element $(t_0 x_0, t_1 x_1, \ldots) \in X_G$ the component $t_i \in [0, 1]$ and*

$$g_i : f_i^{-1}(0, 1] \to G, \forall i \geq 0,$$

*which assigns to the element $(t_0 x_0, t_1 x_1, \ldots)$ the component $x_i \in G$. Then $x_i$ cannot be uniquely defined outside $f_i^{-1}(0, 1]$ in a natural way. Because, if $f \in X_G$ and $g \in G$, then there exist the following relations between the action of $G$ and the maps $f_i$ and $g_i : g_i(fg) = g_i(\alpha)g$ and $f_i(fg) = f_i(\alpha)$. Endow the smallest topology on the set $X_G$ such that every map $f_i : X_G \to [0, 1]$ and $g_i : f^{-1}(0, 1] \to G$ is continuous, with $f_i^{-1}(0, 1]$ the subspace topology. From the definition of the action sigma, it follows that $X_G$ is a G-space with the G-set structure map*

$$\sigma : X_G \times G \to X_G$$

*continuous. Denote*

*(i)   the orbit space by $X_G$ mod $G$ by $B_G$,*
*(ii)  the quotient map by $p : X_G \to B_G$ and*
*(iii) the resulting bundle $\omega_G = (X_G, p, B_G)$.*

*This bundle $\omega_G = (X_G, p, B_G)$ constructed by Milnor is a **numerable principal G-bundle**.*

Example 5.9.2 illustrates Milnor construction at some concrete situation to obtain a numerable principal G-bundle $\omega_G$, where the topological groups $G$ are $\mathbf{Z}_2$, $S^1$ and $S^3$.

*Example 5.9.2*   **(Numerable principal G-bundle)**

(i)  Consider the n-sphere $S^n$ in $\mathbf{R}^{n+1}$ for $n \geq 1$ and the antipodal map

$$f : S^n \to S^n, x \mapsto -x.$$

Then $f^2 = f \circ f$ is identity map $1_{S^n}$ and the group $G = \{1_{S^n}, f\} \cong \mathbf{Z}_2$, which is a subgroup of the group of homeomorphisms of $S^n$. Here the space $X_G = S^n$. Define the action $\sigma$ of $G$ on $X_G = S^n$

$$\sigma : S^n \times G \to S^n, (x, g) \mapsto \begin{cases} x \text{ if } g = 1_{S^n} \\ -x \text{ if } g = f. \end{cases}$$

This asserts that the quotient space $X_G$ mod $G = B_G = \mathbf{R}P^n$ and the resulting bundle $w_G = (S^n, p, \mathbf{R}P^n)$ is a numerable principal G-bundle of dimensions $\leq n - 1$.

(ii)  Let $G = S^1$ be the unit circle in the complex plane $\mathbf{C}$ and $S^{2n+1}$ be the $(2n + 1)$-sphere in $\mathbf{R}^{2n+2}$ ( identified with $\mathbf{C}^{n+1}$). This topological group $G = S^1$ is regarded as a subgroup $S^1 \subset \mathbf{C} - \{0\}$ with usual multiplication of nonzero complex numbers. Consider the right action $\sigma$ of $S^1$ on $S^{2n+1}$

$$\sigma : S^{2n+1} \times S^1 \to S^{2n+1}, ((z_0, z_1, \ldots, z_n), e^{i\theta}) \mapsto (e^{i\theta}z_0, e^{i\theta}z_1, \ldots, e^{i\theta}z_n).$$

Let $X_G = G * G * \cdots * G$ be the infinite join. Then $X_G$ mod $G = B_G = \mathbf{C}P^n$ is the complex $n$-dimensional projective space. This implies that $w_G = (S^{2n+1}, p, \mathbf{C}P^n)$, is a principal numerable G-bundle of dimensions $\leq 2n$.

(iii)  Let $G = S^3$. This topological group is regarded as a subgroup $S^3 \subset \mathbf{H} - \{0\}$ with usual multiplication of nonzero quaternionic numbers. Let $X_G = G * G * \cdots * G$ be the infinite join. Then proceeding as before, it follows that $w_G = (S^\infty, p, \mathbf{H}P^\infty)$, where $S^\infty$ denotes the infinite dimensional sphere in infinite dimensional quaternionic space $\mathbf{H}^\infty$ with weak topology, and $\mathbf{H}P^\infty$ denotes infinite dimensional quaternionic projective space.

The concept of classifying spaces introduced by Milnor in 1956 is formulated in Definition 5.9.3.

**Definition 5.9.3** (*Classifying space of $K_G$*) Given a topological group $G$, the functor

$$K_G : \mathcal{H}tp \to \mathcal{S}et$$

is said to be a representation of the functor $K_G$, if there exist a space $B_G$, called the classifying space of $K_G$ and a bundle $\xi_G = (X_G, p_G, B_G)$, called universal bundle in $K_G(B_G)$, such that there is a natural equivalence

$$\psi : K_G \to [-, B_G]$$

of functors defined from the category $\mathcal{H}tp$ to the category $\mathcal{S}et$.

**Remark 5.9.4**  It follows from Definition 5.9.3 that

(i)  for any topological group $G$, the map

$$\psi(B) : [B, B_G] \to K_G(B), [f] \mapsto [f^*(\xi_G)])$$

  is a bijection and
(ii)  for any topological space which is homotopy equivalent to $B_G$ is also a classifying space for $K_G$.

**Theorem 5.9.5**  (Milnor) *Let $G$ be a topological group. Then the $G$-bundle $\omega_g = (X_G, p, B_G)$ is a numerable principal $G$-bundle, and this bundle is a universal $G$-bundle.*

**Proof**  It follows from Theorem 5.9.1 by using Milnor method of construction of $\omega_g = (X_G, p, B_G)$.  □

## 5.10  Applications and Computations

This section presents an application of Galois correspondence arising from the function that assigns to each covering space $p : (X, x_0) \to (B, b_0)$ the subgroup $p_*(\pi_1(X, x_0))$ of $\pi_1(B, b_0)$ and communicates applications of covering to compute fundamental groups of some important spaces such as real projective spaces, Klein bottle, lens space and figure-eight.

### 5.10.1  Application of Galois Correspondence Theorem

This subsection communicates an application of Galois correspondence $\psi$ formulated in Theorem 5.5.8 by using the action $\sigma$ of fundamental group of the base space of a covering space on its fiber formulated in Theorem 5.10.2.

**Theorem 5.10.1**  *Let $(X, p)$ be a covering space over a connected, locally path connected and semilocally simply connected base space $B$. Then*

(i)  *the components of the total space $X$ are in bijective correspondence with orbits obtained by the action of the fundamental group $\pi_1(B, b_0)$ on the fiber $p^{-1}(b_0)$ over $b_0 \in B$ and*
(ii)  *the subgroup assigned to the component of $X$ containing a given lifting $\tilde{b}_0 = x_0$ of the point $b_0$ by the Galois correspondence $\psi$ is the stabilizer group $G_{x_0}$ of $x_0 \in X$, where the action $\sigma$ on the fiber keeps $x_0$ fixed.*

**Proof**  By hypothesis, $(X, p)$ is a covering space over a connected, locally path-connected and semilocally simply connected base space $B$. Let $x_0, x_1 \in p^{-1}(b_0)$ be two arbitrary points.

(i)  Suppose that the points $x_0$ and $x_1$ lie in the same component of $X$. Then there exists a path

$$\beta : \mathbf{I} \to X : \beta(0) = x_0 \text{ and } \beta(1) = x_1.$$

Then $[p \circ \beta]$ is an element of $\pi_1(B, b_0)$ such that

$$\sigma : p^{-1}(b_0) \times \pi_1(B, b_0), (x_0, [p \circ \beta]) \mapsto x_1.$$

This asserts that the action $\sigma$ is transitive. On the other hand, if $x_0, x_1 \in p^{-1}(b_0)$ are points lying in different components of $X$, then there exists no path connecting them. Consequently, there exists a bijective correspondence between the components of the total space $X$ and the orbits obtained by the action $\sigma$ of the fundamental group $\pi_1(B, b_0)$ on the fiber $p^{-1}(b_0)$, because the set of elements of $p^{-1}(b_0)$ in a given component forms an orbit. Hence it follows that this correspondence is a bijection.

(ii)  Suppose that the point $x_0 \in p^{-1}(b_0)$ is an arbitrary points and $x_0$ lies in some component $C_{x_0}$ of $X$. Then by Galois correspondence $\psi$, the subgroup of $\pi_1(B, b_0)$ corresponding to the component $C_{x_0}$ of $X$ is the image $p_*(G)$ of the group $G = \pi_1(C_{x_0}, x_0)$. Since every loop $\beta \in p_*(G)$ lifts back to a loop in $C_{x_0}$ by the unique lifting property, it follows that $\beta$ carries the $x_0$ to itself and hence it is an element of the stabilizer group $G_{x_0}$ of $x_0$ under the action $\sigma$. Conversely, suppose $f \in \pi_1(B, b_0)$ is in the stabilizer group $G_{x_0}$ of $x_0$ under the action $\sigma$. Then the lift $\tilde{f}$ of $\alpha$ and hence $\tilde{f}$ is a loop in $X$ based at the $x_0$. $\tilde{\alpha} \in G$, which implies $\alpha \in p_*(G)$. This proves that $p_*(G)$ is the stabilizer group $G_{x_0}$ of $x_0$. $\square$

## 5.10.2   Actions of Fundamental Groups on Fibers of Covering Spaces

This subsection considers action of the fundamental group of the base space of a covering space on a fiber. This action plays an important role in the study of the covering space.

Theorem 5.10.2 proves some important results obtained by action of the fundamental group of the base space of a covering space on its fiber.

**Theorem 5.10.2** *Let $(X, p)$ be a covering space of $B$, and $F_{b_0} = Y = p^{-1}(b_0)$ be the fiber over $b_0 \in B$. If $X$ is path connected, then the map*

$$\sigma : Y \times \pi_1(B, b_0) \to Y, \ (x, [f]) \mapsto \tilde{f}(1),$$

*(where $\tilde{f}$ is the unique lifting of $f : (\mathbf{I}, 0) \to (B, b_0)$ such that $\tilde{f}(0) = x$ )*

*satisfies the following properties:*

*(i)  the group $\pi_1(B, b_0)$ acts transitively on $Y$;*

*(ii)* *If $y_0 \in Y \subset X$, then the isotropy group $G_{y_0} = p_* \pi_1(X, y_0)$; and*
*(iii)  the cardinality $|Y| = [\pi_1(B, b_0) : p_* \pi_1(X, y_0)]$.*

**Proof** We first show that

$$\sigma : Y \times \pi_1(B, b_0) \to Y, \ (x, [f]) \mapsto \tilde{f}(1)$$

is a right action. Clearly, $\sigma$ is well-defined, because

(i) $\tilde{f}$ is the unique lifting of

$$f : (\mathbf{I}, 0) \to (B, b_0) : \tilde{f}(0) = x$$

and

(iii) $\sigma$ is independent of the choice of the representative of the class $[f]$ by the Monodromy Theorem 5.4.7.

Next, we show that it satisfies the conditions of an action. If $f$ is a constant path at $b_0$, then $\tilde{f}$ is also a constant path at $x \in Y$ such that

$$\sigma(x, [f]) = x \cdot [f] = \tilde{f}(1) = x.$$

If $[f], [g] \in \pi_1(B, b_0)$, and $\tilde{f}$ is the lifting of $f$ with $\tilde{f}(0) = x$ and $\tilde{g}$ is the lifting of $g$ with $\tilde{g}(0) = \tilde{f}(1)$, then $\tilde{f} * \tilde{g}$ is a lifting of $f * g$ that begins at $x$ and ends at $\tilde{g}(1)$. This asserts that $x \cdot [f * g] = (x \cdot [f])[g]$. Hence it follows that $\sigma : Y \times \pi_1(B, b_0) \to Y$ is an action.

(i) To show that $\sigma$ is transitive, take a fixed point $y_0 \in Y$ and any point $x \in Y$. By hypothesis, $X$ is path connected. Hence there is a path $\tilde{\alpha}$ in $X$ from $y_0$ to $x$ such that $p \circ \tilde{\alpha}$ is a closed path in $B$ at $b_0$ having lifting with initial point $y_0$ is $\tilde{\alpha}$. Hence it follows that

$$[p \circ \tilde{\alpha}] \in \pi_1(B, b_0) \ and \ y_0 \cdot [p \circ \tilde{\alpha}] = \tilde{\alpha}(1) = x.$$

This asserts that $\pi_1(B, b_0)$ acts transitively on $Y$.

(ii) To prove (ii), take any closed path $f$ based at the point $b_0 \in B$ and $G = \pi_1(B, b_0)$. If $\tilde{f}$ is the lifting of $f$ with $\tilde{f}(0) = y_0$, then the corresponding isotropy group of $\sigma$ at $y_0$ is given by

$$G_{y_0} = \{[f] \in \pi_1(B, b_0) : y_0 \cdot [f] = y_0\}.$$

This implies that if $[f] \in G_{y_0}$, then $y_0 \cdot [f] = y_0$ and $\tilde{f}(1) = y_0 = \tilde{f}(0)$. Hence it follows that $\tilde{f} \in \pi_1(X, y_0)$. This shows that $G_{y_0} \subset p_* \pi_1(X, y_0)$. On the other hand, if $[f] = [p \circ \tilde{g}]$ for some $[\tilde{g}] \in \pi_1(X, y_0)$, then $\tilde{f} = \tilde{g}$, because both are liftings of $f$ having the same initial point $y_0$. This implies that

$$\tilde{f}(1) = \tilde{g}(1) \Rightarrow y_0 \cdot [f] = \tilde{f}(1) = y_0 \Rightarrow [f] \in G_{y_0}$$

and hence $p_* \pi_1(X, y_0) \subset G_{y_0}$. This proves that the isotropy group $G_{y_0} = p_* \pi_1(X, y_0)$.

(iii) Since the group $\pi_1(B, b_0)$ acts transitively on $Y$, with isotropy group $G_{y_0} = p_* \pi_1(X, y_0)$ at $y_0 \in Y$, it follows that the cardinality $|Y|$ of $Y$ is given by

$$|Y| = [\pi_1(B, b_0) : p_* \pi_1(X, y_0)]. \qquad \square$$

**Corollary 5.10.3** *Let $X$ be a path-connected space and $(X, p)$ be a universal covering space of $B$, and $b_0 \in B$. Then the cardinality of the fiber $Y = p^{-1}(b_0)$ is given by*

$$|Y| = |\pi_1(B, b_0)|.$$

**Proof** Since by the given condition, $\pi_1(X, x_0) = 0$, it follows by using from Theorem 5.10.2(iii)that

$$|Y| = |\pi_1(B, b_0)|. \qquad \square$$

**Corollary 5.10.4** *Let $X$ be a path-connected space. Given a covering space $(X, p)$ of $b_0 \in B$, $x_0 \in p^{-1}(b_0)$, if the induced homomorphism*

$$p_* : \pi_1(X, x_0) \to \pi_1(B, b_0)$$

*is onto, then the map $p : X \to B$ induces an isomorphism of groups*

$$p_* : \pi_1(X, x_0) \to \pi_1(B, b_0).$$

**Proof** $p_*$ is a monomorohism by Corollary 5.4.8, and it is also an epimorphism , since by hypothesis, $p_*$ is onto. This proves that $p_*$ is an isomorphism. $\qquad \square$

### 5.10.3  Fundamental Groups of Orbit Spaces

This subsection computes the fundamental groups of some important spaces which are obtained as orbit spaces. For example, we compute the fundamental groups of projective spaces, lens spaces, figure-eight and Klein's bottles by representing them as orbit spaces.

**Definition 5.10.5** A topological group $G$ with identity $e$ acting on a topological space $X$ is said to satisfy the condition **(A)** :

if for every point $x \in X$, there exists a nbd $U_x$ of $x$ such that

$$\Phi_g(U_x) \cap U_x \neq \emptyset$$

implies $g = e$, where

$$\Phi_g : X \to X, x \mapsto gx$$

is a homeomorphism.

The special group action of the group $G$ of homeomorphisms of $X$ satisfying the condition **(A)** reformulated in Definition 5.10.6 is said to be properly discontinuous.

**Definition 5.10.6**  An action $\sigma$ of a topological group $G$ on a topological space $X$

$$\sigma : G \times X \to X$$

is said to be **properly discontinuous,** if $X$ has a nbd $U$ such that the nbds $gU$ and $kU$ are disjoint for every pair of distinct elements $g, k \in G$.

For the particular case, when $G$ is a discrete topological group is also interesting. For example, consider Example 5.10.7.

**Example 5.10.7**  Let $G_n$ be the finite group of $n$ elements which are precisely the $n$ $n$-th roots of unity and $\mathbf{C}^* = \mathbf{C} - \{0\}$. The action

$$\sigma : G_n \times \mathbf{C}^*, \ (g, z) \mapsto gz \ ( \text{usual multiplication of complex numbers} )$$

is properly discontinuous.

**Example 5.10.8**  The automorphism group $\mathcal{A}ut(X/B)$ of the covering spaces $(X, p)$ of a fixed base space $B$ satisfies the condition **(A)** of Definition 5.10.5.

**Definition 5.10.9**  (*Regular covering space*) Let $(X, p)$ be a covering space of $B$. Then it is said to be regular if $p_* \pi_1(X, x_0)$ is a normal subgroup of $\pi_1(B, b_0)$.

**Example 5.10.10**  Let $B$ be a connected, locally path-connected space and $G$ satisfies the condition **(A)** on $X$ then $(X, p)$ is a regular covering space of $X$ mod $G$, where $p : X \to \mathrm{mod}\,G, x \mapsto Gx$ is the natural projection.

**Theorem 5.10.11**  *Let $\sigma : G \times X \to X$ be an action of a topological group $G$ on $X$ satisfying the condition* **(A)** *of Definition 5.10.5.*

(i)  *For any path-connected space $X$, the group $G$ is the group of automorphisms of the covering space*
$$p : X \to X \text{ mod } G, \ x \mapsto Gx$$

(ii)  *For any is path connected and locally path-connected space $X$, the group $G$ is isomorphic to the quotient group $\pi_1(X \text{ mod } G)/p_*\pi_1(X)$.*

(iii)  *For any simply connected space $X$, the groups $\pi_1(X \text{ mod } G)$ and $G$ are isomorphic.*

**Proof**  (i)  By hypothesis, $X$ is path connected. The automorphism group contains $G$ as a subgroup and equals this group, because, if $f$ is any covering (deck) transformation, then given any point $x \in X$, $x$ and $f(x)$ are in the same orbit and hence there is some $g \in G$ having the property $g(x) = f(x)$. This implies that $f = g$, since covering transformations of a connected covering space are uniquely determined at this situation.

(ii) By hypothesis , $X$ is a path connected and locally path-connected space. Hence this part follows.

(iii) By $X$ is a simply connected space. Then for any point $y \in X$ mod $G$, the fundamental group $\pi_1(X, x_0) = \{e\}$, $\forall x_0 \in p^{-1}(y)$. This implies that $p_* \pi_1(X, x_0) = \{e\}$. Hence part (5.10.11) follows from part (5.10.11). $\qquad\square$

### 5.10.4  Computing the Fundamental Group of $\mathbf{R}P^n$

This subsection computes by the fundamental groups of $\mathbf{R}P^n$ for $n \geq 2$. Theorem 5.10.15 computes it by using group action and shows that $\pi_1(\mathbf{R}R^n) \cong \mathbf{Z}_2$, $\forall n \geq 2$. For the particular case, when $n = 2$, it is computed in an alternative method in Theorem 5.10.18. The $n$-dimensional real projective n-space $\mathbf{R}P^n$ is obtained from the $n$-sphere $S^n = \{x \in \mathbf{R}^{n+1} : ||x|| = 1\}$ by identifying its antipodal points. It is formulated in Definition 5.10.12.

**Definition 5.10.12**  Let $A : S^n \to S^n$, $x \mapsto -x$ be the antipodal map. Then $A \circ A = 1_{S^n}$ and hence $A$ generates an action of the two element group $G = \{+1, -1\}$ defined by the relation $(+1)x = x$ and $(-1)x = -x$ for all $x \in S^n$. This action on $S^n$ has the orbit space $S^n$ mod $G$, denoted by $\mathbf{R}P^n$, called the **real $n$-dimensional projective space.**

Recall the following two theorems for their next applications.

**Theorem 5.10.13**  $(S^2, p)$ is a covering space of $\mathbf{R}P^2$, and $\mathbf{R}P^2$ is a surface.

**Proof**  See Chap. 4. $\qquad\square$

**Theorem 5.10.14**  $(S^n, p)$ is a covering space of $\mathbf{R}P^n$, where $p$ is the map identifying antipodal points of $S^n$ for $n > 1$.

**Proof**  See Chap. 4. $\qquad\square$

**Theorem 5.10.15**  $\pi_1(\mathbf{R}R^n) \cong \mathbf{Z}_2$ for $n \geq 2$.

**Proof**  Consider the action of the $G = \{+1, -1\}$ on $S^n$ given in Definition 5.10.12 and the covering space $(S^n, \mathbf{R}P^n)$ of multiplicity 2. Since $S^n$ is simply connected for $n \geq 2$ ( see Chap. 2), it follows by the Theorem 5.10.11 that the fundamental group of orbit space is $G$. Since $S^n$ mod $G = \mathbf{R}P^n$, it follows that

$$\pi_1(S^n \text{ mod } G) = \pi_1(\mathbf{R}P^n) = G \cong \mathbf{Z}_2 \, \forall n \geq 2. \qquad\square$$

**Remark 5.10.16**  (i) Geometrically, a generator of the group $\pi_1(\mathbf{R}P^n)$ is any loop obtained by projecting a continuous path on the sphere $S^n$, which connects two antipodal points.

(ii)  The above action is free in the sense that $gx = x \Rightarrow g = e$. Does there exist any other finite group $G$ acting freely on $S^n$ and defining covering space $S^n \rightarrow S^n$ mod $G$? The answer is $\mathbf{Z}_2$ is the only nontrivial group that can act freely on $S^n$ if $n$ is even.

Theorem 5.10.17 provides an alternative approach for computing $\pi_1(\mathbf{R}R^n)$ for $n \geq 2$.

**Theorem 5.10.17** $\pi_1(\mathbf{R}R^n) \cong \mathbf{Z}_2$ for $n \geq 2$.

**Proof** **Proof I** As $S^n$ is simply connected for $n \geq 2$, so from the covering space $p : S^n \rightarrow \mathbf{R}P^n$ it follows by the Theorem 5.10.11 that the fundamental group of orbit space is $G$. Thus $\pi_1(S^n \text{ mod } G) = G \Rightarrow \pi(\mathbf{R}^n) = G \cong \mathbf{Z}_2$ for $n \geq 2$.

**Proof II** Consider the universal covering space $(S^n, q)$ of $\mathbf{R}P^n$ where $q$ identifies the antipodal points of $S^n$. Then by using Exercise 5.24.1 of Sect. 5.24, it follows that

$$|\pi_1(\mathbf{R}P^n)| = 2$$

and hence it is proved that $\pi_1(\mathbf{R}P^n) \cong \mathbf{Z}_2$.  □

Theorem 5.10.18 computes $\pi_1(\mathbf{R}P^2, y)$ in a little different method.

**Theorem 5.10.18** $\pi_1(\mathbf{R}P^2, y) \cong \mathbf{Z}_2$.

**Proof** Consider the projection map

$$p : S^2 \rightarrow \mathbf{R}P^2.$$

It is a covering map by Theorem 5.10.13. Since $S^2$ is simply connected, we apply Theorem 5.4.25 to obtain a bijective map $\psi$ between $\pi_1(\mathbf{R}P^2, y)$ and the set $p^{-1}(y)$. This proves that $\pi_1(\mathbf{R}P^2, y) \cong \mathbf{Z}_2$, because $p^{-1}(y)$ is a two-element set and $\pi_1(\mathbf{R}P^2, y)$ is a group of order 2 and any group of order 2 is isomorphic to $\mathbf{Z}_2$.  □

## 5.10.5  Computing the Fundamental Group of Klein's Bottle

This subsection computes the fundamental group of Klein's bottle, which is a well-known topological space constructed in Volume I of the present book series of Basic Topology. Let $G$ be the group of transformations of the Euclidean plane $\mathbf{R}^2$ generated by $\alpha$ and $\beta$. Consider the action $\sigma$ of $G$ on $\mathbf{R}^2$

$$\sigma : G \times \mathbf{R}^2 \rightarrow \mathbf{R}^2, (\alpha, (x, y)), (\beta, (x, y)) \mapsto (x + 1, y), (1 - x, y + 1).$$

This implies that

$$\alpha^{-1}(x, y) = (x - 1, y) \text{ and } \beta^{-1}(x, y) = (1 - x, y - 1).$$

Since $\mathbf{R}^2$ is simply connected and the action $\sigma$ satisfies condition **(A)** of Definition 5.10.5, then by Theorem 5.10.11, it follows that

$$\pi_1(\mathbf{R}^2 \bmod G) \simeq G.$$

The quotient space $\mathbf{R}^2 \bmod G$ is the Klein's bottle obtained by the action of $\sigma$ on $\mathbf{R}^2$ and gives a representation of Klein's bottle as an orbit space whose fundamental group is generated by $\alpha$ and $\beta$, because

$$\begin{aligned}
\beta^{-1}\alpha\beta(x, y) &= \beta^{-1}\alpha(1 - x, y + 1) \\
&= \beta^{-1}(2 - x, y + 1) = (1 - 2 + x, y) \\
&= (x - 1, y) = \alpha^{-1}(x, y), \ \forall\,(x, y) \in \mathbf{R}^2
\end{aligned}$$

asserts that $\beta^{-1}\alpha\beta = \alpha^{-1}$. Hence it follows that the fundamental group of the Klein bottle is generated by $\alpha$ and $\beta$.

### 5.10.6   *Computing the Fundamental Groups of Lens Spaces*

This subsection computes the fundamental group of lens spaces $(L(m, p))$ constructed by H. Tietze (1888–1971) in 1908. They are three-dimensional manifolds and form an important class of topological spaces for the study of algebraic topology.

   **Construction of lens space** $(L(m, p))$: Consider $S^3 = \{(z_1, z_2) \in \mathbf{C}^2 : |z_1|^2 + |z_2|^2 = 1\} \subset \mathbf{C}^2$.

   Let $m > 1$ be an integer space and $p$ be an integer relatively prime to $m$ and $S^3 = \{(z_1, z_2) \in \mathbf{C}^2 : |z_1|^2 + |z_2|^2 = 1\} \subset \mathbf{C}^2 = \mathbf{R}^4$. Let $\rho = e^{\frac{2\pi i}{m}}$ be a primitive $m$-th root of unity.

Define a continuous map

$$f : S^3 \to S^3, (z_1, z_2) \mapsto (\rho z_1, \rho^p z_2) = (e^{\frac{2\pi i}{m}} z_1, e^{\frac{2\pi i p}{m}} z_2).$$

This implies that $f : S^3 \to S^3$ is a homeomorphism of period $m$ in the sense that $f^m = 1_{S^3}$. Consider an action $\sigma$ of $\mathbf{Z}_m$ on $S^3$ induced by the homeomorphism $f$

$$\sigma : \mathbf{Z}_m \times S^3 \to S^3, \ (k, (z_1, z_2)) \mapsto f^k(z_1, z_2).$$

Geometrically, this action $\sigma$ is generated by the rotation $z \mapsto e^{\frac{2\pi i}{m}} z$ of the unit sphere $S^3$, and hence it has no fixed point. Analytically, it is proved that the action $\sigma$ has no fixed point , because given an integer $r : 0 < r < m$, the equation $z = e^{\frac{2\pi i r}{m}} z$ has a solution $z = 0$ but $z = 0$ is not a point of $S^3$.

The orbit spaces $S^3$ mod $\mathbf{Z}_m$ thus obtained is called a **lens space** denoted by $L(m, p)$. Thus the lens space is the quotient space $S^3/\sim$, where

$$(z_1, z_2) \sim (z'_1, z'_2) \Leftrightarrow (z'_1, z'_2) = f^k(z_1, z_2) \quad \text{for some } k \in \mathbf{Z}_m.$$

**Construction of generalized lens space** $L(m, p_1, \ldots, p_{n-1})$: Consider

$$S^{2n-1} = \{(z_1, z_2, \ldots, z_n) \in \mathbf{C}^n : |z_1|^2 + |z_2|^2 + \cdots + |z_n|^2 = 1\} \subset \mathbf{C}^n.$$

Given an integer $m > 1$ and integers $p_1, p_2, \ldots, p_{n-1}$ relatively prime to $m$, let $\rho = e^{\frac{2\pi i}{m}}$ be a primitive $m$-th root of unity. Define a continuous map

$$f : S^{2n-1} \to S^{2n-1}, (z_1, z_2, \ldots, z_n) \mapsto (\rho z_1, \rho^{p_1} z_2, \ldots, \rho^{p_{n-1}} z_n) = (e^{\frac{2\pi i}{m}} z_1, e^{\frac{2\pi i p_1}{m}} z_2, \ldots, e^{\frac{2\pi i p_{n-1}}{m}} z_n)$$

This implies that $f : S^{2n-1} \to S^{2n-1}$ is a homeomorphism and the homeomorphism $f$ induces an action $\sigma$ of $\mathbf{Z}_m$ on $S^{2n-1}$

$$\sigma : \mathbf{Z}_m \times S^{2n-1} \to S^{2n-1}, \ (k, (z_1, z_2, \ldots, z_n)) \mapsto f^k(z_1, z_2, \ldots, z_n).$$

**Theorem 5.10.19** $\pi_1(L(m, p)) \cong \mathbf{Z}_m$.

**Proof** Since the group $\mathbf{Z}_m$ is finite and the space $S^3$ is Hausdorff, the action $\sigma$ of $\mathbf{Z}_m$ on $S^3$ satisfies condition **(A)** of Definition 5.10.5. Hence $\mathbf{Z}_m \simeq \pi_1(S^3 \text{ mod } \mathbf{Z}_m) = \pi_1(L(m, p))$. $\qquad\square$

**Theorem 5.10.20** $\pi_1(L(m, p_1, \ldots, p_{n-1})) \cong \mathbf{Z}_m$ of the generalized lens space $L(m, p_1, \ldots, p_{n-1})$.

**Proof** Since by construction the lens space $(L(m, p_1, \ldots, p_{n-1})$ is obtained as the orbit spaces $S^{2n-1}$ mod $\mathbf{Z}_m$, proceed as in Theorem 5.10.19 to prove that

$$\pi_1(L(m, p_1, \ldots, p_{n-1})) \cong \mathbf{Z}_m. \qquad\square$$

**Corollary 5.10.21** $\pi_1(\mathbf{R}P^2)) \cong \mathbf{Z}_2$.

**Proof** Consider the particular the group $\mathbf{Z}_2 = \{1_{S^2}, A\}$, where $1_{S^2} : S^2 \to S^2$ is the identity map and $A : S^2 \to S^2, x \mapsto -x$ the antipodal map. Hence it follows that the orbit space $S^2$ mod $\mathbf{Z}_2 = L(2, 1)$ obtained by the action of $\mathbf{Z}_2$ on $S^2$ is the real projective plane $\mathbf{R}P^2$. This proves that $\pi_1(\mathbf{R}P^2)) \cong \mathbf{Z}_2$. $\qquad\square$

### 5.10.7  Computing the Fundamental Group of Figure-Eight

This subsection computes the fundamental group of figure-eight in two methods. This group is not abelian. The computation of this group by van Kampen theorem is available in Chap. 2.

**Theorem 5.10.22** *The fundamental group of the figure-eight $F_8$ is not abelian.*

*Proof* **Proof I: Geometric method.** Consider the universal covering space $(X, p)$ of figure-eight and the path $\tilde{f} : \mathbf{I} \to X : t \mapsto (t, 0)$ which goes along the x-axis from the origin $(0, 0)$ to the point $(1, 0)$. Let $\tilde{g} : \mathbf{I} \to X, t \mapsto (0, t)$ which goes along the y-axis from the origin $(0, 0)$ to the point $(0, 1)$. If $f = p \circ \tilde{f}$ and $g = p \circ \tilde{g}$, then $f$ and $g$ are loops on the figure-eight $F_8$ based at $x_0$, going around the circles $C_1$ and $C_2$, respectively. To prove that $f * g$ and $g * f$ are not path homotopic, lift each of them to a path in $X$ beginning at the origin. Then the path $f * g$ lifts to a path such that it goes along the x-axis from the origin to $(1, 0)$ and goes thereafter once around the circle tangent to the x-axis at $(1, 0)$. On the other hand, the path $g * f$ lifts to a path in $X$ that goes along the y-axis from the origin to $(0, 1)$ and then goes once around the circle tangent to the y-axis at $(0, 1)$. Consequently, it follows that $[f * g] \neq [g * f]$, because the lifted paths do not end at the same point. This implies that $[f] \circ [g] \neq [g] \circ [f]$.

  **Proof II: Graph theoretic method.** Let $G$ be a free group on two letters $\alpha$ and $\beta$ and $X = \text{Graph}(G, \alpha, \beta)$ be the graph constructed as follows: The vertices of $X$ are precisely the elements of $G$. This implies that the vertices of $X$ are the reduced words $\alpha$ and $\beta$. The edges of $X$ are of the two types: $(g, g\alpha) : g \in G$ and $(g, g\beta) : 1g \in G$. Again $(g, g\alpha), (g, g\beta), (g\alpha^{-1}, g)$ and $(g, g\beta^{-1})$ are the only four edges corresponding to the vertex $g$. Define a map

$$\sigma : G \times X \to X, (h, x) \mapsto \begin{cases} hg, & \text{for the vertex } x = g \in G \\ (hg, hg\alpha), & \text{for the edge } x = (g, g\alpha) \\ (hg, hg\beta), & \text{for the edge } x = (g, g\beta) \end{cases}$$

Denote $\sigma(h, x)$ by $hx$. The definition of $\sigma$ asserts an action of $G$ on $X$. Because, for the identity element $1_G \in G$,

$$1_G \cdot g = g, \ 1_G \cdot (g, g\alpha) = (g, g\alpha) \ \text{ and } \ 1_G \cdot (g, g\beta) = (g, g\beta),$$

and since $h_2 g \in X$ is a vertex, then

$$(h_1 h_2) \cdot g = h_1 h_2 \cdot g, \ h_1 \cdot (h_2 \cdot g) = h_1 \cdot (h_2 \cdot g) = h_1 h_2 \cdot g, \ \forall h_1, h_2 \in G$$

and finally, for all $g, h_1, h_2 \in G$.

$$h_1 \cdot (h_2 \cdot (g, g\alpha)) = h_1 \cdot (h_2 g, h_2 g\alpha) = h_1 \cdot (h_2 g, h_1 (h_2 g\alpha)) = (h_1 h_2) \cdot (g, g\alpha)$$
$$= h_1 \cdot (h_2 \cdot (g, g\beta)) = h_1 \cdot (h_2 g, h_2 g\beta)$$
$$= h_1 \cdot (h_2 g, h_1 (h_2 g\beta)) = h_1 h_2 \cdot (g, g\beta).$$

The orbit space $X$ mod $G$ thus obtained by the action is the figure-eight space. Its two loops are the images of the edges $(g, g\alpha)$ and $(g, g\beta)$. Since $X$ is simply connected, it follows from Theorem 5.10.11 that $\pi_1(X$ mod $G, *) \cong G$, where $G$ is the free group on two generators. ☐

## 5.11 The Relative Homotopy Groups

This section introduces the concept of the relative homotopy groups $\pi_n(X, A, x_0)$ by generalizing both the concepts of the fundamental group $\pi_1(X, x_0)$ and absolute homotopy groups $\pi_n(X, x_0)$ studied in Chap. 2. Homotopy groups play an important role in the study of fiber bundles. The concept of higher homotopy groups was born to study the problem to classify homotopically the continuous maps of an $n$-sphere into a given space. For $n = 1$, this group is known as the fundamental group $\pi_1(X, x_0)$ of a pointed topological space $(X, x_0)$. The same technique works in defining the absolute homotopy groups $\pi_n(X, x_0)$ for $n > 1$. Both the fundamental groups and absolute homotopy groups are already studied in Chap. 2. The close connection between homotopy and homology groups and the existence of relative homology groups $H_n(X, A)$, studied in Chap. 3, stimulate to define the relative homotopy groups $\pi_n(X, A, x_0)$ for any triplet of topological spaces $(X, A, x_0)$ and provides an important system, highly related to homology theory except for certain properties:

(i) $\pi_0(X, x_0)$ and $\pi_0(X, A, x_0)$ fail to be ordinarily groups.
(ii) the groups $\pi_1(X, x_0)$ and $\pi_2(X, A, x_0)$ are not ordinarily abelian.
(iii) the excision property for homology (see Chap. 3) fails to hold for homotopy.

### 5.11.1 Standard Notations and Construction of $\pi_n(X, A, x_0)$

The main objective of this subsection is to generalize the concepts of absolute homotopy groups $\pi_n(X, x_0)$ studied in Chap. 2 by defining the relative homotopy groups $\pi_n(X, A, x_0)$. For this purpose, this subsection first explains the standard notations to be used in construction of $\pi_n(X, A, x_0)$.

**Definition 5.11.1** A triplet $(X, A, x_0)$ of three topological spaces consists of a space $X$, a nonempty subspace $A$ of $X$ and a point $x_0 \in A$. In particular, if $x_0$ is the only point of $A$, then the triplet is simply denoted by the pair $(X, x_0)$.

**Example 5.11.2** Let $\mathbf{I}^n = \{t = (t_1, t_2, \ldots, t_n) : t_i \in \mathbf{I}, i = 1, 2, \ldots, n\} \subset \mathbf{R}^n$ be the topological product of the closed unit interval $\mathbf{I} = [0, 1]$ for $n \geq 1$. The space $\mathbf{I}^n$ is called the $n$-cube in the Euclidean space $\mathbf{R}^n$. Its initial $(n-1)$-face given by $t_n = 0$ is identified with the space $\mathbf{I}^{n-1}$. Let $\mathbf{J}^{n-1}$ denote the union of all remaining $(n-1)$-faces of $\mathbf{I}^n$. Consequently, $\partial\,\mathbf{I}^n = \dot{\mathbf{I}}^n = \mathbf{I}^{n-1} \cup \mathbf{J}^{n-1}$ and $\partial\,\mathbf{I}^{n-1} = \mathbf{I}^{n-1} \cap \mathbf{J}^{n-1}$. The ordered triple $(\mathbf{I}^n, \partial\mathbf{I}^n, \mathbf{J}^{n-1})$ of three spaces constitutes a triplet, whose form is more general than the triplet $(X, A, x_0)$.

**Definition 5.11.3** Let $X$ be a topological space and $A$ be a subspace of $X$ with $x_0 \in A$. By a continuous map

$$f : (\mathbf{I}^n, \mathbf{I}^{n-1}, \mathbf{J}^{n-1}) \to (X, A, x_0),$$

we mean a continuous map $f : \mathbf{I}^n \to X$ such that

(i) $f(\mathbf{I}^{n-1}) \subset A$, i.e., $f$ maps $\mathbf{I}^{n-1}$ into $A$, and
(ii) $f(\mathbf{J}^{n-1}) = x_0$, i.e., $f$ maps $\mathbf{J}^{n-1}$ into $x_0$.
    In particular, $f$ sends $\partial\,\mathbf{I}^n$ into $A$ and $\partial\,\mathbf{I}^{n-1}$ on the point $x_0$.
    Let $F^n = F^n(X, A, x_0)$ denote the set of all such maps $f$ topolozied by the compact open topology.

*Example 5.11.4* If $f : \partial\mathbf{I}^n \to A$ is a continuous map such that $f(\partial\,\mathbf{I}^{n-1}) = x_0 \in A$, then $f : (\partial\mathbf{I}^n, \partial\mathbf{I}^{n-1}) \mapsto (A, x_0)$ is a continuous map and $F^n(X, x_0)$ denote the set of all such maps $f$ topolozied by the compact open topology.

**Definition 5.11.5** Let $(X, A, x_0)$ be a given triplet. Then for any pair of maps $f, g \in F^n(X, A, x_0)$ their sum $f * g$ is defined by

$$(f * g)(t) = \begin{cases} f(2t_1, t_2, \ldots, t_i, \ldots, t_n), & \text{if } 0 \le t_1 \le 1/2 \\ g(2t_1 - 1, t_2, \ldots, t_i, \ldots, t_n), & \text{if } 1/2 \le t_1 \le 1 \end{cases}$$

$\forall t = (t_1, t_2, \ldots, t_n) \in \mathbf{I}^n$. Clearly, $f * g$ is continuous, because, for $n \ge 2$ and $t_1 = 1/2$, the above both lines reduce to $x_0$ and hence $f * g \in F^n(X, A, x_0)$ for $n \ge 2$. For $n = 1$, this is also true providing $A = \{x_0\}$.

*Remark 5.11.6* Given any two elements $f, g \in F^n(X, A, x_0)$, $f * g$ can be equally well defined by

$$(f * g)(t) = \begin{cases} f(t_1, t_2, \ldots, 2t_i, \ldots, t_n), & \text{if } 0 \le t_i \le 1/2 \\ g(t_1, t_2, \ldots, 2t_i - 1, \ldots, t_n), & \text{if } 1/2 \le t_i \le 1 \end{cases}$$

This implies that this definition of $f * g$ does not depend on the particular coordinate $t_i$ we use.

**Definition 5.11.7** Let $f, g \in F^n(X, A, x_0)$ be two continuous maps. They are said to be homotopic relative to the system $\{\mathbf{I}^{n-1}, A; \mathbf{J}^{n-1}, x_0\}$ if there exists a continuous map

$$H_t : \mathbf{I}^n \to X, \ \forall\, t \in \mathbf{I}$$

such that

(i)  $H_0 = f$,
(ii)  $H_1 = g$ and
(iii)  $H_t \in F^n(X, A, x_0)$.

It is symbolized as $f \simeq g$ rel $\{\mathbf{I}^{n-1}, A; \mathbf{J}^{n-1}, x_1\}$. Geometrically, it means that $f$ and $g$ can be joined by a continuous curve in the space $F^n(X, A, x_0)$.

**Construction of** $\pi_n(X, A, x_0)$: The homotopy relation on $F^n(X, A, x_0)$ is an equivalence relation, and hence it divides $F^n(X, A, x_0)$ into disjoint equivalence classes, called homotopy classes. The set of homotopy classes of all these maps relative to the system $\{\mathbf{I}^{n-1}, A; \mathbf{J}^{n-1}, x_0\}$) is denoted by $\pi_n(X, A, x_0)$. If $[f]$ denotes the homotopy class of the map $f \in F^n(X, A, x_0)$ and $0$ denotes the homotopy class of the constant map $c(I^n) = x_0$, then for $n = 0$, the set $\pi_0(X, A, x_0)$ denotes the set of all path components of the space $F^n(X, A, x_0)$. In $F^n(X, A, x_0)$, if $f_i \simeq g_i (i = 1, 2)$ rel $\{\mathbf{I}^{n-1}, A; \mathbf{J}^{n-1}, x_1\}$, then the two homotopies can be combined by pasting lemma to provide a homotopy

$$f_1 * f_2 \simeq g_1 * g_2 \text{ rel } \{\mathbf{I}^{n-1}, A; \mathbf{J}^{n-1}, x_1\},$$

This property of $f * g$ in $F^n(X, A, x_0)$ asserts that if $[f], [g] \in \pi_n(X, A, x_0)$, all products $f' * g'$ for $f' \in [f]$ and $g' \in [g]$ will lie in a single homotopy class $[k] \in \pi_n(X, A, x_0)$. This justifies to define a composition, called addition in $\pi_n(X, A, x_0)$ by the rule

$$[f] + [g] = [f * g], \ \forall\, [f], [g] \in \pi_n(X, A, x_0)$$

in the sense that this composition does not depend on the particular choice of the representatives of the homotopy classes.

**Theorem 5.11.8** $\pi_n(X, A, x_0)$ *is an abelian group under the composition '$+$' for for every $n > 1$.*

**Proof** **Associativity law**: To prove this law, it is sufficient to show that for any three elements $f, g, h \in F^n(X, A, x_0)$ the property that $(f * g) * h \simeq f * (g * h)$ holds. For this purpose, take a homotopy of the $t_1$- axis which stretches the closed interval $[0, 1/4]$ into the closed interval $[0, 1/2]$, translates the closed interval $[1/4, 1/2]$ into the closed interval $[1/2, 3/4]$ and contracts the closed interval $[1/2, 1]$ into the closed interval $[3/4, 1]$.

**Zero element**: It is the homotopy class $[c]$ of the constant map defined by $c(\mathbf{I}^n) = x_0$, because the homotopy $c * f \simeq f$ for every $f \in F^n(X, A, x_0)$ is obtained by deforming the $t_1$ interval to shrink the closed interval $[0, 1/2]$ to the point $0$ and expands the closed interval $[1/2, 1]$ into the closed interval $[0, 1]$. Additionally, if a map $f \in F^n(X, A, x_0)$ sends $\mathbf{I}^n$ into $A$ represents the zero element. Because, if $H$ is a homotopy of $\mathbf{I}^n$ onto itself which contracts $\mathbf{I}^n$ into its face $t_n = 1$, then a homotopy is defined by

$$H(t, s) = (t_1, t_2, \ldots, t_{n-1}, (1 - s)t_n + s).$$

This implies that $F(t, s) = f(H(t, s))$ is a homotopy of $f$ into the constant map $c$.

**Existence of negative element**: For any $f$, its inverse element

$$\overline{f}(t) = f(1 - t_1, t_2, \ldots, t_n)$$

is also an element of $F^n(X, A, x_0)$ such that both $f * \overline{f}$ and $\overline{f} * f$ are homotopic to the constant map $c$.

Hence it follows that $\pi_n(X, A, x_0)$ forms a group under the above additive operation $+$. This is called the **n-th relative homotopy group of|**, $X$ modulo $A$ at $x_0$. The class $[c] = 0$ is the group-theoretic zero element of $\pi_n(X, A, x_0)$, and the inverse element of $[f]$ is the element $f \circ \theta$, where

$$\theta : \mathbf{I}^n \to \mathbf{I}^n, t = (t_1, t_2, \ldots, t_n) \mapsto (1 - t_1, t_2, \ldots, t_n).$$

**The group** $\pi_n(X, A, x_0)$ **is abelian for** $n > 2$: If $\mathbf{J}^{n-1}$ is pinched to a point $s_0$, then the triplet $(\mathbf{I}^n, \mathbf{I}^{n-1}, \mathbf{J}^{n-1})$ would admit a configuration, which is equivalent to the triplet $(\mathbf{D}^n, S^{n-1}, s_0)$ of spaces consisting of the unit Euclidean $n$-cell $\mathbf{D}^n$, its boundary $\partial D^n = S^{n-1}$, and a base point $s_0 \in S^{n-1}$. This implies that an element of $\pi_n(X, A, x_0)$ can be equally well defined as a homotopy class (relative to the system $\{S^{n-1}, A; s_0, x_0\}$) of the maps $f : (\mathbf{D}^n, S^{n-1}, s_0) \to (X, A, x_0)$. Since for every $n > 2$, there exists a rotation of $\mathbf{D}^n$ which keeps the point $s_0$ fixed and interchanges two halves of $\mathbf{D}^n$, it follows that the group $\pi_n(X, A, x_0)$ is abelian. This completely proves that $(\pi_n(X, A, x_0), +$ is an abelian group for every integer $n > 1$.  □

Proposition 5.11.9 gives a sufficient condition for existence of a nonconstant function $f \in F_n(X, A, x_0)$ representing the zero element of the group $\pi_n(X, A, x_0)$. The result is useful for our study.

**Proposition 5.11.9** *Let* $f \in F_n(X, A, x_0)$ *be a function such that* $f(\mathbf{I}^n) \subset A$, *then* $[f] = 0$

**Proof** By hypothesis, $f \in F_n(X, A, x_0)$ and $f(\mathbf{I}^n) \subset A$. Define a homotopy

$$H_t(t_1, t_2, \ldots, t_{n-1}, t_n) = f(t_1, \ldots, t_{n-1}, t + t_n - t t_n).$$

Then $H_t \in F_n(X, A, x_0)$, $\forall t \in \mathbf{I}$ and is such that $H_0 = f$, and $H_1(\mathbf{I}^n) = x_0$. This proves that $[f] = 0$.  □

## 5.11.2 Boundary Operator

This subsection defines the boundary operator

$$\partial : \pi_n(X, A, x_0) \rightarrow \pi_{n-1}(A, x_0), \ \forall n \geq 1.$$

It is an important concept in homotopy theory having interesting algebraic properties.

**Definition 5.11.10** Let a given element $\alpha \in \pi_n(X, A, x_0)$ be represented by a continuous map

$$f : (\mathbf{I}^n, \mathbf{I}^{n-1}, \mathbf{J}^{n-1}) \rightarrow (X, A, x_0).$$

Then $\alpha = [f]$. We now consider two possible cases.

    **Case I**: If $n = 1$, then $f$ sends $\mathbf{I}^{n-1}$ to a point of $A$, which gives a path component $\beta$ of $\pi_0(X, A, x_0)$.

    **Case II**: If $n > 1$, consider the restriction $g = f|_{\mathbf{I}^{n-1}}$. Then $g$ is a continuous map

$$g : (\mathbf{I}^{n-1}, \partial \mathbf{I}^{n-1}) \rightarrow (A, x_0).$$

This implies that $g$ represents an element $[g] = \beta \in \pi_{n-1}(A, x_0)$, which is independent of the choice of the map $f$ representing the given element $\alpha$. Consider the assignment $\alpha \mapsto \beta$ by setting $\partial([f]) = [g]$. This assignment induces a transformation

$$\partial : \pi_n(X, A, x_0) \rightarrow \pi_{n-1}(A, x_0), \ \alpha \mapsto \beta, \ \forall n \geq 1,$$

which is called the **boundary operator.** Since

$$\partial(\alpha + \beta) = \partial \alpha + \partial \beta, \ \forall \alpha, \beta \in \pi_n(X, A, x_0)$$

for all $n > 1$, it follows that $\partial$ is a homomorphism, called the **boundary homomorphism** all $n > 1$.

## 5.11.3 The Induced Transformation

This subsection gives the concept of a transformation

$$f_* : \pi_n(X, A, x_0) \rightarrow \pi_n((Y, B, y_0)$$

induced by a continuous map $f : (X, A, x_0) \rightarrow (Y, B, x_0)$ and relates this transformation with the boundary operator $\partial$. Moreover, Definition 5.11.11 formulates the induced transformation

$$f_* : \pi_n(X, A, x_0) \to \pi_n((Y, B, y_0)$$

for $n = 0, n \geq 2$ in general and for $n = 1$ at some particular situation.

**Definition 5.11.11** Let $f : (X, A, x_0) \to (Y, B, y_0)$ be an arbitrary continuous map. Then $f : X \to Y$ is a continuous map such that $f(A) \subset B$ and $f(x_0) = y_0$.

**Case I for** $n = 0$: Since $f$ carries the path components of $X$ into the path components of $Y$, the assignment $f$ induces a transformation $f_* : \pi_0(X, x_0) \to \pi_0(Y, y_0)$, which carries the zero element of $\pi_0(X, x_0)$ into the zero element of $\pi_0(Y, y_0)$.

**Case II for** $n \geq 2$: Take any map $g \in F^n(X, A, x_0)$. Then the composite map $f \circ g \in F^n(Y, B, y_0)$. Hence the assignment $f \mapsto f \circ g$ defines a continuous map

$$f_\square : F^n(X, A, x_0) \to F^n(Y, B.y_0).$$

Since the map $f_\square$ is continuous, it carries the path components of $F^n(X, A, x_0)$ to the path components of $F^n(Y, B, y_0)$. Moreover, if $g \simeq h$ in $F^n(X, A, x_0)$, then $f \circ g \simeq f \circ h$ in $F^n(Y, B, y_0)$. Consequently, $f$ induces a map

$$f_* : \pi_n(X, A, x_0) \to \pi_n(Y, B, y_0)$$

of homotopy classes. Since $f(g * k) = (f \circ g) * (f \circ g), \forall g, h \in F^n(X, A, x_0)$, it follows that $f_*$ is a homomorphism, called the homomorphism induced by $f$.

**Case III for** $n = 1$ In this case, $A = \{x_0\}$, $B = \{y_0\}$. Hence $\pi_n(X, A, x_0)$ and $\pi_n(Y, B, y_0)$ become the fundamental groups $\pi_1(X, x_0)$ and $\pi_1(Y, y_0)$, respectively, with $f_*$ a homomorphism between them.

The above discussion is summarized in a basic result in homotopy theory formulated in Theorem 5.11.12.

**Theorem 5.11.12** *Every continuous map* $f : (X, A, x_0) \to (Y, B, y_0)$ *induces a transformation*

$$f_* : \pi_n(X, A, x_0) \to \pi_n(Y, B, y_0)$$

*of the corresponding homotopy classes such that*

(i) *the transformation* $f_* : \pi_0(X, x_0) \to \pi_0(Y, y_0)$ *sends the zero element of* $\pi_0(X, x_0)$ *into the zero element of* $\pi_0(Y, y_0)$;
(ii) $f_*$ *is a homomorphism of groups for every* $n \geq 2$;
(iii) *if* $A = \{x_0\}$, $B = \{y_0\}$ *and* $n = 1$, *then* $f_* : \pi_1(X, x_0) \to \pi_1(Y, y_0)$ *is a homomorphism of fundamental groups.*

### 5.11.4 Algebraic Properties

For any triplet $(X, A, x_0)$, the homotopy groups, the boundary operator $\partial$ and the induced transformations have seven fundamental properties of which three algebraic properties are proved in this section and others in subsequent sections.

**Proposition 5.11.13** *Let* $1_d : (X, A, x_0) \to (X, A, x_0)$ *be the identity map. Then it induces the identity transformation*

$$1_{d*} : \pi_n(X, A, x_0) \to \pi_n(X, A, x_0), \ \forall n \geq 0.$$

**Proof** It follows from the definition of the induced transformation

$$1_{d*} : \pi_n(X, A, x_0) \to \pi_n(X, A, x_0), \ [f] \mapsto [1_d \circ f] = [f], \ \forall [f] \in \pi_n(X, A, x_0).$$

$\square$

**Proposition 5.11.14** *Let* $f : (X, A, x_0) \to (Y, B, y_0)$ *and* $g : (Y, B, y_0) \to (Z, C, z_0)$ *be two continuous maps of triples. Then their induced transformations have the property:*

$$(g \circ f)_* = g_* \circ f_* : \pi_n(X, A, x_0) \to \pi_n(Z, C, z_0), \ \forall n \geq 0.$$

**Proof** By the given condition, $f$ and $g$ induce transformations of homotopy classes

$$f_* : \pi_n(X, A, x_0) \to \pi_n(Y, B, y_0)$$

and

$$g_* : \pi_n(Y, B, y_0) \to \pi_n(Z, C, z_0),$$

respectively. Hence the proposition follows the definitions of the induced transformations $f_*$ and $g_*$. $\square$

Proposition 5.11.15 relates the boundary operator with the induced transformation.

**Proposition 5.11.15** *Let* $f : (X, A, x_0) \to (Y, B, y_0)$ *be a continuous function and* $g = f|_A = (A, x_0) \to (B, y_0)$ *be the restriction of* $f$ *to its subspace* $A$. *Then*

$$\partial \circ f_* = g_* \circ \partial : \pi_n(X, A, x_0) \to \pi_{n-1}(B, y_0), \ \forall \ n \geq 1.$$

**Proof** To prove the proposition it is sufficient to prove the commutativity of the rectangular diagram as shown in Fig. 5.32 for every $n \geq 1$. The commutativity of the diagram in Fig. 5.32 follows from definitions of $\partial$, the induced transformations $f_*$ and $g_*$ for every $n \geq 1$. $\square$

**Remark 5.11.16** More algebraic properties are available in Sects. 5.11.5 and 5.13.3.

**Fig. 5.32** Commutativity of
the rectangle involving $\partial$ and
induced transformations

$$
\begin{array}{ccc}
\pi_n(X, A, x_0) & \xrightarrow{\ f_*\ } & \pi_n(Y, B, y_0) \\
\partial \downarrow & & \partial \downarrow \\
\pi_{n-1}(A, x_0) & \xrightarrow{\ q_*\ } & \pi_{n-1}(B, y_0)
\end{array}
$$

## 5.11.5  Functorial Property of the Relative Homotopy Groups

This section continues the study homotopy theory by proving the functorial property
of the relative homotopy groups $\pi_n(X, A, x_0)$, $\forall n \geq 2$ and homotopy properties
of maps $f \in F^n(X, A, x_0)$ and also considers the homotopy equivalence of a map
$f \in F^n(X, A, x_0)$.

**Proposition 5.11.17** *For any two homotopic maps* $f, g : (X, A, x_0) \to (Y, B, b_0)$,
*their induced transformations*

$$
f_* : \pi_n(X, A, x_0) \to \pi_n(Y, B, b_0)
$$

*and*

$$
g_* : \pi_n(X, A, x_0) \to \pi_n(Y, B, b_0)
$$

*are the same for every n.*

**Proof** By hypothesis, the maps $f, g : (X, A, x_0) \to (Y, B, b_0, )$ are homotopic.
Then there exists a homotopy

$$
H_t : f \simeq g
$$

in $F^n(X, A, x_0)$. To prove the proposition, it is sufficient to show that $f_*(\beta) = g_*(\beta)$, $\forall \beta \in \pi_n(X, A, x_0)$. We now consider two possible cases.

   **Case I**: Let $\beta \in \pi_n(X, A, x_0)$ be an arbitrary element. Then for $n = 0$, $A = \{x_0\}$
and $B = \{y_0\}$, the element $\beta$ gives a path component of $X$. If $x \in \beta$, then $f_*(\beta)$ and
$g_*(\beta)$ are path components of $Y$ containing the points $f(x)$ and $g(x)$, respectively.
Define a path

$$
\beta : \mathbf{I} \to Y, \ t \mapsto H_t(x).
$$

Then the path $\beta$ joins the point $f(x)$ to the point $g(x)$ and hence $f_*(\beta) = g_*(\beta)$, $\forall \beta \in \pi_n(X, A, x_0)$. This proves that $f_* = g_*$.

   **Case II**: Let $\beta \in \pi_n(X, A, x_0)$ be an arbitrary element. For $n > 0$, take a map

$$
h \in F_n(X, A, x_0) : [h] = \beta.
$$

Then $f_*(\beta)$ and $g_*(\beta)$ are represented by the maps $f \circ h$ and $g \circ h$, respectively.
Consider the homotopy

$$
H_t \circ h : f \circ h \simeq g \circ h.
$$

It implies that $f_*([h]) = g_*([h])$ and hence $f_*(\beta) = g_*(\beta)$, $\forall \beta \in \pi_n(X, A, x_0)$ asserts $f_* = g_*$. $\quad\square$

**Definition 5.11.18** Let $f : (X, A, x_0) \to (Y, B, y_0)$ be a continuous map. It is said to be a **homotopy equivalence** if there exists a continuous map $g : (Y, B, y_0) \to (X, A, x_0)$ such that

(i) $g \circ f$ is homotopic to the identity map on $(X, A, x_0)$ and
(ii) $f \circ g$ is homotopic to the identity map on $(Y, B, y_0)$.

The above discussion is summarized in a basic result formulated in Proposition 5.11.19.

**Proposition 5.11.19** (*Homotopy invariance*) *Every homotopy equivalence* $f : (X, A, x_0) \to (Y, B, b_0)$ *induces an isomorphism*

$$f_* : \pi_n(X, A, x_0) \to \pi_n(Y, B, b_0), \quad \forall n \geq 2.$$

**Corollary 5.11.20** *Let X and Y be two path-connected topological spaces. If they are homotopy equivalent, then the groups*

$$\pi_n(X, A, x_0) \cong \pi_n(Y, B, y_0), \quad \forall n \geq 2.$$

**Remark 5.11.21** If the topological spaces $X$ and $Y$ are path connected and homotopy equivalent, then the groups $\pi_n(X, A, x_0)$ and $\pi_n(Y, B, y_0)$ are isomorphic for every $n > 1$.

**Proof** It follows from Proposition 5.11.19. $\quad\square$

**Corollary 5.11.22** *For every* $n \geq 2$, *the groups* $\pi_n(X, A, x_0)$ *depend on the homotopy type of the triplet* $(X, A, x_0)$ *of topological spaces.*

**Corollary 5.11.23** *If A is a strong deformation retract of X, then the inclusion map* $i : A \hookrightarrow X$ *induces isomorphisms*

$$i_* : \pi_n(A, x_0) \to \pi_n(X, x_0), \quad \forall n \geq 1.$$

**Proof** It follows from Corollary 5.11.22. $\quad\square$

**Theorem 5.11.24** *If* $\mathcal{H}tp^2$ *denotes the homotopy category of triplets and their continuous maps, and* $\mathcal{A}b$ *denotes the category of abelian groups and homomorphisms, then*

$$\pi_n : \mathcal{H}tp^2 \to \mathcal{A}b$$

*is a covariant functor for every integer* $n > 2$.

**Proof** For every integer $n > 2$, we utilize the assignments $(X, A, x_0) \to \pi_n(X, A, x_0)$ and $f \mapsto f_*$ to define

(i)  the object function : $(X, A, x_0) \to \pi_n(X, A, x_0)$ and

(ii)  the morphism function : $f \mapsto f_*$.

Hence it follows that for every integer $n > 2$,

$$\pi_n : \mathcal{H}tp^2 \to \mathcal{A}b$$

is a covariant functor.                                                 □


## 5.11.6   Role of Cells and Spheres in Computing Homotopy Groups

There are some standard homotopy tricks to compute homotopy groups. Certain bundles are used in such computations. This subsection conveys such a technique.

**Definition 5.11.25**  A pair of an $n$-cell and its boundary, denoted by $(E^n, S^{n-1})$, consists of a topological space $E^n$ and a subspace homeomorphic to $(\mathbf{I}^n, \partial \mathbf{I}^n)$.

**Example 5.11.26**  The Euclidean $n$-cell $\mathbf{D}^n$ defined by $\mathbf{D}^n = \{x \in \mathbf{R}^n : ||x|| \le 1\}$ and its boundary $S^{n-1}$ defined by $S^{n-1} = \{x \in \mathbf{R}^n : ||x|| = 1\}$ is a pair of an $n$-cell and its boundary. This pair is considered as the prototype.

**Definition 5.11.27**  An $n$-cell $(E, S)$ is said to be oriented in the sense that there is an orientation of $(E, S)$ obtained by a choice of a generator $\tau_n$ of $H_n(E, S)$.

Given a base point $s_0 \in S$, let $f : (E, S, s_0) \to (X, A, x_0)$ be a continuous map of triplets. We take a continuous map

$$g : (\mathbf{I}^n, \mathbf{I}^{n-1}, \mathbf{J}^{n-1}) \to (E, S, s_0)$$

such that $g_*(u_n) = \tau_n$. We may suppose that $g$ maps $\mathbf{I}^n - @ \, \mathbf{I}^n$ topologically onto $(E - S$maps $\mathbf{I}^n - @ \, \mathbf{I}^n$ topologically onto $(E - S$ and maps $\mathbf{I}^{n-1} - @ \, \mathbf{I}^{n-1})$ topologically onto $(S, s_0)$, because, if $\mathbf{J}^{n-1}$ is pinched to a point, then the quotient space of $(\mathbf{I}^n, \mathbf{I}^{n-1}, \mathbf{J}^{n-1})$ is homeomorphic to $(E, S, s_0)$. This implies that the composite map $f \circ g \in F^n(X, A, x_0)$. If

$$h : (\mathbf{I}^n, \mathbf{I}^{n-1}, \mathbf{J}^{n-1}) \to (E, S, s_0)$$

is a second map satisfying the property of $g$, then $g_*(u_n) = h_*(u_n)$, and hence it follows that both $g$ and $h$ represent the generator of $\pi_n(E, S, s_0)$. This proves that $g \simeq h$ in $F^n(E, S, s_0)$. This asserts that the homotopy class of $f \circ g$ depends only on $f$ and it determines a unique element $c(f) \in \pi_n(X, A, x_0)$. The element $c(f)$ is called the element represented by the map $f$. Since a homotopy of $f$ sending $S$ to $A$ and $s_0$ to $x_0$ gives rise to a homotopy of $f \circ g$, it follows that the element $c(f)$ depends only on the homotopy class of $f$. Given a map $k \in F^n(X, A, x_0)$, let

$f = k \circ g^{-1}$. The map $f$ is single valued and is continuous ( $g^{-1}$ may not be single valued). This implies that $k$ is of the form $f \circ g$. Hence it follows that every element $\sigma \in \pi_n(X, A, x_0)$ is a $c(f)$ for some continuous map $f : (E, S, s_0) \to (X, A, x_0)$.

Summarizing the above discussion a basic result essentially due to Hurewicz is given in Theorem 5.11.28.

**Theorem 5.11.28** *Let $(E, S)$ be an oriented $n$-cell with $s_0 \in S$. Then the homotopy classes of continuous maps $f : (E, S, s_0) \to (X, A, x_0)$ completely determine the homotopy group $\pi_n(X, A, x_0)$ of $(X, A, x_0)$.*

**Corollary 5.11.29** *Let $S^n$ be oriented by a choice of a generator $\tau_n$ of $H_n(S^n)$, where $S^n$ is fixed $n$-sphere. Then the homotopy classes of continuous maps $f : (S^n, s_0) \to (X, x_0)$ completely determine the absolute homotopy group $\pi_n(X, x_0)$ of $(X, x_0)$.*

**Remark 5.11.30** Original definition of the absolute homotopy group $\pi_n(X, x_0)$ given by Hurewicz in 1935 is embedded in Corollary 5.11.29. Its elements are homotopy classes of continuous maps $f : (S^n, s_0) \to (X, x_0)$ completely determine the absolute homotopy group $\pi_n(X, x_0)$ of $(X, x_0)$.

# 5.12  Isomorphism on $\pi_n(X, x_0)$ Induced by a Curve and Role of Base Point

This section studies the role of base points of the homotopy groups $\pi_n(X, x_0)$ and the properties of the homomorphisms induced by a curve on $X$. Its generalization in the context of isomorphisms of $\pi_n(X, A, x_0)$ induced by continuous curve $\sigma : I \to A$ and the role played by the base point $x_0$ is available in 5.22.2.

**Proposition 5.12.1** *Let $X$ be any topological space and $\psi : I \to X$ be a path connecting two given points $x_0, x_1 \in X$. Then $\psi$ induces the identity map*

$$\psi_0 : \pi_0(X, x_1) \to \pi_0(X, x_0).$$

**Proof** By hypothesis, $\psi(0) = x_0$ and $\psi(1) = x_1$. Since $\pi_0(X, x_1)$ denotes the set of all path components of $X$ containing the point $x_1$ and similarly $\pi_0(X, x_0)$ denotes the set of all path components of $X$ containing the point $x_0$ and the points $x_0, x_1$ lie in the same path component, the neutral elements of both $\pi_0(X, x_1)$ and $\pi_0(X, x_0)$ are the same. Denote by

$$\psi_0 : \pi_0(X, x_1) \to \pi_0(X, x_0)$$

the identity map on $\pi_0(X, x_1) = \pi_0(X, x_0)$.  ☐

**Theorem 5.12.2** *Let $X$ be any topological space and given any two points $x_0, x_1 \in X$, let $\psi : I \to X$ be a path from $x_0$ to $x_1 \in X$. Then $\psi$ induces an isomorphism*

**Fig. 5.33** Commutativity of
the rectangle involving $\partial$ and
induced transformations

$$
\begin{array}{ccc}
\pi_n(X, x_1) & \xrightarrow{\psi_n} & \pi_n(X, x_0) \\
f_* \downarrow & & \downarrow f_* \\
\pi_n(Y, y_1) & \xrightarrow{\phi_n} & \pi_n(Y, y_0)
\end{array}
$$

$$\psi_* = \psi_n : \pi_n(X, x_1) \to \pi_n(X, x_0), \ \forall n \geq 1,$$

which depends only on the homotopy class of the path $\psi$ relative to its end points.
Moreover,

(i)  if $\psi : X \to x_0$ is the degenerate path, then

$$\psi_* = \psi_n : \pi_n(X, x_0) \to \pi_n(X, x_0), \ \forall n \geq 1$$

is the identity automorphism;

(ii)  if $\phi : \mathbf{I} \to X$ is another path such that $\phi(0) = \psi(1)$, then

$$(\psi * \phi)_* = (\psi * \phi)_n = \psi_n \circ \phi_n,$$

where $\psi * \phi$ denotes of product path in $X$ from $x_0$ to $x_1$.

(iii)  For any path $\psi : \mathbf{I} \to X$ and every continuous map $f : X \to Y$, the rectangular
diagram of groups and homomorphisms as shown in Fig. 5.33 is commutative
for every $n \geq 1$, where $f(\psi) = \phi$, $f(x_0) = y_0$ and $f(x_1) = y_1$.

**Proof** By hypothesis, $\psi(0) = x_0$ and $\psi(1) = x_1$. Given an arbitrary element $\sigma \in \pi_n(X, x_1)$, let $\sigma = [f]$ for some representative

$$f : (\mathbf{I}^n, \partial \mathbf{I}^n) \to (X, x_1)$$

of $\sigma$. Define a partial homotopy of $f$

$$H_t : \partial \mathbf{I}^n \to X, \ \forall t \in \mathbf{I}$$

by taking

$$H_t(\partial \mathbf{I}^n) = \psi(1 - t), \ \forall t \in \mathbf{I}.$$

Then $H_t$ has a continuous extension by homotopy extension property of $\partial \mathbf{I}^n$ in $\mathbf{I}^n$.

$$\tilde{H}_t : \mathbf{I}^n \to X, \ \forall t \in \mathbf{I} : \tilde{H}_0 = f.$$

The homotopy $\tilde{H}_t$ is called a **homotopy of** $f$ **along the curve** $\psi$. Since $\tilde{H}_1$ sends $(\partial \mathbf{I}^n)$ into the point $\psi(0) = x_0$, it represents an element $\alpha \in \pi(X, x_0)$. This element depends only on $\sigma$. Consider the transformation

$$\psi_n : \pi_n(X, x_1) \to \pi_n(X, x_0), \ \forall n \geq 1, \sigma \mapsto \alpha,$$

where the element $\alpha \in \pi(X, x_0)$ is defined as above. It depends only on the homotopy class of the path $\psi$ on $X$ by using exercise 5.24.1 of Sect. 5.24. We claim that $\psi_n$ is an isomorphism.

$\psi_n$ **is a homomorphism.** Let $\beta, \gamma \in \pi_n(X, x_1)$ be two arbitrary elements be represented by two maps

$$f, h : (\mathbf{I}^n, \partial \mathbf{I}^n) \to (X, x_1)$$

and

$$F_t, H_t : \mathbf{I}^n \to X, \ \forall t \in \mathbf{I}$$

be homotopies of $f$ and $h$, respectively, along the curve $\psi$. Then $F_1$ represents $\psi_n(\beta)$ and $H_1$ represents $\psi_n(\gamma)$.                                                          □

## 5.13  Homotopy Sequence and Its Basic Properties

This section conveys the concept of homotopy sequence of a triplet and proves its exactness property. The homotopy groups $\pi_n$ and their homomorphisms such as $\partial$ and $f_*$ have basic properties in homotopy theory analogous to those enjoyed by homology groups and their homomorphisms in homology theory (see Chap. 3). Such properties provide powerful tools for the study of both homotopy and homology theories, specially for computing homotopy and homology groups of some important spaces. This section proves also some immediate consequences of the homotopy exact sequences.

### 5.13.1  Homotopy Sequence of Triplets

This subsection defines homotopy sequence of the triplet $(X, A, x_0)$. Consider the inclusion maps

$$i : (A, x_0) \hookrightarrow (X, x_0)$$

and

$$j : (X, x_0) \hookrightarrow (X, A, x_0).$$

They induce transformations

$$i_* : \pi_n(A, x_0) \to \pi_n(X, x_0),$$

and

$$j_* : \pi_n(X, x_0) \to \pi_n(X, A, x_0).$$

Then $i_*$, $j_*$ and the boundary operator $\partial$ constitute together a beginningless sequence

$$\cdots \to \pi_{n+1}(X, x_0) \xrightarrow{j_*} \pi_{n+1}(X, A, x_0) \xrightarrow{\partial} \pi_n(A, x_0) \xrightarrow{i_*} \pi_n(X, x_0) \xrightarrow{j_*} \pi_n(X, A, x_0)$$

$$\xrightarrow{\partial} \cdots \pi_1(X, A, x_0) \xrightarrow{\partial} \pi_0(A, x_0) \xrightarrow{i_*} \pi_0(X, x_0)$$

$$(5.1)$$

This sequence denoted by $\pi(X, A, x_0)$ is known as the **homotopy sequence** of the triplet $(X, A, x_0)$.

Every set in the sequence (5.1) has its zero element, and every transformation in the sequence (5.1) (5.1) carries the zero element into the zero element.

**Definition 5.13.1** (*Exactness property homotopy sequence*) The homotopy sequence (5.1) of any triplet $(X, A, x_0)$ is said to be exact if at every term of the sequence (5.1) except the last one, the kernel of every homomorphism on the right coincides with the image of the preceding homomorphism on the left. In other words, the homotopy sequence (5.1) is said to be exact if for $n \geq 0$,

(i)   Im $j_*$ = ker $\partial$ in $\pi_{n+1}(X, A, x_0)$;
(ii)  Im $\partial$ = ker $i_*$ in $\pi_n(A, x_0)$;
(iii) Im $i_*$ = ker $j_*$ in $\pi_n(X, x_0)$.

**Remark 5.13.2** The exactness property of the homotopy sequence (5.2) of any triplet $(X, A, x_0)$ is proved in Theorem 5.13.8.

## 5.13.2   Elementary Properties

This subsection proves some basic properties formulated in Propositions 5.13.3–5.13.7.

**Proposition 5.13.3** *If $f : (X, A, x_0) \to (X, A, x_0)$ is the identity map, then*

$$f_* : \pi_n(X, A, x_0) \to \pi_n(X, A, x_0)$$

*is the identity transformation.*

**Proof** It follows from Proposition 5.11.13.                                         □

**Proposition 5.13.4** *For any two continuous maps $f : (X, A, x_0) \to (Y, B, y_0)$ and $g : (Y, B, y_0) \to (Z, C, z_0)$, their induced transformations*

$$f_* : \pi_n(X, A, x_0) \to \pi_n(Y, B, b_0)$$

*and*

$$g_* : \pi_n(Y, B, y_0) \to \pi_n(Z, C, z_0)$$

*are such that they satisfy the property*

$$(g \circ f)_* = g_* \circ f_* : \pi_n(X, A, x_0) \to \pi_n(Z, C, z_0)$$

*for each dimension n.*

**Proof** It follows from Proposition 5.11.14. □

**Proposition 5.13.5** *For any continuous map* $f : (X, A, x_0) \to (Y, B, y_0)$, *if k is its restriction, i.e., if*

$$k = f|_A : (A, x_0) \to (B, y_0),$$

*then*

$$\partial \circ f_* = k_* \circ \partial.$$

**Proof** It follows from Proposition 5.11.15. □

**Proposition 5.13.6** *For any two continuous maps* $f : (X, A, x_0) \to (Y, B, y_0)$ *and* $g : (Y, B, y_0) \to (Y, B, y_0)$, *if they are connected by a homotopy* $H_t$ *which maps* $A \times \mathbf{I}$ *into B and carries* $\{x_0\} \times \mathbf{I}$ *into* $y_0$, *then* $f_* = g_*$ *for every n.*

**Proof** It follows from Proposition 5.11.17. □

**Proposition 5.13.7** *If* $X = \{x_0\}$ *consists of single point, then* $\pi_n(X, x_0)$ *consists of only the zero element at each dimension.*

**Proof** It follows from Definition of $\pi_n(X, x_0)$. □

### 5.13.3 Exactness Property

This subsection proves the exactness property of the homotopy sequence $\pi(X, A, x_0)$ of the triplet $(X, A, x_0)$.

**Theorem 5.13.8** (Exactness property) *The sequence*

$$\cdots \to \pi_{n+1}(X, x_0) \xrightarrow{j_*} \pi_{n+1}(X, A, x_0) \xrightarrow{\partial} \pi_n(A, x_0) \xrightarrow{i_*} \pi_n(X, x_0) \xrightarrow{j_*} \pi_n(X, A, x_0)$$

$$\xrightarrow{\partial} \cdots \pi_1(X, A, x_0) \xrightarrow{\partial} \pi_0(A, x_0) \xrightarrow{i_*} \pi_0(X, x_0)$$

$$(5.2)$$

*is exact.*

**Proof** To prove the exactness of the given homotopy sequence, it is sufficient to show that for every $n \geq 1$,

(i) $\text{Im } j_* = \ker \partial$ in $\pi_{n+1}(X, A, x_0)$, which proves the exactness at the term $\pi_{n+1}(X, A, x_0)$;

(ii) $\text{Im } \partial = \ker i_*$, which proves the exactness at the term $\pi_n(A, x_0)$;

(iii) $\text{Im } i_* = \ker j_*$, which proves the exactness at the term $\pi_n(X, x_0)$.

(i) **Proof of (i)** To show that $\text{Im } j_* = \ker \partial$ in $\pi_{n+1}(X, A, x_0)$, take an element $\alpha \in \pi_{n+1}(X, x_0)$. Suppose $f \in F^{n+1}(X, x_0)$ represents the element $\alpha$. Then $f$ is an element in $F^{n+1}(X, x_0)$ such that it represents the element $j^*(\alpha)$. By definition of $F^{n+1}(X, x_0)$, it follows that $f$ carries $\mathbf{I}^n$ into the point $x_0$. This implies that the restricted map $\partial f = f|_{\mathbf{I}^n}$ sends $\mathbf{I}^n$ into the point $x_0$ and hence $\partial f$ represents the zero element of $\pi_n(A, x_0)$. This asserts that $(\partial \circ j_*)(\alpha) = 0$ showing $Im j_* \subset \ker \partial$. For the reverse inclusion, suppose $f \in F^{n+1}(X, A, x_0)$ represents an element $\alpha \in \pi_{n+1}(X, A, x_0)$ such that $\partial(\alpha) = 0$. Then the map $\partial f$ is homotopic to the constant map. Hence there exists a homotopy

$$H_t : \mathbf{I}^n \to A : H_0 = f|_{\mathbf{I}^n}, H_1(\mathbf{I}^n) = x_0 \text{ and } H_t(\partial \mathbf{I}^n) = x_0, \forall t \in \mathbf{I}.$$

Construct a map

$$F_t : \partial \mathbf{I}^{n+1} \to A, s \mapsto \begin{cases} H_t(s), & \text{if } s \in \mathbf{I}^n \\ x_0, & \text{if } s \in \mathbf{J}^n \end{cases}$$

for all $t \in \mathbf{I}$. Then $F_0 = f|_{\partial \mathbf{I}^{n+1}}$. Hence by homotopy extension property (HEP), the homotopy $F_t$ has a extension $\widetilde{F}_t : \mathbf{I}^{n+1} \to X$ such that $\widetilde{F}_0 = f$. Again since $\widetilde{F}_1(\partial \mathbf{I}^{n+1}) = F_1(\partial \mathbf{I}^{n+1}) = x_0$, $\widetilde{F}_1$ represents an element $[g] \in \pi_{n+1}(X, x_0)$. Since $\widetilde{F}_t \in F^{n+1}(X, A, x_0)$, it follows that $j_*([g]) = \alpha$. If $n = 1$, $[f]$ is represented by a path $f : \mathbf{I} \to X$ such that $f(0) \in A$ and $f(1) = x_0$. The given condition $\partial(\alpha) = 0$ shows that $f(0)$ is contained in the same path component of $A$ as $x_0$. Hence there exists a homotopy $F_t : \mathbf{I} \to X$ such that $F_0 = f, F_t(0) \in A, F_t(1) = x_0$ and $F_1(0) = x_0$. Consequently, $F_1$ represents an element $[g] \in \pi_1(X, x_0)$, and hence the homotopy $F_t$ shows that $j_*([g]) = \alpha$. Thus. $\ker \partial \subset \text{Im } j_*$. Combining the above inclusions, it follows that

$$\text{Im } j_* \subset \ker \partial \subset \text{Im } j_*.$$

This implies that $Im j_* = \ker \partial$ and hence it completes the proof of exactness of the given homotopy sequence at the term $\pi_{n+1}(X, A, x_0)$.

(ii) **Proof of (ii)** To show that $\text{Im } \partial = \ker i_*$, take an element $\alpha \in \pi_n(X, A, x_0)$ and suppose $f \in F^n(X, A, x_0)$ represents the element $\alpha$ and hence $\alpha = [f]$. Then the element $(i_* \circ \partial)(\alpha)$ is represented by $g = f|_{\mathbf{I}^{n-1}}$. Construct a homotopy

$$G_t : \mathbf{I}^{n-1} \to X, (t_1, \ldots, t_{n-1}) \mapsto f(t_1, t_2, \ldots, t_{n-1}, t), \forall t \in \mathbf{I}$$

Then $G_0 = g, G_1(\mathbf{I}^{n-1}) = x_0$ and $G_t \in F^{n-1}(X, x_0)$ for $n > 1$. Hence $(i_* \circ \partial)(\alpha) = 0$ implies $i_* \circ \partial = 0$. This shows that $\operatorname{Im} \partial \subset \ker i_*$.

For the reverse inclusion, first let $n > 1$ and $\alpha \in \pi_n(A, x)$ be represented by $f \in F^{n-1}(A, x_0)$ such that $i_*(\alpha) = 0$. Then there exists a homotopy $F_t : \mathbf{I}^{n-1} \to X$ such that $F_0 = f$, $F_1(\mathbf{I}^{n-1}) = x_0$ and $F_t(\partial \mathbf{I}^{n-1}) = x_0$. Define a map

$$g : I^n \to X, (t_1, t_2, \ldots, t_{n-1}, t_n) \mapsto F_{t_n}(t_1, t_2, \ldots, t_{n-1}).$$

Then $g \in F^n(X, A, x_0)$ and represents an element $[g] \in \pi_n(X, A, x_0)$. Since $g|_{\mathbf{I}^{n_1}} = f$, it follows that $\partial([g]) = \alpha$.

For the case, when $n = 1$, the element $\alpha$ is a path component of $A$ and $i_*(\alpha) = 0$ implies that $\alpha$ is contained in the path component of $X$ that contains the point $x_0$. Take any point $x \in X$ and define a path

$$f : \mathbf{I} \to X : f(0) = x \ \text{ and } f(1) = x_0.$$

Then this path $f$ represents an element $\beta \in \pi_1(X, A, x_0)$. Since $f(0)$ is a point of $\alpha$, it follows that $\partial(\beta) = \alpha$. Combining the above inclusions, it shows that

$$\operatorname{Im} \partial \subset \ker i_* \subset \operatorname{Im} \partial.$$

This implies that $Im \ \partial = \ker 1_*$ and hence it completes the proof of exactness of the given homotopy sequence at the term $\pi_n(A, x_0)$.

.

(iii) **Proof of (iii)** $\operatorname{Im} i_* = \ker j_*$: First we show that $\operatorname{Im} i_* \subset \ker j_*$. For $n \geq 1$, take any element $\alpha \in \pi_n(A, x_0)$ and choose a representative $f \in F^n(A, x_0)$ of the element $\alpha$. Then the element $(j_* \circ i_*)(\alpha) \in \pi_n(X, A, x_0)$ is represented by the map $j \circ i \circ f \in F^n(X, A, x_0)$. Since $(j \circ i \circ f)(\mathbf{I}^n) \subset A$, it follows by Proposition 5.11.9 that $(j_* \circ i_*)(\alpha) = 0$, since $\alpha \in \pi_n(A, x_0)$ is an arbitrary element. This shows that $j_* \circ i_* = 0$. Hence $\operatorname{Im} i_* \subset \ker j_*$.

For the reverse inclusion, take an element $\alpha \in \pi_n(X, x_0)$ be such that $j_*(\alpha) = 0$ and choose a function $f \in F_n(X, x_0)$ to represent the element $\alpha$. Then $j_*(\alpha) = 0$ shows that there is a homotopy $F_t : \mathbf{I}^n \to X$ such that $F_0 = f$, $F_1(\mathbf{I}^n) = x_0$, and $F_t \in F^n(X, A, x_0)$. Define a homotopy

$$G_t : \mathbf{I}^n \to X, (t_1, t_2, \ldots, t_{n-1}, t_n) \mapsto \begin{cases} F_{2t_n}(t_1, t_2, \ldots, t_{n-1}, 0), & \text{if } 0 \leq 2t_n \leq t \\ F_t(t_1, t_2, \ldots, t_{n-1}, \frac{2t_n - t}{2 - t}), & \text{if } t \leq 2t_n \leq 1 \end{cases}$$

Then $G_0 = f$, $G_1(\mathbf{I}^n) \subset A$ and $G_t(\partial \mathbf{I}^n) = x_0$ for all $t \in I$. Hence $G_1$ represents an element $[g] \in \pi_n(A, x_0)$ and the homotopy $G_t$ shows that $i_*([g]) = \alpha$. Hence $\ker j_* \subset \operatorname{Im} i_*$.

Combining the above inclusions, it follows that

$$\text{Im } i_* \subset \text{ker } j_* \subset \text{Im } i_*.$$

This implies that $\text{Im } i_* = \text{ker } j_*$ and hence it completes the proof of the exactness of the given homotopy sequence at the term $\pi_n(X, x_0)$.    □

Definition 5.13.9 gives a very useful extension of the concept of homotopy sequence of $(X, A, x_0)$ for the triple $(X, A, B)$, where $X$, $A$ and $B$ are three topological spaces such that $X \supset A \supset B$ and $x_0 \in B$ is a base point.

**Definition 5.13.9** (*Generalized homotopy sequence of* $(X, A, B)$) This is the sequence

$$\cdots \to \pi_{n+1}(X, B) \xrightarrow{j_*} \pi_{n+1}(X, A) \xrightarrow{\partial} \pi_n(A, B) \xrightarrow{i_*} \pi_n(X, B) \xrightarrow{j_*} \pi_n(X, A) \to$$
$$\cdots$$
$$(5.3)$$

where $i : A \hookrightarrow X$ and $j : B \hookrightarrow A$ are inclusion maps. The operator $\partial$ is the composite

$$\pi_{n+1}(X, A) \xrightarrow{\partial} \pi_n(A) \xrightarrow{k_*} \pi_n(A, B)$$
$$(5.4)$$

where $k$ is the inclusion map. The sequence (5.3) ends at the term $\pi_2(X, A)$ and reduces to the sequence of a pair when $B = \{x_0\}$. This sequence is called **generalized homotopy sequence** of $(X, A, B)$.

**Theorem 5.13.10** (Exactness property of generalized homotopy sequence) *The sequence (5.3) is exact.*

**Proof** Proceed as in Theorem 5.13.8.    □

### 5.13.4 Homomorphism and Isomorphism of Homotopy Sequences

This subsection studies homotopy sequences induced by a continuous map of triplets.

**Definition 5.13.11** Let $f : (X, A, x_0) \to (Y, B, y_0)$ be a continuous map. If $f_1 : (X, x_0) \to (Y, y_0)$ and $f_2 : (A, x_0) \to (B, y_0)$ are continuous maps induced by $f$, then the maps $f$, $f_1$ and $f_2$ induce homomorphisms of their corresponding homotopy groups as shown in the diagram in Fig. 5.34.

**Proposition 5.13.12** *The diagram in Fig. 5.34 is commutative.*

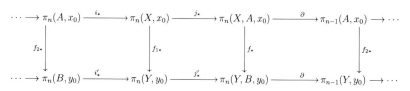

**Fig. 5.34** Homomorphism of homotopy sequences

**Proof** Propositions 5.13.4 and 5.13.5 together prove the commutativity of each rectangle in the diagram. □

**Remark 5.13.13** The commutative diagram in Fig. 5.34 is called the **homomorphism of the homotopy sequence** of $(X, A, x_0)$ into the homotopy sequence of $(Y, B, y_0)$. This two sequences are connected by the homomorphisms by the induced homomorphisms $f*$, $f_{1*}$ and $f_{2*}$ making each rectangle in the diagram commutative. In particular, if $f$ is a homeomorphism, then $f$ induces an **isomorphism of the homotopy** of $(X, A, x_0)$ into the homotopy sequence of $(Y, B, y_0)$.

## 5.14 Homotopy Groups of Cells and Spheres

This section studies isomorphism induced by a homotopy equivalence of one homotopy sequence into the homotopy sequence of the other.

**Definition 5.14.1** Let $f : (X, A, x_0) \to (Y, B, y_0)$ be a continuous map. It is said to be a homotopy equivalence if there is a continuous map $g : (Y, B, y_0) \to (X, A, x_0)$ such that

(i) $g \circ f$ is homotopic to the identity map of $(X, A, x_0)$; and
(ii) $f \circ g$ is homotopic to the identity map of $(Y, B, y_0)$. The homotopies shift the subspaces $A$ and $B$ on themselves and keep the points $x_0$ and $y_0$ fixed.

**Proposition 5.14.2** *Every homotopy equivalence $f : (X, A, x_0) \to (Y, B, y_0)$ induces an isomorphism of the homotopy sequence of $(X, A, x_0)$ onto the homotopy sequence of $(Y, B, y_0)$.*

**Proof** By hypothesis, $f : (X, A, x_0) \to (Y, B, y_0)$ is a homotopy equivalence. Hence it follows from Proposition 5.13.6 that $f$ induces an isomorphism of the homotopy sequence of $(X, A, x_0)$ onto the homotopy sequence of $(Y, B, y_0)$. □

**Proposition 5.14.3** *Let $f : (X, A, x_0) \to (X, A, x_0)$ be the identity map. If $f$ is homotopic to the constant map $c : (X, A, x_0) \to x_0$, then every homotopy group of $(X, A, x_0)$ contains only the zero element.*

**Proof** By hypothesis, the identity map $f : (X, A, x_0) \to (X, A, x_0)$ is homotopic to the constant map $c : (X, A, x_0) \to x_0$. Hence it follows by using Proposition 5.13.7 that every homotopy group of $(X, A, x_0)$ contains only the zero element. □

**Proposition 5.14.4** *Let $E$ be an $m$-cell (open or closed) and $x_0 \in E$ be any point. Then $\pi_n(E, x_0) = 0$ at every dimension $n$.*

**Proof** Since every cell $E$ is contractible to any of its points, the identity map of $E$ is homotopic to constant map. Hence the proposition follows from Proposition 5.14.3.                                                                        □

Let $(E, S)$ be a closed $m$-cell in the sense that it is homeomorphic to $(\mathbf{I}^m, \partial\mathbf{I}^n)$ and let $x_0 \in S$ be any point. Consider the part (5.5) of the homotopy sequence of $(E, S, x_0)$

$$\pi_{n+1}(E, x_0) \xrightarrow{j_*} \pi_{n+1}(E, S, x_0) \xrightarrow{\partial} \pi_n(S, x_0) \xrightarrow{i_*} \pi_n(E, x_0)$$
$$\cdots \quad (5.5)$$

Since $\pi_{n+1}(E, x_0) = 0$, the image $Im\, j_* = 0$ and hence $ker\,\partial = 0$ by exactness property of the sequence (5.5 ). Again, since $\pi_n(E, x_0) = 0$, the kernel $ker\, i_* = \pi_n(S, x_0)$. Hence by the exactness property of the sequence, it follows that $Im\, \partial = \pi_n(S, x_0)$. This implies that

$$\partial : \pi_{n+1}(E, S, x_0) \cong \pi_n(S, x_0).$$

The above discussion is summarized in a basic more general result given in Proposition 5.14.5.

**Proposition 5.14.5** *If every third term of an exact sequence of groups and homomorphisms is zero, then the remaining adjacent pairs of groups are isomorphic.*

**Remark 5.14.6** Two basic theorems of homotopy theory embodied in Theorems 5.14.7 and 5.14.8 have been proved in Chap. 2 by applying Freudenthal suspension theorem saying that

$$\sigma_n : \pi_n(S^n) \to \pi_{n+1}(S^{n+1}), \quad [\alpha] \mapsto [\tilde{\alpha}]$$

is an isomorphism for $m < 2n - 1$ and is surjective for $m \leq 2n - 1$, where $\tilde{\alpha} : S^{n+1} \to S^{n+1}$ is an continuous extension of the continuous map $\alpha : S^n \to S^n$ determined uniquely upto homotopy. On the other hand, the proof of the first theorem by simplicial approximation theorem and the second theorem by using Hurewicz natural homomorphism

$$\psi : \pi_n(X, A, x_0) \to H_n(X, A)$$

are available in [Steenrod, 1951]. Hurewicz homomorphism and relation between homology and homotopy groups have already been studied in Chap. 3. The importance of the natural homomorphism between homotopy and homology theory is realized from the statement of Theorem 5.14.9.

**Theorem 5.14.7** *The homotopy groups $\pi_m(S^n) = 0$ for $m < n$.*

**Proof** See Chap. 2 or see proof of Theorem 5.18.10.                                     □

**Theorem 5.14.8** *The first nontrivial homotopy groups of the n-sphere $S^n$ is the $\pi_n(S^n)$, and it is the infinite cyclic group.*

**Proof** See Chap. 2. For an alternative proof see [Steenrod, 1951, p. 78].           □

Theorem 5.14.9 gives a sufficient condition for equality of the groups $\pi_n(X, A, x_0)$ and $H_n(X, A)$ upto an isomorphism.

**Theorem 5.14.9** *(Isomorphism theorem of Hurewicz )* *Let $(X, A, x_0)$ be a triplet such that*

(i)  *the subspace A is simply connected;*
(ii)  *X is also simply connected.*
(iii)  *Moreover, if $\pi_m(X, A, x_0) = 0$ for $2 \leq m < n$, then*

$$\psi : \pi_n(X, A, x_0) \rightarrow H_n(X, A)$$

*is an isomorphism.*

**Proof** The proof is very long (see [Hu, 1933] ).                                        □

**Corollary 5.14.10** *If the triplet $(X, A, x_0)$ satisfies the condition of Theorem 5.14.9, then $H_m(X, A) = 0$ for $1 \leq m < n$.*

**Proof** By hypothesis, $\pi_n(X, A, x_0) = 0$ for $2 \leq m < n$. Hence the Hurewicz homomorphism $\psi$ maps isomorphically onto $H_n(X, A)$. Conversely, if $(X, A, x)$ satisfies the first two conditions and if $H_m(X, A) = 0$ for $2 \leq m < n$, then by iteration, it can be proved that $\pi_n(X, A, x_0) = 0$ for $n = 2$ followed by $m = 2, 3, \ldots, n$. Hence it follows that the condition (iii) can be replaced by the condition saying that $H_m(X, A) = 0$ for $2 \leq m < n$. Under the conditions (i), (ii) and (iii'), the result still holds.                                                                                        □

**Corollary 5.14.11** *The first nontrivial homology group and the first nontrivial homotopy group have the same dimension and are isomorphic under the natural isomorphism.*

**Corollary 5.14.12** $\pi_n(S^n) \cong H_n(S^n) \cong \mathbf{Z}$.

**Corollary 5.14.13** *If $(E^n, S^{n-1})$ is an n-cell with boundary $S^{n-1}$ and $x_0 \in S^{n-1}$, then*

$$\pi_n(E^n, S^{n-1}, x_0) \cong H_n(E^n, S^{n-1}) \cong \mathbf{Z}.$$

## 5.15  Stable Homotopy Groups

This section discusses stable homotopy groups based on the **Freudenthal suspension theorem** saying that suspension homomorphism

$$E : \pi_m(S^n) \rightarrow \pi_{m+1}(S^{n+1})$$

is an isomorphism for $m < 2n - 1$ and is onto for $m \leq 2n - 1$. [Freudenthal, 1938], which has been discussed in Chap. 2. It is a natural generalization of Freudenthal suspension theorem. The term 'stable' is used in topology if a phenomenon occurs essentially in the same way which is independent of dimension provided that the dimension is sufficiently large. Stable homotopy theory is an important theory in algebraic topology, and it witnessed its greatest development in the late 1950s. It is applied to develop spectral homology and cohomology theory and specially $K$-theory by Atiyah and Hirzebruch. R. Thom (1923–2002) used this theory to reduce the problem of classifying manifolds up to cobordism to a solvable problem in stable homotopy theory.

For stable homotopy theory with applications in homology and cohomology theories, the paper [Adams, 1974], the books [Adhikari, 1916] and [Gray, 1975] are referred.

### 5.15.1  Stable Homotopy Groups

Consider an $n$-connected $CW$-complex and the suspension map

$$\Sigma : \pi_r(X) \rightarrow \pi_{r+1}(X)$$

is an isomorphism for $r < 2n + 1$. In particular, for $r \leq n$ the suspension map $\Sigma$ is an isomorphism and $\Sigma X$ is an $(n + 1)$-connected CW-complex. Now, consider the sequence ( 5.6 ) of groups and homomorphisms

$$\pi_r(X) \rightarrow \pi_{r+1}(\Sigma X) \rightarrow \cdots \rightarrow \pi_{r+m}(\Sigma^m X) \rightarrow \cdots \qquad (5.6)$$

The homomorphism

$$\pi_{r+m}(\Sigma^m X) \rightarrow \pi_{r+m+1}(\Sigma^{m+1} X)$$

is an isomorphism for $r + m < 2(n + m) + 1$, i.e., for $m > r - 2n + 1$, because, $\Sigma^m X$ is $(n + m)$-connected. This implies that for fixed $n$ and $r$ and sufficiently large enough $m$, all the groups in the sequence of groups and homomorphisms (5.6) are isomorphic. This leads to the concept of the stable homotopy group denoted by $\pi_r^s(X)$ of $X$, for which the sequence (5.6 ) is stabilized.

**Definition 5.15.1** (*Stable homotopy group*) Given an $(n-1)$-connected space $X$, for $r \geq 0$, the $r$-th stable homotopy group of $X$, denoted by $\pi_r^s(X)$, is defined to be the group

$$\pi_{n+m}(\Sigma X), \forall m > r - 2n + 1.$$

This is well-defined, since, adding any finite number of groups and homomorphisms to the beginning of (5.6) does not affect the resulting stable homotopy group.

## 5.15.2 Some Examples

*Example 5.15.2* Consider the homotopy exact sequence (5.8) of the fibration: $p : S^3 \to S^2$ :

$$\cdots \to \pi_3(S^1, s_0) \to \pi_3(S^3, s_0) \xrightarrow{p_*} \pi_3(S^2, s_0) \to \pi_2(S^1, s_0) \to \cdots \quad (5.7)$$

The exactness property of the above sequence (5.7) of groups and homomorphisms asserts that the homomorphism $p_*$ induced by $p$

$$p_* : \pi_3(S^3, s_0) \to \pi_3(S^2, s_0)$$

is an isomorphism, because $\pi_3(S^1, s_0) \cong \pi_2(S^1, s_0) = 0$. This implies that $\pi_3(S^2, s_0) \cong \mathbf{Z}$. It provided the first example given by H. Hopf, where $\pi_m(S^n, s_0) \neq 0$ for $m > n$. The map $p$ is known the **Hopf map**, studied earlier in Chap. 2.

*Example 5.15.3* For each $q$, consider

$$\pi_{2q+2}(S^{q+2}, s_0) \xrightarrow{\Sigma} \cong \pi_{2q+3}(S^{q+3}, s_0) \xrightarrow{\Sigma} \cong \cdots \xrightarrow{\Sigma} \cong \pi_{q+n}(S^n, s_0) \xrightarrow{\Sigma} \cong \cdots$$

We denote the common group $\pi_{n+q}(S^n, s_0)$, by $\pi_q^S$. It is called the *k***th stable homotopy group** for the fibration: $p : S^3 \to S^2$.

**Theorem 5.15.4** (Hopf) $\pi_3(S^2) \cong \pi_2(S^2) \cong \mathbf{Z}$.

*Proof* Consider the Hopf fibration

$$S^1 \hookrightarrow S^3 \to S^2$$

and its corresponding homotopy exact sequence (5.8):

$$\cdots \to \pi_3(S^1, s_0) \to \pi_3(S^3, s_0) \xrightarrow{p_*} \pi_3(S^2, s_0) \to \pi_2(S^1, s_0) \to \cdots \quad (5.8)$$

Proceed as Example 5.15.2 to prove the theorem using the result that $\pi_2(S^2) \cong \mathbf{Z}$.

□

## 5.16    Consequences of Homotopy Exact Sequence

This section proves some results in homotopy theory as consequences of the exactness property of the homotopy sequence (5.2).

### 5.16.1    Direct Sum Theorems in Higher Homotopy Groups

This section proves **direct sum theorems** as immediate consequences of the exactness property of the homotopy sequence (5.2).

**Theorem 5.16.1** *Let $(X, A, x_0)$ be a triplet and $r : X \to A$ be a retraction of $X$. If $x_0 \in A$ and $i : A \hookrightarrow X$ and $i : (X, A) \hookrightarrow (X, A, x_0)$ are inclusion maps, then*

$$\pi_n(X, x_0) \cong \pi_n(A, x_0) \oplus \pi_n(X, A, x_0)$$

*for any $n \geq 2$ and the inclusion map $i : A \hookrightarrow X$ induces a monomorphism*

$$i_* : \pi_n(A, x_0) \to \pi_n(X, x_0), \ \forall n \geq 1.$$

**Proof** By hypothesis, $r \circ i = 1_A$. Hence the induced homomorphism

$$r_* \circ i_* : \pi_n(A, x_0) \to \pi_n(A, x_0)$$

is the identity automorphism on $\pi_n(A, x_0)$ for every integer $n \geq 1$. This asserts that $i_* : \pi_n(A, x_0) \to \pi_n(X, x_0)$ is a monomorphism and $r_* : \pi_n(X, x_0) \to \pi_n(A, x_0)$ is an epimorphism for every $n \geq 1$. Since the group $\pi_n(X, x_0)$ is abelian. Again for $n \geq 2$, the relation $r_* \circ i_* = 1_d$ implies that the group $\pi_n(X, x_0)$ decomposes into the direct sum $\pi_n(X, x_0) = G \oplus K$, where $G = \text{Im}\, i_*$ and $K = \ker r_*$. Clearly. $G \cong \pi_n(A, x_0)$, because, $i_*$ is a monomorphism. The exactness of the homotopy sequence (5.2) of the triplet $(X, A, x_0)$ asserts that $j_* : \pi_n(X, x_0) \to \pi_n(X, A, x_0)$ is an epimorphism for every integer $n \geq 2$ and $\ker j_* = \text{Im}\, i_* = G$ and $j_*$ maps $K$ isomorphically onto $\pi_n(X, A, x_0)$ and hence $K \cong \pi_n(X, A, x_0)$.    □

**Corollary 5.16.2** *Let $(X, A, x_0)$ be a triplet, $r : X \to A$ be a retraction of $X$. If $x_0 \in A$ and $i : A \hookrightarrow X$ and $j : (X, A) \hookrightarrow (X, A, x_0)$ are inclusion maps, then the group $\pi_n(X, x_0)$ decomposes into direct sum*

$$\pi_n(X, x_0) \cong G \oplus K, \ \forall n \geq 2,$$

*where $G$ is the image of $i_*$ in $\pi_n(X, x_0)$, and $K$ is the kernel of the homomorphism $r_* : \pi_n(X, x_0) \to \pi_n(A, x_0)$.*

**Proof** It follows from the proof of direct sum theorem 5.16.1.    □

**Remark 5.16.3** For a given triplet $(X, A, x_0)$, if $A$ is a retract of $X$, then the group $\pi_2(X, A, x_0)$ is abelian.

**Definition 5.16.4** Let $(X, A, x_0)$ be a triplet. Then the topological space $X$ is said to be deformable into its subspace $A$ relative to a point $x_0 \in A$, if there exists a homotopy $H_t : X \to X$ satisfying the following conditions

$$H_0(x) = x, H_1(x) \in A \text{ and } H_t(x_0) = x_0, \forall x \in X, t \in \mathbf{I}.$$

**Theorem 5.16.5** Let $(X, A, x_0)$ be a triplet and the topological space $X$ be deformable into its subspace $A$ relative to a point $x_0 \in A$. If $i : A \hookrightarrow X$ be the inclusion map, then $\pi_n(A, x_0)$ decomposes as a direct sum

$$\pi_n(A, x_0) \cong \pi_n(X, x_0) \oplus \pi_{n+1}(X, A, x_0), \forall n \geq 2$$

and

$$i_* : \pi_n(A, x_0) \to \pi_n(X, x_0) \text{ is an epimorphism, } \forall n \geq 1.$$

**Proof** By the given conditions, there exists a homotopy

$$H_t : X \to X : H_0(x) = x, H_1(x) \in A \text{ and } H_t(x_0) = x_0, \forall x \in X, t \in \mathbf{I}.$$

Then the continuous map

$$f : (X, x_0) \to (A, x_0), x \mapsto H_1(x)$$

is such that $i \circ f = H_1 \simeq H_0$ rel $x_0$. This implies that $i_* \circ f_* : \pi_n(X, x_0) \to \pi_n(X, x_0)$ is the identity automorphism on $\pi_n(X, x_0)$. Consequently, $f_*$ is a monomorphism, and $i_*$ is an epimorphism for every integer $n \geq 1$. Since for every integer $n \geq 2$, the group $\pi_n(A, x_0)$ is abelian, it follows from $i_* \circ f_* = 1_d$, the group $\pi_n(A, x_0)$ decomposes as direct sum

$$\pi_n(A, x_0) = \text{Im } f_* \oplus \ker i_*,$$

where $i_*$ is an epimorphism and $f_*$ is a monomorphism for any $n \geq 1$. This implies that $\text{Im } f_* \cong \pi_n(X, x_0)$. Finally, the exactness of the homotopy sequence (5.2) of the triplet $(X, A, x_0)$ asserts that

$$\partial : \pi_{n+1}(X, A, x_0) \to \pi_n(A, x_0)$$

is a monomorphism and hence $\ker i_* = \text{Im } \partial \cong \pi_{n+1}(X, A, x_0)$. $\qquad \square$

**Corollary 5.16.6** Let $(X, A, x_0)$ be a triplet and $i : A \hookrightarrow X$ be the inclusion map. If the continuous map $f : (X, x_0) \to (A, x_0)$ is such that the maps $i \circ f \simeq 1_X : (X, x_0) \to (X, x_0)$, then the group $\pi_n(A, x_0)$ decomposes as a direct sum

$$\pi_n(A, x_0) \cong \pi_n(X, x_0) \oplus \pi_{n+1}(X, A, x_0) = G \oplus K, \ \forall n \geq 2,$$

where $G = Im f_*$ and $K = ker \ i_*$ are subgroups of $\pi_n(A, x_0)$.

**Proof** It follows from the proof of Theorem 5.16.5.    □

## 5.16.2    Characterization of n-Connected Spaces

This subsection characterizes $n$-connected spaces in terms of homomorphisms of homotopy groups induced by inclusion maps by using homotopy exact sequence.

**Definition 5.16.7** A pair $(X, A)$ of topological spaces is said to be

(i) **0-connected** if every path component of $X$ intersects $A$;
(ii) **$n$-connected** if $(X, A)$ is 0-connected and $\pi_k(X, a) = 0$ for $1 \leq k \leq n$ and $\forall a \in A$.

Proposition 5.16.8 gives a characterization of $n$-connected spaces by homotopy exact sequence.

**Proposition 5.16.8** *A pair $(X, A)$ of topological spaces with inclusion map $i : (A, x_0) \to (X, x_0)$ is $n$-connected $(n \geq 0)$ iff the induced map $i_* : \pi_k(A, x_0) \to \pi_k(X, x_0)$*

*(i)  is a bijection for $k < n$; and*
*(ii)  it is a surjection for $k = n$ and $\forall x_0 \in A$.*

**Proof** It follows from the homotopy sequence (5.2) of the triplet $(X, A, x_0)$.    □

## 5.17    Homotopy Sequence of Fibering and Hopf Fibering of Spheres

This section is devoted to the study of the homotopy sequence of fibering. Moreover, it describes Hopf fibering:

$$p : S^{2n-1} \to S^n \ for \ n = 2, 4, 8,$$

which are the early examples of bundles spaces. They are used in computing higher homotopy groups of certain topological spaces. H. Hopf (1894–1975) described various fiberings of spheres by spheres in his papers [Hopf, 1931, 1935]. The map $p : S^3 \to S^2$ of the 3-sphere on the 2-sphere defined by Hopf in 1935, known as a Hopf map, is the simplest.

## 5.17.1   Homotopy Sequence of Fibering

Consider the projection $p : X \to B$. Let $b_0 \in B$ be the base point of $B$ and $F = p^{-1}(b_0) \neq \emptyset$ be the fiber space of $p$. For $x_0 \in F$, this subsection studies the homotopy sequence of the triplet $(X, F, x_0)$.

Define a continuous map

$$q : (X, F, x_0) \to (B, b_0) : p = q \circ j,$$

where $j : (X, x_0) \hookrightarrow (X, F, x_0)$ is the inclusion map. Then $q$ induces a bijection

$$q_* : \pi_n(X, F, x_0) \to \pi_n(B, b_0)$$

for every $n \geq 1$ by Exercise 5.24.1 of Sect. 5.24. Hence $q_*^{-1}$ is an isomorphism. Consider the homomorphism

$$k_* : \partial \circ q_*^{-1} : \pi_n(B, b_0) \to \pi_{n-1}(F, x_0)$$

for ever integer $n \geq 1$. This produces a beginninless sequence

$$\cdots \xrightarrow{p_*} \pi_{n+1}(B, b_0) \xrightarrow{k_*} \pi_n(F, x_0) \xrightarrow{i_*} \pi_n(X, x_0) \xrightarrow{p_*} \pi_n(B, x_0)$$
$$\xrightarrow{k_*} \cdots \xrightarrow{p_*} \pi_1(B, b_0) \xrightarrow{k_*} \pi_0(F, x_0) \xrightarrow{i_*} \pi_0(X, x_0)$$
$$(5.9)$$

Theorem 5.17.1 proves the exactness of the sequence (5.9), which is known as the homotopy sequence of the fibering $p : X \to B$ based at the point $x_0$.

**Theorem 5.17.1** (Homotopy sequence of the fibering) *The sequence (5.9) is exact.*

**Proof** The exactness of the sequence (5.9) follows from the exactness of the sequence (5.2) proved in Theorem 5.13.8. □

**Proposition 5.17.2** *Let the fiber space $F$ in the homotopy sequence (5.9) of fibering is totally disconnected. Then the induced homomorphism*

$$p_* : \pi_n(X, x_0) \to \pi_n(B, b_0)$$

*(i)  is an isomorphism for $n \geq 2$; and*
*(ii) $p_*$ is a monomorphism for $n = 1$.*

**Proof** By the given condition, the groups $\pi_n(F, x_0) = 0$, $\forall n \geq 1$. Using exactness property of the homotopy sequence (5.9), the proposition is proved. □

**Theorem 5.17.3** *If $p : X \to B$ be a fibering with a cross section $s : B \to X$, such that for every $b_0 \in B$ the element $x_0 = s(b_0) \in F = p^{-1}(b_0)$, then*

(i) $\pi_n(X, x_0) \cong \pi_n(B, b_0) \oplus \pi_n(F, x_0)$, $\forall n \geq 2$, and
(ii) $p_* : \pi_n(X, x_0) \to \pi_n(B, b_0)$ is an epimorphism for every $n \geq 1$.

**Proof** By hypothesis $s : B \to X$ is a cross section. Hence $p \circ s = 1_B$ induces the identity automorphism

$$p_* \circ s_* : \pi_n(B, b_0) \to \pi_n(B, b_0).$$

This implies that for $n \geq 1$,

(i) $s_* : \pi_n(B, b_0) \to \pi_n(X, x_0)$ is a monomorphism; and
(ii) $p_* : \pi_n(X, x_0) \to \pi_n(B, b_0)$ is an epimorphism.

For the particular case, when $n \geq 2$, the group $\pi_n(X, x_0)$ is abelian. Since $p_* \circ s_* = 1_d$, it follows that $\pi_n(X, x_0)$ is the direct sum of the groups

$$\pi_n(X, x_0) = \text{Im } s_* \oplus \ker p_*.$$

Then Im $s_* \cong \pi_n(B, b_0)$, because $s_*$ is a monomorphism. Finally, it is proved by using the exactness property of the sequence (5.9) that $i_*$ is a monomorphism for every $n \geq 1$, because, $p_*$ is an epimorphism for every $n \geq 1$. This implies that $\ker p_* = \text{Im } i_* \cong \pi_n(F, x_0)$. $\qquad\square$

**Theorem 5.17.4** *If $(X, A, x_0)$ is a triplet such that the subspace $A$ is contractible in $X$ relative to a point $x_0 \in A$, then*

(i) *$\pi_n(X, A, x_0) \cong \pi_n(X, x_0) \oplus \pi_{n-1}(A, x_0)$, $\forall n \geq 3$;*
(ii) *the inclusion map $i : A \hookrightarrow X$ induces homomorphisms $i_*$ mapping $\pi_n(A, x_0)$ into the zero element of $\pi_n(X, x_0)$ for every $n \geq 1$.*

**Proof** By hypothesis, $A$ is contractible in $X$ relative to a point $x_0 \in A$. Hence there exists a homotopy

$$F_t : A \to X : F_0 = i, \ F_1(A) = x_0, \text{ and } F_t(x_0) = x_0.$$

This implies that $i_*$ maps $\pi_n(A, x_0)$ into the zero element of $\pi_n(X, x_0)$ for every $n \geq 1$. This proves (ii). Next we consider the case for $n \geq 2$. Since $i_* = 0$, by using the exactness property of the homotopy sequence (5.2) of $(X, A, x_0)$, it follows that $\partial$ is an epimorphism and $j_*$ is a monomorphism. Consider $\pi_n(X, A, x_0)$ as an extension of $\pi_n(X, x_0)$ by $\pi_{n-1}(X, x_0)$.
Let $\beta \in \pi_{n-1}(A, x_0)$ be represented by a continuous map $g : (\mathbf{I}^{n-1}, \partial \mathbf{I}^{n-1}) \to (A, x_0)$. Define a continuous map

$$k : (\mathbf{I}^n, \mathbf{I}^{n-1}, \mathbf{J}^{n-1}) \to (X, A, x_0), \ (t_1, t_2, \ldots, t_n) \mapsto (F_{t_n} \circ g)(t_1, t_2, \ldots, t_{n-1})$$

and a homomorphism

$$f_* : \pi_{n-1}(A, x_0) \to \pi_n(X, A, x_0), \beta \mapsto \alpha = [k], \ \forall n \geq 2.$$

Then the homomorphism $\partial \circ f_* = 1_d$ is the identity automorphism of $\pi_{n-1}(A, x_0)$, because, $k|_{I^{n-1}} = g$. This implies that $f_*$ is a monomorphism for each $n \geq 2$. If $n \geq 3$, then the group $\pi_n(X, A, x_0)$ is abelian and hence $\partial \circ f_* = 1_d$ asserts by group theory that the group $\pi_n(X, A, x_0)$ decomposes into the direct sum

$$\pi_n(X, A, x_0) = \operatorname{Im} \ f_* \oplus \ker \partial$$
$$\cong \pi_{n-1}(A, x_0) \oplus \pi_n(X, x_0),$$

because $f_*$ is a monomorphism implies $Im \ f_* \cong \pi_{n-1}(A, x_0)$ and $j_*$ is a monomorphism implies $\ker \partial = \operatorname{Im} j_* \cong \pi_n(X, x_0)$.  □

**Corollary 5.17.5** *Let $X$ be a fiber space over a base space $B$ with projection $p$ : $X \to B$. If $x_0 \in B$ is a base point such that the fiber $F = p^{-1}(b_0)$ is contractible in $X$, then the group $\pi_n(B, x_0)$ decomposes into the direct sum*

$$\pi_n(B, x_0) \cong \pi_n(X, x_0) \oplus \pi_{n-1}(F, x_0).$$

**Proof** Consider the homotopy sequence of the fibering $p : X \to B$. Then the corollary follows by using Theorem 5.17.4.  □

**Theorem 5.17.6** *If $F \hookrightarrow X \to B$ is a fiber bundle such that the inclusion map $i : F \hookrightarrow X$ is homotopic to a constant map. Then the long exact homotopy sequence (5.9) of the fibering $p : X \to B$ based at $x_0$ is split into short exact sequences giving isomorphisms*

$$\pi_n(B) \cong \pi_n(X) \oplus \pi_{n-1}(F).$$

**Proof** By hypothesis the inclusion map $i : F \hookrightarrow X$ is homotopic to a constant map. Hence its induced maps $i_* : \pi_n(F, x_0) \to \pi_n(X, x_0)$ appearing in the long exact sequence (5.9) have the property $i_* = 0$. This produces a short exact sequence for every $n \geq 1$

$$0 \to \pi_n(X, x_0) \ \xrightarrow{p_*} \ \pi_n(B, b_0) \ \xrightarrow{k_*} \ \pi_{n-1}(F, x_0) \to 0.$$

Finally, since $p : X \to B$ has the homotopy lifting property (HLP) with respect to all Euclidean disks, there exists a splitting map $\psi : \pi_n(B, b_0) \to \pi_n(X, x_0)$ in the sense that $p_* \circ \psi$ is the identity map on $\pi_n(B, b_0)$. Hence the map splits for every $n \geq 1$ the short exact sequence

$$0 \to \pi_n(X, x_0) \ \xrightarrow{p_*} \ \pi_n(B, b_0) \ \xrightarrow{k_*} \ \pi_{n-1}(F, x_0) \to 0.$$

This asserts by group theory that the group $\pi_n(B, x_0)$ decomposes into the direct sum

$$\pi_n(B, x_0) \cong \pi_n(X, x_0) \oplus \pi_{n-1}(F, x_0).$$  □

**Corollary 5.17.7**  *For the Hopf fiberings*

$$p : S^{2n-1} \to S^n : n = 2, 4, 8$$

*the groups $\pi_m(S^n)$ decomposes into the direct sum*

(i)

$$\pi_m(S^n) \cong \pi_m(S^{2n-1}) \oplus \pi_{m-1}(S^{n-1}) : n = 2, 4, 8 \quad and \quad for \ all \ m \geq 2$$

    *and in particular*

(ii)

$$\pi_m(S^4) \cong \pi_m(S^7) \oplus \pi_{m-1}(S^3) \ and \ \pi_m(S^8) \cong \pi_m(S^{15}) \oplus \pi_{m-1}(S^7), \ \forall \, m \geq 2.$$

***Proof***  Let $F$ be the fiber in each of the Hopf fiberings $p : S^{2n-1} \to S^n$ for $n = 2, 4, 8$. Since the fiber $F$ is an $(n-1)$-sphere $S^{n-1}$ and is contractible in $(2n-1)$-sphere $S^{2n-1}$, the corollary follows by using Corollary 5.17.5.    □

**Corollary 5.17.8**  *The Hopf bundles $p : S^3 \to S^2$, $S^7 \to S^4$ and $S^{15} \to S^8$ produce isomorphisms*

(i) $\pi_m(S^2) \cong \pi_m(S^3)$, $\forall \, m \geq 3$.
(ii) $\pi_m(S^4) \cong \pi_{m-1}(S^3)$, *if* $2 \leq m \leq 6$.
(iii) $\pi_m(S^8) \cong \pi_{m-1}(S^7)$, *if* $2 \leq m \leq 14$.
(iv) $\pi_7(S^4) \cong \mathbf{Z} \oplus \pi_6(S^3)$.
(v) $\pi_{15}(S^8) \cong \mathbf{Z} \oplus \pi_{14}(S^7)$.

***Proof***  It follows from Corollary 5.17.7.    □

**Corollary 5.17.9**  *The groups $\pi_7(S^4)$ and $\pi_{15}(S^8)$ contain $\mathbf{Z}$ summands.*

***Proof***  Using the results given in Corollary 5.17.8 that $\pi_7(S^4) \cong \mathbf{Z} \oplus \pi_6(S^3)$ and $\pi_{15}(S^8) \cong \mathbf{Z} \oplus \pi_{14}(S^7)$, the corollary follows.    □

### 5.17.2   Fiberings of Spheres Over Projective Spaces

This subsection considers the fiberings of spheres over projective spaces.

**Theorem 5.17.10**  (Fiberings of spheres over projective spaces) *Let $\mathbf{C}P^n$ denote the $n$-dimensional complex space. Then*

(i) $\pi_1(\mathbf{C}P^n) = 0$,
(ii) $\pi_2(\mathbf{C}P^n) \cong \mathbf{Z}$,
(iii) $\pi_m(\mathbf{C}P^n) \cong \pi_m(S^{2n-1})$ *for* $m > 2$.

***Proof*** Let $S$ denote the unit sphere in the complex $n$-space $\mathbf{C}^n$ and $\mathbf{C}P^n$ be the $n$-dimensional complex space and $q : \mathbf{C}^n \to \mathbf{C}P^n$ be the natural projection. Consider the fibering $p = q|_S : S \to \mathbf{C}P^n$. Then $S$ is an $(2n - 1)$ sphere, and the fiber $F$ is a 1-sphere. For $n = 1$, $\mathbf{C}P^n$ consists of a single point, and for $n > 1$, the exactness property of the homotopy sequence of the fibering $p : S \to \mathbf{C}P^n$ asserts that

(i) $\pi_1(\mathbf{C}P^n) = 0$;
(ii) $\pi_2(\mathbf{C}P^n) \cong \mathbf{Z}$;
(iii) $\pi_m(\mathbf{C}P^n) \cong \pi_m(S^{2n-1})$ for $m > 2$. □

**Theorem 5.17.11** *Let $S^q \to S^m \to S^n$ is a fiber bundle. Then*

$$q = n - 1 \quad and \quad m = 2n - 1.$$

***Proof*** Suppose $n \le m$ and $q \le m$ and $q + n = m$. If $q = m$, then $n = 0$, and $S^0$ is not connected. Since $S^m \to S^n$ is a surjection, this gives a contradiction and hence $q < m$, and $S^q \to S^m$ is homotopic to a constant map. Consequently, $\pi_i(S^n) \cong \pi_i(S^m) \oplus \pi_{i-1}(S^q)$, $\forall i > 0$. This asserts that $q > 0$ and $m > n$. Then for $i = 1, 2, \ldots, n$, it follows that $\pi_i(S^q) = 0$ if $i < n - 1$ and $\pi_{n-1}(S^q) \cong \mathbf{Z}$. This implies that $q = n - 1$ and hence it follows that $m = 2n - 1$. □

## 5.17.3 Fiberings of Spheres by Spheres

This subsection is devoted to study Hopf fibering: $p : S^{2n-1} \to S^n$, for $n = 2, 4, 8$. There was a long-standing problem: whether a continuous map $p : S^k \to S^n$ for $k > n > 1$ is necessarily nullhomotopic? H. Hopf (1894–1971) solved this problem in 1935 through his discovery of famous map $p : S^3 \to S^2$, named after him. For the fibering: $p : S^{2n-1} \to S^n$, for $n = 2, 4, 8$, its fiber space $F$ is $S^{n-1}$ for $n = 2, 4, 8$. The simplest of Hopf fiberings is the one with the Hopf map $p : S^3 \to S^2$, where

$$S^3 = \{(z, w) \in \mathbf{C}^2 : z\bar{z} + w\bar{w} = 1\},$$

and $S^2$ represents the complex projective line consisting of pairs $[z, w]$ of complex numbers, not both zero, determined by the equivalence relation

$$[z, w] \sim [\lambda z, \lambda w] : \lambda \ne 0$$

and the projection map

$$p : S^3 \to S^2, (z, w) \mapsto [z, w].$$

The map is onto, because, every pair $[z, w]$ can be normalized on division by

$$\lambda = (z\bar{z} + w\bar{w})^{\frac{1}{2}}.$$

The map $p$ is well-defined, because for $|\lambda| = 1$, if $(z, w) \in S^3$, then $(\lambda z, \lambda w) \in S^3$ and their image points are the same, i.e., $p(z, w) = p(\lambda z, \lambda w)$. Conversely, if $p(z, w) = p(z', w')$, then $(z', w') = (\lambda z, \lambda w)$ for some $\lambda$ such that $|\lambda| = 1$. Consequently, the inverse image of a point of $S^2$ under the map is given by the inverse image of any point $S^2$ on multiplication by $e^{i\theta} : 0 \leq \theta \leq 2\pi$. **Geometrically, this** inverse image represents a great circle of $S^3$, and hence $S^3$ is decomposed into a family of great circles having $S^2$ as a decomposition space. Clearly, $p$ is continuous.

**Definition 5.17.12** (*Hopf map*) The continuous onto map

$$p : S^3 \to S^2, (z, w) \mapsto [z, w]$$

is called the Hopf map.


## 5.18  More Study on Hopf Map $p : S^3 \to S^2$

This section generalizes the Hopf map $p : S^3 \to S^2$ through the study of some spaces that arise in projective geometry. The constructions of Hopf fiberings $p : S^7 \to S^4$ from the quaternions and $p : S^{15} \to S^8$ from Calyley numbers are similar to that of $p : S^3 \to S^2$, and they are described below. Recall from Chap. 4 that for $\mathbf{F} = \mathbf{R}, \mathbf{C}$ or $\mathbf{H}$, the right vector space $\mathbf{F}^n$ consists of elements, which are ordered sets of $n$ elements of $\mathbf{F}$. If $x = (x_1, x_2, \ldots, x_n) \in \mathbf{F}^n$ and $\alpha \in \mathbf{F}$, then $x\alpha = (x_1\alpha, \ldots, x_n\alpha)$.

Using the usual inner product $x$ and $y$ in $\mathbf{F}^n$ by $\langle x, y \rangle = \sum_1^n \overline{x}_i y_i$, where $\overline{x}_i$ is the conjugate of $x_i$, it is proved that

(i) $\langle y, x \rangle = \overline{\langle x, y \rangle}$;
(ii) $\langle x\alpha, y \rangle = \overline{\alpha}\langle x, y \rangle$;
(iii) $\langle x, (y\alpha) \rangle = \langle x, y \rangle\alpha$.

**Remark 5.18.1** The orthogonality relation is symmetric, because, $\langle x, y \rangle = 0$ iff $\langle y, x \rangle = 0$. Let $S$ be the unit sphere in $\mathbf{F}^n$ defined by the locus $\langle x, x \rangle = 1$ and $G_n$ be the orthogonal, unitary or sympletic group according as $\mathbf{F} = \mathbf{R}, \mathbf{C}$ or $\mathbf{H}$, then each $G_n$ is a compact Lie group. Let $\mathbf{F}P^n$ be the projective space associated with $\mathbf{F}$, and it is topolozized by considering it as a quotient space of $\mathbf{F}^{n+1} - \{0\}$. It can be thought of the set of all lines through the origin in $\mathbf{F}^{n+1} = \overbrace{\mathbf{F} \oplus \mathbf{F} \oplus \cdots \oplus \mathbf{F}}^{n+1}$, since, every point of $\mathbf{F}^{n+1} - \{0\}$ determines a line through the origin 0 and if $x$ and $y$ are nonzero elements of $\mathbf{F}^{n+1}$, then $x \sim y$ iff there is an element $\lambda(\neq 0) \in F$ such that $y = x\lambda$. This is an equivalence relation and defines $\mathbf{F}P^n$ as the quotient set of equivalence classes endowed with the quotient topology. In particular, $\mathbf{R}P^n$, $\mathbf{C}P^n$ and $\mathbf{H}P^n$ are called $n$-dimensional real, complex and quaternionic projective spaces. The natural projection map $\mathbf{F}^{n+1} - \{0\} \to \mathbf{F}P^n$, $w \mapsto [w]$ is continuous, and defines maps on restriction to the unit sphere of $\mathbf{F}^n$

$$p_n : S^n \to \mathbf{R}P^n,$$

$$q_n : S^{2n+1} \to \mathbf{C}P^n,$$

and

$$r_n : S^{4n+3} \to \mathbf{H}P^n.$$

Usually, the common notation $p$ is used instead of $p_n, q_n$ or $r_n$, unless there is any confusion.

**Remark 5.18.2** (i) $p : S^7 \to S^4$ is a Hopf fibering with fibers 3-spheres and (ii) $p : S^{15} \to S^8$ is a Hopf fibering with fibers 7-spheres.

Example 5.18.3 plays a key role in computing the homotopy groups of sphere (results are only partly known), and hence it reflects the importance of bundle theory.

**Example 5.18.3 (Real, complex and quaternionic Hopf bundles)**

(i) (Real Hopf bundle) $\xi = (S^n, p, \mathbf{R}P^n, \mathbf{Z}_2)$ is a locally trivial fiber bundle with fiber $\mathbf{Z}_2$.
(ii) (Complex Hopf bundle) $\eta = (S^{2n+1}, p, \mathbf{C}P^n, S^1)$ is a trivial fiber bundle with fiber $S^1$.
(iii) (Quaternionic Hopf bundle) $\gamma = (S^{4n+3}, p, \mathbf{H}P^n, S^3)$ is a locally trivial fiber bundle with fiber $S^3$.

**Definition 5.18.4** A continuous map $f : X \to Y$ is said be inessential if there exists a homotopy $H_t : X \to Y$ such that $H_0 = f$ and $H_1(X) = y_0 \in Y$, otherwise, the map is said to be essential.

**Proposition 5.18.5** *Let $B$ be a topological space with more than one point. If $S^n$ is a fiber space over the base space $B$, then the projection $p : S^n \to B$ is an essential map in the sense of Definition 5.18.4.*

**Proof** It is proved by the method of contradiction. Suppose that $p : S^n \to B$ is an inessential map. Then there exists a homotopy $H_t : S^n \to B$ for all $t \in \mathbf{I}$ such that $H_0 = p$ and $H_1(S^n) = b_0 \in B$. If $1_{S^n} : S^n \to S^n$ is the identity map, then $p \circ 1_{S^n} = p$. Since $p : S^n \to B$ has the bundle property, it has the covering homotopy property by Theorem 5.20.2. Hence there exists a homotopy $F_t : S^n \to S^n$ such that $F_0 = 1_{S^n}$ and $p \circ F_t = H_t$, $\forall t \in \mathbf{I}$. By hypothesis, since $B$ has more than one point, $F_1$ sends $S^n$ into the fiber $p^{-1}(b_0)$, which is a proper subset of $S^n$. This implies that the map $F : S^n \times \mathbf{I} \to S^n$ such that $F(x, t) = F_t(x)$, $\forall x \in X$, $\forall t \in \mathbf{I}$ is inessential. This asserts that the identity map $1_S^n$ is homotopic to a constant map. But it is not possible by degree theorem (see Chap. 3). This proves that $p : S^n \to B$ is an essential map. $\square$

### 5.18.1  *Problems of Computing* $\pi_m(S^n)$

This subsection communicates the problems of computing the homotopy groups $\pi_m(S^n)$, the simplest noncontractable spaces. The homotopy groups $\pi_m(S^n)$ are not completely known. So, computing the homotopy groups completely is one of the major unsolved problems in homotopy theory. The homotopy group $\pi_m(S^n)$ for $m \leq n$ are known. It has been proved that

(i) $\pi_m(S^n) = 0 = \{0\}$, for $m < n$ by Theorem 5.18.10.
(ii) $\pi_m(S^1) = 0 = \{0\}$, for $m > 1$, by Theorem 5.4.27.
(iii) $\pi_1(S^1) \cong \mathbf{Z}$ (see Chap. 2 ).
(iv) $\pi_n(S^n) \cong \mathbf{Z}$ (see Chap. 2).

**Remark 5.18.6** There is a natural question: is $\pi_m(S^n) = \{0\}$ for every integer $m > n$. H. Hopf first solved this problem in 1931 by showing that $\pi_3(S^2)$ is not trivial by Hopf Theorem 5.18.7. The sample Table 5.21 shows that there are other examples for $\pi_m(S^n)$ (for $m > n$) are known for particular pair of integers $m$ and $n$ but not known in all possible cases.

### 5.18.2  *More Theorems on Hopf Maps*

This subsection continues the study of Hopf maps started in Chap. 2. In general, continuous maps $p : S^{2n-1} \to S^n$ for $n = 2, 4, 8$ introduced by H. Hopf in 1935 [Hopf, 1935] are now called Hopf maps, which are early examples of bundle spaces with the property that $p$ is not homotopic to a constant map. The aim of his study was to investigate certain homotopy groups of spheres. Historically, it was not known until 1930 whether a given continuous map $p : S^m \to S^n$ for $m > n > 1$ is not homotopic to a constant map. Hopf presented in 1930 the first example of a continuous map $p : S^3 \to S^2$ which is not homotopic to a constant map by proving that $\pi_3(S^2) \neq 0$ ( see Theorem 5.18.7).

### 5.18.3  *Hopf Theorem*

Theorem 5.18.7, known as Hopf theorem and proved by Hopf in 1930 exhibits first example of a continuous map $p : S^3 \to S^2$ which is not homotopic to a constant map by proving that $\pi_3(S^2) \neq \{0\}$ (see Theorem 5.18.7).

**Theorem 5.18.7** $\pi_3(S^2) \neq \{0\}$.

**Proof** Consider the 3-sphere $S^3 = \{(z, w) \in \mathbf{C} \times \mathbf{C} : |z|^2 + |w|^2 = 1\}$ and the Hopf map $p : S^3 \to S^2$. Define an equivalence relation $\sim$ on $S^3$ :

**Fig. 5.35** Commutative
diagram involving Hopf map
$p : S^3 \to S^2$

$$(z, \omega) \sim (z', \omega') \Leftrightarrow (z, \omega) = (\alpha z', \alpha \omega') : \alpha \in \mathbf{C} \text{ and } |\alpha| = 1.$$

(i) If $X = S^3 / \sim$ is the quotient space topologized by the quotient topology and
(ii) $p : S^3 \to X, \ (z, \omega) \mapsto [(z, \omega)]$ is the projection map,

then $p^{-1}[(z, \omega)]$ called the fiber over $[(z, \omega)]$, is a great circle of $S^3$. Since $X$ is
homeomorphic to $S^2$, by replacing $X$ by $S^2$, we have the Hopf map $p : S^3 \to S^2$. This
implies that $S^3$ is decomposed into a family of great circles with $S^2$ as a quotient space.
To prove the theorem, it is sufficient to show that $p$ is not homotopic to a constant
map. We prove it by method of contradiction. Suppose that there exists a homotopy
$F : S^3 \times I \to S^2$ between $p$ and a constant map $c$. Then it defines a homotopy
$\tilde{F} : S^3 \times I \to S^3$ such that the triangle in diagram Fig. 5.35 is commutative.

The map $\tilde{F}$ is continuous and gives a homotopy between the identity map $1_{S^3}$ on
$S^3$ and a constant map $c$. This implies that $S^3$ is contractible, but it is not true. This
contradiction proves that $\pi_3(S^2) \neq \{0\}$.  □

Theorem 5.18.8 proves that the Hopf map $p : S^7 \to S^4$ is not homotopic to any
constant map.

**Theorem 5.18.8**  $\pi_7(S^4) \neq \{0\}$.

**Proof**  Consider the 7-sphere $S^7$ as the topological space

$$S^7 = \{(z, \omega) \in \mathbf{H} \times \mathbf{H} : ||z||^2 = 1\},$$

where $\mathbf{H}$ denotes the division ring of quaternions. Let $Y$ denote the unit disk in $\mathbf{H}$
given by the topological space

$$Y = \{z \in \mathbf{H} : ||z|| \leq 1\}.$$

Let $X$ be the quotient space obtained by identifying the boundary $\partial Y$ of $Y$ to a single
point. Then $X$ is homeomorphic to $S^4$, since the real dimension of $Y$ is 4. Proceed
as in Theorem 5.18.8 to prove that $\pi_7(S^4) \neq \{0\}$.  □

**Theorem 5.18.9**  $\pi_{15}(S^8) \neq \{0\}$.

**Proof**  Consider the Hopf map

$$p : S^{15} \to S^8.$$

To prove the theorem proceed performing likewise construction in $\mathbf{R}^{16}$ as in Theorem 5.18.7 and in Theorem 5.18.8.  □

### 5.18.4   Hurewicz Theorem

Theorem 5.18.10 known as Hurewicz and proved by Hurewicz in 1935 gives a sufficient condition imposed on $m$ and $n$ such that $\pi_m(S^n) = \{0\}$. This theorem is proved
by simplicial approximation theorem saying that if $K$ and $L$ are finite simplicial com-
plexes and $f : |K| \rightarrow |L|$ be a continuous map between their polyhedra, and then for
a chosen sufficiently large $r$, there exists a simplicial approximation $g : |K^r| \rightarrow |L|$
to $f : |K^r| \rightarrow |L|$ and $f \simeq g$ [Adhikari, 2016].

**Theorem 5.18.10**   (Hurewicz) $\pi_m(S^n) = \{0\}$, $\forall\, m, n$ $satsfying$ $0 < m < n$.

**Proof** Let $[f] \in \pi_m(S^n)$ be an arbitrary element. Suppose that $[f]$ is represented by
a continuous map

$$f : (S^m, 1) \rightarrow (S^n, 1).$$

Since $S^m$ can be represented as the boundary complex of simplexes of dimension
$m + 1$ and similarly, $S^n$ as the boundary complex of simplexes of dimension $n + 1$,
the map

$$f : (S^m, 1) \rightarrow (S^n, 1)$$

has a simplicial approximation $g$ that cannot map a simplex onto a simplex of higher
dimension. This implies that the map $g$ is continuous that cannot be onto. Then there
exists a point $s_0 \in S^n$ such that $s_0 \notin Im\, g$ and $g$ is a map such that $Im\, g = g(S^m)$ is
contained in a contractible space. Hence it follows that $g$ is homotopic to a constant
map $c : S^m \rightarrow S^n$. This implies that $[f] = [g] = [c] = 0$. Since $[f]$ is an arbitrary
element of $\pi_m(S^n)$, it is proved that $\pi_m(S^n) = \{0\}$, $\forall\, m, n$ satisfying $0 < m < n$.
□

**Theorem 5.18.11** For any $r > 1$, $\pi_r(S^1, s_0) = 0$.

**Proof** Consider the exponential map

$$p : \mathbf{R} \rightarrow S^1, \; t \mapsto e^{2\pi i t}.$$

Then $p : \mathbf{R} \rightarrow S^1$ is a covering map. Since $S^r$ is simply connected, every continuous map $f : S^r \rightarrow S^1$ has a lifting to a continuous map $\tilde{f} : S^r \rightarrow \mathbf{R}$ by the lifting
property such that $\tilde{f}$ is homotopic to a constant map $\tilde{c}$. Hence there exists a homotopy

$$\tilde{H}_t : \tilde{f} \simeq \tilde{c}.$$

Project this homotopy $\tilde{H}_t$ on $S^1$ to obtain a homotopy

$$H_t : f \simeq c.$$

This proves that $\pi_r(S^1, s_0) = 0$, $\forall\, r > 1$.
□

**Fig. 5.36** Rectangle
involving fiber map and its
induced map

### 5.18.5  Fiber Maps and Induced Fiber Spaces

Given two fiberings $p : X \to B$ and $q : Y \to A$, a continuous map $f : X \to Y$ is
said to be a fiber map if $f$ sends fibers into fibers. This concept is formulated in
Definition 5.18.12.

**Definition 5.18.12** Let $p : X \to B$ and $q : Y \to A$ be two fiberings. A continuous
map $f : X \to Y$ is said to be a **fiber map** if for every point $b \in B$, there exists a
point $a \in A$, such that
$$f(p^{-1}(b)) \subset q^{-1}(a).$$

Every fiber map induces a map on base spaces.

**Definition 5.18.13** Let $p : X \to B$ and $q : Y \to A$ be two fiberings and $f : X \to Y$
be a fiber map. Then $f$ induces a map

$$f_* : B \to A, b \mapsto (q \circ f)(p^{-1}(b)).$$

Then for an arbitrary subset $V \subset A$

$$f_*(V) = (p \circ f^{-1})(q^{-1}(V))$$

asserts that $f_*$ is continuous.

**Remark 5.18.14** For bundles $p : X \to B$ and $q : Y \to A$, let $f : X \to Y$ be a fiber
map. Then $p$ is open and hence the induced map $f_* = g : B \to A$ is a continuous
map such that the diagram in Fig. 5.36 is commutative in the sense that $q \circ f = g \circ p$.

**Example 5.18.15**  (i) **(Trivial fibering)** The fibering $q : Y \to A$ is said to be trivial
over the base space $B$, with the projection $p : X \to B$ as a fiber map, where
$A = B = Y$, $q = 1_B$ (identity map on B) and the induced map $g = 1_B$.
(ii) Let $p : X \to B$ be the trivial fibering over the base space $B$. Then every map
$f : X \to Y$ is a fiber map, where $B = X$ and $p = 1_X$.

## 5.19  Homotopy Sequence of Bundles

This section proves Theorem 5.19.1, which is basic result in bundle theory and defines
homotopy sequence of a bundle in Sect. 5.19.2. For its exactness property and direct
sum theorems related to bundles, see Exercises 5.24.1 and 5.24.1 of Sect. 5.24. Direct

sum theorems related to bundles are analogous to Theorem 5.17.3 plays a key role providing a very strong necessary condition for the existence of a cross section for homotopy groups of dimensions $\geq 2$. An exceptional behavior occurs for $n = 1$. For example, consider the Klein bottle. It is bundle with base space is the circle $B$, obtained from a line segment $L$ by identifying its end points. Its fiber $F$ is also a circle, and it admits a cross section. If the fundamental group $\pi_1(X, x_0)$ of the Klein bottle were a direct sum of $\pi_n(B, b_0) \oplus \pi_n(F, x_0)$, then the fundamental group would be an abelian group, because both the groups $\pi_1(B, b_0)$ and $\pi_1(F, x_0)$ are infinite cyclic. But this is not true in this case, because the fundamental group of the Klein bottle is a group on two generators $\alpha$ and $\beta$ having the only one relation $\alpha\beta = \beta^{-1}\alpha$.

Möbius band is a bundle obtained from the product space $L \times F$ by identifying the two ends of the cylinder $L \times F$

### 5.19.1  Fundamental Property

Theorem 5.19.1 proves a fundamental property of $p_*$ in homotopy theory which is not enjoyed in homology theory.

**Theorem 5.19.1** (Fundamental property) *Let* $\xi = (X, p, Y)$ *be a bundle over* $Y$ *and* $B \subset Y$ *be a subspace. If* $A = p^{-1}(B)$, $x_0 \in A$ *and* $y_0 = p(x_0)$, *then the map* $p : (X, A, x_0) \to (Y, B, y_0)$ *induces an isomorphism*

$$p_* : \pi_n(X, A, x_0) \to \pi_n(Y, B, y_0), \ \forall n \geq 2.$$

**Proof** The map $p : (X, A, x_0) \to (Y, B, y_0)$ induces a homomorphism by Theorem 5.11.12

$$p_* : \pi_n(X, A, x_0) \to \pi_n(Y, B, y_0), \ \forall n \geq 2.$$

$p_*$ **is a monomorphism**: It is proved by showing that kernal of $p_*$ is zero. Let $\alpha \in \pi_n(X, A, x_0)$ be an element such that $p_*(\alpha) = 0$. If $\alpha$ is represented by $f \in F^n(X, A, x_0)$, then there exists a homotopy

$$H_t \in F^n(Y, B, y_0) : H_0 = p \circ f \text{ and } H_1 = c,$$

where $c$ is a constant map. Then there a covering homotopy $H_t'$ of $f$ such that $H_t'(\mathbf{J}^{n-1}) = x_0$, because $H_t(\mathbf{J}^{n-1}) = y_0$. Again, $H_t'(\mathbf{I}^{n-1}) \subset A$, because, $H_t(\mathbf{I}^{n-1}) \subset B$. This implies that $H_t'$ is a homotopy in $F^n(X, A, x_0)$ of $f$ into a map

$$f' : (\mathbf{I}^n, \mathbf{I}^{n-1}, \mathbf{J}^{n-1}) \to (X_0, X_0, x_0),$$

where $X_0$ is the fiber over $y_0$. Define a continuous map

$$F_t : \mathbf{I}^n \to \mathbf{I}^n, \ (t_1, t_2, \ldots, t_n) \mapsto (t_1, t_2, \ldots, t_{n-1}, (1-t)t_n + t), \ \forall t \in \mathbf{I}.$$

Then $F_t$ is a homotopy of $\mathbf{I}^n$ over itself into its face $t_n = 1$. This face stays in the subspace $\mathbf{J}^{n-1}$, which is also deformed into itself. Define

$$F'_t : \mathbf{I}^n \to \mathbf{I}^n, \ (t_1, t_2, \ldots, t_n) \mapsto f'(F_t(t_1, t_2, \ldots, t_{n-1}, t_n)) \ \forall t \in \mathbf{I}.$$

This implies that $F'_t$ is a homotopy in $F^n(X, A, x_0)$ of $f'$ into a constant map. This proves that $\alpha = 0$ and hence kernal of $p_*$ is zero.

$p_*$ **is an epimorphism**: Take an arbitrary element $\beta \in (Y, B, y_0)$ and let $f \in F^n(Y, B, y_0)$ represent the element $\alpha$. Define

$$G_t : \mathbf{I}^n \to \mathbf{I}^n, \ (t_1, t_2, \ldots, t_n) \mapsto f(F_t(t_1, t_2, \ldots, t_{n-1}, t_n)) \ \forall t \in \mathbf{I}.$$

This implies that $G_t$ is a homotopy of $f$ into a constant map. Let $g$ be the map, which carries $\mathbf{I}^n$ into $x_0$. Then there exists an element $\alpha \in \pi_n(X, A, x_0)$ such that $p_*(\alpha) = \beta$.
This completes the proof of the theorem.                                              $\square$

**Corollary 5.19.2** *Let $\xi = (X, p, Y)$ be a bundle over $Y$ and $y_0 \in Y$ be a base point. If $F = p^{-1}(y_0)$, $x_0 \in F$ and $y_0 = p(x_0)$, then the map $p : (X, F, x_0) \to (Y, y_0)$ induces an isomorphism*

$$p_* : \pi_n(X, F, x_0) \to \pi_n(Y, y_0), \ \forall n \geq 2.$$

## 5.19.2  Homotopy Sequence of a Bundle

This subsection defines homotopy sequence of a bundle. For its exactness property, see Exercise 5.24.1 of Sect. 5.24.

Let $\xi = (X, p, Y)$ be a bundle over $Y$ and $y_0 \in Y$ be a base point. Let $F = p^{-1}(y_0)$, $x_0 \in F$ and $y_0 = p(x_0)$. Suppose

$$i : (F, x_0) \hookrightarrow (X, x_0), \ x \mapsto x$$

and

$$j : (X, x_0, x_0) \hookrightarrow (X, F, x_0), \ x \mapsto x$$

are the inclusion maps. Then the homotopy sequence of the triplet $(X, F, x_0)$ is

$$\cdots \longrightarrow \pi_n(F, x_0) \xrightarrow{\ i_* \ } \pi_n(X, x_0) \xrightarrow{\ j_* \ } \pi_n(X, F, x_0) \xrightarrow{\ \partial \ } \pi_{n-1}(F, x_0) \longrightarrow \cdots$$

Let $q$ denote $p$ considered as a map $q : (X, F, x_0) \to (Y, y_0, y_0)$. Then $q \circ j :$ $(X, x_0) \to (Y, y_0)$ is the same as the map $p : (X, x_0) \to (Y, y_0)$. Use Corollary 5.19.2 which asserts that the map $p : (X, F, x_0) \to (Y, y_0)$ induces an isomorphism

$$q_* : \pi_n(X, F, x_0) \to \pi_n(Y, y_0), \ \forall n \geq 2$$

to define

$$\Delta = \partial \circ q_*^{-1} : \pi_n(Y, y_0) \to \pi_n(F, x_0), \ \forall n \geq 2.$$

Then the sequence of groups and homomorphisms

$$\cdots \longrightarrow \pi_n(F, x_0) \xrightarrow{i_*} \pi_n(X, x_0) \xrightarrow{p_*} \pi_n(Y, y_0) \xrightarrow{\Delta} \pi_{n-1}(F, x_0) \to \cdots$$

$$\cdots \longrightarrow \pi_2(Y, y_0) \xrightarrow{\Delta} \pi_1(F, x_0) \xrightarrow{i_*} \pi_1(X, x_0) \xrightarrow{p_*} \pi_1(Y, y_0) \to \cdots$$

is called the **homotopy sequence of the bundle** $\xi$ based at $x_0$.

## 5.20  Bundle Space and Bundle Property

**Definition 5.20.1**  A continuous $p : X \to B$ is said have the **bundle property** if there exists a topological space $D$ such that for every $b \in B$, there exists an open nbd $U$ of the point $b$ and a homeomorphism $\psi_U$ such that

$$\psi_U : U \times D \to p^{-1}(U) : p\psi_U(y, z) = y, \ \forall \, y \in U, z \in D.$$

Then $X$ is called the **bundle space,** $B$ is called the **base space**, and $D$ is called the **director space** relative to the projection $p : X \to B$. The nbd $U$ is called a decomposing nbd (space), and the homeomorphism $\psi_U$ is called the decomposing function.

**Theorem 5.20.2**  (Covering homotopy theorem)  *If a map $p : X \to B$ has the bundle property, then it has the covering homotopy property for every paracompact Hausdorff space.*

**Proof**  See [p. 50, Steenrod, 1951].  □

**Corollary 5.20.3**  *Let $X$ be a bundle space over the base space $B$ relative to projection $p : X \to B$. Then $X$ is fiber space over the space $B$ relative to projection $p$.*

**Proof**  It follows from Theorem 5.20.2  □

**Theorem 5.20.4**  *The Hopf map*

$$p : S^3 \to S^2, (z, w) \mapsto [z, w]$$

*is a 1-sphere bundle.*

**Proof** Represent 1-sphere as $S^1 = \{z \in \mathbf{C} : |z| = 1\}$. For the points $z_1 = [1, 0]$ and $z_2 = [0, 1]$ of $S^2$, consider the open sets

$$U = S^2 - \{z_1\} \text{ and } V = S^2 - \{z_2\}.$$

Hence it follows that $U$ and $V$ forms an open covering of $S^2$ and each point in $U$ is represented by a pair $[z, 1]$. Consider the map

$$\psi_U : U \times S^1 \to S^2, ([z, 1], w) \mapsto (\frac{wz}{(z\overline{z} + 1)^{\frac{1}{2}}}, \frac{w}{(z\overline{z} + 1)^{\frac{1}{2}}}).$$

Then $\psi_U$ maps $U \times S^1$ homeomorphically onto $p^{-1}(U)$ such that $p\psi_U(u, d) = u$, $\forall u \in U$ and $d \in D$ some topological space $D$ (called **director space**).     □

**Corollary 5.20.5** *The 3-sphere is a bundle space over the 2-sphere $S^2$ relative to the Hopf map $p : S^3 \to S^2$.*

**Proof** It follows from the proof of Theorem 5.20.4.     □

The above discussion is summarized in a result formulated in Theorem 5.20.6.

**Theorem 5.20.6** *The 3-sphere $S^3$ is decomposed into a family of great circles, called fibers of the Hopf map $f : S^3 \to S^2$ with the 2-sphere $S^2$ as a decomposition space.*

## 5.21    Table of $\pi_i(S^n)$ for $1 \leq i, n \leq 8$

This section exhibits Table 5.1 displaying a small sample of the values of the groups $\pi_i(S^n)$ extracted from the paper [Toda, 1962]. This table is a consequence of Freudenthal suspension theorem formulated by Freudenthal in 1937 (see Sect. 5.15). An analogue table is also given in Chap. 2.

**Table 5.1** Sample table of $\pi_i(S^n)$ for $1 \leq i, n \leq 8$

| | i=1 | 2 | 3 | 4 | 5 | 6 | 7 | 8 |
|---|---|---|---|---|---|---|---|---|
| n=1 | $\mathbf{Z}$ | 0 | 0 | 0 | 0 | 0 | 0 | 0 |
| 2 | 0 | $\mathbf{Z}$ | $\mathbf{Z}$ | $\mathbf{Z}_2$ | $\mathbf{Z}_2$ | $\mathbf{Z}_{12}$ | $\mathbf{Z}_2$ | $\mathbf{Z}_2$ |
| 3 | 0 | 0 | $\mathbf{Z}$ | $\mathbf{Z}_2$ | $\mathbf{Z}_2$ | $\mathbf{Z}_{12}$ | $\mathbf{Z}_2$ | $\mathbf{Z}_2$ |
| 4 | 0 | 0 | 0 | $\mathbf{Z}$ | $\mathbf{Z}_2$ | $\mathbf{Z}_2$ | $\mathbf{Z} \times \mathbf{Z}_{12}$ | $\mathbf{Z}_2 \times \mathbf{Z}_2$ |
| 5 | 0 | 0 | 0 | 0 | $\mathbf{Z}$ | $\mathbf{Z}_2$ | $\mathbf{Z}_2$ | $\mathbf{Z}_{24}$ |
| 6 | 0 | 0 | 0 | 0 | 0 | $\mathbf{Z}$ | $\mathbf{Z}_2$ | $\mathbf{Z}_2$ |
| 7 | 0 | 0 | 0 | 0 | 0 | 0 | $\mathbf{Z}$ | $\mathbf{Z}_2$ |
| 8 | 0 | 0 | 0 | 0 | 0 | 0 | 0 | $\mathbf{Z}$ |

## 5.22　Action of $\pi_1$ on $\pi_n$ and $n$-simplicity

This section studies action of $\pi_1$ on $\pi_n$ and conveys an important action. The fundamental group $\pi_1(X, x_0)$ acts on $\pi_n(X, x_0)$ as a group of automorphisms for every $n \geq 1$.

### 5.22.1　Automorphism Induced by a Closed Curve

For the study of this subsection, we use Theorem 5.12.2, which asserts that for any path-connected space $X$, the group $\pi_n(X, x_0)$ is independent of the choice of the base point $x_0 \in X$ in the sense that any path $\alpha$ in $X$ from $x_0$ to $x_1$ induces an isomorphism

$$\alpha_* : \pi_n(X, x_1) \to \pi_n(X, x_0).$$

This isomorphism $\alpha_*$ satisfies the following properties for any two points $x_0$, $x_1 \in X$

(i) if $\alpha : \mathbf{I} \to X$ is a path from $x_0$ to $x_1$ and $\beta : \mathbf{I} \to X$ is a path from $x_1$ to $x_2$, then their product path $\alpha * \beta : \mathbf{I} \to X$ induces an isomorphism $(\alpha * \beta)_*$ such that

$$(\alpha * \beta)_* = \alpha_* \circ \beta_*.$$

(ii) if $\alpha \simeq \gamma : \mathbf{I} \to X$ are two paths from $x_0$ to $x_1$, (keeping their end points fixed), then they induce the same isomorphism, i.e., $\alpha_* = \gamma_*$.

The above discussion is summarized in a basic result formulated in Theorem 5.22.1.

**Theorem 5.22.1** *Let X be a pointed topological space with base point $x_0 \in X$. Then for every $n \geq 1$, the fundamental group $\pi_1(X, x_0)$ acts on $\pi_n(X, x_0)$ as a group of automorphisms.*

**Remark 5.22.2** For $n = 1$, the action $\psi$ of $\pi_1$ on itself is by inverse automorphism. For $n > 1$, the action endows the abelian group $\pi_n(X, x_0)$ a module structure over the group ring $\mathbf{Z}[\pi_1(X, x_0)]$. It is said that $\pi_n$ is a $\pi_1$-module instead of $\mathbf{Z}[\pi_1(X, x_0)]$-module.

### 5.22.2　Isomorphism of $\pi_n$ Induced by a Curve

This subsection shows that if the space $A$ of the triple $(X, A, x_0)$ is path-connected, then the groups $\pi_n(X, A, x_0)$ are independent of the choice of the base point $x_0 \in A$, $\forall n \geq 2$. Theorem 5.22.3 gives a generalization of Theorem 5.12.2.

**Theorem 5.22.3** *Let* $(X, A, x_0)$ *be a triple of spaces and* $A$ *be path-connected. Then* $\pi_n(X, A, x_0)$ *is independent of the choice of base point* $x_0 \in A$.

**Proof** Let $\psi : \mathbf{I} \to X$ be a path in $A$ from the point $x_0$ to the point $x_1$ in $A$. To prove the theorem we have to prove that $\psi$ induces an isomorphism

$$\psi_* = \psi_n : \pi_n(X, A, x_1) \to \pi_n(X, A, x_0), \ \forall n \geq 2$$

such that

(i) if $\theta$ is a path from $x_0$ to $x_1$ in $A$ and $\eta$ is a path from $x_1$ to $x_2$ in $A$ then

$$(\theta * \eta)^* = (\theta * \eta)_n = \theta_n \circ \eta_n = (\sigma * \phi)_n,$$

where $\theta * \eta$ denotes product path in $A$ from $x_0$ to $x_1$.

(ii) if $\psi$ and $\phi$ are two homotopic paths in $A$ from $x_0$ to $x_1$ relative to end points, then

$$\psi_n = \phi_n, \ \forall n \geq 2.$$

**Geometric proof** Given an element $\sigma \in \pi_n(X, A, x_0)$, let $f$ represent the element $\sigma$, construct a homotopy $F_t$ of $f$ such that

(i) $F_t$ moves $\mathbf{J}^{n-1}$ along the inverse path $\psi^{-1}$ into $x_0$ keeping the image of $J^{n-1}$ a point at every step and
(ii) $F_t$ deforms $\mathbf{I}^{n-1}$ over $A$.

Hence, it follows that the composite map represents an element $\beta \in \pi_n(X, A, x_0)$. The assignment $\sigma \to \beta$ defines an isomorphism

$$\psi_n : \pi_n(X, A, x_1) \to \pi_n(X, A, x_0), \sigma \to \beta, \ \forall n \geq 2.$$

**Analytical proof** Proceed as in Theorem 5.12.2 for an analytical proof or see [Steenrod, 1951]. $\qquad\square$

**Corollary 5.22.4** *If $X$ is any topological space and $\psi : \mathbf{I} \to X$ is any path connecting two given points $x_0, x_1 \in X$. Then $\psi$ induces an isomorphism*

$$\psi_n : \pi_n(X, x_1) \to \pi_n(X, x_0), \ \forall n \geq 1,$$

*which depends only on the homotopy class of the path $\psi$ relative to its end points.*

### 5.22.3 *n-simplicity*

**Definition 5.22.5** A group $G$ is said to act simply on a group $H$, if $g \cdot h = h, \ \forall g \in G$ and $\forall h \in H$. This action is also called a trivial action of $G$ on $H$.

**Definition 5.22.6** A topological space $X$ is said to be $n$-simple, if for every base point $x_0 \in X$, its fundamental group $\pi_1(X, x_0)$ acts simply on the group $\pi_n(X, x_0)$.

**Proposition 5.22.7** *Let $X$ be a path-connected space. Then it is n-simple, iff there exists a point $x_0 \in X$ such that the fundamental group $\pi_1(X, x_0)$ acts simply on $\pi_n(X, x_0)$.*

**Proof** It follows by using Definition 5.22.6.                                      □

**Example 5.22.8** Every simply connected space is $n$-simple for any $n \geq 1$.

**Example 5.22.9** Every path-connected topological group is $n$-simple for any $n \geq 1$ by Proposition 5.22.11.

Theorem 5.22.10 characterizes $n$-simplicity of a topological space by specified pairs of homotopic maps.

**Theorem 5.22.10** *Let $X$ be a topological space and $S^n$ be the n-sphere with a base point $s_0 \in S^n$. Then $X$ is n-simple iff for any point $x_0 \in X$ and for any pair of continuous maps $f$ and $g$ such that*

$$f, g : S^n \to X : f(s_0) = g(s_0) = x_0,$$

*the homotopy relation $f \simeq g$ implies $f \simeq g$ rel $s_0$.*

**Proof** First suppose that the space $X$ is $n$-simple and $f \simeq g$. We claim that $f \simeq g$ rel $s_0$. Then the fundamental group $\pi_1(X, x_0)$ acts simply on the group $\pi_n(X, x_0)$ by Proposition 5.22.7. Since by hypothesis, $f \simeq g$, there exists a homotopy

$$H_t : S^n \to X : H_0 = f \text{ and } H_1 = g.$$

Let $[f] = \alpha \in \pi_n(X, x_0)$ and $[g] = \beta \in \pi_n(X, x_0)$. Then $\alpha, \beta \in \pi_n(X, x_0)$ are the elements represented by $f$, $g$, respectively. Define a path in $X$

$$\gamma : \mathbf{I} \to X, \ t \mapsto H_t(s_0).$$

Then

$$\gamma(0) = H_0(s_0) = f(s_0) = x_0 = \gamma(1)$$

implies $\gamma$ represents an element $u \in \pi_1(X, x_0)$ such that $\alpha = u \cdot \beta$. Since by hypothesis, $\pi_1(X, x_0)$ acts simply on the group $\pi_n(X, x_0)$, it follows that $u \cdot \beta = \beta$ and hence it follows that $\alpha = \beta$. This asserts that $f \simeq g$ rel $s_0$. To prove its converse, suppose that $f \simeq g$ implies $f \simeq g$ rel $s_0$. Let $\alpha \in \pi_n(X, x_0)$ be an arbitrary element represented by a continuous map $f : S^n \to X$ such that $f(s_0) = x_0$. Hence the element $u\alpha \in \pi_n(X, x_0)$ is represented by a continuous map $g : S^n \to X$ such that $g(s_0) = x_0$. Then $f \simeq g$ and hence by hypothesis, $f \simeq g$ rel $s_0$. This asserts that $u \cdot \alpha = \alpha$ and hence it is proved that $X$ is $n$-simple.          □

**Proposition 5.22.11** *Let $G$ be a path-connected topological group. Then it is $n$-simple for every $n \geq 1$.*

**Proof** Let $S^n$ be the $n$-sphere with a base point $s_0 \in S^n$. By hypothesis, $G$ is a path-connected topological group with its identity element $g_0 = e$. Let $f, g : S^n \to G$ be two maps such that $f \simeq g$ and $f(s_0) = g_0 = g(s_0)$. Then there exists a homotopy

$$H_t : S^n \to G : H_0 = f \quad \text{and} \quad H_1 = g.$$

Define a continuous map

$$F_t : S^n \to G, \ s \mapsto (H_t(s_0))^{-1} H_t(s).$$

Then $F_0 = f$, $F_1 = g$ and $F_t(s_0) = g_0$, $\forall t \in \mathbf{I}$ implies that $F_t : f \simeq g$ rel $s_0$. Hence it follows by Theorem 5.22.10 that $G$ is $n$-simple for every $n \geq 1$.  □

**Theorem 5.22.12** *Let $G$ be a Lie group and $H$ be a closed connected subgroup of $G$. Then $G/H$ is $n$-simple for every integer $n \geq 1$.*

**Proof** Consider the natural projection

$$p : G \to G/H, \ g \mapsto gH.$$

If $p(H) = x_0 \in G/H$, then to prove the theorem, it is sufficient to show that the group $\pi_1(G/H, x_0)$ acts trivially on $\pi_n(G/H, x_0)$, since $G$ acts transitively on $G/H$. Define

$$F^n(G/H, x_0) = \{f : (\mathbf{I}^n, \partial \, \mathbf{I}^n) \to (G/H, x_0) \text{ such that } f \text{ is continuous}\}.$$

Let $\alpha : \mathbf{I} \to X/A$ be a closed curve based at the point $x_0$. If $\alpha(t)$ is regarded as a homotopy of $x_0$, then a covering homotopy determines a curve $\alpha'(t)$ such that

$$\alpha'(0) = e \text{ and } p\alpha(t) = \alpha'(t).$$

This implies that $\alpha'(t) \in H$. Adjoin a curve in $H$ from the point $\alpha'(1)$ to the point $e$ to obtain a closed curve $\beta'(t)$ such that $p \circ \beta' = \beta \simeq \alpha$. Then

$$H : \mathbf{I}^n \times \mathbf{I} \to G/H, (t, s) \mapsto \beta'(1 - s) f(t)$$

is a homotopy of $f$ around the curve $\beta^{-1}$ back into $f$. Since $f \in F^n(G/H, x_0)$ is an arbitrary map, the theorem follows.  □

**Remark 5.22.13** The action of $\pi_1$ on $\pi_n$ is used in algebraic topology such as to prove the homological version of Whitehead theorem which asserts that if $X$ and $Y$ are both simply connected $CW$-complexes and if a continuous map $f : X \to Y$ induces isomorphisms

$$f_* : H_n(X) \to H_n(Y)$$

for all homology groups, then $f$ is a homotopy equivalence [Adhikari, 2016, p. 526].

## 5.23   Further Applications

Theorem 5.23.1 studies infinite earring initiated in Chap. 4 from the homotopy viewpoint and commutes its fundamental group.

**Theorem 5.23.1** *Let X be the union of a countably infinite family of circles with subspace topology in the plane* $\mathbf{R}^2$, *called the 'infinite earring or shrinking wedge of circles' in the plane* $\mathbf{R}^2$. *Then*

(i) $\pi_1(X) \neq \{0\}$;
(ii) *X has no universal covering.*

**Proof** Let $C_n$ be the circle of radius $1/n$ in $\mathbf{R}^2$ with center at $(1/n, 0)$, for each $n \geq 1$ and $X$ be the subspace of $\mathbf{R}^2$ which is the union of these circles as shown in Fig. 5.37.

Hence $X$ is the union of a countably infinite family of circles, which is called the **infinite earring** or 'shrinking wedge of circles' in the plane $\mathbf{R}^2$.

(i) Let $U$ be a nbd of the origin $x_0$ in $X$. Consider the inclusion map

$$i : U \hookrightarrow X.$$

Then the homomorphism

$$i_* : \pi(U, x_0) \to \pi_1(X, x_0)$$

between the corresponding fundamental groups induced by the inclusion

$$i : U \hookrightarrow X$$

is not trivial. Because, for any integer $n$, there is a retraction $r : \tilde{X} \to C_n$ which sends each circle $C_i$ for $i \neq n$ to the point $x_0$. Take $n$ sufficiently large such that $C_n \subset U$. Then the inclusion maps $j : U \hookrightarrow X$ and inclusion $k : U \hookrightarrow U$ induce monomorphism $j_*$ and $k_*$ in the triangle of groups and homomorphisms in Fig. 5.38 commutative. This implies that the homomorphism $j_*$ can not be trivial. This proves the part (i) of the theorem.
(ii) The part (ii) follows from part (i).     □

Theorem 5.23.2 is a basic theorem in homotopy theory. It is now proved by using homotopy sequence of a fibration.

**Fig. 5.37** Infinite earring

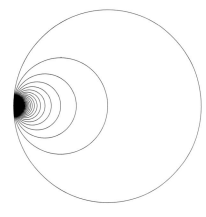

**Fig. 5.38** Commutative
diagram induced by
inclusion maps

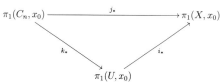

**Theorem 5.23.2** *Let X be a path-connected space and $x_0 \in X$ be an arbitrary point. If $\Omega_{X_{x_0}}$ be the space of all loops in X based at $x_0$ endowed with compact open topology. Then*

*(i)* $\pi_n(\Omega_{X_{x_0}}) \cong \pi_{n+1}(X)$;
*(ii) The group $\pi_1(\Omega_{X_{x_0}})$ is abelian;*
*(iii) The groups $\pi_n(X)$ are all abelian for $n \geq 2$.*

**Proof** $E$ be the space of all paths $\alpha$ in $X$ starting at the point $x_0$ endowed with compact open topology and ending at different pints $x \in X$. Then $\alpha : \mathbf{I} \to X$ is a continuous map such that $\alpha(0) = x_0$ and $\alpha(1) = x$. If

$$p : E \to X, \ \alpha \mapsto \alpha(1) = x,$$

then $\xi = (E, p, X)$ is a fibration represented as

$$\Omega_{X_{x_0}} \to E \to X$$

with fiber $\Omega_{X_{x_0}}$, total space $E$ and base space $X$. Since each $\alpha$ is contractible on itself to the point $x_0$, it follows that $E$ is also contractible and hence $\pi_n(E) = 0$. Consider the homotopy sequence of the fibration $\xi = (E, p, X)$

$$\Omega_{X_{x_0}} \to E \to X$$

$$\cdots \longrightarrow \pi_{n+1}(E) \longrightarrow \pi_{n+1}(X) \longrightarrow \pi_n(\Omega_{X_{x_0}}) \longrightarrow \pi_n(E) \cdots$$

Since the above sequence is exact and $\pi_{n+1}(E) = 0$, it follows that $\pi_n(\Omega_{X_{x_0}}) \cong \pi_{n+1}(X)$. Hence, the theorem follows.  □

**Proposition 5.23.3** *If $(X, A)$ be a pair of topological spaces such that both $X$ and $A$ are path connected, then*

(i)  *the set $\pi_1(X, A, x_0)$ and*
(ii) *the set of cosets $aH$ of the subgroup $H$ of $\pi_1(X, x_0)$ represented by loops in $A$ at $x_0$*

*are equivalent as sets.*

**Proof** By the given condition, $\pi_1(X, A, x_0)$ represents the set of homotopy classes of paths in $X$ starting from an arbitrary point in $A$. Construct a map

$$\phi : \pi_1(X, x_0) \to \pi_1(X, A, x_0)$$

by taking a loop at $x_0$ as an element of $\pi_1(X, A, x_0)$. We claim that $\phi$ is a bijection.

$\phi$ **is surjective** By hypothesis, $A$ is path connected, and hence every element of $\pi_1(X, A, x_0)$ is homotopic to a loop based at the point $x_0$. This implies that $\phi$ is surjective.

$\phi$ **is injective** Since any two loops $\alpha, \beta \in \pi_1(X, x_0)$ are homotopic rel $A$ iff $[\alpha^{-1} * \beta]$ is represented by a loop in $A$. This implies that $\phi$ is injective.  □

**Proposition 5.23.4** *For a given triplet $(X, A, x_0)$, if $A$ is a strong deformation retract of $X$ and the inclusion, $i : (A, x_0) \to (X, x_0)$ is the inclusion, them its induced homomorphism*

$$i_* : \pi_n(A, x_0) \to \pi_n(X, x_0)$$

*is an isomorphism for every $n > 0$.*

**Proof** By hypothesis, $A \subset X$ is a strong deformation retract of $X$. Then there is a retraction $r : X \to A$ such that $i \circ r \simeq 1_X$ rel $A$. Then the map $i$ a homotopy equivalence and consequently, its induced homomorphism $i_* : \pi_n(A, x_0) \to \pi_n(X, x_0)$ is an isomorphism for all $n > 1$.  □

**Corollary 5.23.5** *For a given triplet, $(X, A, x_0)$ if $A$ is a strong deformation retract of $X$, then $\pi_n(X, A, x_0) = 0$ for every integer $n > 0$.*

**Proof** Consider the inclusion map

$$i : (A, x_0) \hookrightarrow (X, x_0).$$

Then its induced homomorphism

$$i_* : \pi_n(A, x_0) \to \pi_n(X, x_0)$$

in the corresponding homotopy groups is an isomorphism for every $n > 0$ by Proposition 5.23.4. Hence the corollary follows from the exactness property of the sequence (5.2) of the triplet $(X, A, x_0)$. $\qquad\qquad\qquad\qquad\qquad\qquad\qquad\qquad\qquad\qquad\qquad\square$

**Definition 5.23.6**  A continuous map $p : X \to B$ is said to have the **polyhedra covering homotopy property (PCHP)** if it has the covering homotopy property for every triangulable space $Y$. If $p$ has PCHP, then $p$ is said to be a fibering.

**Proposition 5.23.7**  *Given a continuous map $f : (X, A, x_0) \to (Y, B, y_0)$ of triplets, if $f : X \to Y$ is a fibering and $A = f^{-1}(B)$, then the induced homomorphism*

$$f_* : \pi_n(X, A, x_0) \to \pi_n(Y, B, y_0)$$

*is an isomorphism for each $n > 1$.*

**Proof  $f_*$ is an epimorphism:** Let $\beta \in \pi_n(Y, B, y_0)$ be an arbitrary element represented by a map $g : (\mathbf{I}^n, \mathbf{I}^{n-1}, \mathbf{J}^{n-1}) \to (Y, B, y_0)$. Since the subspace $\mathbf{J}^{n-1}$ is a strong deformation retract of the space $\mathbf{I}^n$, there is a continuous map

$$k : \mathbf{I}^n \to X : f \circ h = g \text{ and } k(\mathbf{J}^{n-1}) = x_0.$$

This defines a continuous map of triplets

$$k : (\mathbf{I}^n, \mathbf{I}^{n-1}, \mathbf{J}^{n-1}) \to (X, A, x_0),$$

because $A = f^{-1}(B)$, $f \circ k = g$, $k(\mathbf{I}^{n-1}) \subset A$. Then $f_*([k]) = \beta$ asserts that the homomorphism $f_*$ is an epimorphism. $f_*$ **is a monomorphism** : Let $\beta, \gamma \in \pi_n(X, A, x_0)$ be two elements such that $f_*(\alpha) = f_*(\beta)$ represented by $g, k \in \pi_n (X, A, x_0)$. Then maps $f \circ g$ and $f \circ k$ represent the same element of $\pi_n(Y, B, y_0)$. Hence , there exists a map

$$F : (\mathbf{I}^n \times \mathbf{I}, \mathbf{I}^{n-1} \times \mathbf{I}, \mathbf{J}^{n-1} \times \mathbf{I}) \to (Y, B, y_0) : F(w, 0) = (f \circ g)(w) \text{ and } H(w, 1) = (f \circ k)(w), \forall w \in \mathbf{I}^n.$$

If $C = (\mathbf{I}^n \times \{0\}) \cup (\mathbf{J}^{n-1} \times \mathbf{I}) \cup (\mathbf{I}^n \times \{1\})$, then $C$ is a closed subspace of $\mathbf{I}^n \times \mathbf{I}$. Construct a continuous map

$$H : C \to X, (w, t) \mapsto \begin{cases} g(w), & \text{if } w \in \mathbf{I}^n, t = 0 \\ x_0, & \text{if } w \in \mathbf{J}^{n-1}, t \in \mathbf{I} \\ k(w), & \text{if } w \in \mathbf{I}^n, t = 1. \end{cases}$$

Then $f \circ H = F|_C$. Clearly, $H$ has an extension

$$\widetilde{H} : \mathbf{I}^n \times \mathbf{I} \to X : f \circ \widetilde{H} = F,$$

because, $C$ is a strong deformation retract of $\mathbf{I}^n \times \mathbf{I}$. This implies that $\widetilde{H}(\mathbf{I}^{n-1} \times \mathbf{I}) \subset A$. Hence the map

$$\widetilde{H} : (\mathbf{I}^n \times \mathbf{I}, \mathbf{J}^{n-1} \times \mathbf{I}, \mathbf{J}^{n-1} \times \mathbf{I}) \to (X, A, x_0)$$

satisfies the properties

(i) $\widetilde{H}(w, 0) = g(w)$, $\forall\, w \in \mathbf{I}^n$ and
(ii) $\widetilde{H}(w, 1) = k(w)$, $\forall\, w \in \mathbf{I}^n$.

     This shows that the maps $g$ and $h$ represent the same element of $\pi_n(X, A, x_0)$. This proves that if $f_*(\alpha) = f_*(\beta)$, then $\alpha = \beta$.    $\square$

**Proposition 5.23.8** *Let $X = \{x_0\}$ be a one-point space consisting of a single point $x_0$. Then $\pi_n(X, x_0) = 0$ for every $n \geq 0$.*

**Proof** By hypothesis, $X = \{x_0\}$. then for each $n$. Since the map $f : \mathbf{I}^n \to X$ is the only map of $\mathbf{I}^n$ onto $X$, which is a constant map for every $n \geq 0$, it follows that $\pi_n(X, x_0) = 0$ for every $n \geq 0$.    $\square$

## 5.24 Exercises and Multiple Choice Exercises

As solving exercises plays an essential role of learning mathematics, various types of exercises and multiple choice exercises are given in this section. They form an integral part of the book series.

### 5.24.1 Exercises

1. Let $p : (X, x_0) \to (B, b_0)$ be a fibration having fiber $F = p^{-1}(b_0) \subset X$ and the **mapping fiber** $F_p = \{(x, \alpha) \in X \times B^{\mathbf{I}} : \alpha(0) = b_0 \text{ and } \alpha(1) = p(x)\}$ with product topology. Show that the fiber $F$ and the mapping fiber $F_p$ are homotopy equivalent spaces.
2. Given any fibration $p : X \to B$ and a continuous map $f : A \to B$, let $\mathcal{H}_1 = \{[h : A \to X : p \circ h = f]\}$ and $\mathcal{H}_2 = \{[\tilde{h} : A \to X : p \circ \tilde{h} \simeq f]\}$ be two sets of homotopy classes. Show that there exists a bijection

$$\psi : \mathcal{H}_1 \to \mathcal{H}_2.$$

3. Given a normal space $X$ and its subspace $A$, show that the inclusion

$$i : A \hookrightarrow X$$

is a cofibration iff the inclusion

$$j : A \hookrightarrow V$$

is a cofibration for some open neighborhood $V$ of $A$ in $X$.

4. Given a a continuous map : $A \to X$ with mapping cylinder $M_f$, and the inclusion

$$i : A \to M_f, x \mapsto [x, 0],$$

show that the inclusion $i$ is a cofibration.

5. Let $B$ be a path-connected space and $p : X \to B$ be a fibration having fiber $F = p^{-1}(b_0)$. If

$$i : F \hookrightarrow X$$

is the inclusion, show that for any topological space $Z$, the sequence of homotopy sets

$$[Z, F] \xrightarrow{i_*} [Z, X] \xrightarrow{p_*} [Z, B]$$

is exact.

6. Let $A$ be be a subspace of the topological space $X$ and $q : X \to X/A$ the natural projection map. If $i : A \hookrightarrow X$ is a cofibration, with cofiber $X/A$, show that for any topological space $Z$, the 3 terms sequence of homotopy sets

$$[X/A, Z] \xrightarrow{q^*} [X, Z] \xrightarrow{i^*} [A, Z]$$

is exact.

7. Let $\xi$ and $\eta$ be two principal G-bundles over the same space $B$ and $\psi = (f, \tilde{f})$ : $\xi \to \eta$ be a morphism of principal G-bundles. If $\tilde{f} : B \to B$ is the identity map $1_B$, show that $\psi$ is an equivalence of principal $G$-bundles.

8. Let $\xi = (X, p, B, \mathbf{F}^n)$ be an $n$-dimensional vector bundle. A **Gauss map** of $\xi$ in $\mathbf{F}^m (n \leq m \leq \infty)$ is a continuous map $f : X \to \mathbf{F}^m$ such that $f|p^{-1}(b)$ : $p^{-1}(b) \to \mathbf{F}^n$ is a linear monomorphism. Prove the following statements:

   (i) corresponding to every $n$-dimensional vector bundle $\xi$ over a paracompact space $B$, there exists Gauss map $f : X \to \mathbf{F}^m$ for $\xi$,

   (ii) corresponding to an open covering $\{U_i : 1 \leq i \leq m\}$ of $B$, such that $\xi|U_i$ is trivial, then $\xi$ has a Gauss map $f : X \to \mathbf{F}^{mn}$.

9. Show that

   (i) given a covering space $(\tilde{X}, p)$ of $X$ and $x_0 \in X$, $\tilde{x}_0 \in p^{-1}(x_0)$, the induced homomorphism

   $$p_* : \pi_1(\tilde{X}, \tilde{x}_0) \to \pi_1(X, x_0)$$

   is a monomorphism.
   [ Hint: Suppose that $[\tilde{f}], [\tilde{g}] \in \pi_1(X, x_0)$ and $[\tilde{f}] \neq [\tilde{g}]$. This implies that $p_*([\tilde{f}]) = [p \circ \tilde{f}]$ and $p_*([\tilde{g}]) = [p \circ \tilde{g}]$. Hence it follows that

$$p \circ \tilde{f} \simeq p \circ \tilde{g} \text{ rel } \dot{I} \Leftrightarrow \tilde{f} \simeq \tilde{g} \text{ rel } \dot{I}.$$

On the other hand

$$\tilde{f} \not\simeq \tilde{g} \text{ rel } \dot{I} \Leftrightarrow p \circ \tilde{f} \not\simeq p \circ \tilde{g} \text{ rel } \dot{I}$$

because, otherwise, a contradiction would be arrived by Theorem 5.4.7. This proves that $p_*$ is well defined and it is a monomorphism. ]

(ii) homotopic closed path in the punctured Euclidean plane $\mathbf{R}^2 - \{0\}$ has the same winding number.

(iii) every covering of a rectangle ( may be open or closed) is trivial.
[ Hint: Use homotopy lifting theorem.]

10. Let $X$ be a connected covering space of a path-connected space $B$ with covering projection $p : (X, x_0) \to (B, b_0)$ such that $p(x_0) = b_0$. Show that

(i) the induced homomorphism $p_* : \pi_n (X, x_0) \to \pi_n(B, b_0)$ is an isomorphism for any $n \geq 2$.

(ii) $\pi_n(S^1) = \{0\}$ for any $n \geq 2$.

[ Hint: $p_* : \pi_n(\mathbf{R}) \to \pi_1(S^1)$ is an isomorphism for any $n \geq 2$ by (a). Since all the groups of the contractible space $\mathbf{R}$ are

$$X,$$

(ii) follows.]

11. Prove the following statements:

(i) Given a covering space $(X, p)$ of $B$ and two points $b_0, b_1 \in B$, if $F_0$ is the fiber over $b_0$ and $F_1$ is the fiber over $b_1$, the fibers $F_0$ and $F_1$ are the homeomorphic.

[Hint: Since each fiber is discrete and any two fibers have the same cardinal numbers by Theorem 5.10.2(iii), they are homeomorphic.]

(ii) Given a path-connected topological group $G$ and a discrete normal subgroup $N$ of $G$, the natural homomorphism,

$$p : G \to G/N, g \mapsto gN,$$

forms a covering space of $G/N$.

(iii) The map

$$p : S^1 \to S^1, z \mapsto z^2$$

is a covering map having its generalization to the map

$$p : S^1 \to S^2, z \mapsto z^n.$$

(iv) The antipode preserving continuous

$$f : S^1 \to S^1$$

is not nullhomotopic.

12. Given any manifold $M$, let $N$ be submanifold of $M$ such that $N$ is a closed subset of $M$. If $\xi = (X, p, M)$ is a vector bundle over $M$, show that every smooth section $s$ of the restricted bundle $\xi|N$ can be extended to a smooth section of $\xi$.
   [ Hint: Consider the smooth section $s$ as a map with values in a vector space and an open nbd $U$ of for every point $p \in N$ and a section $\tilde{s}$ of $\xi|U$ such that $\tilde{s} = s$ on $U \cap N$. Use Smoothing theorem (ssee Chap. 3, Volume II) and partition of unity subordinate to an open covering of $M$ consisting of such open nbds $\{U_i\}$ in $N$ together with the open set $M - N$. ]

13. Let $\xi = (X, p, M)$ and $\xi' = (X', p', M)$ be two vector bundles of same dimension over a manifold $M$ and $N$ be a closed submanifold of $M$. Show that every isomorphism

$$\psi : \xi|N \to \xi'|N$$

has an extension to an isomorphism

$$\tilde{\psi} : \xi|U \to \xi'|U$$

over an open nbd $U$ of $N$.

14. Show that the 4-manifold $S^2 \times S^2$ is simply connected, but it is not homeomorphic to $S^4$.
   [Hint: Use the results : $\pi_2(S^2 \times S^2) \cong \mathbf{Z} \oplus \mathbf{Z}$ and $\pi_2(S^4) = 0$.]

15. Let $(X, x_0)$ and $(B, b_0)$ be pointed topological spaces. Consider the homotopy exact sequence (5.9) of the fibering $p : X \to B$. Prove the following statements:

   (a) If $F$ is a retract of $B$, then $\pi_n(X, x_0) \cong \pi_n(B, b_0) \oplus \pi_n(F, x_0)$ for every $n \geq 2$ and $p_*$ is an epimorphism for every $n \geq 1$.
   (b) If $X$ is deformable into $F$, then $\pi_n(F, x_0) \cong \pi_n(X, x_0) \oplus \pi_{n+1}(B, b_0)$ for every $n \geq 2$ and $p_* = 0$ for every $n \geq 1$.
   (c) If $F$ is contractible in $X$, then $\pi_n(B, b_0) \cong \pi_n(X, x_0) \oplus \pi_{n-1}(F, x_0)$ for every $n \geq 2$ and $p_*$ is a monomorphism for every $n \geq 1$.

   [ Hint : As $n \geq 2$, use Propositions 5.16.1 & 5.16.5 and exactness property of the homotopy sequence of a fibering $p : X \to B$.]

16. (a) Show that for any triplet $(X, A, x_0)$ the formula $a + b - a = (\partial a)b$ holds for all $a, b \in \pi_2(X, A, x_0)$, where $\partial : \pi_2(X, A, x_0) \to \pi_1(A, x_0)$ is the usual boundary operator, and $(\partial a)b$ denotes the action of $\partial a$ on $b$.
   (b) Deduce from (a) that the image of the map $j_* : \pi_2(X, x_0) \to \pi_2(X, A, x_0)$ lies in the entire of $\pi_2(X, A, x_0)$.

17. Show that a continuous map $f : (\mathbf{D}^n, S^{n-1}, s_0) \to (X, A, x_0)$ defines the zero element in $\pi_n(X, A, x_0)$ iff $f \simeq g$ rel $S^{n-1}$ for some $g : (\mathbf{D}^n, S^{n-1}, s_0) \to (X, A, x_0)$ such that $g(\mathbf{D}^n) \subset A$.

18. Let $p : X \to B$ be a weak fibration with $p(x_0) = b_0$ If $b_0 \in A \subset B, x_0 \in p^{-1}(b_0)$, $Y = p^{-1}(A)$, show that the induced transformation $p_* : \pi_n(X, Y, y_0) \to \pi_n(B, A, b_0)$ is a bijection for every $n \geq 1$.
[Hint. Use mathematical induction on $n$ starting from $n = 1$.]

19. Let $p : E \to B$ be a locally trivial fiber bundle and $b_0 \in B$. If $F = p^{-1}(b_0)$ and $f_0 \in F$, show that for every $n > 1$, $p_* : \pi_n(E, F, f_0) \to \pi_n(B, b_0, b_0)$ is an isomorphism.

20. Let $p : X \to B$ be a covering of $X$ with discrete fiber $F$. Suppose $b_0 \in B$ and $x_0 \in p^{-1}(b_0)$, show that

   (i) $p_* : \pi_n(X, x_0) \to \pi_n(B, b_0)$ is an isomorphism for all $n > 1$ and a monomorphism for $n = 1$;
   (ii) if $X$ is 0-connected, then the points of $F$ are in 1-1 correspondence with the cosets of $p_*(\pi_n(X, x_0))$ in $\pi_1(B, b_0)$.
   [Hint. Since $F$ is discrete, $\pi_n(F, x_0) = \pi_n(\{x_0\}, x_0) = 0$ for all $n \geq 1$.]

21. If $B$ is locally path connected and $(X, p)$ is regular, then for $x_0 \in p^{-1}(b_0)$,

$$\mathcal{A}ut(X/B) \cong \pi_1(B, b_0)/p_*\pi_1(X, x_0)$$

[ Hint: Use Monodromy theorem of the regular covering space].

22. Let $O(n, \mathbf{R})$ be the topological group of real orthogonal $n \times n$ matrices and $SO(n, \mathbf{R})$ be the subspace of $O(n, \mathbf{R})$ of real orthogonal matrices of determinant 1. Show that the inclusion map $i : SO(n, \mathbf{R}) \hookrightarrow O(n, \mathbf{R})$ induces an isomorphism

$$i_* : \pi_n(SO(n, \mathbf{R}), 1) \to \pi_n(O(n, \mathbf{R}), 1) \text{ for } n \geq 1.$$

[Hint. Consider the exact homotopy sequence

$$\cdots \to \pi_{n+1}(\mathbf{Z}_2, 1) \to \pi_n(SO(n, \mathbf{R}), 1) \xrightarrow{i_*} \pi_n(O(n, \mathbf{R}), 1) \to \pi_n(\mathbf{Z}_2, 1),$$

where $\pi_n(\mathbf{Z}_2, 1) = 0$ for $n \geq 1$.]

23. Suppose there exist fiber bundles $S^{n-1} \to S^{2n-1} \to S^n$, for all $n$. Show that the groups $\pi_i(S^n)$ would be finitely generated free abelian groups computable by induction, and nonzero for $i \geq n \geq 2$.

24. Let $p : S^3 \to S^2$ be the Hopf bundle and $q : T^3 \to S^3$ be the quotient map collapsing the complement of a ball in the 3-dimensional torus $T^3 = S^1 \times S^1 \times S^1$ to a point. Show that $p \circ q : T^3 \to S^2$ induces the trivial map $(p \circ q)_* : \pi_n(T^3) \to \pi_n(S^2)$, but not homotopic to a constant map.

25. Let $X$ be a path-connected space with a base point $x_0 \in X$ and $f : S^n \to X$ be a continuous map such that $f(s_0) = x_0$, where $s_0$ is a base point of $S^n$. If

$Y = X \bigcup_f D^{n+1}$, and $i : X \hookrightarrow Y$ is inclusion, show that induced homomorphism
$i_* ; \pi_m(X, x_0) \to \pi_m(Y, y_0)$

(i)  is an isomorphism if $m < n$;
(ii)  is an epimorphism if $m = n$ and
(iii)  ker $i_*$ is generated by $\alpha^{-1}[f]\alpha \in \pi_n(X, x_0)$, where $\alpha \in \pi_1(X, x_0)$.

26. Let $\xi = (X, p, A, G)$ be a principal G-bundle. Show that $\xi$ has an H-structure iff there exists a map $\tilde{f} : A \to B_H$ such that $B\alpha \circ \tilde{f} \simeq f_\xi$, where $f_\xi : A \to B_G$ is the unique map(upto homotopy) such that $f_\xi^*(\xi_G) \cong \xi$.

27. Let $\xi$ be a numerable principal G-bundle over $B \times \mathbf{I}$ and $f$ be the map

$$f : B \times \mathbf{I} \to B \times \mathbf{I}, (b, t) \mapsto (b, 1)$$

Show that

(i)  there exists a G-morphism $(g, f) : \xi \to \xi$, and
(ii)  the principal G-bundles $\xi$ and $f^*(\xi)$ over $B \times \mathbf{I}$ are isomorphic.

28. Given a covering space $(\tilde{X}, p)$ of $X$ and a connected space $Y$, if $f, g : Y \to \tilde{X}$ are two continuous maps such that $p \circ f = p \circ g$, and if they agree at some point $y_0 \in Y$, then show that $f = g$.

29. Prove the following statements :

(i)  If $B$ is a paracompact space and $p : X \to B$ is the projection of a fiber bundle, then $p$ is a fibration.
(ii)  Given a vector bundle $\xi$ over $B \times \mathbf{I}$, there exists an open covering $\{U_i\}_{i \in A}$ of $B$ such that $\xi|(U_i \times \mathbf{I})$ is trivial.
(iii)  Given a paracompact space $B$, a continuous map

$$f : B \times \mathbf{I} \to B \times \mathbf{I}, (b, t) \mapsto (b, 1)$$

and a vector bundle
$$\xi = (X, p, B \times \mathbf{I}, \mathbf{F}^n),$$

there exists a continuous map $g : X \to X$ with a morphism of vector bundles $(g, f) : \xi \to \xi$ and a linear isomorphism $g$ on each fiber
(iv)  Given a vector bundles $\xi$ over $B \times \mathbf{I}$, the vector bundles $f^*(\xi|(B \times \{1\})$ and $\xi$ are isomorphic.
(v)  The two vector bundles $\xi$ and $\xi|((B \times \{1\}) \times \mathbf{I})$ over the same base space $B \times \mathbf{I}$ are isomorphic.
(vi)  There exists an isomorphism of vector bundles

$$(f, g) : \xi|(B \times \{0\}) \to \xi|(B \times \{1\})$$

30. Let $\gamma_n^\infty$ represent the $n$-dimensional vector bundle over the Grassmann manifold $G_n(\mathbf{F}^m)$. Prove the following statements:

   (i) if $B$ is a paracompact space, then each $n$-dimensional vector bundle over $B$ is B-isomorphic to the induced vector bundle $f^*(\gamma_n^\infty)$ for some continuous map $f : B \to G_n(\mathbf{F}^\infty)$;

   (ii) For any two continuous maps $f, g : B \to G_n(\mathbf{F}^m)$ such that their induced vector bundles $f^*(\gamma_n{}^m)$ and $g^*(\gamma_n{}^m)$ are B-isomorphic, if

$$i : G_n(\mathbf{F}^m) \hookrightarrow G_n(\mathbf{F}^{2m}) \quad \text{for } 1 \leq m \leq \infty$$

   is the natural inclusion, then

$$i \circ f \simeq i \circ g.$$

   (iii) for any paracompact space $B$, the fiber bundle $p : X \to B$ is a fibration.

31. Show that for every $n$-dimensional vector bundle $\xi = (X, p, B, \mathbf{F}^n)$ over a para-compact space $B$, there exists a continuous function $f : B \to G_n(\mathbf{F}^\infty)$ such that the vector bundles $\xi$ and $f^*(\{\gamma_n^\infty\})$ are B-isomorphic.
   [ Hint: See Chap. 4. ]

32. Let $f, g : (\mathbf{I}^n, \partial \mathbf{I}^n) \to (X, x_0)$ be two homotopic maps relative to $\partial \mathbf{I}^n$ and $\psi, \phi : \mathbf{I} \to X$ are two homotopic paths relative to end points. If $F_t, G_t : \mathbf{I}^n \to X : t \in \mathbf{I}$ are homotopies of $f$ along the path $\psi$ and that of $g$ along the path $\phi$, respectively, show that $F_1$ and $G_1$ are homotopic relative to $\partial \mathbf{I}^n$.

33. Let $p : X \to B$ be fibration and $b, b' \in B$ are two points such that they have the same path component. Show that

$$p^{-1}(b) \simeq p^{-1}(b').$$

34. Let

$$F \to X \to B$$

   be a fibration sequence with fiber $F$, total space $X$ and contractible base space $B$. Show that

   (i) $X \simeq F \times B$ and
   (ii) the inclusion $F \hookrightarrow X$ is a homotopy equivalence.

35. Prove the following statements:

   (i) An inclusion $A \hookrightarrow X$ of topological spaces is a cofibration iff the inclusion

$$(A \times \mathbf{I}) \cap (X \times 0) \hookrightarrow X \times \mathbf{I}$$

is a retraction.

(ii) Let $A$ and $X$ be two topological spaces such that $A$ is contractible and closed in $X$. If the inclusion $A \hookrightarrow X$ is a cofibration, then the quotient map

$$p : X \to X/A$$

is a homotopy equivalence.

36. Let $\xi = (X, p, B)$ be a covering space. Then $\xi$ is said to be **regular** if the subgroup $p_* \pi_1(B, b_0)$ of $\pi_1(B, b_0)$ is normal. Let $G$ be the group of covering transformations of $\xi = (X, p, B)$. Prove the following statements:

(i) If $G$ is a properly discontinuous group of homeomorphisms of a space $X$, then the projection $p : X \to X$ mod $G$ is a covering projection.

(ii) If $X$ is connected, then this covering space $\xi = (X, p, X, \mod G)$ is regular, and $G$ is its group of covering transformations

(iii) If a topological space $X$ is simply connected and $G$ is a properly discontinuous group of homeomorphisms of $X$, then the groups $G$ is isomorphic to the fundamental group $\pi_1(X \mod G)$.

37. Let $\eta$ be a numerable principal $G$-bundle over $B \times \mathbf{I}$ and $f : B \times \mathbf{I} \to B \times \mathbf{I}$, $(b, t) = (b, 1)$ be a map. Show that

(i) there exists a $G$-morphism $(g, f) : \eta \to \eta$ and

(ii) the principal $G$-bundles $\eta$ and $f^*(\eta)$ are isomorphic over $B \times \mathbf{I}$.

38. Prove the following statements:

(i) Let $(X, p)$ be a universal covering space of $B$, where $B$ is locally path connected. Then for any point $b_0 \in B$, the groups $\mathcal{A}ut(X/B) \cong \pi_1(B, b_0)$.

(ii) A principal G-bundle $\xi = (X, p, B, G)$ is a universal $G$-bundle iff its total space $X$ is contractible.

(iii) Let $A$ be a contractible space such that it is a closed subspace of a topological space $X$. If $i : A \hookrightarrow X$ is cofibration, then the natural quotient map

$$p : X \to X/A$$

is a homotopy equivalence.

(iv) Let $B$ and $B'$ be two base spaces of two universal $G$-bundles. Then the spaces $B$ and $B'$ are homotopy equivalent.

(v) $p : X \to B$ be a fibration such that the points $b$, $b' \in B$ lie in the same path component of $B$. Then their fibers $p^{-1}(b)$ and $p^{-1}(b')$ in $X$ are homotopy equivalent.

39. If a map $f : X \to Y$ satisfies the covering homotopy property, show that the inverse images of the points of $Y$ are homotopy equivalent.

40. Given a continuous map $p : (X, A, x_0) \to (Y, B, b_0)$, let $p : X \to Y$ be a fibering and $A = p^{-1}(B)$ be a subspace of $X$. Prove that $p$ induces a bijection

$$p_* : \pi_n(X, A, x_0) \to \pi_n(Y, B, y_0), \ \forall n > 0.$$

41. Let $p : X \to Y$ be a fibration and $x_0 \in X$, $y_0 \in Y$ be two points such that $p(x_0) = y_0$. If $y_0 \in B \subset Y$ and $A = p^{-1}(B) \subset X$, show that $p$ induces an isomorphism

$$p_* : \pi_n(X, A, x_0) \to \pi_n(Y, B, y_0), \ \forall n > 0.$$

42. Let $\xi = (X, p, B)$ be a principal G-bundle over $B$ and $f, g : A \to B$ are homotopic maps. Show that the induced bundles $f^*(\xi)$ and $g^*(\xi)$ are isomorphic G-bundles over $A$.

43. Show that the homotopy sequence of any bundle based at a point defined in Sect. 5.19.2 is exact,
    [Hint: Using notations of Sect. 5.13.3, proceed as in exactness property of homotopy sequence of any triplet given in Theorem 5.13.8 for the triplet $(X, F, x_0)$ in this case].

44. (**Direct sum theorems for bundles**) Let $\xi = (X, p, B)$ be a bundle over $B$ and $b_0 \in B$ be a base point. If $F = p^{-1}(b_0)$, $x_0 \in F$ and $\xi$ admits a cross section, show that

    (i) $\pi_n(X, x_0) \cong \pi_n(B, b_0) \oplus \pi_n(F, x_0), \ \forall n \geq 2$;
    (ii) for $n = 1$, the fundamental group $\pi_1(X, x_0)$ contains two subgroups $G$ and $H$ such that $G$ is invariant and isomorphic to the fundamental group $\pi_1(F, x_0)$ and $p_*$ maps $H$ isomorphically onto the fundamental group $\pi_1(B, b_0)$ and every element of $\pi_1(B, b_0)$ is uniquely determined as the product of an element of $G$ with an element of $H$;
    (iii) $p_* : \pi_n(X, x_0) \to \pi_n(B, b_0)$ is an epimorphism for every $n \geq 1$.

    Hence prove that

$$\pi_n(B \times F, (b_0, x_0)) \cong \pi_n(B, b_0) \oplus \pi_n(F, x_0), \ \forall n \geq 1.$$

    [ Hint: For the first part, use the result that the product space is a bundle and it admits a cross section and hence proceed as in Theorem 5.17.3 for $n \geq 2$. For $n = 1$, use (ii). ]

45. (**Long exact sequence for a fibration**) Let $p : X \to B$ be a fibration and $b_0 \in B$ be a base point. If $F = p^{-1}(b_0)$, $x_0 \in F$, show that $p$ induces an a homomorphism $p_*$ such that

    (i)

$$p_* : \pi_n(X, F, x_0) \to \pi_n(B, b_0, b_0)$$

    is an isomorphism for all $n > 0$ and
    (ii) for any path-connected base space $B$, there is a long exact sequence

$$\cdots \longrightarrow \pi_n(F, x_0) \xrightarrow{\ i_* \ } \pi_n(X, x_0) \xrightarrow{\ p_* \ } \pi_n(B, b_0) \xrightarrow{\ \partial \ } \pi_{n-1}(F, x_0) \longrightarrow \cdots$$

46. Show that a covering projection $p : X \to B$ is a principal $G$-bundle for the group $G$ of covering transformations with the discrete topology.

47. Let $X$ be a $G$-space $X$. Show that the set of all automorphisms of the trivial $G$-bundle with projection

$$p_2 : X \times B \to B$$

are in bijective correspondence with the set of all continuous maps

$$f : B \to G.$$

48. Show that a fiber bundle $\xi : X \xrightarrow{\ p\ } B$ is trivial if the base $B$ is contractible.
49. Prove that a fibration $p : X \to B$ is a Serre fibration iff its the base $B$ is paracompact.
50. Let $G_{n,k} = G_k(\mathbf{R}^n)$ denote the Grassmann manifold of $k$-planes through the origin in $\mathbf{R}^n$. Show that

   (i) $O(n, \mathbf{R})$ acts transitively on $G_{n,k}$;
   (ii) $G_{n,k} \simeq O(n, \mathbf{R})/O(k, \mathbf{R}) \times O(n - k, \mathbf{R})$.
       See Bredon p 464

51. Let $(X, p)$ be the universal covering space of $B$ and $Aut(X/B)$ be the group of all automorphisms of (X,B).
    Prove that

   (i) the automorphism group $Aut(X/B)$ is isomorphic to the fundamental group $\pi_1(B)$ of $B$ and
   (ii) if $|\pi_1(B)|$ is the order of the group $\pi_1(B)$, then $|\pi_1(B)|$=number of sheets of the universal covering space.

## 5.24.2   Multiple Choice Exercises

Identify the correct alternative (s) ( there may be more than one ) from the following list of exercises:

1. Let $(X, p, B)$ be a universal covering space of a connected space $B$. If $b_0 \in B$, $x_0 \in p^{-1}(b_0)$, then the induced homomorphism

$$p_* : \pi_1(X, x_0) \to \pi_n(B, b_0)$$

   (i) is an isomorphism for $n = 1$;
   (ii) is an isomorphism for all $n > 1$;
   (iii) is an isomorphism for $n > 5$.

2. Let $\mathbf{R}P^m$ be the real projective space and $\mathbf{S}^m$ be the unit sphere in $\mathbf{R}^{m+1}$. Then

$$\pi_n(\mathbf{R}P^m) \cong \pi_n(\mathbf{S}^m)$$

(i) if $n = 1$;
(ii) only if $n > 1$;
(iii) only if $n > 6$.

3. Let $\xi = (X, p, B)$ be a principal G-bundle over $B$ and $f, g : Y \to B$ are homotopic maps.

(i) The induced bundles $f^*(\xi)$ and $g^*(\xi)$ are isomorphic G-bundles over $Y$.
(ii) the induced bundles $f^*(\xi)$ and $g^*(\xi)$ are isomorphic G-bundles over $B$.
(iii) The induced bundles $f^*(\xi)$ and $g^*(\xi)$ are neither isomorphic G-bundles over $B$ nor $Y$.

4. Consider a fibration sequence

$$F \to X \to B$$

with fiber $F$, total space $X$ and contractible base space $B$. Then

(i) the spaces $X$ and the product space $F \times B$ are homotopy equivalent;
(ii) the inclusion $i : F \hookrightarrow X$ is a homotopy equivalence;
(iii) the inclusion $i : F \hookrightarrow X$ is a homeomorphism.

5. Let $f : (X, A, x_0) \to (Y, B, y_0)$ be a continuous map of triplets. If $f : X \to Y$ is a fibering and $A = f^{-1}(B)$, then the induced homomorphism

$$f_* : \pi_n(X, A, x_0) \to \pi_n(Y, B, y_0)$$

(i) is a monomorphism but it is not an epimorphism;
(ii) is an epimorphism but it is not an monomorphism;
(iii) is an isomorphism.

6. Let $\mathcal{H}tp^2$ denote the homotopy category of triplets and their continuous maps, and $\mathcal{A}b$ denote the category of abelian groups and homomorphisms. Then

$$\pi_n : \mathcal{H}tp^2 \to \mathcal{A}b$$

(i) is a covariant functor for every integer $n > 2$.
(ii) is a covariant functor for every integer $n \geq 2$.
(iii) is a contravariant functor for every integer $n > 2$.

# References

Adams JF. Stable homotopy and generalized homology. University of Chicago Press; 1974.

Adhikari A, Adhikari MR. Basic topology, vol 1: metric spaces and general topology. India: Springer; 2022a.

Adhikari A, Adhikari MR. Basic topology, vol 2: topological groups, topology of manifolds and lie groups. India: Springer; 2022b.

Adhikari MR, Adhikari A, Groups, rings and modules with applications. Hyderabad: Universities Press; 2003.

Adhikari MR, Adhikari A, Textbook of linear algebra: an introduction to modern algebra. New Delhi: Allied Publishers; 2006.

Adhikari MR, Adhikari A. Basic modern algebra with applications. New Delhi, New York, Heidelberg: Springer; 2014.

Adhikari MR. Basic algebraic topology and its applications. India: Springer; 2016.

Arkowitz M. Introduction to homotropy theory. New York: Springer; 2011.

Bredon GE. Topology and geometry. New York: Springer; 1993.

Dugundji J. Topology. Newtown, MA: Allyn & Bacon; 1966.

Hopf H. Ueber die Abbildungen von Sphären niedriger Dimension. Fund Math. 1935;25:427–40.

Hu ST. Homotropy theory. New York: Academic Press; 1959.

Mukherjee A. Differential topology. New Delhi: Hindustan Book Agency; 2015.

Steenrod N. The topology of fibre bundles. Prentice: Prentice University Press; 1951.

# Chapter 6
# Geometric Topology and Further Applications of Algebraic Topology

Geometric topology primarily studies manifolds and their embeddings in other manifolds. A particularly active area is low-dimensional topology, which studies manifolds of four or fewer dimensions. This includes knot theory, which makes a study of mathematical knots. This chapter gives a brief study of geometric topology by communicating the concepts of knots and knot groups. It also gives further applications of topological concepts and results discussed in earlier chapters with a view to understand the beauty, power and scope of the subject topology. Moreover, it provides alternative proofs of some results proved in the previous chapters such as Brouwer–Poincaré theorem, Van Kampen theorem, Borsuk–Ulam theorem for any finite dimension. It proves Ham Sandwich theorem and Lusternik–Schnirelmann theorem.

For this chapter the books [Adams, 1958, 1960, 1972], [Adhikari and Adhikari, 2014, 2016, 2022a, 2022b], [Aguilar et al. 2002], [Arkowitz and Martin, 2011], [Armstrong, 1983], [Bredon, 1993], [Basak, 2017], [Dold and Thom, 1958], [Hu, 1966], [Steenrod, 1951] and some others are referred in the Bibliography.

## 6.1 Geometric Topology: Embedding Problem of the Circle in $\mathbf{R}^3$ with Knot and Knot Groups

The main aim of **geometric topology** is to study manifolds and their maps including embeddings of one manifold into another. This section studies the embedding problems of geometric topology such as various embeddings of the circle $S^1$ in the Euclidean 3-space $\mathbf{R}^3$ or in the 3-sphere $\mathbf{S}^3$. Such problems arise in geometry leading to the concept of knots. These problems make a return to geometry and are

© The Author(s), under exclusive license to Springer Nature Singapore Pte Ltd. 2022
M. R. Adhikari, *Basic Topology 3*,
https://doi.org/10.1007/978-981-16-6550-9_6

mainly studied in **low-dimensional topology**. For possible solution of such problems, knot groups are studied through the concepts of fundamental group and the **one-point compactification of $\mathbf{R}^3$**. This study displays an interplay among geometry, topology and algebra. Physicists and bio-scientists use knot theory in their study. **Historically,** H. Tietze (1880–1964) laid the foundations of knot theory.

**Definition 6.1.1** A **knot** $K$ is a homeomorphic image of an embedding $f : S^1 \to \mathbf{R}^3$, and the fundamental group

$$\pi_1(\mathbf{R}^3 - K)$$

of the complement of the knot $K$ in $\mathbf{R}^3$ is called the **knot group of** $K$.

***Example 6.1.2*** As a knot $K$ is a subspace of $\mathbf{R}^3$, which is homeomorphic to the circle, it is represented by its projection in the plane of the paper. For example, trivial knot, figure-eight knot and square knots and some others are interesting. For their geometrical representation see Remark 6.1.4.

(i)  The **trivial knot** or unknot is defined by the standard embedding

$$i : S^1 \hookrightarrow \mathbf{R}^3$$

It is called the **circle knot,** which is the simplest knot and consists of the unit circle in the $xy$-plane with its knot group **Z**.

(ii)  For trefoil knot see Fig. 6.1.

(iii)  For figure-eight knot see Fig. 6.2.

(iv)  For square knot see Fig. 6.3.

***Remark 6.1.3*** Since a knot $K$ is represented by its projection in the plane of the paper, 'trivial knot' or 'unknot' consists of the unit circle in the $xy$-plane.

***Remark 6.1.4*** For representing a knot **geometrically** and working with it relatively comfortable, the standard practice is to project it into the plane in such a way that its projection only crosses itself at a finite number of points, at most two pieces of the knot meet at such crossing, and it does so at 'right angles.' For example, the knots represented in Figs. 6.1, 6.2 and 6.3 known as 'trefoil knot,' 'figure-eight knot' and 'square knot,' respectively.

**Definition 6.1.5** Let $K_1$ and $K_2$ be two knots. They are said to **equivalent as knots** if there exists a homeomorphism

$$f : \mathbf{R}^3 \to \mathbf{R}^3 : f(K_1) = K_2.$$

In other words, two knots are said to be equivalent or same if there exists a homeomorphism

**Fig. 6.1**  Trefoil knot

**Fig. 6.2**  Figure-eight knot

**Fig. 6.3**  Square knot

$$f : \mathbf{R}^3 \to \mathbf{R}^3$$

that sends one knot onto the other knot.

**Remark 6.1.6**  Geometrically, two knots are equivalent if one knot can be deformed into the other knot by a continuous deformation.

**Definition 6.1.7**  Let $S^m$ be the $m$-dimensional sphere in $\mathbf{R}^{m+1}$ and $S^n$ be the $n$-dimensional sphere in $\mathbf{R}^{n+1}$. If $m < n$ and

$$f : S^m \to S^n$$

is an embedding, then $f(S^m)$ is called the  $m$-**dimensional knot in** $S^n$. It is also called an higher-dimensional knot.

**Remark 6.1.8**  The higher-dimensional knots given in Definition 6.1.7 are studied through the generalized Jordan curve theorem given in Exercise 8 of Sect. 6.4.1 by using  **homology and cohomology theories** [Basak, 2017].

**Remark 6.1.9**  The 3-sphere is considered as the one-point compactification of the Euclidean space $\mathbf{R}^3$, the latter space $\mathbf{R}^3$ is not compact (see **Basic Topology, Volume 1**). This result is utilized in Theorem 6.1.10.

**Theorem 6.1.10** *Given a knot $K$, the inclusion map*

$$i : (\mathbf{R}^3 - K) \hookrightarrow (S^3 - K)$$

*induces an isomorphism between the corresponding fundamental groups*

$$i_* : \pi_1(\mathbf{R}^3 - K) \to \pi_1(S^3 - K).$$

**Proof** By definition, $K$ is a compact subset of $\mathbf{R}^3$. Again, the subspace $S^3 - K$ is the union of the open set $\mathbf{R}^3 - K$ and an open ball $D$ obtained by including the compactification point and the complement of a large closed ball in $\mathbf{R}^3$ containing the knot $K$. Since both the spaces $\mathbf{D}$ and $\mathbf{D} \cap (\mathbf{R}^3 - K)$ are simply connected and $\mathbf{D} \cap (\mathbf{R}^3 - K)$ is homeomorphic to $S^2 \times \mathbf{R}$, it follows that the inclusion map

$$i : (\mathbf{R}^3 - K) \hookrightarrow (S^3 - K)$$

induces an isomorphism.

$$i_* : \pi_1(\mathbf{R}^3 - K) \to \pi_1(S^3 - K).$$

$\square$

**Definition 6.1.11** (Torus knot) Given two relatively prime positive integers $(p, q)$, the torus knot $K_{p,q}$

$$K = K_{p,q} \subset \mathbf{R}^3$$

is the image of the embedding

$$h : S^1 \to S^1 \times S^1 \subset \mathbf{R}^3, \; z \mapsto (z^p, z^q),$$

with the torus $S^1 \times S^1$ having the natural embedding in $\mathbf{R}^3$.

**Remark 6.1.12** **Geometrically**, the torus knot of the form $K_{p,q} \subset \mathbf{R}^3$ is the image in the torus of the line with the equation $px = qy$ in $\mathbf{R}^3$ is a knot that winds $p$ times around the torus one way and it winds $q$ times around the other way. More precisely, the torus knot $K = K_{p,q}$ winds the torus a total of $p$ times in the longitudinal direction and $q$ times in the meridian direction.

**Definition 6.1.13** Let $f : \mathbf{R}^3 \to \mathbf{R}^3$ be a homeomorphism. Then $f$ is called **isotopic to the identity map** if there is a homotopy

$$F : \mathbf{R}^3 \times I \to \mathbf{R}^3$$

having the property

$$F_t : \mathbf{R}^3 \to \mathbf{R}^3, x \mapsto F(x, t)$$

is a homeomorphism such that $F_0 = 1_{\mathbf{R}^3}$ ( identity map on $\mathbf{R}^3$) and $F_1 = f$.

**Remark 6.1.14** Given two knots $K_1$ and $K_2$, let $f : \mathbf{R}^3 \to \mathbf{R}^3$ be a homeomorphism isotopic to the identity map such that $f(K_1) = K_2$.

(i) Then $H_t(K_1)$ provides a family of continuous maps which move gradually from the knot $K_1$ to the knot $K_2$ as $t$ increases from 0 to 1.
(ii) Since $S^3$ is the one-point compactification of $\mathbf{R}^3$, a homeomorphism $f : \mathbf{R}^3 \to \mathbf{R}^3$ has a unique extension to a homeomorphism $\tilde{f} : S^3 \to S^3$ ( see Chap. 5).

**Definition 6.1.15** Let $f : \mathbf{R}^3 \to \mathbf{R}^3$ be a homeomorphism. It is called **orientation preserving (or orientation reversing)** if its extension homeomorphism

$$\tilde{f} : S^3 \to S^3$$

preserves (or reverses) the orientation of $S^3$.

**Example 6.1.16** Every homeomorphism $f : \mathbf{R}^3 \to \mathbf{R}^3$, which is isotopic to the identity map $1_{\mathbf{R}^3}$ on $\mathbf{R}^3$ is orientation preserving, since every homeomorphism

$$F_t : \mathbf{R}^3 \to \mathbf{R}^3$$

has a continuous extension which is the homeomorphism

$$\tilde{F}_t : S^3 \to S^3$$

because homotopic maps have the same degree by Hopf's classification theorem (see Chaps. 2 and 3).

**Example 6.1.17** Reflection in a plane which is a homeomorphism of $\mathbf{R}^3$, transforms every knot to its mirror image. Since it is orientation reversing, it is not isotopic to the identity. On the other hand every orientation-preserving homeomorphism of $\mathbf{R}^3$ is isotopic to the identity.

**Proposition 6.1.18** *Equivalent knots have homeomorphic complements in $\mathbf{R}^3$.*

**Proof** Suppose that $K_1$ and $K_2$ are two equivalent knots. Then there exists a homeomorphism $f : \mathbf{R}^3 \to \mathbf{R}^3$ such that $f(K_1) = K_2$. Hence, its restriction map $f|_{(\mathbf{R}^3 - K_1)}$ determines a homeomorphism

$$\overline{f} : \mathbf{R}^3 - K_1 \to \mathbf{R}^3 - K_2.$$

This asserts that the knots $K_1$ and $K_1$ have homeomorphic complements in $\mathbf{R}^3$. □

**Remark 6.1.19** As the knot group $\pi_1(\mathbf{R}^3 - K)$ of $K$ is the complement of $K$ in $\mathbf{R}^3$, it is utilized to classify various knots.

**Definition 6.1.20** Let $K$ be a given knot. It is said to be **untied** if there is an isotopy of $\mathbf{R}^3$ that sends $K$ to the standard circle $S^1 \subset \mathbf{R}^3$.

*Remark 6.1.21*  **The circle knot** is a trivial knot having its knot group isomorphic to $\mathbf{Z}$, which is the infinite cyclic group. This asserts that if a given knot $K$ has the non abelian knot group, then the knot $K$ cannot be a trivial knot, and hence, this knot $K$ cannot be untied.

*Remark 6.1.22*  If there is a homeomorphism $h : \mathbf{R}^3 \to \mathbf{R}^3$ which is isotopic to the identity such that $h(K_1) = K_2$ for two knots $K_1$ and $K_2$, then the knots $H_t(K_1)$ give a continuous family of maps which move gradually from $K_1$ to $K_2$ as $t$ increases from 0 to 1. Since $S^3$ is the one-point compactification of $\mathbf{R}^3$, a homeomorphism $h : \mathbf{R}^3 \to \mathbf{R}^3$ has a unique extension to a homeomorphism $\tilde{h} : S^3 \to S^3$.

**Definition 6.1.23** A homeomorphism $h : \mathbf{R}^3 \to \mathbf{R}^3$ is said to be orientation preserving (or orientation reversing) if its extension homeomorphism $\tilde{h} : S^3 \to S^3$ preserves (or reverses) the orientation of $S^3$.

*Example 6.1.24*  A homeomorphism which is isotopic to the identity is orientation preserving, because we can extend each homeomorphism $H_t : \mathbf{R}^3 \to \mathbf{R}^3$ to the homeomorphism $\tilde{H}_t : S^3 \to S^3$, since homotopic maps have the same degree.

*Example 6.1.25*  Reflection in a plane is a homeomorphism of $\mathbf{R}^3$ and transforms a knot to its mirror image. It is orientation reversing and cannot be isotopic to the identity.

*Remark 6.1.26*  Any orientation-preserving homeomorphism of $\mathbf{R}^3$ is isotopic to the identity.

**Definition 6.1.27** A knot $K$ is said to be untied if there is an isotopy of $\mathbf{R}^3$ that would take $K$ to the standard circle $S^1 \subset \mathbf{R}^3$.

*Remark 6.1.28*  Circle knot is a trivial knot. If a knot $K$ is trivial, then the fundamental group of its complement (which is homeomorphic to the solid torus) is the infinite cyclic group. Hence the knot group of $K$ is abelian. This shows that if the knot group of a knot $K$ is not abelian, then $K$ cannot be a trivial knot which means that $K$ cannot be untied.

*Remark 6.1.29*  For some sort of reasonable presentation for a knot group in terms of generators and relations the book Armstrong [Armstrong, 1983] is referred.


## 6.2   Further Applications of Topology

This section gives further applications of topological concepts and results discussed in earlier chapters with a view to understand the beauty, power and scope of the subject topology.

### 6.2.1 Isotopy and Its Applications

This subsection studies the concept of isotopy, which is closely related to the concept of embedding.

**Definition 6.2.1** Let $X$ and $Y$ be two topological spaces and $H : X \times I \to Y$ be a continuous map. Then the family of continuous maps

$$\{H_t : X \to Y, \ x \mapsto H(x, t)\}$$

is said to be an **isotopy,** if every $H_t : X \to Y$ is an embedding.

**Definition 6.2.2** Let $f, g : X \to Y$ be two embeddings. They are said to be **isotopic** denoted by $f \cong g$ if there exists an isotopy

$$\{H_t : X \to Y\} \ \text{such that} \ H_0 = f, \ \text{and} \ H_1 = g.$$

**Definition 6.2.3** Let $f : X \to Y$ be an embedding. It is said to be an **isotopic equivalence** if there exists an embedding $g : Y \to X$ such that

$$f \circ g \cong 1_Y : Y \to Y \ \text{and} \ g \circ f \cong 1_X : X \to X.$$

**Example 6.2.4** Every homeomorphism is an isotopic equivalence, but its converse is not. For example, consider the closed interval $\mathbf{I}$ and its open interval $X = \{t : 0 < t < 1\} \subset \mathbf{I}$. Then the inclusion map $i : X \hookrightarrow \mathbf{I}$ is an isotopic equivalence but it is not a homeomorphism. The family of homeomorphisms given in Exercise 1 of Sect. 6.4.1 forms of a family of isotopic equivalences.

### 6.2.2 Application of Topology to Theory of Numbers

It is well known that the set of prime integers is infinite. This result is proved in this section by using topological tools.

**Proposition 6.2.5** *Let $\Omega$ be a family of subsets of $\mathbf{N}$ consisting of the empty set $\emptyset$ and all those subsets of $\mathbf{N}$ which are expressible in the form*

$$X_n = \{n, n+1, n+2, n+3, \ldots : n \in \mathbf{N}\}.$$

*Then $\Omega$ forms a topology on $\mathbf{N}$.*

**Proof** $X_1 = \mathbf{N}$ shows that the whole set $\mathbf{N}$ is in $\Omega$. Again since, $\Omega$ is totally ordered by set inclusion, it follows that intersection of any two sets in $\Omega$ is also in $\Omega$. Let $\mathbf{S}$ be a subfamily of $\Omega - \{\emptyset, \mathbf{N}\}$ in the sense that $\mathbf{S} = \{X_n : n \in J \subset \mathbf{N}\}$. As $J$ is a subset of positive integers, it has a smallest positive integer $p$. Hence

$$\bigcup\{X_n : n \in J\} = \{p, p+1, p+2, p+3, \dots\} = X_p \in \Omega.$$

Hence it follows that $\Omega$ forms a topology on $\mathbf{N}$.                                      □

**Proposition 6.2.6** (i) *Given a nonempty subset $X \subset \mathbf{N}$, let there exist a positive integer $n_X$ such that $X$ contains no arithmetic progressions (AP) of length greater than $n_X$. Then subsets of $\mathbf{N}$ having the property together with $\emptyset$ and the set $\mathbf{N}$ form a collection of closed set for some topology on $\mathbf{N}$.*
    *[ Use Van der Waerden's theorem which asserts that given an integer $n \in \mathbf{N}$, there is an integer $n_0$ such that for any subset $X \subset \{1, 2, \dots, n_0\} = Y$, either $X$ or $Y - X$ contains an AP.*
    *Let $A$ and $B$ be two subsets of $\mathbf{N}$ which contain no AP of length at least the given $n$. If $A \cup B$ contains a sufficiently long AP, then $A$ or $B$ contains an AP of length more than $n$ implies a contradiction.]*
 (ii) *The collection of all infinite APs in $\mathbf{N}$ forms a base for some topology on $\mathbf{N}$.*
    *[ Use the result that the finite intersection of APs in $\mathbf{N}$ is also an AP.]*
(iii) *Using this topology on $\mathbf{N}$ show that the set of prime integers is infinite.*
    *[ Hint: The sets $A(k, d) = \{k, k+d, k+2d, \dots : k = 1, 2, \dots, d\}$ are open, pairwise disjoint, and form a covering of $\mathbf{N}$. Hence, it follows that each of them is closed. As a particular situation, for each prime integer $p$, the sets of the form $\{p, 2p, 3p, \dots, \}$ forms a covering of $\mathbf{N} - \{1\}$. This shows that the set of prime integers cannot be finite; otherwise, if the set were finite, then the set $\{1\}$ would be open. This shows that it is not a union of arithmetic progressions. This concludes that the set of prime integers cannot be finite and hence it is infinite.]*

**Proposition 6.2.7** *There are infinitely many primes in $\mathbf{Z}$.*

**Proof** First a topology is defined on $\mathbf{Z}$ by using doubly infinite APs. Given $k, d \in \mathbf{Z}, d \neq 0$, define a set $A(k, d) = \{k + nd : n \in \mathbf{Z}\}$. Call a nonempty subset $U \subset \mathbf{Z}$ to be open if it is a union of sets of the form $A(k, d)$ or $U = \emptyset$. Then $\mathbf{Z}$ is an open set and an arbitrary union of sets of the form $A(k, d)$ is also of the same form. Let $U$ and $V$ are of the same form and $a \in U \cap V$ be an arbitrary element. Let $a \in A(k, d_1)$ and $a \in A(k, d_2)$ be two APs containing $k$ such that

$$a \in A(k, d_1) \subset U$$

and

$$a \in A(k, d_2) \subset V.$$

Then $a \in A(k, d_1 d_1) \subset A(k, d_1) \cap A(k, d_2) \subset U \cap V$. This asserts that $U \cap V$ is an open set and every nonempty open set is infinite. $A(k, d)$ is also a closed set. $A(k, d)$ can be expressed as

$$A(k, d) = \mathbf{Z} - \bigcup_{r=1}^{d-1} A(k+r, d).$$

As any integer $m \in \mathbf{Z} - \{-1, 1\}$ has at least one prime divisor $p$, the integer $m$ is contained in $A(0, p)$. Hence it follows that

$$\mathbf{Z} - \{-1, 1\} = \bigcup_{p \text{ is prime}} A(0, p).$$

If possible, suppose there are finitely many primes in $\mathbf{Z}$. Then the finite union of closed sets

$$\bigcup_{p \text{ is prime}} A(0, p)$$

being a closed set in $\mathbf{Z}$, the above equality implies that the subset $\{-1, 1\}$ is open in $\mathbf{Z}$, which is not true, as it is finite. This implies a contradiction. $\qquad\square$

## 6.3 Borsuk–Ulam Theorem with Applications

The aim of this section is to prove Borsuk–Ulam theorem 6.3.1 in a general form by using homology theory for all finite dimensions by generalizing this theorem for two-dimensional case proved in Chap. 2 by homotopy theory. Finally, this section provides some of its applications such as Ham Sandwich theorem and Lusternik–Schnirelmann theorem obtained as direct consequences of Borsuk–Ulam theorem 6.3.1. Historically, S. Ulam (1909–1984) formulated the Borsuk–Ulam theorem without any correct proof but K. Borsuk (1905–1982) proved it first in 1933. Since then, different alternative proofs are found in literature was a conjecture posed by Borsuk.

### 6.3.1 Borsuk–Ulam Theorem

This subsection proves Borsuk–Ulam theorem by using homology theory.

**Theorem 6.3.1** (Borsuk–Ulam Theorem ) *Given any pair of integers $m$, $n$ with $m > n \geq 0$, there does not exist any continuous map*

$$h : S^m \to S^n,$$

*which preserves antipodal points in the sense $h(x) = h(-x)$, $\forall x \in S^m$.*

**Proof** This theorem is proved by method of contradiction. Suppose there exists a continuous map

$$h : S^m \to S^n : h(x) = h(-x), \forall x \in S^m$$

and $S^n$ is obtained from $S^m$ by taking the last $m - n$ coordinates equal to zero. Let

$$i : S^n \hookrightarrow S^m$$

be the usual inclusion map. Consider the composite map

$$i \circ h : S^m \to S^m.$$

This map preserves the antipodal points. This implies that deg $(i \circ h)$ must an odd integer. Again consider the induced composite homomorphism

$$(i \circ h)_* = i_* \circ h_* : H_m(S^m) \xrightarrow{h_*} H_m(S^n) \xrightarrow{i_*} H_m(S^m)$$

is the trivial homomorphism in the homology theory because it is a sequence covariant functors and the homology group $H_m(S^n) = 0$, $for\ m > n$. This proves that deg $(i \circ h) = 0$, which produces a contradiction.                                        $\square$

**Remark 6.3.2** As a direct consequence of Borsuk–Ulam theorem 6.3.1, it follows that geometrically, every continuous map $h : S^n \to \mathbf{R}^n$ identifies a pair of antipodal points of $S^n$.

**Corollary 6.3.3** *Let $h : S^n \to \mathbf{R}^n$ be continuous map such that it preserves the antipodal points of $S^n$. Then there exists a point $x \in S^n$ such that $h(x) = 0$.*

**Proof** Suppose there exists no point $x \in S^n$ such that $h(x) = 0$. Then

$$h(x) \neq 0, \ \forall x \in S^n.$$

Define a map

$$f : S^n \to S^{n-1}, x \mapsto \frac{h(x)}{||h(x)||}.$$

Then $f$ is well-defined and continuous. By hypothesis, $h$ preserves antipodal points. This implies that the continuous map $f$ also preserves antipodal points. This contradicts the Borsuk–Ulam theorem 6.3.1.                                        $\square$

**Corollary 6.3.4** *The $n$-sphere $S^n$ is not embeddable in $\mathbf{R}^n$.*

**Proof** Since the sphere $S^n$ is not homeomorphic to a subspace of $\mathbf{R}^n$ by Remark 6.3.2, it is proved that the sphere $S^n$ cannot be embedded in $\mathbf{R}^n$.                                        $\square$

### 6.3.2  Ham Sandwich Theorem

This subsection proves Ham Sandwich theorem 6.3.5 as a direct consequence of Borsuk–Ulam theorem 6.3.1.

**Theorem 6.3.5** (Ham Sandwich Theorem) *Given $n$ bounded convex subsets $X_1$, $X_2, \ldots, X_n$ of $\mathbf{R}^n$, there exists a hyperplane bisecting all of the $X_i$'s simultaneously.*

**Proof** **Case I:** First suppose that $n = 3$ and consider the continuous map

$$f : S^3 \to \mathbf{R}^3, (f_1(x), f_2(x), f_3(x)),$$

where $f_i : S^3 \to \mathbf{R}$ is a continuous map such that for every $x \in S^3$, the numerical value $f_i(x)$ denotes the volume of the part of $X_i$ lying on the same side of the hyperplane $H_x$ at the point $x$ and passing through the point $(0, 0, 0, 1/2)$ for $i = 1, 2, 3$. Then there exists a point $s_0 \in S^3$ such that $f(s_0) = f(-s_0)$ by Borsuk–Ulam theorem 6.3.1. This asserts that $f_1(s_0) = f_1(-s_0)$, $f_2(s0) = f_2(-s0)$ and $f_3(s_0) = f_3(-s_0)$.

**Case II:** For $n > 3$, proceed as in **Case I**.                              □

### 6.3.3   Lusternik–Schnirelmann Theorem for Higher Dimension

Borsuk–Ulam theorem 6.3.1 is applied to prove Lusternik–Schnirelmann theorem 6.3.6 for $S^n$, which generalizes Lusternik–Schnirelmann for dimension 2, proved in Chap. 2.

**Theorem 6.3.6** (Lusternik–Schnirelmann theorem for $S^n$) *If $S^n$ is covered by $n + 1$ closed sets $X_1, X_2, \ldots, X_{n+1}$ of $S^n$, then one of them contains a pair of antipodal points.*

**Proof** By the given condition, $\displaystyle\bigcup_{i=1}^{n+1} X_i = S^n \subset \mathbf{R}^{n+1}$. Consider $\mathbf{R}^n$ as the Euclidean space with usual metric $d$. Let $d(x, X_i)$ be the distance of $x$ from the closed set $X_i$. Construct a continuous map

$$f : S^n \to \mathbf{R}^n, \ x \mapsto (d(x, X_1), \ldots, d(x, X_n)).$$

Then Borsuk–Ulam theorem 6.3.1 implies that $f$ must identify a pair of antipodal points of $S^n$. This implies that there exists a point $s_0 \in S^n$ such that $d(s_0, X_i) = d(-s_0, X_i)$ for $0 \leq i \leq n$. Consider the two possible cases.

**Case I**: If $d(s_0, X_i) = 0$ for some $i$, then the pair of antipodal points $s_0, -s_0 \in X_i$. Because every $X_i$ being a closed set by hypothesis, both the points $s_0, -s_0 \in X_i$.

**Case II**: If $d(s_0, X_i) > 0$ for all $i = 1, 2, \ldots, n$, then the pair of antipodal points $s_0, -s_0 \in X_{n+1}$ because $X_i$'s form a cover of $S^n$ by hypothesis.                              □

### 6.3.4  Van Kampen Theorem: An Application of Graph Theory

This subsection proves Van Kampen theorem 6.3.16 by using graph-theoretic results together with algebraic concept of free product of two groups. This theorem provides a technique for computing the fundamental groups of topological spaces which are decomposed into simpler spaces having their fundamental groups already known and expressing the edge group as a set of generators and relations. This theorem is also known as **Seifert–Van Kampen theorem.**

**Definition 6.3.7** A tree $T$ is a 1-dimensional subcomplex of a complex having its polyhedron both path connected and simply connected.

**Definition 6.3.8** A tree $M_T$ is said to be maximal if for any tree $T^*$ containing $M_T$, the trees $T^* = M_T$.

**Definition 6.3.9** The $r$-skeleton $K^r$ of a simplicial complex $K$ is the subcomplex of $K$ which consists of all $n$-faces of simplexes of $K$ having $n \leq r$.

**Definition 6.3.10** Given a map $f : |K| \to |L|$ between polyhedra and a point $x$ in $|K|$, the point $f(x) \in |L|$ belongs to the interior of a unique simplex of $L$, called the carrier of $f(x)$.

**Definition 6.3.11** Let $f : |K| \to |L|$ be a continuous map between polyhedra. Then a simplicial map $s : K \to L$ is said to be **simplicial approximation of** $f$, if $s(x)$ is a point of the carrier of $f(x)$ for every point $x$ in $K$.

**Proposition 6.3.12** *A maximal tree of a complex contains all its vertices.*

**Proof** Let $M_T$ be a maximal tree of a complex $K$ which does not contain all its vertices. There exists at least one vertex $v$ of $K$ such that $v$ is in $K - M_T$. Since $K$ is path connected for any vertex $w$ of $M_T$, the two vertices $w$ and $v$ can be joined by a path in $|K|$. Then this path can be replaced by an edge path $w v_1 v_2 \cdots v_m v$ by simplicial approximation theorem. Let $v_k$ be the last vertex of this edge lying in $M_T$. Now, a new subcomplex $S$ of $K$ is constructed by including the vertex $v_{k+1}$ and the edge generated by $v_k v_{k+1} \cdots$ . Then $|S|$ is a deformation retract of $|M_T|$, and hence, $S$ is also tree. But it contradicts the assumption that $M_T$ is a maximal tree. This contradiction asserts that the maximal tree $M_T$ does contain all the vertices of $K$. $\square$

**Definition 6.3.13** Let $K$ be a complex and $S$ be a subcomplex of $K$ with $|S|$ be simply connected. List the vertices of $K$ as $v = v_0, v_1, v_2, \ldots, v_m$ and denote the generators $g_{ij}$ for every pair of vertices $v_i$ and $v_j$ of $K$ as follows:

(i)  $g_{ij} = 1$, if $v_i$ and $v_j$ span a simplex of $S$,
(ii) $g_{ij} g_{jk} = g_{ik}$ if $v_i$, $v_j$ and $v_k$ span a simplex of $K$,

(iii)  $g_{ii} = 1$,
(iv)  $g_{ij}^{-1} = g_{ji}$.

The group generated by $g_{ij}$ is denoted by $G(K, S)$.

Construct another group $E(K, S)$ called **the edge group of $K$ based at a vertex of $K$** in Definition 6.3.14.

**Definition 6.3.14**  Define another group $E(K, S)$ called **the edge group of $K$ based at a vertex of $K$** as follows:
An edge path in $K$ is a sequence $v_0, v_1, \ldots, v_n$ of vertices of $K$ such that every consecutive pairs $v_i$ and $v_{i+1}$ spans a simplex of $K$ with the possibility $v_i = v_{i+1}$. Define a simplicial version of homotopy by defining two edge paths of $K$ to be equivalent if one can be obtained from the other by a finite number of operations of the forms

(i)  if $v_{j-1} = v_j$, then replace $\cdots v_{j-1} v_j \cdots$ by $\cdots v_j \cdots$ and conversely
(ii)  or, if $\{v_{j-1}, v_j, v_{j+1}\}$ spans a simplex of $K$, which may not be a 2-simplex, replace $\cdots v_{j-1} v_j v_{j+1} \cdots$ by $\cdots v_{j-1} v_{j+1} \cdots$ and conversely (like two sides of a triangle can be replaced by the third side and vice versa).

This is an equivalence relation between the edge paths. The class corresponding to the edge path $v_0 v_2 \cdots v_j$ is denoted by $[v_0 v_2 \cdots v_j]$. The set of the equivalence classes of edge loops at the vertex $v$ of $K$, denoted by $v$ forms a group under the binary operation (juxtaposition)

$$[v v_1 v_2 \cdots v_n v] \cdot [v w_1 w_2 \cdots w_n v] = [v v_1 v_2 \cdots v_n v v w_1 w_2 \cdots w_n v]$$

with the identity element $[v]$ and the inverse of its element $[v v_1 v_2 \cdots v_n v]$ is the element $[v v_n \cdots v_2 v_1 v]$.

This is called the **edge group of $K$ based at its vertex $v$ and is denoted by $E(K, v)$**.

Theorem 6.3.15 proves that the groups $E(K, S)$ and $E(K, v)$ are isomorphic.

**Theorem 6.3.15**  *Let $S$ be a subcomplex of $K$ such that $|S|$ be simply connected. Then the groups $G(K, S)$ and $E(K, v)$ are isomorphic.*

**Proof**  To prove the theorem first define a homomorphism

$$\alpha : G(K, S) \to E(K, v), \quad g_{ij} \mapsto [e_i v_j e_j^{-1}].$$

Define another homomorphism

$$\beta : E(K, v) \to G(K, S), \quad [v v_k v_t v_m \ldots v_n v] \mapsto g_{0k} g_{kt} g_{tm} g_{n0}.$$

Then

$$\alpha \circ \beta = 1_{E(K,v)} \text{ and } \beta \circ \alpha = 1_{G(K,S)}$$

assert that $\alpha$ is an isomorphism with $\beta$ as its inverse isomorphism.     □

Given two simplicial complexes $S$, $K$ in the same Euclidean space, if they intersect in a common subcomplex and $|S|, |K|, |S \cap K|$ are all path-connected spaces with their known fundamental groups, then the fundamental group $\pi_1(|S \cup K|)$ is calculated by **Van Kampen theorem** 6.3.16.

Consider the two possibilities:

(i) If the simplicial complexes $S$ and $K$ intersect in a single vertex, then any edge loop in $S \cup K$ based at this vertex is a product of loops, each of which is in either $S$ or $K$. This facilitates to calculate the free product $\pi_1(|S|) * \pi_1(|K|)$ for the fundamental group of $|S \cup K|$.

(ii) For the general situation, an analogous arguments holds, except that the free product $\pi_1(|S|) * \pi_1(|K|)$ effectively counts the homotopy classes of these loops lying in $|S \cap K|$ twice (one in each of $\pi_1(|S|)$, $\pi_1(|K|)$). So, in this case, some extra relations are required as given in Van Kampen theorem 6.3.16.

**Theorem 6.3.16** (Van Kampen Theorem) *Let $S$, $K$ be two simplicial complexes in the same Euclidean space with $|S|, |K|$ and $|S \cap K|$ are path-connected spaces. Suppose*

$$i : |S \cap K| \hookrightarrow |S|,$$

$$j : |S \cap K| \hookrightarrow |K|$$

*are inclusion maps and $v$ is a vertex of $S \cap K$, which is taken as a base point of $S \cap K$. Then the fundamental group $\pi_1(|S \cup K|, v)$ is the free product $\pi_1(|S|, v) * \pi_1(|K|, v)$ with the relations $i_*(x) = j_*(x)$ for every $x \in \pi_1(|S \cap K|, v)$.*

***Proof*** Let $M_T$ be a maximal tree in $S \cap K$. Extend it to a maximal tree $M_{T_1}$ in $S$ and a maximal tree $M_{T_2}$ in $K$. Then $M_{T_1} \cup M_{T_2}$ is a maximal tree in $S \cup K$. By proposition 6.3.12 and theorem 6.3.15, it follows that the group $\pi_1(|S \cup K|)$ is generated by elements $g_{ij}$ corresponding to edges of $S \cup K - M_{T_1} \cup M_{T_2}$, with relations $g_{ij}g_{jk} = g_{ik}$ provided by the triangles of $S \cup K$. But this is precisely the group obtained by taking a generator $b_{ij}$ for each edge of $S - M_{T_1}$, a generator $c_{ij}$ for each edge of $K - M_{T_2}$, with relation of the form $b_{ij}b_{jk} = b_{ik}, c_{ij}c_{jk} = c_{ik}$ corresponding to the triangles of $S$, $K$ with additional relations $b_{ij} = c_{ij}$, whenever $b_{ij}$ and $c_{ij}$ correspond to the same edge of $S \cap K$. Since the edges of $S \cap K - T_M$, considered as edges of $S$ give a set of generators for $i_*(\pi_1(|S \cap K|))$. Similarly the same edges, considered also as edges of $K$, give a set of generators for $j_*(\pi_1(|S \cap M|))$.     □

## 6.3.5   Proof of Jordan Curve Theorem by Homology Theory

This subsection continues the study of Jordan curve theorem initiated in Chap. 2 and studies it and its generalization from the viewpoint of homology theory. Recall that a homeomorphic image of a circle in the plane $\mathbf{R}^2$ is a Jordan curve $\mathbf{J}$. This theorem says that the complement in the plane $\mathbf{R}^2$ of a Jordan curve $J$ consists of two open components, each of which as $J$ as its boundary. It is one of the most classical theorems in topology, and it is also one of the oldest problems of a purely topological nature. It is related to connectedness and continuum theory. Intuitively, this theorem is simple asserting that a Jordan curve $\mathbf{J}$ ( which is a subspace of $\mathbf{R}^2$ homeomorphic to $S^1$) separates $\mathbf{R}^2$ into two complementary components. But its proof is not obvious. The first correct proof was given by O. Veblem in 1905 [Veblem, 1905]. This subsection proves Jordan curve theorem and its generalization by using homology theory.

**Lemma 6.3.17**   (i) *If* $X$ *is subspace of* $S^n$ *homeomorphic to* $\mathbf{D}^k$ *for some* $k \geq 0$, *then the reduced homology groups*

$$\widetilde{H}_i(S^n - X) = 0, \quad \text{for every } i;$$

(ii) *If* $A$ *is a subspace of* $S^n$ *homeomorphic to* $S^k$ *for some* $k$ *with* $0 \leq k < n$, *then*

$$\widetilde{H}_i(S^n - A) \cong \begin{cases} \mathbf{Z}, & \text{if } i = n - k - 1 \\ 0, & \text{otherwise.} \end{cases}$$

**Proof**   (i) Apply induction on $k$. For $k = 0$, the proof is trivial, because the space $S^n - X$ is homeomorphic to $\mathbf{R}^n$. Next, if $f : \mathbf{I}^k \to X$ is a homeomorphism, then the open sets $D = S^n - f(\mathbf{I}^{k-1} \times [0, 1/2])$ and $S = S^n - f(\mathbf{I}^{k-1} \times [1/2, 1])$, then $D \cap S = S^n - X$ and $D \cup S = S^n - f(\mathbf{I}^{k-1} \times \{1/2\})$. By induction $\widetilde{H}_i(D \cup S) = 0$ for all $i$. This asserts by Mayer–Vietoris sequence that there are isomorphisms

$$\psi : \widetilde{H}_i(S^n - X) \to \widetilde{H}_i(D) \oplus \widetilde{H}_i(S) \text{ for every } i.$$

Since the two components of $\psi$ are induced by the inclusions $S^n - X \hookrightarrow D$ and $S^n - X \hookrightarrow S$, it follows that there exists an $i$-dimensional cycle $\beta$ in $S^n - X$ that is not a boundary in $S^n - X$. This implies that $\beta$ is also not a boundary in at least one of $D$ and $S$. For, $i = 0$, 'cyclic' is considered augmented chain complexes, which are studied in reduced homology. In an analogous way, subdivide the last $\mathbf{I}$ factor of $\mathbf{I}^k$ into quarters, eights, ... to get a nested sequence of closed subintervals $\mathbf{I}_1 \supset \mathbf{I}_2 \supset \cdots$ with intersection one point $p \in \mathbf{I}$, such that $\beta$ is not a boundary in $S^n - f(\mathbf{I}^{k-1} \times \mathbf{I}_m)$ for any $m$. By induction on $k$, $\beta$ is the boundary of a chain $\alpha$ in $S^n - f(\mathbf{I}^{k-1} \times \{p\})$. This implies that $\alpha$ is a finite linear combination of singular simplices with compact image in $S^n - f(\mathbf{I}^{k-1} \times \{p\})$. This asserts by compactness $\alpha$ is a chain in $S^n - f(\mathbf{I}^{k-1} \times \mathbf{I}_m)$ for some $m$. This contradiction implies that $\beta$ is a boundary in $S^n - X$. Hence (i) follows by induction on $k$.

(ii) Apply induction on $k$. For $k = 0$ the case is trivial because in this case $S^n - A \approx S^{n-1} \times \mathbf{R}$. Represent the space $A$ as a union of two subspaces $X_1 \cup X_2$, where $X_1$ and $X_2$ are homeomorphic to $D^k$ and $X_1 \cap X_2$ is homeomorphic to $S^{k-1}$. Apply Mayer–Vietoris sequence for $C = S^n - X_1$ and $B = S^n - X_2$, both of which have trivial reduced homology groups by (i). Hence there exist isomorphisms

$$\widetilde{H}_i(S^n - A) \cong \widetilde{H}_{i+1}(S^n - (X_1 \cap X_2)), \; \forall i.$$

This implies (ii).                                                                                        □

**Theorem 6.3.18** (Jordan curve ) *The complement of a Jordan curve* $\mathbf{J}$ *in the plane* $\mathbf{R}^2$ *consists of two open components, each of which as* $\mathbf{J}$ *has its boundary.*

**Proof** Use (ii) of Lemma 6.3.17 to prove that a subspace of $S^2$ homeomorphic to $S^1$ separates $S^2$ into two complementary open complements, since open subsets of $S^n$ are locally path connected. Finally, use $\mathbf{R}^2$ in place of $S^2$ to complete the proof because deleting a point from an open set in $S^2$ does not change its connectedness,                                                                        □

Theorem 6.3.19 gives a generalization of the Jordan curve theorem 6.3.18 and its proof is analogous.

**Theorem 6.3.19** (Generalized Jordan curve theorem) *Every subspace of* $S^n$ *homeomorphic to* $S^{n-1}$ *separates it into two components, and these components have the same homology group same as the homology group of a point. Moreover, both the complementary regions are homeomorphic to open balls.*

**Proof** Proceed as in proof of Theorem 6.3.18.                                                    □

## 6.3.6  Homology Groups of $\bigvee_{i \in \mathbf{A}} S_i^n$

This subsection computes the homology groups of the wedge product $\bigvee_{i \in \mathbf{A}} S_i^n$ of any family $\{S_i^n : i \in \mathbf{A}\}$ of the $n$-spheres.

**Theorem 6.3.20** *Let* $\mathbf{A}$ *be an indexing set, and* $S_i^n$ *be a copy of the $n$-sphere for each* $i \in \mathbf{A}$. *Then the reduced homology groups*

$$\widetilde{H}_m\left(\bigvee_{i \in \mathbf{A}} S_i^n\right) \equiv \begin{cases} \bigoplus \mathbf{Z}(i), & \text{if } m = n \\ 0, & \text{otherwise,} \end{cases}$$

where $\bigoplus_{i \in \mathbf{A}} \mathbf{Z}(i)$ *is a free abelian group with generators* $i \in \mathbf{A}$.

**Proof** Consider the topological spaces $\Sigma(\bigvee_{i \in \mathbf{A}} S_i^n)$ and $\bigvee_{i \in \mathbf{A}} \Sigma S_i^n = \bigvee_{i \in \mathbf{A}} S_i^{n+1}$. Since they are homotopy equivalent, the theorem follows. $\qquad\square$

### 6.3.7 More Application of Euler Characteristic

Euler characteristic has various applications. For example, it is proved using Euler characteristic that there are only five regular simple polyhedra (see Chap. 2). This subsection studies Euler characteristic of a finite $CW$-complex.

**Theorem 6.3.21** *Let* $X$ *be a finite CW-complex such that* $\chi(X) \neq 0$. *If* $\psi_t : X \to X$ *is a flow, then there exists a fixed point of* $\psi_t$ *for every* $t \in \mathbf{R}$.

**Proof** By hypothesis, $\psi_t : X \to X$ is a flow. If $\wedge_{\psi_t}$ and $\wedge_{1_X}$ denote the Lefschetz numbers of the map $\psi_t$ and the identity map $1_X$ on $X$ respectively, then

$$\wedge_{\psi_t} = \wedge_{1_X} = \chi(X) \neq 0.$$

Hence by Lefschetz fixed-point theorem, it follows that there exists a fixed point $x_0(t)$ of $\psi_t$. Construct a sequence of subspaces

$$X_n = \{x \in X : \psi_{1/2^n}(x) = x\}$$

for $n = 1, 2, \ldots,$ . This implies that $X_n \supset X_{n+1}$, and every space $X_n$ is a nonempty closed set such that

$$X_\infty = \bigcap_n X_n \neq \emptyset.$$

This implies that $X_\infty$ is a set of points fixed under all rational numbers of dyadic form $k/2^n$. Since the set of rational numbers are dense is the real number space $\mathbf{R}$, every element in $X_\infty$ is a fixed point of $\psi_t$ for any $t \in \mathbf{R}$. $\qquad\square$

### 6.3.8 Application of Algebraic Topology to Algebra

In general, algebraic topology involves algebraic techniques to obtain topological information but some algebraic results can be proved conveniently in a way where the direction is reversed. For example, Theorem 6.3.22 in algebra is proved by using tools of algebraic topology given in Exercise 4 of Sect. 6.4.1.

**Theorem 6.3.22** *Any subgroup* $G$ *of a free group* $F$ *is free.*

***Proof*** By hypothesis, $G$ is an arbitrary subgroup of a free group $F$. Let $\mathcal{B}$ be a basis of $F$. Then corresponding to the basis $\mathcal{B}$, there exists a graph $G_{\mathcal{B}}$ by taking $G_{\mathcal{B}}$ the wedge of circles such that $\pi_1(G_{\mathcal{B}}) \cong F$. Hence there exists a covering space $p : E \to G_{\mathcal{B}}$ by Exercise 4 of Sect. 6.4.1 such that

(i)  the induced homomorphism

$$p_* : \pi_1(E) \to \pi_1(G_{\mathcal{B}})$$

is a monomorphism and
(ii)  $p_*(\pi_1(E)) = G.$

Since $p_*$ is a monomorphism, $\pi_1(E) \cong G$. Finally, it follows that the group $G \cong \pi_1(E)$ is free, because $E$ is a graph by using Exersise 4 of Sect. 6.4.1.  □

***Remark 6.3.23*** It has been proved in an earlier chapter that the antipodal map

$$A : S^n \to S^n, \ x \mapsto -x$$

generates an action of $\mathbf{Z}_2$ on $S^n$ with the projective space $\mathbf{R}P^n$ the corresponding orbit space $S^n \ mod \ \mathbf{Z}_2$. It is a compact connected manifold of dimension $n$, and this action is free. Hence a natural question arises: does there exist any other finite group which acts freely on $S^n$? Theorem 6.3.25 gives its partial answer which asserts that $\mathbf{Z}_2$ is the only nontrivial group that can act freely on $S^n$ for an even integer $n$. To prove this result, we first prove Proposition 6.3.24.

**Proposition 6.3.24** *If a topological group $G$ acts on a topological space $X$, then every $g \in G$, defines a homeomorphism*

$$\psi_g : X \to X, \ x \to gx.$$

***Proof*** For every $g \in G$, the map

$$\psi_g : X \to X, \ x \mapsto gx$$

is continuous and $\psi_g$ is such that

(i)  $\psi_g \circ \psi_{g^{-1}} = 1_X$ and
(ii)  $\psi_g \circ \psi_{g^{-1}} = 1_X.$

Hence it follows that $\psi_g$ is a homeomorphism.  □

**Theorem 6.3.25** *If $n$ is an even integer, $\mathbf{Z}_2$ is the only nontrivial group that can act freely on $S^n$.*

***Proof*** Consider an action of an arbitrary topological $G$ on $S^n$

$$\psi : G \times S^n, (g, x) \mapsto gx.$$

Then for every $g \in G$, the map

$$\psi_g : S^n \to S^n, x \mapsto gx$$

is a homeomorphism by Proposition 6.3.24. Since the degree of a homeomorphism is $\pm 1$, the action of $G$ on $S^n$ determines a degree function

$$d : G \to \{\pm 1\}, \ g \mapsto \deg \psi_g.$$

The degree function $d$ is a homomorphism. If the action is free, then $d$ maps every nontrivial element of $G$ to $(-1)^{n+1}$, since $\psi_g$ has no fixed point. Consequently, if $n$ is even, $d$ has trivial kernel and hence $G \subset \mathbf{Z}_2$. This proves the theorem.  ☐

**Remark 6.3.26** Sect. 6.4.1 provides for more results of algebra that can be proved by using algebraic topology.

## 6.3.9 *Whitehead Theorem and Its Applications*

Whitehead Theorem 6.3.29 is a basic result in algebraic topology proved by J. H. C. Whitehead in [Whitehead, 1949], where he introduced the concept of $CW$ complex.

**Definition 6.3.27** Let $\mathcal{T}op_*$ be the category of pointed topological spaces and their continuous maps. Then a map $f : (X, x_0) \to (Y, y_0) \in \mathcal{T}op_*$ with $y_0 = f(x_0)$ is said to be a weak homotopy equivalence if its induced map

(i) $f_* : \pi_0(X, x_0) \to \pi_0(Y, y_0)$ is a bijection and
(ii) $f_* : \pi_n(X, x_0) \to \pi_n(Y, y_0)$ is an isomorphism of groups for all $n \geq 1$.

On the other hand, $f$ is said to be an $m$ equivalence for some integer $m \geq 1$ if

(i) $f_* : \pi_m(X, x_0) \to \pi_m(Y, y_0)$ is an isomorphism of groups for all $0 < m < n$ and
(ii) $f_* : \pi_m(X, x_0) \to \pi_m(Y, y_0)$ is an epimorphism for $m = n$.

A weak homotopy equivalence $f : (X, x_0) \to (Y, y_0) \in \mathcal{T}op_*$ with $y_0 = f(x_0)$ is sometimes written in brief as $f : X \to Y$.

**Proposition 6.3.28** *Let $f : X \to Y$ be a weak homotopy equivalence between $CW$ complexes and $K$ be a $CW$ complex with base point $k_0$ a 0-cell. Then the induced map*

$$f_* : [K, X] \to [K, Y], [\alpha] \mapsto [f \circ \alpha]$$

*is a bijection.*

**Proof** Let $M_f$ be the mapping cylinder of the given map $f : X \to Y \in \mathcal{T}op_*$ with inclusion $i : X \hookrightarrow M_f$ and $g : M_f \to Y$ be a homotopy equivalence. Then $f = g \circ i$. This implies that $i$ is also a weak homotopy equivalence. This asserts that

$$i_* : [K, X] \to [K, Y]$$

is a bijection. This proves that $f_* = g_* \circ i_*$ is a bijection, since $g_*$ is a bijection.   $\Box$

Every homotopy equivalence is a weak homotopy equivalence by definition, but its converse is true under certain conditions prescribed in Whitehead theorem 6.3.29. Theorem 6.3.34 provides a sufficient condition under which the concepts a weak homotopy equivalence and homotopy equivalence coincide.

**Theorem 6.3.29** (Whitehead) *Let $f : X \to Y$ be a weak homotopy equivalence between CW complexes X and Y. Then it is also a homotopy equivalence.*

**Proof** By hypothesis, $f : X \to Y$ is a weak homotopy equivalence between $CW$ complexes. Then by Proposition 6.3.28, it follows that

$$f_* : [Y, X] \to [Y, Y], [\alpha] \mapsto [f \circ \alpha]$$

is a bijection. Then there exists a continuous map $g : Y \to X$ such that $f \circ g \simeq 1_Y$. This implies that $g$ is also a weak homotopy equivalence. Similarly, it can be proved that $g \circ f \simeq 1_X$. Consequently, $f$ is a homotopy equivalence with $g$ its homotopy inverse.                                                                $\Box$

**Corollary 6.3.30** (Another form of Whitehead theorem) *Let $f : X \to Y$ be a weak homotopy equivalence between CW complexes X and Y such that its induced homomorphism*

$$f_* : \pi_n(X) \to \pi_n(Y)$$

*is an isomorphism for every integer $n \geq 1$.*

## 6.3.10   Eilenberg–MacLane Spaces and Their Applications

Eilenberg–MacLane spaces form an important family of topological spaces having only one nontrivial homotopy groups. Such spaces introduced by S. Eilenberg (1915–1998) and S. MacLane (1909–2005) in 1945 are named after them. The importance of Eilenberg–MacLane spaces is twofold. Because Eilenberg–MacLane spaces

(i)  develop homotopy theory and
(ii)  they closely link the study of cohomology theory with homotopy theory.

**Definition 6.3.31** Let $G$ be an arbitrary group. Given a positive integer $n$, an Eilenberg–MacLane space of type $K(G, n)$ is a pointed $CW$-complex $X$ such that

(i)  $X$ has only one nontrivial homotopy group;

(ii) $\pi_n(X) = G$ and all other homotopy groups of $X$ (i.e., in all dimensions except $n$) vanish;

(iii) the group $G$ is to be abelian for all $n > 1$.

The concept of an Eilenberg–MacLane space $K(G, n)$ is well-defined for all $n \geq 1$, because there is only one space of type $K(G, n)$ upto homotopy equivalence.

**Definition 6.3.32** An Eilenberg–MacLane space is of the form

(i)  $K(G, 1)$ is a path-connected space having fundamental group isomorphic to a given group $G$ and a contractible universal covering space.

(ii)  $K(G, 0)$ is defined to be the group $G$ with the discrete topology.

***Example 6.3.33*** Consider the homotopy groups of the infinite dimensional real projective space $\mathbf{RP}^\infty$, of the infinite dimensional complex projective space $\mathbf{CP}^\infty$ and of unit circle $S^1$ in $\mathbf{C}$ :

$$\pi_i(\mathbf{RP}^\infty) = \begin{cases} \mathbf{Z}_2, & \text{if } i = 1, \\ 0, & \text{if } i \neq 1. \end{cases}$$

$$\pi_i(\mathbf{CP}^\infty) = \begin{cases} \mathbf{Z}, & \text{if } i = 2, \\ 0, & \text{if } i \neq 2. \end{cases}$$

$$\pi_i(S^1) = \begin{cases} \mathbf{Z}, & \text{if } i = 1, \\ 0, & \text{if } i \neq 1. \end{cases}$$

Hence it follows that

(i)  $K(\mathbf{Z}_2, 1) = \mathbf{RP}^\infty$ (infinite dimensional real projective space).

(ii)  $K(\mathbf{Z}, 2) = CP^\infty$ (infinite dimensional complex projective space).

(iii)  $K(\mathbf{Z}, 1) = S^1$ (unit circle in $\mathbf{C}$), but $S^2$ is not an Eilenberg–MacLane space of type $K(\mathbf{Z}, 2)$.

Theorem 6.3.34 provides a sufficient condition under which the concepts of a weak homotopy equivalence and homotopy equivalence coincide, which are in general different.

**Theorem 6.3.34** (Whitehead) *For every abelian group $G$, there exists a weak homotopy equivalence*

$$\psi_n : K(G, n) \to \Omega K(G, n+1), \quad \forall n \geq 1,$$

*which is also a homotopy equivalence.*

***Proof*** For every integer $n \geq 1$, using the earlier results such as

(i)  $\pi_n(\Omega K(G, n+1)) \cong \pi_{n+1}(K(G, n+1)) \cong G$ and

(ii)  $\pi_n(\Omega K(G, n+1)) \cong \pi_n(K(G, n)) \cong G$.

Hence it follows that there exists a continuous map

$$\psi_n : K(G, n) \to \Omega K(G, n + 1)$$

for every $n \geq 1$ and its induced homomorphism

$$\psi_{n*} : \pi_n(K(G, n)) \to \pi_n(\Omega K(G, n + 1))$$

is an isomorphism for every $n \geq 1$. This implies that

$$\psi_n : K(G, n) \to \Omega K(G, n + 1)$$

is a weak homotopy equivalence for every $n \geq 1$, since all other homotopy groups are trivial. Finally, since the loop space $\Omega K(G, n + 1)$ is in the homotopy type of a $CW$-complex, it follows that $\psi_n$ is a homotopy equivalence.     □

**Theorem 6.3.35** *Let $X$ be a $CW$-complex and $K(G, n)$ be an Eilenberg- MacLane space. Then the set $[X, K(G, n)]$ of homotopy classes of maps $f : X \to K(G, n)$ admits a group structure for every $n \geq 1$.*

**Proof** By Whitehead theorem 6.3.34 the Eilenberg–MacLane space $K(G, n)$ is homotopy equivalent to the $H$-space $\Omega K(G, n + 1)$, $\forall n \geq 1$. This proves that the set $[X, K(G, n)]$ admits a group structure ( see Chap. 2).     □

Dold–Thom theorem 6.3.36 establishes a close relation between the groups $\pi_i(SP^\infty(X))$ and $H_i(X; \mathbf{Z})$ for every $CW$-complex $X$ in $\mathcal{T}op_*$.

**Theorem 6.3.36** (Dold Theorem) *The functor*

$$F_1 : \mathcal{T}op_* \to \mathcal{G}rp, \ X \mapsto \pi_i(SP^\infty(X))$$

*and the functor*
$$F_2 : \mathcal{T}op_* \to \mathcal{G}rp, \ X \mapsto H_i(X; \mathbf{Z})$$

*coincide for every $i \geq 1$.*

**Proof** See [Dold and Thom, 1958]     □

**Corollary 6.3.37** *For a connected $CW$-complex $X$, there is a natural isomorphism*

$$\psi : \pi_n(SP^\infty(X)) \to H^n(X; \mathbf{Z}),$$

*for every $n \geq 1$.*

**Proof** It follows from Dold–Thom theorem 6.3.36.     □

**Corollary 6.3.38** $SP^\infty(S^n)$ *is a $K(\mathbf{Z}, n)$.*

**Proof** Take in particular, $X = S^n$ in Dold–Thom theorem 6.3.36 to prove the corollary. $\qquad\square$

**Example 6.3.39** (i) The infinite symmetric product space $SP^\infty(S^n)$ of $S^n$ is an Eilenberg–MacLane space $K(\mathbf{Z}, n)$ for every integer $n \geq 1$ because it is a CW-complex having only one nontrivial homotopy group such as $\pi_n(K(\mathbf{Z}, n)) \cong \mathbf{Z}$.
(ii) The $n$-symmetric product space $SP^n(S^2) \approx \mathbf{C}P^n$.
(iii) The infinite symmetric product space $SP^\infty(S^2) \approx \mathbf{C}P^\infty$.

**Remark 6.3.40** For the inclusion map

$$X = SP^1(X) \hookrightarrow SP^\infty X,$$

its induced homomorphism

$$\pi_n(X) \to \pi_n(SP^\infty(X)) = H_n(X; \mathbf{Z})$$

is the Hurewicz homomorphism (see Exercise 7 of Sect. 6.4.1). In particular for $X = S^1$, the map

$$SP^n(S^n) \hookrightarrow SP^\infty(S^n)$$

induces on $\pi_1$ by Hurewicz theorem, an isomorphism

$$\sigma_1 : \mathbf{Z} \to \mathbf{Z}.$$

### 6.3.11 Adams Theorem on Vector Field Problem

Adams theorem 6.3.43 solved a long a standing problem on vector field in 1960 saying that there exists a continuous map

$$f : S^{2n-1} \to S^n$$

having Hopf invariant one only when $n = 2, 4$ and 8. This theorem is named after J. F. Adams who solved this problem in his papers [Adams, 1958, 1960].

**Definition 6.3.41** Let $f : S^{2n-1} \to S^n$ be a continuous map for $n \geq 1$. Then there exists a unique integer depending only on the homotopy class of $f$. This integer denoted by $\mathcal{H}(f)$ is called the **Hopf invariant** of the map $f$.

**Remark 6.3.42** The assignment

$$\psi : \pi_{2n-1}(S^n) \to \mathbf{Z}, \ f :\to \mathcal{H}(f)$$

is a homomorphism such that $\mathcal{H}(f) = 1$ only when $n = 2, 4$ and $8$ by Theorem 6.3.43.

Adams theorem 6.3.43 proves an algebraic result saying that **R**, **C** and **H** are the only nontrivial real division algebra. This proof is beyond the scope of the book and only referred.

**Theorem 6.3.43** *(Adams)* there exists a continuous map

$$f : S^{2n-1} \to S^n$$

having Hopf invariant one only when $n = 2, 4$ and $8$.

***Proof*** See Adams [(Adams, 1958, 1960)].                                          □

## 6.4  Exercises and Multiple Choice Exercises

As solving exercises plays an essential role in learning mathematics, various types of exercises and multiple choice exercises are given in this section. They form an integral part of the book series.

### 6.4.1  Exercises

1. Let $\mathbf{R}^n$ be the $n$-dimensional Euclidean space. Prove the following statements:

   (i) every disk $\mathbf{D}_r = \{x \in \mathbf{R}^n : ||x|| = r\}$ of an arbitrary radius $r$ is homeomor-phic to $\mathbf{D}^n$;
   (ii) the unit dis $\mathbf{D}^n = \{x \in \mathbf{R}^n : ||x|| = 1\}$ in $\mathbf{R}^n$ is homeomorphic to $\mathbf{R}^n$;
   (iii) every disk $\mathbf{D}_r = \{x \in \mathbf{R}^n : ||x|| = r\}$ of an arbitrary radius $r$ is also home-omorphic to $\mathbf{R}^n$.
   (iv) the above family of homeomorphisms forms of a family of isotopic equiva-lences.

2. Let a connected graph be embedded in a sphere such that any face has exactly three edges. Show that for this graph

   (i) $\mathbf{E} = 3\mathbf{V} - 6$;
   (ii) $\mathbf{F} = 2\mathbf{V} - 4$

   where **V**, **E**, and **F** denotes the number of vertices, edges and faces of the graph, respectively.

3. Let $K$ be a simplicial complex in $\mathbf{R}^n$. Show that

(i) $|K|$ is a closed bounded subset of $\mathbf{R}^n$;

(ii) $|K|$ is a closed compact space;

(iii) every point of $|K|$ belongs to the interior of exactly one simplex of $K$;

(iv) $|K|$ can be obtained by taking the simplexes of $K$ and giving their union the identification topology;

(v) if $|K|$ is connected, then it is also path connected.

4. Prove the following statements:

   (i) If $G$ is a connected graph with maximal tree $T$, then the fundamental group $\pi_1(G)$ of $G$ is a free group having a basis $[e_k]$ corresponding to the edges $e_k$ of $G$ - $T$.

   (ii) If $G$ is any graph, then every covering space $X$ of $G$ having vertices and edges are the corresponding lifts of vertices and edges of the graph $G$.

   (iii) If $B$ is a path connected, locally path connected and also is a semilocally simply connected space, then corresponding to any subgroup $H$ of the group $G = \pi_1(B)$, there exists a covering space $p : E \to B$ and a point $e_0 \in E$ such that $H = p_*\pi_1(E, e_0)$.

5. **(Simplicial approximation theorem)** Let $f : |K| \to |L|$ be a continuous map between polyhedra. Show that for sufficiently large $r$, there is a simplicial approximation $s : |K^r| \to |L|$ of $f$.

6. Let $X = S^2 \bigvee S^1$ be the one-point union of the 2-sphere $S^2$ and the circle $S^1$, and $Y = S^2 \cup l$, where $l$ denotes the line segment joining the north and south poles of $l$. Show that their fundamental groups $\pi(X)$ and $\pi(Y)$ are isomorphic. [ Hint: The spaces $X$ an $Y$ are homotopically equivalent.]

7. ( **Hurewicz theorem**) Let $(X, x_0)$ and $(Y, y_0)$ be two path-connected pointed spaces and $f : (X, x_0) \to (Y, y_0)$ be a base point preserving continuous map such that its induced homomorphism in the homotopy theory

$$f_* : \pi_m(X, x_0) \to \pi_m(Y, y_0)$$

is an isomorphism for every integer $m \leq n - 1$ and an epimorphism for $m = n$. Show that its induced homomorphism in the homology theory $\mathcal{H}$ with integral coefficients also such that

$$f_* : H_m(X, x_0) \to H_m(Y, y_0)$$

is an isomorphism for every integer $m \leq n - 1$ and an epimorphism for $m = n$. [Hint: See Chaps. 2 and 3. ]

8. **(Generalized Jordan curve theorem)** If $X \subset S^n$ is homeomorphic to the sphere $S^m$ ( $m < n$), show that the homology groups of the complement $S^n - X$ with coefficient group $\mathbf{Z}$ are

$$H_i(S^n - X; \mathbf{Z}) = \begin{cases} \mathbf{Z} \oplus \mathbf{Z}, & \text{if } m = n - 1 \text{ or } i = 0 \\ \mathbf{Z}, & \text{if } m < n - 1 \text{ and } i = 0 \text{ or } i = n - m - 1 \\ 0, & \text{otherwise.} \end{cases}$$

9. Let $S^3$ denote the topological group of quaternions of unit modulus. Consider the 2-sphere $S^2 = \{(a, b, c, d) \in S^3 : a = 0\}$ as a subspace of $S^3$. Show that

   (i) for every $y \in S^3$, the map

$$\psi_y : S^3 \to S^3, \quad x \mapsto yxy^{-1}$$

   is a continuous map such that it maps $S^2$ into $S^2$;

   (ii) there is a map $f : S^3 \to SO(3, \mathbf{R})$ which induces a homeomorphism

$$f_* : \mathbf{R}P^3 \to SO(3, \mathbf{R});$$

   (iii) the fundamental group $\pi_1(SO(n, \mathbf{R})) \cong \mathbf{Z}_2 \ \forall n \geq 3$;

   (iv) the two-dimensional homotopy group $\pi_1(SO(n, \mathbf{R})) = \{0\}, \ \forall n \geq 3$.

10. Prove the following statements:

    (i) If $X$ is a path connected, commutative, associative $H$-space $X$ with a strict identity element, then it is in a weak homotopy type of a product of Eilenberg–MacLane spaces.

    (ii) If $\mathcal{C}_0$ is the category of pointed topological spaces having homotopy type of $CW$ complexes, then functor $SP^\infty$ defines Eilenberg–MacLane spaces on $\mathcal{C}_0$.

       [Hint: Use Exercise 10(i) because if $X$ is a CW-complex, then $SP^\infty(X)$ is path connected and has the weak homotopy type of $\prod_n K(H_n(X), n)$.]

    (iii) The Eilenberg–MacLane $K(\mathbf{Z_m}, 1)$ is an infinite dimensional lens space $l^\infty(m) = S^\infty \bmod \mathbf{Z}_m$, that cannot be replaced by any finite dimensional $CW$-complex.

    (iv) If $X \neq \emptyset$ be a closed connected subspace of $S^3$, then the complement $S^3 - X$ is an Eilenberg–MacLane space $K(\mathbf{Z}, 1)$.

    (v) If $X$ is a torus knot, then $S^3 - X$ is an Eilenberg–MacLane space $K(\mathbf{Z}, 1)$.

## 6.4.2  Multiple Choice Exercises

Identify the correct alternative(s) (there may be more than one) from the following list of exercises:

1. Consider the following topological spaces.

   (i) The 2-sphere $S^2$ has the fixed-point property.

    (ii) The one-point union $S^1 \vee S^1$ of two circles has the fixed-point property.

    (iii) The torus $S^1 \times S^1$ has the fixed-point property.

2.  (i) The 3-sphere $S^3$ has no fixed-point property.

    (ii) There is a continuous unit tangent vector field over 3-sphere $S^3$.

    (iii) Every rotation of 4-sphere $S^4$ has a fixed point.

3.  (i) For every point $x$ of the 5-dimensional complex projective space $\mathbf{C}P^5$, its inverse image $p^{-1}$ of the identification map

$$p : S^{11} \to \mathbf{C}P^5, \ z \mapsto [z]$$

    is a great circle of the 11-dimensional sphere $S^{11}$.

    (ii) Let $\mathbf{H}$ be the division ring of quaternions. Then every polynomial over $\mathbf{H}$ has a root in $\mathbf{H}$.

    (iii) The lens space $L(2, 1)$ is homeomorphic to the real projective space $\mathbf{R}^3$.

4.  (i) The antipodal map $f : S^{11} \to S^{11}$, $x \mapsto -x$ is of degree $-1$.

    (ii) The antipodal map $f : S^{16} \to S^{16}$, $x \mapsto -x$ is of degree $+1$.

    (iii) If a continuous map $f : S^{15} \to S^{15}$ has continuous extension over the 16-dimensional Euclidean space $\mathbf{R}^{16}$, where $S^{15}$ is the boundary of $\mathbf{R}^{16}$, then $f$ is of degree 0.

5.  (i) The action of $O(n, \ \mathbf{R})$ on the $n$-dimensional Euclidean plane $\mathbf{R}^n$

$$\psi : O(n, \ \mathbf{R}) \times \mathbf{R}^n, (A, \ x) \mapsto Ax$$

    is not transitive.

    (ii) This action $\psi$ is transitive.

    (iii) In particular, this action on $S^{n-1}$ is transitive having orbit space of the point $x$ is the spheres of radius $||x||$.

6. Let $X(x, y) = -y \, \partial/\partial x + x \, \partial/\partial y$ be a vector field in the Euclidean plane $\mathbf{R}^2$. Then, the action

$$\psi : \mathbf{R} \times \mathbf{R}^2 \to \mathbf{R}^2, \ (t, (x, y)) \mapsto (x \cos t - y \sin t, \ x \sin t + y \cos t)$$

is a flow generated by the vector field $X$ satisfying the following properties:

    (i) the flow through $(x, y)$ is the circle having the center at the origin;

    (ii) if $\psi_t \equiv \psi(t, \ -)$ then $\psi_t = \psi_{2n\pi+t}$

    (iii) one-parameter group $\{\psi_t = \psi(t, \ -)\}$ is isomorphic to $SO(2, \mathbf{R})$ or to the circle group $S^1 \cong U(1, \ \mathbf{R})$.

# References

Adams JF. On the nonexistence of elements of Hopf invariant one. Bull Amer Math Soc. 1958;64:279–82.

Adams JF. On the nonexistence of elements of Hopf invariant one. Ann Math. 1960;72:20–104.

Adams JF. Algebraic topology: A student's guide. Cambridge: Cambridge University Press; 1972.

Adhikari A, Adhikari MR. Basic topology, vol. 1: metric spaces and general topology. India: Springer; 2022a.

Adhikari A, Adhikari MR, Basic topology, vol. 2: topological groups, topology of manifolds and lie groups. India: Springer; 2022b.

Adhikari MR. Basic Algebraic topology and its applications. India: Springer; 2016.

Adhikari MR, Adhikari. Basic modern algebra with applications. Springer, New Delhi, New York, Heidelberg; 2014.

Aguilar, Gitler, S, Prieto, C. Algebraic topology from a homotopical view point. Springer, New York, 2002.

Arkowitz M. Introduction to homotopy theory. New York: Springer; 2011.

Armstrong A. Basic topology. New York: Springer; 1983.

Basak S. A study of some aspects of knots and links, Ph.D Thesis, University of Calcutta, India; 2017.

Bredon G. Topology and Geometry, Springer, GTM 139; 1993.

Dold A, Thom R. Quasifaserungen und unendliche symmetrische Produkte. Ann Math. 1958;67(2):239–81.

Steenrod N, The topology of fibre bundles. Princeton: Princeton University Press; 1951.

Whitehead JHC. Combinatorial homotopy I. Bull. Am. Math. Soc. 1949;55(5):213–245.

# Chapter 7
# Brief History of Algebraic Topology: Motivation of the Subject and Historical Development

This chapter conveys **the history of emergence** of the concepts leading to the development of algebraic topology as a subject with their motivations. Just after the concept of homeomorphisms is clearly defined, the subject of topology begins to study those properties of geometric figures which are preserved by homeomorphisms with an eye to classify topological spaces up to homeomorphism, which stands the ultimate problem in topology, where a geometric figure is considered to be a point set in the Euclidean space $\mathbf{R}^n$. But this undertaking becomes hopeless, when there exists no homeomorphism between the two given topological spaces.

(i) The concept of **topological property** such as compactness and connectedness introduced in general topology solves this problem in a very few cases which is studied in Basic Topology Volume 1. A study of the subspaces of the Euclidean plane $\mathbf{R}^2$ gives an obvious example.

(ii) On the other hand, the subject algebraic topology was born to solve the problems of impossibility in many cases with a shift of the problem by associating **invariant objects** in the sense that homeomorphic spaces have the same object (up to equivalence). Initially, these objects were integers, and subsequent research reveals that more fruitful and interesting results can be obtained from the algebraic invariant structures such as groups and rings. For example, homology and homotopy groups are very important **algebraic invariants (they are also called topological invariants)** which provide strong tools to study the structure of topological spaces. These algebraic objects are assigned to topological spaces in such a way that **natural operations** on the latter correspond to **natural operations** on the former in the sense continuous maps correspond to group homomorphisms and homeomorphisms to isomorphisms, etc., (its converse is not necessarily true). This approach of assignment in the language of category theory is called **functorial**. In this way, it is often possible to distinguish between different topological spaces by demonstrating that certain assigned alge-

© The Author(s), under exclusive license to Springer Nature Singapore Pte Ltd. 2022   449
M. R. Adhikari, *Basic Topology 3*,
https://doi.org/10.1007/978-981-16-6550-9_7

braic objects are not isomorphic. Algebraic topology is now used to invade many problems of contemporary mathematics.

(iii) The homology groups are algebraic invariants that stem from homology theory inaugurated by Heny Poincaré (1854–1912) in 1895. Development of homology theory discussed in Chap. 3 starts from its invention by Heny Poincaré in 1895 to the approach formulating axiomatization of homology, announced in 1952 by S. Eilenberg (1913–1998) and N. Steenrod (1910–1971), now known as **Eilenberg and Steenrod axioms.** This approach simplifies the proofs of many results by escaping avoidable difficulties to promote active learning in homology and cohomology theories, which is the most important contribution to algebraic topology after the invention of homotopy and homology by Poincare' in 1895. This functorial approach facilitates in variety of cases to solve topological problems through the sovability of corresponding algebraic problems. The motivation of the study of algebraic topology comes from the study of geometric properties of topological spaces from the algebraic viewpoint.

(iv) The homotopy groups are also important algebraic invariants studied in Chap. 2 and they stem from homotopy theory. The early development of homotopy theory was found through the work H. Poincaré, L. E. J. Brouwer (1881–1966), H. Hopf (1894–1971), W. Hurewicz (1904–1956), H. Freudenthal (1905–1990) and some others.

(v) The concepts born in the development of homology and homotopy theories to solve topological problems have found outstanding applications to other areas of mathematics leading to the starting points of many theories such as category theory, homological algebra, K-theory, to mention a few (see Adhikari, 2016). This is a remarkable feature in the history of topology.

The systematic study of algebraic topology as a subject began with precise formulations and correct proofs at the turn of the nineteenth to twentieth century (1895–1904) through the work of Henri Poincaré (1854–1912) in his land-marking 'Analysis situs,' Paris, 1895. Unfortunately, his deep insight did not invite sufficiently attraction of mathematicians until the 1920s, when the situation began to change with applications in many mathematical theories. For example, the importance of homotopy invented by Poincaré was first established by H. Hopf (1895–1971) in 1835 with his discovery of a new continuous map, now known as **Hopf map.** The fibration

$$p : S^3 \to S^2$$

is also known as **Hopf fibration** plays a central role in both geometry and algebraic topology. He proved a surprising result that $\pi_3(S^2) \neq 0$. Geometrically, it means that there is a 'three-dimensional hole' in $S^2$ that cannot be filled. Many mathematicians consider the discovery of Hopf map as the **starting point of modern homotopy theory.** The development of the ideas of homotopy and fiber bundles after 1935 closely links to the study of differential topology.

The exponential growth of algebraic topology both in theory and applications has been found in 1940. Algebraic topologists consider **H. Poincaré founder and H.**

**Hopf as co-founder of algebraic topology.** As many fundamental ideas of algebraic topology, specially in homotopy theory, were born through the work of W. Hurewicz (1904–1956) in 1935–1936, he is also considered co-founder of algebraic topology, at least for homotopy theory.

A characteristic of a topological space which is shared by homeomorphic spaces is called a topological invariant in the sense that it is an invariant which is preserved by a homeomorphism. The main objective in algebraic topology is to create and study topological invariants, which are algebraic in nature and they are also algebraic invariants. Fundamental groups, higher homotopy groups, homology and cohomology groups (ring) are central topics of study in algebraic topology. The concept of topological invariant is utilized in classification of topological spaces up to homeomorphism. On the other hand, the motivation of combinatorial topology is to study a topological space by representing it as a union of simple pieces with a specified arrangement, called combinatorial, such that the properties of the original space depend on how the spitted pieces are arranged.

## 7.1 Motivation of the Study of Topology

This section begins with the motivation of the study of topology. Two natural questions arise:

(1) **What is the subject topology**?

(2) **Why we study this subject?**

(1) There are many different answers of (1). **One may call the subject topology as a qualitative study of geometry without reference to distance** in the sense that if one geometric object is obtained from another geometric object by a continuous deformation, then these two geometric objects are considered to be topologically same, called **homeomorphic**. So it is also called a rubber sheet geometry. Accordingly, the geometric objects such as a circle, ellipse and a square are topologically the same, though they are geometrically different.

(2) **There are also many different answers** of (2). The simplest answer is topology is both highly elegant and useful which come from beauty, scope and power of the subject. Its beauty comes from both its various interesting geometric constructions. Its usefulness comes from the basic properties of continuous functions and geometric objects with their applications in mathematics and also beyond mathematics. For example, it facilitates a study of practically all branches of mathematics, including algebra, real analysis, complex analysis, functional analysis, graph theory, number theory, dynamical systems, and differential equations and many more.

**What is the main problem of study in topology?** The main problem in topology is the classification problem of topological spaces up to homeomorphism. To solve this classification problem, given two topological spaces, either we have to find an explicit expression of a homeomorphism between these two spaces or we have to show that it is not possible to construct such a homeomorphism. **Algebraic and differential topology were born to prove this impossibility**. The usual technique is to assign 'invariant' objects which are shared by homeomorphic spaces (i.e., same for homeomorphic spaces). The earliest invariant objects were Euler characteristics which are integers. Subsequently, integral invariants are generalized by inventing algebraic invariants such as groups, rings and modules which offer more information about the structure of the concerned topological spaces. For example, **fundamental group, homotopy and homology groups provide deep insight into the structure of the topological spaces.** Homology underwent developed first since its invention by H. Poincaré in 1895; on the other hand, homotopy did not develop until 1930. Since then, there has been an explosive development of homotopy theory and its connection with homology theory has become a central theme of topology. Many concepts initially introduced in homotopy and homology theories such as $K$-theory, Brouwer fixed-point theorem and so on have found surprising applications to other areas of mathematics and also beyond mathematics.

## 7.2 Analysis Situs of Henri Poincaré

Topology is now one of the most exciting and powerful fields of research in modern mathematics because of its importance. Its origins may be traced back several hundred years. Historically, topology emerged as a distinct field of mathematics with the publication famous memoirs **Analysis Situs and its Five Supplements of Henri Poincaré (1854–1912 ) from 1895 onward (1895–1904).** Historically, before 1895, several topological ideas were found in mathematics during the previous century and a half. Mathematical terminology changes with time. For example, the Latin terminology **Analysis Situs** is essentially due to G. Leibniz (1646–1716) based on the concept limit and continuity and the German terminology **Topologie** is due to J. B. Listing (1802–1882) in 1847. The Latin terminology **Analysis Situs** means 'analysis of position.' Poincaré preferred the terminology Analysis Situs of his above memoirs and O. Veblen (1880–1960) used this terminology in his famous Colloquium volume of 1922. In English language, the term topology was first used in 1883 in mathematical sense. But confusion arose because of using the same term in botanical sense also since 1659. However, the other alternative older terms of topology were superseded by the term **topology**, perhaps, because of simplicity of its derived terms such as **topological** and **topologist**, etc.

A new geometry was born through Analysis Situs of Poincaré. He remarked in 1912.

**"Geometers usually distinguish two kinds of geometry, the first of which they qualify as metric and the second as projective. … . But it is a third. …; this is analysis situs. In this discipline, two figures are equivalent whenever one can pass from one to the other by a continuous deformation; whatever else the law of this deformation may be, it must be continuous. Thus a circle is equivalent to**

**an ellipse or even to an arbitrary closed curve, but it not equivalent to a straight line segment since this segment is not closed. A sphere is equivalent to a convex surface; it is not equivalent to a torus since there is a hole in a torus and in a sphere there is not."**

## 7.2.1  Brief History of Algebraic Topology

This section highlights the emergence of the ideas leading to development of algebraic topology and communicates the contributions of some mathematicians who inaugurated new concepts and new theories or proved basic results of fundamental importance in algebraic topology starting from the creation of fundamental group and homology group by H. Poincaré in 1895, which are the first fundamental and powerful inventions in algebraic topology. This section also communicates the motivation of the study of algebraic topology with historical development of the subject starting from invention of Euler characteristic in 1752, which is a numerical topological invariant followed by invention of other topological invariants which are algebraically groups such as fundamental groups (may be nonabelian) and higher homotopy groups (abelian) associated with pointed topological spaces together with different types of homology and cohomology groups (always abelian) for arbitrary topological spaces.

## 7.2.2  Motivation of Study of Algebraic Topology

The main aim of algebraic topology is to devise methods to construct topological invariants. This subsection conveys the concepts of several topological invariants with their invention. The main areas of algebraic topology include a study of homotopy groups, homology groups and cohomology groups (rings). The first homotopy group, known as the fundamental group, provides information about loops in a topological spare which facilitates intuitively to know the basic shape, or holes, of a topological space. The concepts of higher homotopy groups are the generalization of the fundamental group which provide a sequence of topological invariants. Homology groups assigned to a topological space provide a sequence of abelian groups defined on a chain complex. Cohomology groups are dual concepts of topological groups defined on co-chain complex associated with a topological space also provide a sequence of abelian groups which are topological invariants. The role of topological invariants is to reformulate statements about topological spaces and continuous maps into statements about groups and homomorphism to have a better chance for solution. This transformation of topological problems into algebraic ones is done through homotopy, homology and cohomology theories by using topological invariants such as fundamental groups, higher homotopy groups, homology and cohomology groups. It is easier to define homotopy groups than homology groups but it is difficult to compute homotopy groups, ever for simple spaces. Moreover, the fundamental group of many topological spaces are nonabelian. An older name for the algebraic topology was combinatorial topology in the sense that the investigating topological spaces were constructed from simpler topological spaces by some technique. Topologists were successful in investigating during 1920s and 1939s to convert topological problems to algebraic problems, which led to rename algebraic topology of combinatorial topology.

## 7.3  Development of the Basic Topics in Algebraic Topology

The following subsections convey the motivation and historical development of the basic topics discussed in this book titled **Basic Topology: Volume 3**, starting from Euler's polyhedral formula.

### 7.3.1  Euler's Polyhedral Formula

**Euler's land-marking polyhedral formula**

$$\mathbf{V} - \mathbf{E} + \mathbf{F} = 2$$

in combinatorial qualities of polyhedra is considered the first important topological invariant. **The Euler characteristic** invented by L. Euler (1703–1783) in 1752 is an integral invariant, which distinguishes nonhomeomorphic spaces. For topological and homotopical invariance of Euler characteristics, see Sect. 7.6.4. The search of other invariants has established connections between topology and modern algebra in such a way that homeomorphic spaces have isomorphic algebraic structures.

**Remark 7.3.1** Theorem 7.3.5 saying that there are only five different types of platonic solids is a classical result proved in Chap. 2 as an interesting application of the Euler characteristic in the theory of convex polyhedra. Its proof is based on considering the surface of a convex polyhedron as glued together a finite number of convex polygons with respect to identity map on edges glued.

**Definition 7.3.2** For a given polyhedron $P$, if $n$ edges meet at each vertex and each face is a convex $m$-gon, then the polyhedron $P$ is said to be of type $[m, n]$. In particular, $P$ is said to regular if every $m$-gon is regular.

**Remark 7.3.3** If the type $[m, n]$ of a polygon $P$ is known, then the number $\mathbf{V}$ of the vertices, the numbers $\mathbf{E}$ of the edges and the number $\mathbf{F}$ of the faces of $P$ can be calculated.

**Definition 7.3.4** A **platonic solid** is a polyhedron such that its faces are congruent regular polygons and each vertex lies in the same number of edges, it is also called a **regular simple polyhedron.**

**Theorem 7.3.5** *There are only five platonic solids which are precisely of types:*

$$[3, 3], [4, 3], [3, 4], [5, 3], \ and \ [3, 5].$$

*Remark 7.3.6* The **platonic solids**

(i) $[3, 3]$ represents geometrically **tetrahedron** , where $V = 4, E = 6, F = 4$;
(ii) $[4, 3]$ represents geometrically **cube** , where $V = 8, E = 12, F = 6$;
(iii) $[3, 4]$ represents geometrically **octahedron**, where $V = 6, E = 12, F = 8$;
(iv) $[5, 3]$ represents geometrically **dodecahedron**, where $V = 12, E = 30, F = 20$; and
(v) $[3, 5]$ represents geometrically **icosahedron**, where $V = 20, E = 30, F = 12$.

**These are the only five different types of platonic solids.**

## 7.3.2   *Beginning of Algebraic Topology*

This subsection highlights the emergences of the ideas leading to algebraic topology and communicates the contributions of some mathematicians who inaugurated new concepts and new theories or proved basic results of fundamental importance in algebraic topology starting from the creation of fundamental group and homology group by H. Poincaré in 1895, which are the first fundamental and powerful inventions in algebraic topology. Actually, **Algebraic Topology** was born as a subject through the work of H. Poincaré based on the idea of dividing a topological space into geometric elements corresponding to the vertices, edges, and faces of polyhedra, and their higher-dimensional analogues. Such investigation presents many topological invariants including the Euler characteristic.

   **Historically,** fundamental group and homology groups are the first important topological invariants of homotopy and homology theories which came from such a search embodded in the work of H. Poincaré (1854–1912) in his land-marking 'Analysis situs,' Paris, 1895. He invented homology theory, now called, **simplicial homology in 1895** with an aim to study geometric properties of a topological space by converting topological problems to algebraic ones for the first time in the history of topology. The term '**Homotopy**' was first used by M. Dehn and P. Heegaard in 1907. L. E. J. Brouwer (1881–1967) gave the precise definition of continuous deformation by using the concept of homotopy of continuous maps. **The Jordan Curve Theorem** stated by Jordan in 1892 is a classical theorem. Its first rigorous proof given by Oswald Veblen (1880–1960) in 1905 is one of the remarkable development of algebraic topology. W. Hurewicz made significant contributions to algebraic topology. **The invention of the higher homotopy groups** $\pi_n$ **by W. Hurewicz in 1935–1936** is a natural generalization of the fundamental group to higher-dimensional analogue of the fundamental group. More precisely, $\pi_n$ is a sequence of covariant functors defined by Hurewicz from topology to algebra by extending the concept of fundamental group formulated by $\pi_n(X) = [S^n, X]$. Lens spaces defined by H. Tietze (1880–1964) in 1908 form an important class of 3-manifolds in the study of their homotopy classification.

## 7.4  Historical Note on Homology Theory

Before formal of invention homology by Henri Poincaré in 1895, this concept was found in the work of B. Riemann (1826–1866) during 1850–1860 on the notion of the connectivity order of geometrical objects (now called Riemann surfaces). A surface $X$ is called $(r + 1)$—connected if there is a family of pairwise $(r + 1)$ closed paths, known as cobounding curves, on $X$ such that all of them taken together bound a region in $X$ but no proper subfamily of this family can do it. Such family of cobounding curves is called maximal. For example the connectivity order of the sphere is 1. The connectivity order is independent of the choice of the maximal cobounding curves, which is proved by Riemann. For example, the connectivity order of the torus $T$ is 2, because a single circle on its surface can not separate it but any pair of disjoint closed curves can do it. E. Betti (1823–1892) developed the concept of connectivity in 1870s for submanifolds in the Euclidean spaces $\mathbf{R}^n$. The concept of connectivity order based on the cofounding curves on a region of a manifold leads to the concept of homology theory invented by H Poincaré (1854–1912) in 1895. He assigned to the $n$-cells of an oriented simplicial complex $K$ a matrix $M(n, \mathbf{Z})$, called the $n$-th incidence matrix having entries 0,-1 $or$ $+1$ depending on the nature of the orientations on its $n - 1$-cells induced by the given $n$-cells. He reduced the incidence matrix to normal form to obtain the Riemann's connectivity order, called Betti number of the given oriented surface. He also utilized tosion coefficients the nature of the closed curves in nonorientable surfaces not bounding a region but their multiples can bound it. To define homology groups, Poincaré started with a geometric object (a topological space) which is given by combinatorial data (a simplicial complex) and then he constructed homology groups by using the linear algebra and boundary relations by these data. His homology theory is called **the classical simplicial homology theory**.

   It involves of tedious discussion on the concepts of triangulablity of the topological spaces, orientations of simplexes, incidence numbers, subdivisions, simplicial approximation and also the topological invariance of the simplicial homology groups.

   Emmy Noether (1882–1935) formulated an algebraic approach corresponding to the geometric approach of homology theory invented by Poincaré.

## 7.5    Historical Note on Homotopy Theory

The early development of homotopy theory was found through the work H. Poincaré, L. E. J. Brouwer (1881–1967), H. Hopf (1894–1971), W. Hurewicz (1904–1956), H. Freudenthal (1905–1990) and some others. But **historically,** a systematic study of homotopy theory in the premises of algebraic topology with precise formulations and correct proofs began at the turn of the nineteenth to twentieth century (1885–1904) through the work of Henri Poincare in his 'Analysis situs,' Paris, 1895. But his deep insight did not attract mathematicians sufficiently until the 1920s, when the situation began to change with applications in many mathematical theories. L. E. J. Brouwer (1881–1967) gave the precise definition of continuous deformation by using the concept of homotopy of continuous maps. Homotopy theory excepting fundamental group was first applied as a secondary tool of homology theory. This section addresses the contribution of Brouwer, Hopf, Hurewicz and Freudenthal toward development of homotopy theory since its inauguration in 1895 by H. Poincaré.

### 7.5.1    Inauguration of Homotopy Theory in 1895 by Poincaré

The intuitive idea of continuous deformation which led to the concept of homotopy is found in Lagrange's method in calculus of variation. This idea was also available in the mathematical work of many many mathematicians of the nineteenth century. But it was Brouwer who first formally defined the concepts of homotopy between two continuous maps in 1911, though Poincaré introduced the idea of fundamental group $\pi(X, x_0)$ of a pointed topological space in his paper of 1895 consisting of its elements: each element is a class of a loop that can be considered as a continuous map

$$\alpha : [0, 1] \to X : \alpha(0) = \alpha(1) = x_0.$$

The study of Poincaré on the fundamental group (also called Poincaré's group) is considered the beginning of homotopy theory. The concepts of fundamental groups and homotopy were born through the work of Henri Poincaré in his land-marking 'Analysis situs,' Paris, 1895. This work officially inaugurated homotopy theory along with homology theory. These two theories form the basic parts of algebraic topology. For this reason, Poincaré is called the founder of algebraic topology. His study to solve classification problems in mathematical analysis and Euclidean geometry by creating topological invariants, which are also algeraicgical invariants. He devised ways to assign to a topological space a group or module in such a way that homeomorphic spaces have isomorphic algebraic objects.

### 7.5.2  Brouwer Fixed-Point Theorem and Degree of Spherical Map

L. E. J. Brouwer (1881–1967) gave the precise definition of continuous deformation by using the concept of homotopy of continuous maps. He initiated significant work in 1912 connecting homology and homotopy groups of certain spaces and proved that two continuous maps of a two-dimensional sphere into itself can be continuously deformed into each other if and only if they have the same degree (i.e., if they are equivalent from the view point of homology theory). His definition of the degree of a spherical map is more intuitive than its definition from the viewpoint of homology theory. He defined $deg f$, the degree of $f$ as the number of times of the domain sphere wraps around the range sphere and proves its homotopy invariance. He showed that for self maps of $S^n$, the homotopy class of a continuous map is characterized by its degree. His definition shows that if $f : S^1 \to S^1, z \mapsto z^n$, then $\deg f = n$. The most basic results of Brouwer proved in this volume are

(i)  **Brouwer no retraction theorem**  saying that there exists no continuous onto map $f : \mathbf{D}^n \to S^{n-1}$ which leaves every point of $S^{n-1}$ fixed for every integer $n \geq 1$.

(ii)  **Brouwer degree theorem**  asserting that if $f, g : S^n \to S^n$ are two continuous maps such that $f \simeq g$, then $deg\ f = deg\ g$.

(iii)  the **topological invariance of dimension of**  $\mathbf{R}^n$ saying that if $m$ and $n$ are two distinct positive integers, then the Euclidean $m$-space $\mathbf{R}^m$ and the Euclidean $n$-space $\mathbf{R}^n$ cannot be homeomorphic.

(iv)  **Brouwer fixed-point theorem**  asserts that every continuous map $f : \mathbf{D}^n \to \mathbf{D}^n$ has a fixed point for every integer $n \geq 0$.

These results of Brouwer has made a key role in laying the foundations of algebraic topology.

**Brouwer fixed-point theorem** (named after his name) for $\mathbf{D}^n$ in the $n$-dimensional Euclidean space $\mathbf{R}^n$ was first proved and studied by L. E. J. Brouwer during 1910–2012. It is a celebrated theorem in topology. This theorem is also proved by using the homology or homotopy groups. But Brouwer used neither of them. Instead, he used the notion of degree of spherical maps $f : S^n \to S^n$. An important application of Brouwer fixed-point theorem for $\mathbf{D}^n$ to algebra is the **Perron–Frobenius theorem** in $\mathbf{R}^n$. It asserts that any square matrix with positive entries has a unique eigenvector with positive entries (up to a multiplication by a positive constant), and the corresponding eigenvalue has multiplicity one and is strictly greater than the absolute value of any other eigenvalue.

**Historically**, Brouwer first established a close link between homotopy and homology by showing in 1912 that two continuous mappings of a two-dimensional sphere $S^2$ into itself can be continuously deformed into each other if and only if they have the same degree, the concept defined by Brouwer himself from the viewpoint of homology, i.e., if they are equivalent from the viewpoint of homology theory invented by Poincaré in 1895. The papers of H. Poincaré during 1895–1904 can be considered as blueprints for theorems to come. The results of Brouwer during 1910–1912 may be considered the first one of the proofs in algebraic topology. He proved the celebrated theorem **Brouwer fixed-point theorem** 7.5.1 by using the concept of degree of a continuous spherical map defined by Brouwer himself.

**Theorem 7.5.1** (Brouwer fixed-point theorem for dimension $n$) Every continuous map $f : \mathbf{D}^{n+1} \to \mathbf{D}^{n+1}$ has a fixed point for every finite $n \geq 0$.

### 7.5.3 Brouwer–Poincaré Theorem

Brouwer-Poincaré Theorem 7.5.4 asserts that there is a continuous nonvanishing vector field $f : S^n \to S^n$ ($n \geq 1$), iff $n$ is odd. On the other hand, Corollary 7.5.5 shows that for all even integers $n \geq 1$, there is no vector field $f : S^n \to S^n$.

**Definition 7.5.2** **A vector field on** $S^n$ is a continuous map $v : S^n \to \mathbf{R}^{n+1}$ ($n \geq 1$) such that the inner product $< x, v(x) >= 0$, $\forall x \in S^n$, i.e., the vector $v(x)$ is orthogonal to the vector $x$ in $\mathbf{R}^{n+1}$ for every $x \in S^n$. Moreover, if $v(x) \neq 0$ for all $x \in S^n$, we say that the **vector field is nonvanishing**.

**Remark 7.5.3** (*Geometrical Interpretation*) Definition 7.5.2 implies that a vector field $v$ on $S^n$ is a continuous map which assigns to every vector $x$ of unit length in $\mathbf{R}^{n+1}$, a unit vector $v(x)$ in $\mathbf{R}^{n+1}$ such that $x$ and $v(x)$ are orthogonal, i.e., $x \perp v(x)$, $\forall x \in S^n$. If we consider the vector $v(x)$ starting from the point $x \in S^n$, then this vector $v(x)$ must be tangent to $S^n$ at each point $x$ of $S^n$. If $x$ moves in $S^n$, then endpoint point of the vector $v(x)$ varies continuously in $\mathbf{R}^{n+1}$.

**Theorem 7.5.4** (Brouwer–Poincaré) *The n-sphere $S^n$ admits a continuous nonvanishing vector field iff n is odd.*

**Corollary 7.5.5** *The n-sphere $S^n$ admits no continuous nonvanishing vector field if n is even.*

**Remark 7.5.6** If $n$ is odd, the difficult problem of determining the maximum number of linearly independent nowhere vanishing vector fields on $S^n$ was solved by J.F Adams in 1962 by using $K$-theory Adams (1962).

### 7.5.4 Freudenthal Suspension Theorem

Freudenthal Suspension Theorem 7.5.8 proved in 1938 by H. Freudenthal (1905–1990) is a basic theorem in algebraic topology. For each pair of positive integers, $m$ and $n$, there is a natural homomorphism $\sigma_m : \pi_m(S^n) \to \pi_{m+1}(S^{n+1})$. This homomorphism is called the Freudenthal suspension homomorphism defined by H. Freudenthal in 1937. He observed that the suspension operation on topological spaces shifts by one their low-dimensional homotopy groups. This observation was important in understanding the special behavior of homotopy groups of spheres, because every sphere can be formed topologically as a suspension of a lower-dimensional sphere and this subsequently forms the basis of stable homotopy theory. So, it said that the founding result of stable homotopy theory is the Freudenthal suspension theorem proving that homotopy groups become stable in the sense of isomorphic after performing sufficiently many iterated suspensions. This result leads to the concept of stable homotopy groups of a topological space. Its direct consequences are enormous. For example, Corollary 7.5.9 proved by Hurewicz says that the homotopy groups $\pi_m(S^n) = 0$ for $0 < m < n$.

**Definition 7.5.7** *(Suspension homomorphism)* The natural homomorphism

$$\sigma : \pi_n(S^n) \to \pi_{n+1}(S^{n+1}), [\alpha] \mapsto [\tilde{\alpha}]$$

is a homomorphism, called the suspension homomorphism.

H. Freudenthal proved the following suspension theorem in 1938. In his honor, this suspension theorem is known as **the Freudenthal suspension theorem.**

**Theorem 7.5.8** (The Freudenthal suspension theorem) *The suspension homorphism*

$$\sigma : \pi_m(S^n) \to \pi_{m+1}(S^{n+1})$$

*is an isomorphism for $m < 2n - 1$ and is onto for $m \leq 2n - 1$.*

An immediate consequence of the Theorem 7.5.8 is the following corollaries:

**Corollary 7.5.9** (Hurewicz) *The homotopy groups $\pi_m(S^n) = 0$ for $0 < m < n$.*

**Corollary 7.5.10** (Hopf) *For every integer $n \geq 1$, $\pi_n(S^n) \cong \mathbf{Z}$. (This result is known as **Hopf degree theorem**).*

## 7.5.5   Stable Homotopy Groups

This section studies stable homotopy groups based on the Freudenthal suspension theorem saying that suspension homomorphism

$$E : \pi_m(S^n) \to \pi_{m+1}(S^{n+1})$$

is an isomorphism for $m < 2n - 1$ and is onto for $m \leq 2n - 1$. Freudenthal (1938) has been discussed in Chap. 2. It is a natural generalization of Freudenthal suspension theorem. In algebraic topology we use the word 'stable' when a phenomenon occurs essentially in the same way independent of dimension provided perhaps that the dimension is sufficiently large. The importance of stable homotopy theory was reinforced by two related developments in the late 1950s. One is the introducing of spectral homology and cohomology theory and specially $K$-theory by Atiyah and Hirzebruch. The other one is the work of Thom which reduces the problem of classifying manifolds up to cobordism to a problem, a solvable problem in stable homotopy theory. Moreover, this section studies homotopy groups of a spectrum. Historically, stable phenomena found implicitly before 1937 in the study of reduced homology and cohomology functors that are invariant under suspension without limitations on dimension. Stable homotopy theory appeared as an important topic of algebraic homotopy with Adam's introduction of his spectral and conceptual use of the concept of stable phenomena in his solution to the Hopf invariant problems. Every homology and cohomology theory can be constructed by using homotopy-theoretical techniques only. This establishes a close relation of homology and cohomology theories with stable objects in homotopy theory.

**Construction of the stable homotopy group**: Consider an $n$-connected $CW$-complex and the suspension map

$$\Sigma : \pi_r(X) \to \pi_{r+1}(X)$$

is an isomorphism for $r < 2n + 1$. In particular, for $r \leq n$ the suspension map $\Sigma$ is an isomorphism and $\Sigma X$ is an $(n + 1)$-connected CW-complex. Now, consider the sequence (7.1 ) of groups and homomorphisms

$$\pi_r(X) \to \pi_{r+1}(\Sigma X) \to \cdots \to \pi_{r+m}(\Sigma^m X) \to \cdots \qquad (7.1)$$

The homomorphism

$$\pi_{r+m}(\Sigma^m X) \to \pi_{r+m+1}(\Sigma^{m+1} X)$$

is an isomorphism for $r + m < 2(n + m) + 1$, i.e., for $m > r - 2n + 1$, because, $\Sigma^m X$ is $(n + m)$-connected. This implies that for fixed $n$ and $r$ and sufficiently large enough $m$, all the groups in the sequence of groups and homomorphisms (7.1) are isomorphic. This leads to the concept of the stable homotopy group for which the sequence (7.1 ) is stabilized.

**Definition 7.5.11** *(Stable homotopy group)* Given an $(n - 1)$ -connected space $X$, for $r \geq 0$, the $r$-th stable homotopy group of $X$, denoted by $\pi_r(X)$ is defined to be the group

$$\pi_{n+m}(\Sigma X), \ \forall m > r - 2n + 1.$$

This is well defined, since, adding any finite number of groups and homomorphisms to the beginning of (7.1) does not affect the resulting stable homotopy group.

**Example 7.5.12** Consider the homotopy exact sequence (7.2) of the fibration: $p : S^3 \to S^2$ :

$$\cdots \to \pi_3(S^1, s_0) \to \pi_3(S^3, s_0) \xrightarrow{p_*} \pi_3(S^2, s_0) \to \pi_2(S^1, s_0) \to \cdots \quad (7.2)$$

The exactness property of the sequence (7.2) of groups and hommorphisms asserts that the homomorphism $p_*$ induced by $p$

$$p_* : \pi_3(S^3, s_0) \to \pi_3(S^2, s_0)$$

is an isomorphism, because $\pi_3(S^1, s_0) \cong \pi_2(S^1, s_0) = 0$. This implies that $\pi_3(S^2, s_0) \cong \mathbf{Z}$. It provided the first example given by H.Hopf, where $\pi_m(S^n, s_0) \neq 0$ for $m > n$. The map $p$ is known the **Hopf map**, studied earlier in Chap. 2.

*Example 7.5.13* For each $q$, consider

$$\pi_{2q+2}(S^{q+2}, s_0) \xrightarrow{\Sigma} \cong \pi_{2q+3}(S^{q+3}, s_0) \xrightarrow{\Sigma} \cong \cdots \xrightarrow{\Sigma} \cong \pi_{q+n}(S^n, s_0) \xrightarrow{\Sigma} \cong \cdots$$

We denote the common group $\pi_{n+q}(S^n, s_0)$, by $\pi_q^S$. It is called the $k$**th stable homotopy group** for the fibration : $p : S^3 \to S^2$ :.

## 7.6   The Topology of Fiber Bundles

(i)     The theory of fiber bundles, in particular, vector bundles, establishes a very strong link between algebraic topology and differential topology. The topology of fiber bundles has created general interest and promises for more work, because it is involved of interesting applications of topology to other areas such as algebraic topology, geometry, physics and gauge groups.

(ii)    Whitney gave the first general definition of fibration in 1935 and his own idea on fibration was subsequently developed by him. He studied a special type of fibration, now, called locally trivial fiber space, which is a quadruple $\xi = (X, p, B, F)$ consisting of the total space $X$, base space $B$, fiber space $F$ and the projection map $p : X \to B$.

(iii)   H. Poincaré considered in 1883 a special type of fiber spaces, which were the covering spaces over the open subsets of $\mathbf{C}$ with discrete fibers. Their generalizations are found during the period 1913-1934.

(iv)    Mathematicians realized around 1940 that the covering homotopy property (CHP) of a fiber space $\xi = (X, p, B, F)$ asserts that the projection

$$p : X \to B : p(x_0) = b_0$$

which is a continuous surjective map satisfying the condition that for each point $b \in B$, there exist nbd $U$ and a homeomorphism

$$\psi : U \times F \to p^{-1}U$$

such that

$$p(\psi(x, t)) = x, \ \forall x \in U, t \in F$$

which induces an isomorphism

$$p_* : \pi_n(X, F, x_0) \to \pi_n(B, b_0), \ \forall n \geq 1.$$

This gives rise beginningless an exact homotopy sequence (7.3)

$$\cdots \to \pi_{n+1}(F, x_0) \xrightarrow{j_*} \pi_{n+1}(X, x_0) \xrightarrow{p_*} \pi_{n+1}(B, b_0) \xrightarrow{\partial} \pi_n(F, x_0) \xrightarrow{j_*} \pi_n(X, x_0)$$

$$\xrightarrow{\partial} \cdots \pi_1(F, x_0) \xrightarrow{j_*} \pi_0(X, x_0) \xrightarrow{p_*} \pi_0(B, b_0) \xrightarrow{\partial} 0$$
$$(7.3)$$

which establishes a close link of homotopy groups of the three spaces such as $X$, $B$ and $F$ of $\xi = (X, p, B, F)$.

(v) A real (resp. complex) vector field is a fiber space $\xi = (X, p, B, F)$ with $F$ is a vector space $\mathbf{R}^n$ (resp. $\mathbf{C}^n$).

(vi) The theory of fiber bundles was first recognized during the period 1935–1940 through the work of H. Whitney (1907–1989 ), H. Hopf (1894–1971) and E. Stiefel (1909–1978 ), J. Feldbau (1914–1945) and some others.

(vii) A fiber bundle is a bundle with an additional structure derived from the action of a topological group on the fibers. A fiber bundle is a locally trivial fibration having covering homotopy property. Historically, the first formal definition of fiber bundles was given by Whitney. Chapter 4 of this book studies general theory of bundles and its Chap. 5 is developed based on their homotopy properties. The study topology of fiber bundles has created general interest as it is involved of interesting applications of topology to other areas such as algebraic topology, geometry, physics and gauge groups and addresses the homotopy theory of bundles. Covering spaces provide tools to study the fundamental groups. Fiber bundles provide likewise tools to study higher homotopy groups (which are generalizations of fundamental groups).

(viii) The notion of fiber spaces is the most fruitful generalization of covering spaces. The importance of fiber spaces was realized during 1935–1950 to solve several problems relating to homotopy and homology. The motivation of the study of fiber bundles and vector bundles came from the distribution of signs of the derivatives of the plane curves at each point.

## 7.6.1 Historical Note on Category Theory

**Category theory** plays an important role in the modern study of algebraic topology, because, homotopy, homology and cohomology theories can be conveniently expressed in the language of category theory. This theory is also very important in mathematics to unify different concepts in mathematics and specially in the study of homotopy, homology and cohomology theories and their development in a unified language. The basic objective of category theory is to unify many basic concepts and results of mathematics in an accessible way. Historically, category theory was born through the work of Eilenberg (1913–1998) and S MacLane (1909–2005) during 1942–1945. A category is a certain collection of mathematical objects (possibly with an additional structure) and morphisms which are like mappings agreeing with this structure. The objects may be a set, group, ring, vector space, module, a sequence of abelian groups, topological space, etc. and morphisms are collections of mapping preserving this structure. On the other hand, **a functor** is a natural mapping from one category to the other in the sense that it preserves the identity morphism and composites of well-defined morphisms. It plays a key role in converting a problem of one category to the problem of other category to have a better chance for solution. **A natural transformation** is a certain function from one functor to other one satisfying some specific properties. These concepts together with their dual concepts form the **foundation of category theory**, which provides a convenient language to unify several mathematical results. This language is used throughout the present book.

### 7.6.2  Homology and Cohomology Theories: Eilenberg and Steenrod Axioms

Homology and cohomology theories are basic theories in algebraic topology. While investigating the 3-dimensional and higher dimensional manifolds in 1895, **Henri Poincaré in his 'Analysis Situs' formally introduced the concepts of homotopy, fundamental group, homology groups and Betti numbers**. His monumental work embodied in his 'Analysis Situs' aimed at solving problems on system of differential equations and his research **establishes a surprising connection between analysis and topology**. Historically, homology invented by Henri Poincaré in 1895 was studied by him during 1895–1904. This homology, called simplicial homology is one of the most fundamental powerful inventions in mathematics. He started with a geometric object (a topological space) which is given by combinatorial data (a simplicial complex), then the linear algebra and boundary relations by these data were used to construct homology groups. The shifting of geometric approach of Henri Poincaré to the algebraic approach of Emmy Nother (1882–1935) to homology theory highly motivated P. Alexandroff (1896–1982) and H. Hopf to study homology theory jointly from the algebraic viewpoint in 1935. **The classical simplicial homology theory invented by Poincaré is involved of tedious discussion** on the concepts of triangulalbity of the topological spaces, orientations of simplexes, incidence numbers, subdivisions, simplicial approximation, and also the topological invariance of the simplicial homology groups. Poincaré remarked in 1899 saying that

" **Assume that one can find in $V$ a manifold of $p + 1$ dimension whose boundary consists of $n$ manifolds of $p$ dimension $v_1, v_2, \ldots, v_n$; I will express this fact with the relation $v_1 + v_2 + \cdots + v_n \sim 0$, that I will call it homology.**"

**There are other homology theories constructed by different mathematicians** such as L. Vietories (1891–2002) in 1927, by E.Čech (1893–1960) in 1932 and S. Lefschitz (1884–1972) in 1933 and some others. For example, Vietories constructed homology groups for compact metric spaces; E.Čech constructed homology groups for compact Hausdorff spaces and Lefschitz constructed singular homology groups by generalizing singular homology groups invented by Poincaré. These theories are conveniently used depending on the nature of the topological problems. On the other hand, the axiomatic approach formulated by seven axioms, now called **Eilenberg and Steenrod axioms for homology and cohomology theories**, were announced by S. Eilenberg (1913–1998) and N. Steenrod (1910–1971) in 1945 but first appeared in their celebrated book 'The Foundations of Algebraic Topology' in 1952. This approach classifies and unifies different homology (cohomology groups) and is the most important contribution to algebraic topology since invention of the homology groups by Poincaré in 1895. Eilenberg and Steenrod proved that on the category of compact triangulable spaces all the homology and cohomology theories satisfying their axioms have isomorphic groups. This asserts that there exists only one homology theory and only one cohomology theory on this category.

## 7.6.3  Euler–Poincaré Theorem

**Euler–Poincaré theorem** 7.6.2 is a powerful result in topology. It establishes a close link among geometry, topology and algebra with the help of Euler characteristics of compact polyhedra.

**Euler- Poincaré characteristic is a generalization of Euler characteristic formula**.

**Definition 7.6.1** Let K be a simplicial complex of dimension $n$ and $\beta_p$ denotes the number of p simplexes in K, for $p = 0, 1, 2, ..., n$. Then the alternative sum

$$\kappa(K) = \Sigma_{p=0}^{n}(-1)^p \beta_p \tag{7.4}$$

is called the **Euler characteristic** of $K$ with **Betti numbers** $\beta_p$.

Betti numbers $\beta_p$ in Eq. (7.4) coined by Poincaré (named after E. Betti ), play an important role in algebraic topology to classify topological spaces based on the connectivity of a $p$-dimensional simplicial complex K. The number $\beta_p$ is the same as the rank of the $p$-th homology group $Hp(K;\mathbf{Q})$ with rational coefficients.

**Theorem 7.6.2** (Euler–Poincaré theorem) *The Euler characteristic of an oriented simplicial complex K of dimension n is given by*

$$\kappa(K) = \Sigma_{p=0}^{n}(-1)^p \beta_p = \Sigma_{p=0}^{n}(-1)^p \ rank \ H_p(K;\mathbf{Q}). \tag{7.5}$$

**Corollary 7.6.3** *The Euler characteristic of an oriented simplicial complex K of dimension n is also given by the formula*

$$\kappa(K) = \Sigma_{p=0}^{n}(-1)^{p}\,\beta_{p} = \Sigma_{p=0}^{n}(-1)^{p}\,rank\,H_{p}(K;\mathbf{Z}).$$

**Corollary 7.6.4** *Let X be a compact polyhedron. Then its Euler characteristic is given by the formula*

$$\kappa(X) = \Sigma_{p=0}^{n}(-1)^{p}\,\beta_{p} = \Sigma_{p=0}^{n}(-1)^{p}\,rank\,H_{p}(X;\mathbf{Z}).$$

**Corollary 7.6.5** *The rank of the free abelian part of $H_{p}(K;\mathbf{Q})$ of a finite oriented complex K is the Betti number $\beta_{p}$ of K.*

**Remark 7.6.6** Since homology group is a homotopy as well as a topological invariant, Euler characteristic is also so (see Sect. 7.6.4).

### 7.6.4  Topological and Homotopical Invariance of Euler Characteristics

**Euler–Poincaré theorem** 7.6.2 is applied to prove a powerful results in topology. For example, this theorem applies to prove topological invariance of Euler characteristics in Theorem 7.6.7 in the sense that two homeomorphic compact polyhedra have the same Euler characteristics and homotopical invariance of compact polyhedra in Theorem 7.6.8 in the sense that two homotopy equivalent compact polyhedra have the same Euler characteristics (see Chap. 3).

**Theorem 7.6.7** *Two homeomorphic compact polyhedra have the same characteristics.*

**Theorem 7.6.8** *Two homotopy equivalent compact polyhedra have the same Euler characteristics.*

**Remark 7.6.9** Theorem 7.6.7 establishes a close link among geometry, topology and algebra by Euler characteristics of compact polyhedra, which are integers. For a generalization of theorem 7.6.8 asserting that if the compact polyhedra $X$ and $Y$ are two homotopy equivalent spaces, then $\kappa(X) = \kappa(Y)$. Moreover, $\kappa(X) \in \mathbf{Z}$, which is an algebraic object.

### 7.6.5 Brief History of Combined Aspect of Combinatorial and Set-Theoretic Topologies

**A union of combinatorial and set-theoretic aspects of topology was achieved first by L. E. J. Brouwer** through his investigation of the concept of dimension during 1908–1912. The unified theory was laid on a solid foundation in the period 1915–1930 by J. W. Alexander, P. S. Alexandrov (1896–1982), S. Lefschetz (1884–1972) and others. Until 1930 topology was called 'analysis situs (position analysis).' **Analysis situs conveyed the qualitative properties of geometric figures both in the ordinary space as well as in the space of more than 3-dimensions.** It was Lefschetz who first used and popularized the name topology by publishing a book with this title in 1930.

### 7.6.6 Poincaré Conjecture and Its Solutions

**Poincaré stated a conjecture known as Poincaré conjecture in 1894 or 1904** by saying that a compact smooth $n$-dimensional manifold, which is homotopy equivalent to the $n$-sphere $S^n$ is homeomorphic to $S^n$. Its equivalent statement says: is a compact $n$-manifold homotopically equivalent to $S^n$ homeomorphic to $S^n$? For $n = 3$, **G. Perelman** (1966-) proved this conjecture in 2003 by using Ricci flow. For other values of $n$, it was solved by others before 1994:

(i) For $n = 4$, **M. Freedman** (1951-) proved that the conjecture is true and wins Fields medal for this proof.

(ii) For $n = 5$, **C. Zeeman** (1925–2016 ) demonstrated the conjecture in 1961.

(iii) For $n = 6$, **J. R. Stallings** (1935–2008) proved in 1961 that the conjecture is true.

(iv) For $n \geq 7$, **S. Smale** (1930-) proved that the conjecture is true and also extended his proof for all $n \geq 5$. He wins the Fields medal in 1966 for this work.

### 7.6.7 Interest of Poincaré in Various Scientific Work

**H. Poincaré, founder of algebraic topology** has contributed fundamental work in pure and applied mathematics, celestial mechanics, dynamical system, mathematical physics. He published 30 books and over 500 papers contributing significant work in a variety of areas such as algebra, analysis, differential equation, complex analysis, algebraic geometry, celestial mechanics, relativity theory, mathematical physics, philosophy of mathematics and even in popular science. Like combinatorial or algebraic topology, his original work on chaotic deterministic system has fonded the modern chaos theory. In 1905, he first showed the requirement the gravitational waves emanating from a body and propagating at the speed of light in the Lorentz transformations. His work on perfect invariance of all of Maxwell's equations has lead to the formulation of the theory of special relativity. The Poincaré group introduced by Poincaré used in topology and physics are named after him.

For additional reading the readers may refer to Adams (1972), Adhikari (2016), Adhikari and Adhikari (2003, 2006, 2014, 2022a, 2022b), Barrat (1955), Bredon (1993), Dieudonné (2016), Eilenberg and Steenrod (1952), Freedman (1982), Gray (1975), Hatcher and Allen (2002), Hilton and Wylie (1960), Hu (1966), Hurewicz (1935), Milnor (1962), Poincaré (1895, 2010), Rotman (1988), Spanier (1966), Switzer (1975).

# References

Adams JF. Algebraic topology: a student's guide. Cambridge: Cambridge University Press; 1972.

Adhikari A, Adhikari MR. Basic topology, vol. 1: metric spaces and general topology. India: Springer; 2022a.

Adhikari A, Adhikari MR, Basic topology, vol. 2: topological groups, topology of manifolds and lie groups. India: Springer; 2022b.

Adhikari MR, Adhikari A, Groups, rings and modules with applications. Hyderabad: Universities Press; 2003.

Adhikari MR, Adhikari A, Textbook of linear algebra: an introduction to modern algebra. New Delhi: Allied Publishers; 2006.

Adhikari MR, Adhikari A, Basic modern algebra with applications. New Delhi, New York: Springer; 2014.

Adhikari MR, Basic algebraic topology and its applications. India: Springer; 2016.

Barrat MG. Track groups I. Proc Lond Math Soc. 1955;5(3):71–106.

Bredon GE. Topology and geometry. New York: Springer; 1993.

Dieudonné J, A history of algebraic and differential topology, 1900–1960, Modern Birkhäuser; 1989.

Eilenberg S, Steenrod N. Foundations of algebraic topology. Princeton: Princeton University Press; 1952.

Freedman M. The topology of four-dimensional manifolds. J Diff Geom. 1982;17:357–453.

Gray B. Homotopy theory. An introduction to algebraic topology. New York: Academic Press; 1975.

Hatcher A. Algebraic topology. Cambridge: Cambridge University Press, 2002.

Hilton PJ, Wylie S. Homology theory. Cambridge: Cambridge University Press; 1960.

Hu ST. Homology theory. Oakland CA: Holden Day; 1966.

Hurewicz W. Beitrage de Topologie der Deformationen. Proc K Akad Wet Ser A 1935; 38:112–119, 521–528.

Milnor JW. On axiomatic homology theory. Pacific J Math. 1962; 12:337–341.

Poincaré H. Analysis situs. J de l'Ecole, Polyt. 1895; 1:121

Poincaré H. Papers on topology: analysis situs and its five supplements,Translated by Stillwell, J., History of Mathematics, 2010; 37 Amer Math Soc.

Rotman JJ. An introduction to algebraic topology. New York: Springer; 1988.

Spanier E. Algebraic topology. McGraw-Hill; 1966.

Switzer RM. Algebraic topology-homotopy and homology. Berlin, Heidelberg, New York: Springer; 1975.

Printed in the United States
by Baker & Taylor Publisher Services